Dimitris E. Simos · Varvara A. Rasskazova ·
Francesco Archetti · Ilias S. Kotsireas ·
Panos M. Pardalos (Eds.)

Learning and Intelligent Optimization

16th International Conference, LION 16
Milos Island, Greece, June 5–10, 2022
Revised Selected Papers

 Springer

Editors
Dimitris E. Simos ⓘD
SBA Research
Vienna, Austria

Graz University of Technology
Graz, Austria

Francesco Archetti
Università degli Studi di Milano-Bicocca
Milan, Italy

Panos M. Pardalos ⓘD
University of Florida
Gainesville, FL, USA

Varvara A. Rasskazova ⓘD
Moscow Aviation Institute (National
Research University)
Moscow, Russia

Ilias S. Kotsireas ⓘD
Wilfrid Laurier University
Waterloo, ON, Canada

ISSN 0302-9743   ISSN 1611-3349 (electronic)
Lecture Notes in Computer Science
ISBN 978-3-031-24865-8   ISBN 978-3-031-24866-5 (eBook)
https://doi.org/10.1007/978-3-031-24866-5

This Springer imprint is published by the registered company Springer Nature Switzerland AG
The registered company address is: Gewerbestrasse 11, 6330 Cham, Switzerland

Preface

The sixteenth installment of the international conference series "Learning and Intelligent Optimization" (LION 16) took place in Milos, Greece, during June 5–10, 2022, at the Milos Conference Center with the participation of researchers and academicians from 23 countries, from China to Brazil, across five continents. As the repercussions of COVID-19 continued in some countries and regions, all sessions were digitally broadcasted in real-time. Thus, all participants had the option to join and give their talk in a remote setting. LION 16 featured six invited talks:

- "How does our mind emerge from our brain?", a plenary talk given by Christos Papadimitriou (Columbia University, USA),
- "Crops, Tuples, and Disasters", a tutorial talk given by Bernhard Garn, Klaus Kieseberg, and Dimitris E. Simos (MATRIS Research Group, SBA Research, Austria),
- "A Random Generator of Hypergraphs Ensembles", a tutorial talk presented by Mario Rosario Guarracino, Amor Messaoud, Yassine Msakni, Giovanni Camillo Porzio (University of Cassino and Southern Lazio, Italy, and Ecole Supérieure de Commerce de Tunis, Tunisia),
- "Numerical Infinities and Infinitesimals in Optimization", a tutorial talk given by Yaroslav D. Sergeyev (University of Calabria, Italy),
- "Tourism and Hospitality: Relevant Problems for Machine Learning and Intelligent Optimization", a tutorial talk given by Roberto Battiti (University of Trento and Ciaomanager SRL, Italy), and
- "Distributed Adaptive Gradient Methods for Online Optimization", a tutorial talk given by George Michailidis (University of Florida, USA).

We would like to thank the authors for contributing their work, and the reviewers whose tireless efforts resulted in keeping the quality of the contributions at the highest standards. This volume contains 39 refereed papers carefully selected from 60 submissions using a XX single-blind peer review process, with a minimum of two reviews per paper. Authors of selected papers were invited to revise their work taking into account 36 full papers and 3 short papers for inclusion in this volume.

The editors also express their gratitude to the organizers and sponsors of the LION 16 conference:

- MATRIS Research Group, SBA Research, Austria.
- CARGO Lab, Wilfrid Laurier University, Canada - APM Institute for the Advancement of Physics and Mathematics.

A special thank you goes to the publicity chair of LION 16 (Izem Chaloupka – MATRIS Research Group) for coordinating and managing registrations, communication with participants, the digital media for visibility efforts, online participation, and digital broadcasting, and to the LION 16 volunteers (junior and senior researchers of the MATRIS Research Group: Klaus Kieseberg, Ludwig Kampel, and Bernhard Garn), for the preparation of the conference, managing a scientifically attractive and intriguing booth, developing disaster scenarios and combinatorial black-box testing games, and presenting gifts to two participants on each day of the event. We also thank SBA Research for sponsoring LION 16 and Daniela Freitag-David (Strategic Innovation and Communication, SBA Research) for increasing the conference visibility through online news entries.

Another special thank you goes to the Knuth Prize, Gödel, IEEE-John von Neumann Medal, EATCS, and the IEEE Computer Society Charles Babbage Award holder, Christos Papadimitriou (Columbia University), for contributing to the community through his intriguing plenary talk.

After a two-year-long COVID-19 pandemic and virtual conferences, we were delighted to be able to assemble and reconnect with the vibrant LION community members physically and deliver this LNCS proceedings volume for LION 16, in keeping with the tradition of the four most recent LION conferences [1–4].

June 2022   Dimitris E. Simos
                                            Varvara A. Rasskazova
                                              Francesco Archetti
                                               Ilias S. Kotsireas
                                              Panos M. Pardalos

References

1. Roberto Battiti, Mauro Brunato, Ilias S. Kotsireas, Panos M. Pardalos (Eds.): Learning and Intelligent Optimization - 12th International Conference, LION 12, Kalamata, Greece, June 10–15, 2018, Revised Selected Papers, Lecture Notes in Computer Science, LNCS 11353, Springer, 2019.
2. Nikolaos F. Matsatsinis, Yannis Marinakis, Panos M. Pardalos (Eds.): Learning and Intelligent Optimization - 13th International Conference, LION 13, Chania, Crete, Greece, May 27–31, 2019, Revised Selected Papers, Lecture Notes in Computer Science, LNCS 11968, Springer, 2020.
3. Ilias S. Kotsireas, Panos M. Pardalos (Eds.): Learning and Intelligent Optimization - 14th International Conference, LION 14, Athens, Greece, May 24–28, 2020, Revised Selected Papers, Lecture Notes in Computer Science, LNCS 12096, Springer, 2020.
4. Dimitris E. Simos, Panos M. Pardalos, Ilias S. Kotsireas (Eds.): Learning and Intelligent Optimization - 15th International Conference, LION 15, Athens, Greece, June 20–25, 2021, Revised Selected Papers. Lecture Notes in Computer Science 12931, Springer, 2021.

Organization

General Chairs

Panos M. Pardalos — University of Florida, USA
Francesco Archetti — Consorzio Milano Ricerche and Università degli Studi di Milano-Bicocca, Italy

Program Committee Chairs

Dimitris E. Simos — SBA Research and Graz University of Technology, Austria; Information Technology Laboratory, NIST, USA
Varvara A. Rasskazova — Moscow Aviation Institute, Russia

Local Organizing Committee Chair

Panos M. Pardalos — University of Florida, USA

Publicity Chair

Izem Chaloupka — SBA Research, Austria

Program Committee

Francesco Archetti — Consorzio Milano Ricerche, Italy
Annabella Astorino — ICAR-CNR, Italy
Amir Atiya — Cairo University, Egypt
Rodolfo Baggio — Bocconi University, Italy
Roberto Battiti — University of Trento, Italy
Maude Josée Blondin — Université of Sherbrooke, Canada
Christian Blum — Spanish National Research Council (CSIC), Spain
Juergen Branke — University of Warwick, UK
Mauro Brunato — University of Trento, Italy
Dimitrios Buhalis — Bournemouth University, UK
Sonia Cafieri — Ecole Nationale de l'Aviation Civile, France
Antonio Candelieri — University of Milano-Bicocca, Italy
John Chinneck — Carleton University, Canada
Kostas Chrisagis — City University London, UK
Andre Augusto Cire — University of Toronto, Canada

Andre de Carvalho	University of São Paulo, Brasil
Patrick De Causmaecker	Katholieke Universiteit Leuven, Belgium
Renato De Leone	University of Camerino, Italy
Clarisse Dhaenens	Université Lille 1 - Polytech Lille, CRIStAL, Inria, France
Luca Di Gaspero	University of Udine, Italy
Ciprian Dobre	Politehnica University of Bucharest
Adil Erzin	Sobolev Institute of Mathematics, Russia
Giovanni Fasano	University Ca'Foscari of Venice, Italy
Paola Festa	University of Napoli Federico II, Italy
Antonio Fuduli	Universita' della Calabria, Italy
Martin Golumbic	University of Haifa, Israel
Vladimir Grishagin	Nizhni Novgorod State University, Russia
Mario Guarracino	ICAR-CNR, Italy
Youssef Hamadi	Uber AI, France
Cindy Heo	Ecole hôtelière de Lausanne, Switzerland
Laetitia Jourdan	Inria, LIFL, CNRS, France
Valeriy Kalyagin	Higher School of Economics, Russia
Alexander Kelmanov	Sobolev Institute of Mathematics, Russia
Marie-Eleonore Kessaci	Université de Lille, France
Michael Khachay	Krasovsky Institute of Mathematics and Mechanics, Russia
Oleg Khamisov	Melentiev Institute of Energy Systems, Russia
Zeynep Kiziltan	University of Bologna, Italy
Yury Kochetov	Sobolev Institute of Mathematics, Russia
Ilias Kotsireas	Wilfrid Laurier University, Waterloo, Canada
Dmitri Kvasov	University of Calabria, Italy
Dario Landa-Silva	University of Nottingham, UK
Hoai An Le Thi	Université de Lorraine, France
Daniela Lera	University of Cagliari, Italy
Vittorio Maniezzo	University of Bologna, Italy
Silvano Martello	University of Bologna, Italy
Francesco Masulli	University of Genoa, Italy
Nikolaos Matsatsinis	Technical University of Crete, Greece
Kaisa Miettinen	University of Jyväskylä, Finland
Laurent Moalic	University of Haute-Alsace - IRIMAS, France
Hossein Moosaei	Charles University, Czech Republic
Serafeim Moustakidis	AiDEAS OU, Greece
Evgeni Nurminski	FEFU, Russia
Panos M. Pardalos	University of Florida, USA
Konstantinos Parsopoulos	University of Ioannina, Greece
Jun Pei	Hefei University of Technology, China

Contents

Invited Papers

Optimal Scheduling of the Leaves of a Tree and the SVO Frequencies of Languages

Christos H. Papadimitriou and Denis Turcu[✉][iD]

Columbia University, New York, NY 10027, USA
{christos,d.turcu}@columbia.edu

Abstract. We define and study algorithmically a novel optimization problem related to the sequential scheduling of the leaves of a binary tree in a given order, and its generalization in which the optimum order is sought. We assume that the scheduling process starts at the root of the tree and continues breadth-first in parallel, albeit with possible intervening lock and unlock steps, which define the scheduling cost. The motivation for this problem comes from modeling language generation in the brain. We show that optimality considerations in this problem provide a new explanation for an intriguing phenomenon in linguistics, namely that certain ways of ordering the subject, verb, and object in a sentence are far more common in world languages than others.

Keywords: Binary tree · Optimal leaves scheduling · Language · Basic word order

1   Introduction: The Leaf Scheduling Problem

Consider a binary tree T, e.g. the one in Fig. 1A—where by binary tree we mean a downwards directed tree with $2n - 1$ nodes, one node of degree two (the root), $n - 2$ nodes of degree three, and n nodes (the leaves) of degree one—and suppose that we are also given an order σ of the leaves, say the order subject-verb-object (SVO) in this example. We are interested in assigning integer times to the nodes of the tree according to the following rules:

1. The root is assigned time 0;
2. A non-root node i either is assigned time $t + 1$, where t is the time assigned to its parent, or it is *locked* by its parent;
3. The leaves are assigned times that are strictly increasing in the given order, σ;
4. If a leaf ℓ is assigned a time t, then a locked node i may be assigned time $t + 1$, in which case we say that ℓ *unlocks* i.

We say that an assignment is a *feasible schedule* if it satisfies these rules. Intuitively, this assignment of times formalizes the process in which the nodes of the tree "fire" starting with the root, and the children of a node fire right

D. E. Simos et al. (Eds.): LION 2022, LNCS 13621, pp. 3–14, 2022.
https://doi.org/10.1007/978-3-031-24866-5_1

after their parent did. The exception is that a node may choose to lock one or both of its children at the time of its own firing. A leaf, upon firing, may unlock one locked node. It is clear that, given a tree and an order of its leaves, there is always a feasible schedule: Always lock the child that does not lead to the next leaf in the order, while any firing leaf unlocks the locked ancestor of the next leaf in the order.

Define now the LEAF SCHEDULING PROBLEM to be following: *Given a tree T and an order σ of its leaves, find a feasible schedule that has the smallest cost, where the cost of a schedule is the number of lock commands used (equivalently, the smallest number of unlock commands).* We can also define the *weighted version* of this problem by assigning a weight to every possible lock and unlock command, and minimizing the sum of these weights. We can further define a more complex problem called the *optimum leaf order problem*, in which we are only given a tree T and we seek the order σ that has the smallest scheduling cost.

For example, the optimum leaf scheduling problem for the order SVO in Fig. 1A is the one that assigns 0 to the root, 1 to S and the internal node, 2 to V and 3 to O. That is, the internal node locks O, and V unlocks it. This solution has cost one, since one lock is used, and it is clear that there is no solution with zero cost. In fact, the order S–V–O along with S–O–V are the optimum leaf orders with cost one, while the other four orders have optimum cost two. For a more complicated example, the reader may want to verify that the tree in Fig. 1B, with the leaf order from left to right, has optimum leaf scheduling cost three, while the optimum orders for this tree are the orders ADBC and ADCB with cost one. As we shall see in the next section, both algorithmic problems can be solved by greedy algorithms — with the exception of the optimum leaf order problem with weights, which is NP-hard.

Motivation: Word Orders in Natural Languages

The reason these problems are interesting is because they relate to a classical problem in Linguistics, which we explain next. In English, the subject of a sentence generally comes before the verb while the object, if present, follows both: "dogs chase cats". This ordering is not universal, as other languages adopt any of the six possible orderings, see for example [4]. The same order as in English, denoted SVO, is prevalent in French, Hebrew, modern Greek and Romanian, and overall in about 42% of world languages. The order SOV is slightly more common, accounting for 45% of languages, including Hindi, Urdu, Japanese, Latin, and ancient Greek. The orders VSO (9%), VOS (2%) and OVS (1%) are much less common, while the order OSV (< 1%) is practically disregarded. In English, changing the language's SVO order creates either meaningless sentences ("chase cats dogs") or changes the meaning ("cats chase dogs"). In other languages, such as German, Russian, or modern Greek, deviations from the standard order are tolerated, because nouns have a *case* in these languages, which makes their syntactic role (subject *vs* object) easy to identify independently of position. However, many linguists believe that most languages have a dominant, default word order.

There is extensive literature on justifying the widely varying frequencies of basic word orders, see [1,8,10,11,14,17–19,22]. These past explanations are based on plausible linguistic principles related to the ease of communicating meaning, or the difficulty of learning grammar [8,9,14,22] while more recent explanations consider the mutability and evolution of word orders in languages [17–19]. Here we propose a different explanation based on *the difficulty of articulating sentences in the brain.*

Indeed, one can hypothesize that, in order to generate a sentence such as "cats chase dogs," a speaker must first create, through neuronal circuits in their brain, a tree representation of the sentence as in Fig. 1A,C. There is cognitive evidence [6] suggesting that this tree is binary (that is, there are no nodes with more than two children), and in fact that the three leaves"cats," "chase" and "dogs" are organized as shown in Fig. 1A,C (instead of the alternative, e.g., where S and V are combined first); see Fig. 3 in [20]. Given now this tree, the speaker must articulate it, and this involves selecting and implementing one of the six word orders. To arrive at one of the orders, a neural mechanism of "lock" and

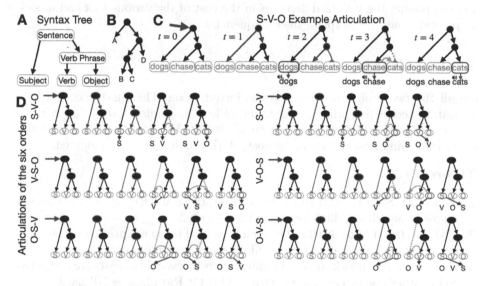

Fig. 1. (A) Basic syntactic tree with a "Verb Phrase" internal node and "Subject", "Verb" and "Object" leaves. **(B)** An alternative binary tree example, more complex than the basic syntactic tree. **(C)** Example articulation from the syntactic tree to sequential speech for the SVO order. Black arrows are inactive. Blue arrows activate the object they point to on the next time step. Red inhibitory signals maintain a lock on the object they point to. Green arrows remove the lock. **(D)** Articulations of all possible basic word orders to sequential speech, starting from the same syntactic tree. Appropriate lock and unlock operations dictate the basic word order. (Color figure online)

"unlock" steps may be used. In Sect. 3 we point out that this can easily be done in the model of brain computation proposed in [20]. It would make sense that all speakers of a language end up using the same fixed order, for reasons of effective information transfer; even though most languages allow, unlike English, more flexibility in articulation orders, many linguists believe that there is a dominant order in each language [5, 8, 17, 19, 22]. But which of the six orders will be chosen as the dominant order? We propose that, the smaller the implementation cost of a word order in the brain, the more likely it should be for the order to be chosen. In the model of brain computation articulated in [20] and explained further in Sect. 3, every node of the tree resides in a different brain area, and long-range inhibitory neurons are used to lock and unlock brain areas and thus articulate the sentence.

The rest of this paper is organized as follows: In Sect. 2 we study algorithmically the leaf ordering problems, while in Sect. 3 we spell out the model of brain computation in [20] and the way it can implement sentence generation. This model ends up providing an explanation for the differences in the probability of word orders: The two most frequent word orders correspond to the two optimal solutions, while the other four lag behind. Adopting a model in the style of statistical mechanics for calculating the frequencies of the orders allows one to even predict the various differences in the cost of the various lock and unlock steps that would best explain these frequencies.

2   The Greedy Algorithm

Recall the two problems defined in the introduction: The LEAF SEQUENCING PROBLEM seeks the optimum scheduling of lock and unlock steps that realizes a given sequence, whereas the OPTIMUM ORDER PROBLEM wants to find the order that minimizes this optimum cost. Both problems can be weighted.

Theorem 1.

1. The LEAF SEQUENCING PROBLEM can be solved in $O(n \log n)$ time through a greedy algorithm; ditto for the weighted case.
2. The OPTIMUM ORDER PROBLEM can be solved by an adaptation of the same greedy algorithm, if all lock and unlock steps have unit cost.
3. However, if the unlock steps have different costs, even if the costs are restricted to be either one or two, the OPTIMUM ORDER PROBLEM is NP-hard.

Proof. (1) We describe the algorithm informally. It entails the sequential firing of all nodes of the tree, starting from the root; the firing propagates from a node to its children down the tree (a breadth-first search implemented by a queue of nodes). Specifically, the root fires at the first parallel time step. At step $t + 1$, the internal nodes whose parents fired at step t will fire. Additionally, *any leaf unlocked* by another leaf at time t will fire at time $t + 1$ (there will be at most one unlocked leaf at any time step). Finally, if one of the internal nodes firing has any leaf children, then each child is locked *unless it is the next leaf to be output.*

To keep track of leaves we maintain a separate heap of *locked leaves* ordered by σ, initially empty, and an index `next`, initially 1. If at some step we encounter a leaf child i of a node being processed, there are two cases: If $\sigma(i) = $ `next`, and no other leaf has been output during this step, then the leaf is output immediately and `next` is increased by 1. Otherwise, $\sigma(i) > $ `next`, and i joins the heap of locked leaves. At the beginning of parallel step t (the round of breadth-first-search processing the nodes of the tree at depth $t - 1$), we check whether the `min` of the heap, call it m, has $\sigma(m) = $ `next`. If so, then we output m and increment `next`. We then proceed with the breadth-first search. The algorithm terminates when both the heap and the queue are empty.

We claim that this algorithm outputs the leaves in the σ order, and that it does so with the fewest lock and unlock operations and in the fewest parallel steps possible. We first claim that every leaf is output as early, in terms of parallel time, as possible. This follows from two things: (a) no leaf i can be output earlier than time $T(i)$, where $T(i)$ satisfies the recurrence $T(i) = \max\{T(\sigma^{-1}(\sigma(i) - 1) + 1, \text{depth}(i)\}$ if i is not the first leaf and $T(i) = \text{depth}(i)$ otherwise; and (b) the algorithm achieves this time, as can be shown by induction on $\sigma(i)$. We also claim that it implements the permutation with the fewest locks, which follows from the two facts that (c) the minimum possible number of locks is $n - 1$ minus the number of *coincidences,* where a coincidence is an i for which the two terms in the recursive definition of T above are equal, and (d) such coincidences are caught and exploited by the algorithm.

(2) For the OPTIMUM ORDER PROBLEM, we start by noticing that every leaf i becomes available to be output at time depth(i). Second, a leaf can be output without lock/unlock steps only if it is output at the precise time it becomes available. Otherwise, if many leaves have the same depth, all but one of them can be feasibly postponed to any time in the future, and unlocked by the leaf that was output immediately before it. Hence the following greedy algorithm achieves the minimum number of lock/unlock steps: We define a one-to one mapping from the n leaves to the time slots $\{d, d + 1, \ldots, D + n\}$, where d and D is the minimum and maximum depth of a leaf of the tree: First, each leaf i is mapped to depth(i), which creates a map which is not one-to-one because of collisions. We then repeatedly go through the time slots, from smaller to larger starting from d and execute the following algorithm: for any time slot t, if it has $\ell > 1$ leaves mapped to it, select $\ell - 1$ of these leaves and assign them to the $\ell - 1$ empty time slots greater than t and closest to t, resolving ties arbitrarily. It is easy to see that this algorithm chooses the permutation of the leaves which has the maximum number of coincidences (leaves fire exactly when they become available), in the sense of the previous paragraph, and thus the minimum possible number of lock and unlock steps.

(3) Finally, for NP-hardness: Imagine that the tree is a full binary tree of depth d — that is, $n = 2^d$ and all leaves arrive simultaneously. Then all permutations are available, and we need to chose the ones that order $\sigma(1), \sigma(2), \ldots, \sigma(n)$ such that $\sum_{i=2}^{n} \text{unlockcost}(\sigma(i-1), \sigma(i))$ is as small as possible. It is easy to see that this is a generic instance of the (open-loop) traveling salesman problem,

which is known to be NP-hard even if the lengths of the edges are either one or two [21]. This completes the proof of Part (3) and of the theorem.

3   Generating Sentences in the Brain

It is by now widely accepted among neuroscientists that, in the brain, information items such as objects, ideas, words, episodes, etc. are represented by large populations of spiking neurons. These populations are called *assemblies*. In [20], a computational system was presented whose basic data item is the assembly of neurons, and its operations include *merge,* the creation of an assembly that has strong synaptic connectivity to and from two already existing assemblies, as well as operations that inhibit and disinhibit brain areas. Notice that by repeated application of the merge operation, trees can be built. Indeed, a simple sentence such as "dogs chase cats" can be generated by first identifying the three assemblies corresponding to the three words in the lexicon — believed to reside in the left medial temporal lobe [7]. Then, these word-assemblies project to create three new assemblies within separate subareas of Wernicke's area in the superior temporal gyrus, corresponding to Subject, Verb and Object brain areas. Next, the Verb and Object assemblies (in this example corresponding to "chase" and "cats", respectively) merge to create a *Verb Phrase* assembly in Broca's area [7]. Finally, the Subject and Verb Phrase assemblies merge to create an assembly representing the Sentence Fig. 1A, in another subarea of Broca's area [7]. A sentence may have many other constituents, such as determiners, adjectives, adverbs, and propositional phrases, but here we focus only on the tree built from its three basic syntactic parts: Subject, Verb, and Object.

Three different binary trees can be built from three leaves, by grouping any two of these leaves first. There is a broad consensus in Linguistics [5,14,17,22], as well as evidence from cognitive experiments [6], supporting the basic tree described above Fig. 1A with an internal Verb Phrase node whose constituents are Verb and Object.

Once the sentence is generated, it may be *articulated*, that is, converted into speech. This can be done by exciting the root of the tree – the Sentence assembly – which then will excite its children in the tree and so on. Eventually, all three leaves will be excited. Each leaf can mobilize motor programs which will articulate each word, but this must be done sequentially. Therefore, *one of the six orders must be selected and implemented*. Perhaps the simplest and most biologically realistic mechanism for implementing a particular order involves two plausible primitives, which we call *lock* and *unlock*. These primitives correspond to the familiar neural processes of inhibition of an area (the activation of a population of inhibitory neurons which will prevent excitatory neurons in this area from firing) and dis-inhibition (the inhibition of the inhibitory population) [3,12,16]. In particular, upon firing, an assembly in the tree can inhibit one of its children from firing. Secondly, any leaf can, upon firing, dis-inhibit any other leaf.

3.1 Scheduling Cost Explains SVO Frequencies

We have already seen that, among the six orders, only two can be implemented by just one lock and one unlock operation, whereas all others require two lock and two unlock operations Fig. 1D. In other words, this simple model immediately predicts "the highest-order bit" of the frequency statistics, namely the prevalence of the SVO and SOV orders. All other orders besides these two require extra inhibition and disinhibition, primitives that are known to require significant brain energy consumption [2,3,15]. Furthermore, extra operations makes the articulation process more complex, and presumably renders this aspect of language more difficult for the learner.

3.2 Leaf Scheduling Cost as Energy

It has been argued in the literature [5,8,17–19] that languages have undergone transitions in their history, in which the word order has changed, and hence the current frequencies reflect a dynamic equilibrium of this dynamic process. This view motivates a naïve statistical-mechanical formulation, treating the frequencies of the basic word orders as a Boltzmann distribution [13], in which states with energy level L are prevalent with probability proportional to $e^{-\beta L}$, for a temperature parameter β. For simplicity, we take $\beta = 1$ in this account (but in our experiments we use a wide range of values for β Fig. 2). The states of our model are the six basic word orders and the associated energies are the number of operations required by each articulation choice. The optimal choices for SVO or SOV have low energies, requiring only two operations (one lock and one unlock), while the other four optimal choices have high energies, requiring four operations (two lock and two unlock) Fig. 1D. The prevalence of the six orders SVO, SOV, VSO, VOS, OSV, and OVS would be proportional to the numbers $e^{-2}, e^{-2}, e^{-4}, e^{-4}, e^{-4}, e^{-4}$, respectively. The orders SVO and SOV would then be expected to be more frequent than the rest by a factor of $e^2 \approx 7.4$, predicting frequencies (.39, .39, .055, .055, .055, .055), a great first-order approximation of the empirical distribution (.45, .42, .09..02, .01, .01).

The true cost of a brain area locking and unlocking another area may differ, depending on the distance between the two brain areas involved and the strength of their neural connections, as well as the duration, in steps, of the locking state of the target area. By introducing such hyper-parameters, in addition to β, and fitting them to the observed data, we can in fact *predict* their values. That is, make predictions about the connectivity, via inhibitory neural connections, between brain areas. It turns out that these predictions are robust to various hyper-parameters, including β. The Boltzmann distribution model provides the basis for estimating the frequency of the six basic word orders. We equate these frequencies [Equation 2] with the empirical observations and numerically solve the system of six non-linear equations. We note that the equations display an analytical degeneracy which is also recovered from the simulations; specifically, four of the six parameters can only be determined up to a common additive constant. This degeneracy is manifest in Eq. 1, in that these four parameters cannot be compared with the other two.

The system of equations does not have an analytical solution, but the six parameters can be approximated using gradient optimization. This method finds the same qualitative results for different values of the coefficient β Fig. 2A and for different values of other hyper-parameters. The results of these calculations are robust enough to support certain predictions about the relative costs of dis-inhibiting one brain area from another Fig. 2B. More specifically, we find that:

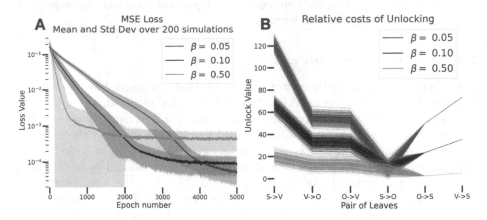

Fig. 2. (A) Loss value plotted during the epochs of gradient descent for various β values. The lines represent the average loss and the shaded areas the standard deviation over 200 initializations. **(B)** Relative costs of the unlock operation from one leaf to another. Colors represent models with varying β. Each line represents the optimized parameters for one model. Note the degeneracy of the solutions for the first four leaf pairs: the lines differ only by an additive constant.

$$U_{S \to V} > U_{V \to O} \gtrless U_{O \to V} > U_{S \to O} \quad \text{and} \quad U_{V \to S} > U_{O \to S}, \tag{1}$$

where $U_{x \to y}$ is the cost to dis-inhibit assembly y from assembly x.

3.3   A Statistical-Mechanical Argument

In statistical mechanics, the probability of a given state of a system depends on its energy and temperature parameter. The Boltzmann distribution provides a way to estimate the thermal equilibrium configuration of all the states of a system. The probability of a state with energy E_i is proportional to $p_i \propto \exp(-\beta E_i)$, where β is a scale factor, inversely proportional to temperature, and E_i depends on the respective unlock costs $U_{x \to y}$ of state i. The system we describe has six states, therefore, the probability of each state is:

$$p_i = \frac{\exp\left(-\beta E_i\right)}{\sum_j \exp\left(-\beta E_j\right)}, \tag{2}$$

for $i, j \in \{SVO, SOV, VSO, VOS, OVS, OSV\}$.

These formulas are simply a heuristic way, aligned with physical principles, of modeling how complexity affects probabilities; however, we also note that in the neuroscience literature (see e.g. [15]) metabolic costs are thoroughly discussed with respect to thermal energy. On this account, we choose the energies of our states to be $E_i \sim 1$ and we assume $\beta \lesssim 1$.

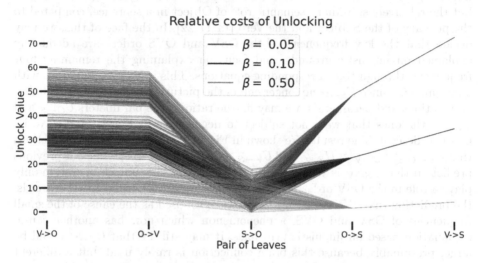

Fig. 3. The predictions if the primacy of Subject and Verb is the cause of the low frequencies of the OSV and OVS orders. Notice that there is no prediction for $U_{S \to V}$.

4   Discussion

Linguistic phenomena should be constantly reinterpreted under the light of new insights, including advancements in our understanding of, or theories about, language processing in the brain. Despite recent progress in this front, articulating the constraints imposed by the neural processes involved in the language function is not easy, due to a large gap, in both scale and focus, between cognitive and systems neuroscience. Our work attempts to bridge this gap using the computational framework of the Assembly Calculus, thus providing a new explanation of the difference in frequencies of the six basic word orders in languages in terms of the difficulty of generating an order from the basic syntax tree of the sentence.

The simplest version of our model qualitatively matches the observed basic word order frequencies, and the most complex version can be tuned to predict the exact frequencies. However, we suspect that the latter calculation may constitute overfitting, as other considerations are likely to enter in the determination of these frequencies, including linguistic considerations of communication efficiency and learnability. These other factors were heretofore the only ones used for this purpose. Our model is not meant to replace these arguments, but add to them and it provides an additional basis for breaking the symmetry of the basic word orders.

We believe that the ultimate explanation of the phenomenon of word orders will integrate both linguistic and neurocomputational evidence, and perhaps learnability considerations, together with more kinds to come. For an example of how this can be done, let us take the linguistic argument that the primary cause of the extreme rarity of orders starting with "O" may not be the difficulty of unlocking Subject or Verb subareas from the Object area according to our model, but the relatively subsidiary semantic role of Object in a sentence, compared to the primacy of the Subject and the Verb [14,17,22]. In the face of this, we may decide that the low frequencies of the OSV and OVS orders are adequately explained on linguistic grounds, and focus on explaining the remaining four frequencies through the corresponding equations. This leads to 4 equations with 5 parameters (since $U_{S \to V}$ no longer enters the picture). To balance the number of equations and parameters, we may fix the ratio of the parameters $U_{O \to S}$ and $U_{V \to S}$ (the ones that were not subject to degeneracy in Fig. 2), and solve by gradient descent. The results are shown in Fig. 3. We notice that our predictions that $U_{V \to O} \gtrsim U_{O \to V} > U_{S \to O}$ and $U_{V \to S} > U_{O \to S}$ are stable, while our previous prediction that $U_{S \to V}$ is very large vanishes because this unlock operation only plays a role in the OSV order, whose frequency we are ignoring. In other words, the prediction that "$U_{S \to V}$ is very large" was proposed as the cause of the small frequencies of OSV and OVS, a phenomenon which now has another causal explanation based on linguistic principles. It may still be that $U_{S \to V}$ would be large, presumably because this brain connection is rarely used, but a different model or experimental evidence may need to be employed to settle this.

References

1. Aitchison, J.: Word order universals. John A. Hawkins Quantitative Analysis of Linguistic Structure Series. Academic Press, New York, London, Sydney, 1983–1986. https://doi.org/10.1016/0024-3841(86)90039-2
2. Attwell, D., Laughlin, S.B.: An energy budget for signaling in the grey matter of the brain. J. Cerebral Blood Flow Metab. **21**(10), 1133–1145 (2001). https://doi.org/10.1097/00004647-200110000-00001. pMID: 11598490
3. Buzsáki, G., Kaila, K., Raichle, M.: Inhibition and brain work Neuron **56**(5), 771–83 (2007). https://doi.org/10.1016/j.neuron.2007.11.008, https://www.sciencedirect.com/science/article/pii/S0896627307009270
4. Dryer, M.: Order of subject, object, and verb. In: Martin, H., Matthew S.D., David, G., Bernard, C. (eds.) The World Atlas Language Structures, chap. 81, pp. 330–333. Oxford University Press, Oxford (2005)
5. Dunn, M., Greenhill, S.J., Levinson, S.C., Gray, R.D.: Evolved structure of language shows lineage-specific trends in word-order universals. Nature **473**(7345), 79–82 (2011). https://doi.org/10.1038/nature09923

6. Frankland, S.M., Greene, J.D.: Two ways to build a thought: distinct forms of compositional semantic representation across brain regions. Cereb. Cortex **30**(6), 3838–3855 (2020). https://doi.org/10.1093/cercor/bhaa001

7. Friederici, A., Chomsky, N.: Language in Our Brain: the Origins of a Uniquely Human Capacity. The MIT Press (2017). https://books.google.com/books?id=MJg-DwAAQBAJ

8. Gell-Mann, M., Ruhlen, M.: The origin and evolution of word order. In: Proceedings of the National Academy of Sciences of the United States of America, vol. 108, no. 42, pp. 17290–17295 (2011). https://doi.org/10.1073/pnas.1113716108, http://starling.rinet.ru/cgi-bin/main.cgi?flags=eygtnnl. Thisarticlecontainssupportinginformationonlineatwww.pnas.org/lookup/suppl/, https://doi.org/10.1073/pnas.1113716108/-/DCSupplemental.www.pnas.org/cgi/doi/10.1073/pnas.1113716108

9. Greenberg, J.: Some universals of grammar with particular reference to the order of meaningful elements. Tech. Rep. 40–70 (1963). http://lear.unive.it//handle/10278/3011

10. Hammarström, H.: The Basic Word Order Typology: an Exhaustive Study. Tech. Rep. (2015)

11. Huber, M., Consortium, T.A.: Order of subject, object, and verb. Atlas of Pidgin and Creole Language Structures Online, pp. 330–333 (2013). http://apics-online.info/parameters/1

12. Jackson, G.M., Draper, A., Dyke, K., Pépés, S.E., Jackson, S.R.: Inhibition, disinhibition, and the control of action in tourette syndrome. Trends in Cognitive Sciences **19**(11), 655–665 (2015). https://doi.org/10.1016/j.tics.2015.08.006, https://www.sciencedirect.com/science/article/pii/S1364661315001837

13. James, S.: Statistical Mechanics: Entropy, Order Parameters and Complexity. In: Oxford Master Series in Physics, OUP Oxford 14 (2006). https://ezproxy.cul.columbia.edu/login?qurl=https

14. Krupa, V.: Syntactic Typology and Linearization. Tech. Rep. **58**(3), 639-645 (1982). https://doi.org/10.2307/413851

15. Laughlin, S.B., de Ruyter van Steveninck, R.R., Anderson, J.C.: The metabolic cost of neural information. Nat. Neurosci. **1**(1), 36–41 (1998). https://doi.org/10.1038/236, http://www.nature.com/articles/nn0598_36

16. Letzkus, J., Wolff, S., Lüthi, A.: Disinhibition, a circuit mechanism for associative learning and memory. Neuron **88**(2), 264–276 (2015). https://doi.org/10.1016/j.neuron.2015.09.024, https://www.sciencedirect.com/science/article/pii/S0896627315008132

17. Maurits, L.: Representation, information theory and basic word order. Tech. Rep. (2011)

18. Maurits, L., Griffiths, T.L.: Tracing the roots of syntax with Bayesian phylogenetics. In: Proceedings of the National Academy of Sciences of the United States of America, vol 111, no. 37, pp. 13576–13581 (2014). https://doi.org/10.1073/pnas.1319042111. https://pubmed.ncbi.nlm.nih.gov/25192934/

19. Maurits, L., Perfors, A., Navarro, D.: Why are some word orders more common than others? A Uniform Information Density account. Tech. rep., vol. 23, pp. 1585–1593(2010)

20. Papadimitriou, C.H., Vempala, S.S., Mitropolsky, D., Collins, M., Maass, W.: Brain computation by assemblies of neurons. In: Proceedings of the National Academy of Sciences **117**(25), 14464–14472 (2020). https://doi.org/10.1073/pnas.2001893117, https://www.pnas.org/content/117/25/14464

21. Papadimitriou, C.H., Yannakakis, M.: The traveling salesman problem with distances one and two. Math. Oper. Res. **18**(1), 1–11 (1993). http://www.jstor.org/stable/3690150
22. Tomlin, R.S.: Basic word order. Functional principles. No. 1, Cambridge University Press (CUP) (1988). https://doi.org/10.1017/s0022226700011646. https://www.cambridge.org/core/journals/journal-of-linguistics/article/abs/russell-s-tomlin-basic-word-order-functional-principles-london-croom-helm-1986-pp-308/7542AFB4A8B28D651F6E109B810F4C04

From Design of Experiments to Combinatorics of Disasters: A Conceptual Framework for Disaster Exercises

Bernhard Garn, Klaus Kieseberg(✉), Berina Celic, and Dimitris E. Simos

SBA Research, 1040 Vienna, Austria
{bgarn,kkieseberg,bcelic,dsimos}@sba-research.org

Abstract. In this paper, we present a conceptual framework for disaster exercises covering the modelling, generation and post-analysis of conducted exercises. Our proposed conceptual framework makes use of a combinatorial approach for actual disaster exercise generation from the literature. In particular, the scenarios in a disaster exercise generated by our framework collectively provide certain combinatorial sequence coverage guarantees and are also minimizing the overall number of scenarios. These coverage guarantees make the created scenarios ideal for the assessment of relief strategies and allow for easy generation of more extensive and effective exercises and training programs. We explain all the individual stages of our proposed framework and how they are mapped to particular steps in the disaster exercise design process.

1 Introduction

Natural as well as man-made disasters are an increasing threat to the safety of modern humanity. Even in our modern society, the damages caused can be so severe, that they cannot be fully coped with. In extreme cases, damages to critical infrastructures and disruptions of critical services can even lead to human fatalities. The global repercussions of disasters can also be severe: Disasters can cause environmental issues like pollution and mass extinction, have a huge impact on the global economy and lead to humanitarian crises such as water and food scarcity as well as displaced populations. In the last decades, natural disasters have been observed more and more frequently [13]. In 2020 alone, 416 natural disasters have been registered worldwide, not including the SARS COVID-19 pandemic [3].

As a result, analyzing disasters or crises in order to prevent them in the future, or at least lessen the damages resulting from them, also has become increasingly important [22]. Disaster exercises are essential instruments in this endeavor, as they can be employed for a multitude of different purposes, such as the assessment of current and future crisis handling processes and relief strategies, the planning of pre-crisis resource requirements and allocation as well as

© Springer Nature Switzerland AG 2022
D. E. Simos et al. (Eds.): LION 2022, LNCS 13621, pp. 15–26, 2022.
https://doi.org/10.1007/978-3-031-24866-5_2

helping train emergency personnel for actual crises [15] in a controlled environment [8]. To be effective, exercises must provide a safe but realistic environment, reproducing all, or at least certain, characteristics of actual crises, and allow for conclusive evaluation.

In this paper, we present a conceptual framework for disaster exercises, including the process of planning, designing and generating disaster scenarios for the use in disaster exercises. Our proposed conceptual framework utilizes a combinatorial approach for disaster exercise generation first described in [1]. The overall structure of our proposed conceptual framework is visualized in Fig. 1. The intention of our presented conceptual framework is to provide for the automated design and generation of disaster exercises featuring scenarios, which collectively for each generated exercise, provide guaranteed coverage for certain permutations of specified length consisting of events that make up a disaster or crisis. This guaranteed coverage of certain event permutations can help with finding problematic sequences of events in a disaster situation that might have otherwise been overlooked and help to better understand deficiencies and weaknesses of current disaster management strategies. Training exercises generated by our proposed framework provide sets of diverse disaster unfolding story lines for participants and, consequently, assured overall training diversity for emergency responders. Additionally, the underlying combinatorics provide the means to minimize the number of scenarios in each generated disaster exercise, leading to efficient and effective training schedules.

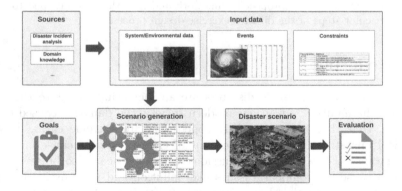

Fig. 1. Structure of the framework for scenario generation.

This paper is structured as follows. In Sect. 2, we provide motivation for our work. We present an overview of our integrated combinatorial framework for disaster exercises in Sect. 3. We conclude this work in Sect. 4, where we also offer an outlook on future work.

2   Motivation

In this section, we contextualize our work to the general domain of disaster management. We give a brief historic perspective on the field of Design of Experiments in Sect. 2.1. In Sect. 2.2, we explain the need for well-designed exercise scenarios in disaster management . We argue how notions of combinatorial sequence coverage can be beneficially made use of in the design of disaster exercises in Sect. 2.3.

2.1   Design of Experiments

Today, the Design of Experiments (DoE) is a well established and recognized branch within statistics, used by both academia and practitioners. DoE provides statistically-based, scientifically objective test strategies for analysis of and conclusions drawn from designed experiments. For a comprehensive treatment of DoE, we refer the interested reader to [19]. Since its inception around 100 years ago, DoE's evolvement and development over the years has had profound impact on test design in various industries. In [19], the following four eras of experimental design have been identified:

1st **Era:** Agricultural experiments spearheaded by work of Sir Ronald A. Fisher [10,11].
2nd **Era:** Industrial era enabled by the formulation of the response surface methodology [5].
3rd **Era:** With the work of Taguchi, the focus shifted towards quality improvements in industrial settings [25].
4th **Era:** Revived activity in statistical design, from both research and application sides.

Modern novel applications of the DoE mentality include combinatorial testing (CT) for software [9,16] and combinatorial security testing (CST) focusing on security testing aspects within information security [24]. These two testing methods combine DoE-inspired testing mentality with appropriate discrete combinatorial structures, which fit better to the considered domains of computer science than *classical* statistical designs from the past. Central to CT and CST is the notion of combinatorial coverage, which also plays a significant role in the research presented in this work. Before we go into the technical details, we take a closer look at the application domain of disaster management next.

2.2   Exercise Scenarios

Disasters and crises can have devastating effects on humankind and therefore pose a great threat to societies everywhere in the world. Because of this, governments and emergency organizations have to be prepared and make efforts to minimize potential damages, to ensure the functionality of important services and to be able to provide crisis relief. Here, proactive measures like exercises can help strengthen the crisis management capability of countries or organizations by providing in-depth analyses of current or new relief strategies and processes, discuss cooperational and jurisdictional aspects and by training emergency personnel in a low-risk environment [8].

Exercises often implement scenarios which specify a possible, but not necessarily probable, context and series of events [26], simulating the events and processes occurring during a disaster or crisis situation. These scenarios can be based on reconstructions of real-life historical disasters or be of completely hypothetical nature and should reproduce the corresponding characteristics and effects of the disaster and feature clear and specific goals or assessment measures [12]. When looking at crisis scenarios on an abstract level, we can differentiate between two building blocks making up a scenario, the *scenario context* and the *crisis*, which are illustrated in Fig. 2. The *scenario context* describes the environment the crisis takes place in [26]. It provides important information such as the geographical and political landscape, the organizations and governmental agencies involved, their structures and relationships, etc. The *crisis* gives a description of the sequence in which crisis-events happen within a specific timeframe [18]. This not only includes events happening during the exercise (*crisis events*), but all events leading up to the crisis (*set-up events*) as well. The events themselves can be of several types with different consequences [18].

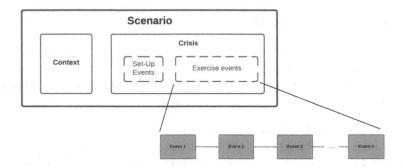

Fig. 2. Abstract structure of a crisis scenario.

Modelling exercise scenarios as sequences of events which are organized in a specific space-time framework [4] provides flexible and highly adaptable scenarios for crisis exercises that make progress easy to monitor and allow for conclusive assessment.

2.3 Event Coverage

In disaster management, the impact a crisis has on a system not only depends on the individual events happening during the crisis and their characteristics, but also on certain parameters of a scenario which specify how the events happen. Amongst these parameters are the overall duration of the scenario, the number of events which make up the scenario, their ordering, the temporal spacing of the events as well as their geographic spread. For example, the order in which events pertaining to the weather, such as changes in weather conditions, happen, might have a tangible effect on the course of a disaster. Using the bushfire-simulation engine SPARK [23], we examined the effects of four different scenarios which feature permutations of event sequences for two types of weather events, wind speed and wind direction, to the spreading of a modelled fire. These four scenarios are based on a historical bushfire disaster which happened in February 2009 near the town of Redesdale in Victoria, Australia, as part of a series of bushfires. The corresponding events have been extracted from weather logs and official disaster records [7] and converted into a time-series. This time-series was then imported alongside other events pertaining to the fire and topographical and land-coverage data into the SPARK engine, where the initial scenario was simulated. Next, three other scenarios were simulated which feature changes in the time series containing the course of the weather events. The results of the simulation of the initial scenario as well as of the three adapted scenarios are depicted in Fig. 3.

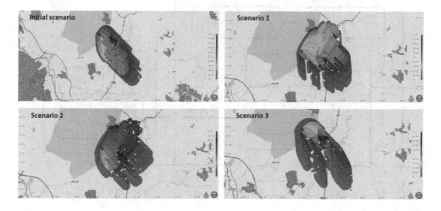

Fig. 3. Simulations pertaining to the 2009 Redesdale fires, Victoria, Australia.

Looking at the results of the simulations, it is evident that the order in which events happen in a scenario has an impact on how the scenario plays out. This means that, to fully test a crisis response plan, testing as many permutations of different scenario parameters as possible is imperative, as the overall response strategy (i.e., regarding the distribution, allocation and coordination of resources and emergency services and personnel) differs on a case-by-case basis and has to be adapted accordingly.

It is obvious that an exhaustive testing approach would be very time consuming and expensive. In fact, for more complex scenarios featuring lots of events it is practically impossible to test out all possible event permutations with exercises. However, utilizing the testing philosophy of *DoE* can help us significantly in this endeavor. Specifically, we make use of certain combinatorial sequence structures from discrete mathematics [6] as means to derive disaster scenarios as outlined in [1], which inherently minimize the overall number of generated scenarios while at the same time guarantee coverage of sub-permutations of selected events.

3　A Conceptual Combinatorial Framework for Disaster Scenario Generation

In this section, we provide an overview of our proposed conceptual framework for combinatorial scenario generation for disaster exercises. We show how the individual concepts and aspects discussed so far come together and are holistically integrated into a single process. In particular, we describe the individual steps of our envisaged process consisting of six stages, which are depicted in Fig. 4.

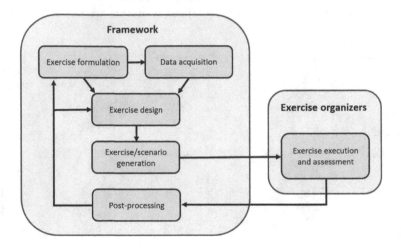

Fig. 4. Process of scenario generation using the framework.

In the first stage, **exercise formulation**, covered in Sect. 3.1, the overall goals of the exercise are discussed and defined. In the stage **data acquisition**, which is considered in Sect. 3.2, sources for information regarding the crisis have to be found and data has to be extracted. This information can include details pertaining to the development of the crisis, the events that occur and the actors that are involved in both the emergence, as well as the combatting, of the crisis. Once data has been acquired the following step, also referred to as the **exercise design** stage, dealt with in Sect. 3.3, is to design the exercise to suit all the desired needs established during the exercise formulation with regards to the information gathered and disaster events identified. In the next stage of the framework, **exercise generation** described in Sect. 3.4, exercise designers have to first select the mathematical (combinatorial) structure most suitable for the previously specified exercise. In particular, translation of events, constraints and properties of the scenario into parameters of the structure and its generation has to be performed. The generated structure containing the scenarios is then implemented into the disaster exercise process, where the exercise can be conducted and the individual scenarios can be evaluated for their impact. This stage is called the **execution and assessment** stage and treated in Sect. 3.5. The findings from this assessment may then be used to alter future exercises by fine-tuning the design or by changing the focus to scenarios similar to ones that performed poorly during the exercise, which can happen in a **post-processing** stage, covered in Sect. 3.6. In the following, we put forth a more in-depth technical description of the individual stages of our proposed conceptual framework.

3.1 Exercise Formulation

Before an exercise can be constructed, exercise designers have to first define the purpose and general goals of the exercise. For example, *crisis management exercises* can be utilized to evaluate crisis relief processes and response measures, whereas *training exercises* focus more on helping emergency services and crisis response units getting acquainted with the processes and measures. Depending on the purpose and goals defined, exercise designers also have to decide on which exercise types suits these purposes best. In this work, we differentiate between three major types of exercises: *tabletop, functional* and *full-scale* exercises [20]. *Tabletop exercises* are discussion-based exercises which are used for the assessment of the relief strategies and processes on an operational and jurisdictional level where participants can familiarize themselves with or discuss existing emergency procedures or develop and test out new ones on a strictly theoretical and strategical basis [17]. *Functional exercises* and *full-scale exercises* are both operational exercises, used to train or validate existing plans, policies and procedures at a more operational level and in a more realistic and real-time environment [15]. While *functional exercises* focus on a single or a small number of very specific functions and processes of a disaster response plan and involve only one or a few organizations, *full scale exercises* involve the whole crisis management as well as multiple agencies, organizations and jurisdictions [14].

The exercise types not only differ in the process and goals of the exercises themselves but also in their characteristics and complexity, as well as in the implementation of the scenarios used. For example, scenarios used in *tabletop exercises* consists of a series of conceptual events discussed from a high-level point of view and are typically presented in a narrative form [21] by a moderator using text handouts accompanied by photos for visualization purposes [12]. *Functional exercises* and *full-scale exercises* feature more realistic scenarios where a lot of focus is also on the operational processes. To successfully convey a certain sense of realism and seriousness, actors are often used simulating victims or other relevant entities like agency officials [14]. Depending on the number of agencies and organizations taking part, different sub-goals also have to be implemented to provide for effective assessment of all the different processes that are part of the relief strategy. Due to the general nature of our framework, the scenarios generated by it can be used for both discussion-based and operational exercises.

3.2 Acquisition of Data

Depending on the purpose and scale of the exercise, a multitude of different information has to be collected in order to generate effective scenarios [2, 13]. This can entail general data regarding the scenario context, such as (systemical) information on the environment and domain the crisis takes place in as well as information regarding specific events affecting the disaster or relief progress.

The acquisition of data usually begins with the sighting and analysis of viable sources of information. In this work, post disaster analysis documents such as official reports and case studies of historical crises can be good starting points. Scanning archives from social media for content created during the time of the crisis can also be a viable source of information, though scenario designers have to be very careful to identify real data from fake. Additionally, crisis events can also be manually identified by experts in the corresponding field the crisis is set in. This is especially important for new technology domains, where there is little historical data available. Once viable data sources have been identified, the information has to be extracted, either manually or by utilizing data-extraction tools, and stored in a knowledge base. Our conceptual framework considers the full scale of modern data mining capabilities and techniques, such as advanced analytics and natural language processing, to enable the extraction of all relevant information and important data points from a multitude of different sources (Fig. 5).

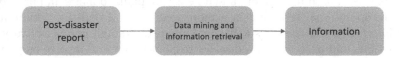

Fig. 5. Illustration of data extraction from a post-disaster report.

3.3 Exercise Design

Once all the information has been successfully allocated and collected, it is systemized with regard to the significance for the overall goals and purposes. This helps identifying the main disaster events of the crisis as well as ones useful for reproducing certain crisis characteristics [18]. An important task here is also to identify event constraints which specify the conditions in which events can occur. The difficulty and scope of the exercise is then determined by setting scenario parameters such as as the number of events occurring, the minimum and maximum length, the desired complexity and constraints of the scenario. Additionally, scenario designers can inject events into the scenario that create urgency and time constraints, add elements of surprise, or present the participants with a dilemma. The exercise complexity, which determines the strength of the sub-permutations that are covered collectively by all scenarios in the exercise, is also set in this stage.

The overall goals determined in the disaster formulation stage can now be fleshed out and specified in order to present exercises participants with achievable and distinct goals [12].

3.4   Exercise Generation

Depending on the properties of the exercise designed in the previous stage, such as scenario length, exercise complexity as well as the occurrence of event and scenario constraints, suitable combinatorial structures have to be identified and the properties translated into the mathematical domain and then mapped to the corresponding parameters of the combinatorial structure. In Fig. 6, which is taken from [1], such a mapping of two exemplary combinatorial structures is shown (for undefined terms and notions we refer to [1] and references there in).

Disaster exercise domain	Combinatorial sequence testing domain
Exercise	Sequence test set (i.e., SCA of strength t or t-way sequence test set with constraints)
Exercise complexity	Strength t
Scenario	Element in sequence test set
Scenario length	Length of test sequence
Story	t-way permutation or target sequence of length t
Events	Event symbols
Multiplicities of events in scenarios	Repetition of event symbols in sequences by constraints
Constraints between events	Constraints between event symbols

Fig. 6. Mapping the scenario properties to the parameters of combinatorial structures.

The combinatorial structures themselves can then be generated by utilizing various mathematical methods and algorithms. After the generation process has finished, the created structure is then transcribed to the desired exercise by mapping the elements of the combinatorial structure to the corresponding disaster scenarios. All artifacts that are constructed during the disaster exercise generation process are stored in the knowledge base for future reference.

3.5   Execution and Assessment

After a disaster exercise has been generated, it is given to the exercise organizers in the form of a set of crisis scenarios, which consist of the disaster events identified earlier and which collectively satisfy the desired strength for the chosen notion of combinatorial sequence coverage. These sets of scenarios can be delivered in different data formats to suit the needs of the management process of the organizations involved.

Depending on the number and complexity of the scenarios, the exercises can be implemented into the disaster management process in a variety of different ways. For example, they might be implemented as smaller regularly occurring *drills* which help with continuously improving the disaster relief processes and proficiency of participants or as a large-scale one-time exercise which could even last for a few days. While conducting the exercise, the progress of the participants as well as the effects on the relief efforts is monitored and evaluated by exercise conductors. This gives a first assessment of the impact the different scenarios have on the system as well as a measure on how the disaster relief strategy and the participants involved are handling them. An illustration of the general assessment procedure for two scenarios is given in Fig. 7.

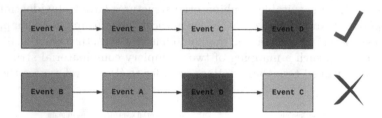

Fig. 7. Illustration of the general assessment procedure for two scenarios.

3.6   Post-processing

Finally the results of the evaluation of the individual scenarios as well as the overall exercise can be used as input for further disaster exercises in the considered domain, as they might be modeled with a bias towards the scenarios that have been identified as *critical* during the assessment. These critical scenarios are then specifically annotated in the knowledge base for further analysis or exercise repetition.

4   Conclusion and Future Work

Based on an analysis of current disaster management approaches, we proposed a DoE-inspired conceptual combinatorial framework for disaster exercises. Our proposed framework is envisioned to be able to parse multiple kinds of data sources via appropriate modern data mining techniques. The integration of domain knowledge enables the semi-automated extraction of relevant data from the input sources. Once a model has been selected, it is translated into a discrete mathematical representation, which then enables the generation of disaster scenarios, that are given to interested organizations and stakeholders, which satisfy required mathematical properties, including guaranteed notions of combinatorial coverage. Once exercises have been conducted, their results can be fed back into the framework for post-processing analysis as well as further exercise tuning. Our framework can be used as part of on overall cycling training schedule.

We see several possibilities for future work. First, we want to evaluate our proposed framework in practice for different kinds of disasters and affected stakeholders. Second, we are interested in broadening our combinatorial modelling and generation framework by analyzing more combinatorial structures for usage in disaster scenario generation to be able to accommodate specialized constraints arising in application domains. Third, the precise mathematical coverage guarantees of disaster exercises generated via our proposed framework could be integrated into general novel vulnerability, risk and impact analysis assessments of disaster threats.

Acknowledgements. SBA Research (SBA-K1) is a COMET Center within the framework of COMET - Competence Centers for Excellent Technologies Program and funded by BMK, BMDW, and the federal state of Vienna. The COMET Program is managed by FFG.

This work was performed partly under the following financial assistance award 70NANB21H124 from U.S. Department of Commerce, National Institute of Standards and Technology.

References

1. Combinatorial sequences for disaster exercise generation. operations research forum currently under review (2021)
2. Ahammed, S., et al: Acquisition, storage, retrieval and dissemination of disaster related data. In: IEEE Region 10 Humanitarian Technology Conference (R10-HTC) (2014). https://doi.org/10.1109/R10-HTC.2014.7026329
3. Benfield, A.: Weather, climate catastrophe insight. 2020 annual report. Aon Benfield. https://www.aon.com/global-weather-catastrophe-natural-disasters-costs-climate-change-2020-annual-report/index.html 1 (2021)
4. Bouget, G., Chapuis, J., Vincent, J.: Conception de scénarios d'attaque de systèmes complexes. In: Workshop 3SGS'09 (2009)
5. Box, G.E.P., Wilson, K.B.: On the experimental attainment of optimum conditions. J. Royal Stat. Soci. Series B (Methodological) **13**(1), 1–38 (1951). https://doi.org/10.1111/j.2517-6161.1951.tb00067.x,https://rss.onlinelibrary.wiley.com/doi/abs/10.1111/j.2517-6161.1951.tb00067.x
6. Colbourne, C., Dinitz, J.: Handbook of Combinatorial Designs, 2nd edn. CRC Press Boca Raton, FL (2007)
7. Commission,V.B.R.: 2009 Victorian Bushfires Royal Commission: final report (2010)
8. for Disease Prevention, E.C., Control: Handbook on simulation exercises in EU public health settings - How to develop simulation exercises within the framework of public health response to communicable diseases. https://www.ecdc.europa.eu/en/publications-data/handbook-simulation-exercises-eu-public-health-settings (2014) (2021)

9. Duan, F., Lei, Y., Kacker, R.N., Kuhn, D.R.: An approach to t-way test sequence generation with constraints. In: 2019 IEEE International Conference on Software Testing, Verification and Validation Workshops (ICSTW), pp. 241–250 IEEE (2019)

10. Fisher, R.A.: The Design of Experiments, 8th ed. Hafner Publishing Company (1966)

11. Fisher, R.A.: Statistical Methods for Research Workers, pp. 66–70. Springer, New York (1992). https://doi.org/10.1007/978-1-4612-4380-9_6. https://doi.org/10.1007/978-1-4612-4380-9_6

12. Grunnan, T., Fridheim, H.: Planning and conducting crisis management exercises for decision-making: the do's and don'ts. EURO J. Decis. Process. 5(1), 79–95 (2017)

13. Guha-Sapir, D.: EM-DAT: The emergency events database. https://www.emdat.bewww.emdat.be

14. of Health, D., Human Services, Health Facilities Administration, S.o.N.H.: Basics of exercise design and administration. https://www.dhhs.nh.gov/oos/bhfa/documents/beda.pdf (2021)

15. Homeland Security, D.: Homeland security exercise and evaluation program (HSEEP). https://www.fema.gov/sites/default/files/2020-04/Homeland-Security-Exercise-and-Evaluation-Program-Doctrine-2020-Revision-2-2-25.pdf (2021)

16. Kuhn, D.R., Kacker, R.N., Lei, Y.: Introduction to combinatorial testing. CRC Press (2013)

17. Leonard Jr, J.J., Roberson, M.G.: Adding realism to tabletop exercises. In: International Oil Spill Conference, vol. 1999, pp. 555–560. American Petroleum Institute (1999)

18. Limousin, P., Tixier, J., Bony-Dandrieux, A., Chapurlat, V., Sauvagnargues, S.: A new method and tools to scenarios design for crisis management exercises. In: CISAP7: 7TH International Conference on Safety and Environment in Process Industry, vol. 53, pp. 319–324 (2016)

19. Montgomery, D.C.: Design and analysis of experiments. John wiley sons (2017)

20. Perry, R.W.: Disaster exercise outcomes for professional emergency personnel and citizen volunteers. J. Contingencies Crisis Manage. 12(2), 64–75 (2004)

21. Peterson, D.M., Perry, R.W.: The impacts of disaster exercises on participants. Disaster Prev. Manag. Int. J. 8(4), 241–255 (1999)

22. Rosenthal, U., Kouzmin, A.: Crises and crisis management: toward comprehensive government decision making. J. Publ. Adm. Res. Theory: J-PART 7(2), 277–304 (1997)

23. Scientific, C., Organisation, I.R.: Spark: A wildfire simulation toolkit. https://research.csiro.au/spark/

24. Simos, D.E., Kuhn, R., Voyiatzis, A.G., Kacker, R., et al.: Combinatorial methods in security testing. Computer 49(10), 80–83 (2016)

25. Taguchi, G.: System of experimental design; engineering methods to optimize quality and minimize costs. Tech. rep. (1987)

26. Walker, W.E.: The use of Scenarios and Gaming in Crisis Management Planning and Training, vol. 7897. Rand Santa Monica, CA (1995)

Separating Two Polyhedra Utilizing Alternative Theorems and Penalty Function

Saeed Ketabchi[1] , Hossein Moosaei[2,3] , Mario R. Guarracino[4(✉)] ,
and Milan Hladík[5]

[1] Department of Applied Mathematics, Faculty of Mathematical Sciences, University of Guilan, Rasht, Iran
sketabchi@guilan.ac.ir

[2] Department of Informatics, Faculty of Science, Jan Evangelista Purkyně University, Ústí nad Labem, Czech Republic
hossein.moosaei@ujep.cz

[3] Prague University of Economics and Business, Prague, Czech Republic

[4] Department of Economics and Law, University of Cassino and Southern Lazio Campus Folcara, Cassino, Italy
mario.guarracino@unicas.it

[5] Department of Applied Mathematics, Faculty of Mathematics and Physics, Charles University, Prague, Czech Republic
hladik@kam.mff.cuni.cz

Abstract. The separation of two polyhedra by a family of parallel hyperplanes is a well-known problem with important applications in operations research,statistics and functional analysis. In this paper, we introduce a new algorithm for constructing a family of parallel hyperplanes that separates two disjoint polyhedra given by a system of linear inequalities. To do this, we consider the alternative system and introduce its dual problem using the alternative theorem. We can find its minimum-norm solution by combining the objective function and constraints into a penalty function. Since our objective function is only once differentiable, we propose an extension of Newton's method to solve the unconstrained objective optimization. The computational outcomes demonstrate the efficacy of the proposed method.

Keywords: Polyhedra · Separating theorem · Theorems of alternative · Penalty function · Generalized

1 Introduction

In computer modeling, a polyhedron is one of the most commonly used geometric solids. For example, polyhedra are used in robotic systems to model obstacles that must be avoided; in computer-aided design, they may represent automobile parts or tools; and in computer graphics or geographical databases, they may

© Springer Nature Switzerland AG 2022
D. E. Simos et al. (Eds.): LION 2022, LNCS 13621, pp. 27–39, 2022.
https://doi.org/10.1007/978-3-031-24866-5_3

represent real-world objects such as mountains or buildings [22]. Classification is a decision-making process that results in the labeling of an observed object as a member of a particular category. Using a training set of labeled observations, it is possible to compute a mathematical model and classify new unlabeled observations. The classification has been utilized in numerous applications, including the prediction of heart disease, lung cancer, or colon tumors, text categorization, computational biology, bioinformatics, and image classification [1,7,9,17,19,28]. A concept related to binary classification is the separation of two polyhedra [3–5]. Separation of parametrized polyhedra was addressed in [15,16].

The method for constructing a family of parallel hyperplanes that separates two disjoint polyhedra was investigated theoretically by Eremin [10] with no computational results. This is a practical version of the Hahn-Banach separating theorem in functional analysis [20,24,26]. A construction of all separating hyperplanes was proposed by Grygarová [12,13]; naturally, it is time expensive to handle all separating hyperplanes.

In this paper, we describe an algorithm for constructing a family of parallel hyperplanes and separating two disjoint polyhedra given by a system of linear inequalities [11]. To find this family, we consider the dual problem of the alternative to this system (Sect. 2). The replacement of the alternative system with the dual problem is advantageous since the dimension of the new variables is less than that of the starting one.

To find the minimum norm solution, we combine the objective function and constraints into a penalty function and obtain an unconstrained quadratic optimization problem. Since the objective function is not twice differentiable, we use the generalized Newton method with a line-search based on the Armijo rule (Sect. 3). The experiment results show the efficiency of our proposed algorithm (Sect. 4). This method is especially efficient for the unconstrained optimization of a piecewise quadratic function [21]. Concluding remarks are given in Sect. 5.

In this paper, vectors are considered as column vectors and we denote the n-dimensional real space by R^n. By $A^\top$, and $\| \cdot \|$, we mean the transpose of matrix A and Euclidean norm, respectively. and, by $(a)_+$ we mean a vector which is obtained from a by replacing the negative components by zero. The symbol f_x stands for the partial derivative of f with respect to x. The complement of a set X is denoted by X^c.

2   Separation of Two Polyhedra

Let $x \in R^n$, and $b \in R^m$, with $\|b\| \neq 0$ be two given vectors and $A \in R^{m \times n}$ ($m > n$) be a rectangular matrix. The linear systems

$$Ax \geq b, \tag{1}$$

and

$$A^\top u = 0_n, \ b^\top u = \rho, \ u \geq 0_m, \tag{2}$$

determine the alternative sets X and U [8,11] defined as:

$$X = \{x \in R^n : Ax \geq b\},$$
$$U = \{u \in R^m : A^\top u = 0_n, \ b^\top u = \rho, \ u \geq 0_m\},$$

where $\rho > 0$ is an arbitrary fixed positive number and 0_i is the zero vector of dimension i.

From (2) we have

$$u^\top (Ax - b) = -\rho. \tag{3}$$

This equality is the key tool for constructing a family of hyperplanes that separate two disjoint and nonempty polyhedra.

Suppose A, b and u are of the form

$$A = \begin{bmatrix} A_1 \\ A_2 \end{bmatrix}, \qquad b = \begin{bmatrix} b_1 \\ b_2 \end{bmatrix}, \qquad u = \begin{bmatrix} u_1 \\ u_2 \end{bmatrix},$$

where A_1 and A_2 are full rank matrices of size $m_1 \times n$ and $m_2 \times n$; $b_1, u_1 \in R^{m_1}$ and $b_2, u_2 \in R^{m_2}$; $m_1 + m_2 = m$. We also suppose that

$$X_1 = \{x \in R^n : A_1 x \geq b_1\}, \qquad X_2 = \{x \in R^n : A_2 x \geq b_2\},$$

are two nonempty sets which determine two polyhedra such that $X = X_1 \cap X_2 = \emptyset$; it means that $Ax \geq b$ is infeasible.

Let $c^\top x - \gamma = 0$ be a hyperplane whose normal vector is $c \in R^n$, with $\|c\| \neq 0$ and γ is a scalar. We say that the hyperplane $c^\top x - \gamma = 0$ separates X_1 and X_2, if $c^\top x - \gamma \geq 0$ for all $x \in X_1$, and $c^\top x - \gamma \leq 0$ for all $x \in X_2$. We can rewrite the systems (1) and (2) and the relation (3) as follows:

$$A_1 x \geq b_1, \quad A_2 x \geq b_2,$$
$$A_1^\top u_1 + A_2^\top u_2 = 0_n, \quad b_1^\top u_1 + b_2^\top u_2 = \rho, \quad u_1 \geq 0_{m_1}, \quad u_2 \geq 0_{m_2}, \tag{4}$$
$$u_1^\top (A_1 x - b_1) + u_2^\top (A_2 x - b_2) = -\rho < 0. \tag{5}$$

Define the linear function $\varphi(x, \alpha) : R^n \times [0, 1] \to R$ with

$$\varphi(x, \alpha) = u_1^\top (A_1 x - b_1) + \alpha \rho. \tag{6}$$

The relation (5) implies that $\varphi(x, \alpha)$ can be equivalently defined as

$$\varphi(x, \alpha) = u_2^\top (b_2 - A_2 x) + (\alpha - 1)\rho. \tag{7}$$

The equality $\varphi(x, \alpha) = 0$, where u_1 and u_2 satisfies (4) and α belongs to $[0, 1]$, determines the hyperplane that separates X_1 and X_2. This means that if $x \in X_1$ then, according to (6), we have $\varphi(x, \alpha) \geq 0$, while if $x \in X_2$ then, according to (7), we have $\varphi(x, \alpha) \leq 0$. The separating hyperplane $\varphi(x, \alpha) = 0$ with $\alpha = 1/2$ was first introduced by Eremin [10,11]. From (6) and (7) the hyperplane $\varphi(x, \alpha) = c^\top x - \gamma = 0$ determines:

$$c = A_1^\top u_1 = -A_2^\top u_2, \qquad \gamma = b_1^\top u_1 - \alpha \rho = -b_2^\top u_2 - (\alpha - 1)\rho,$$

where u_1 and u_2 are arbitrary solutions of the system (4).

For fixed vector $u^\top = [u_1^\top, u_2^\top]$, that satisfies system (4), we define the family of parallel hyperplanes given by the following equivalent definitions:

$$\Gamma(\alpha) = \{x \in R^n : u_1^\top A_1 x - b_1^\top u_1 + \alpha\rho = 0\} = \{x \in R^n : \varphi(x, \alpha) = 0\} \quad (8)$$
$$= \{x \in R^n : -u_2^\top A_2 x + b_2^\top u_2 + (\alpha - 1)\rho = 0\}. \quad (9)$$

The distance between the hyperplanes $\Gamma(1)$ and $\Gamma(0)$ will be called *thickness of the family of hyperplanes.*

We note that the system (4) can be considered as a linear programming problem (its objective function is equal to zero) thus, its dual problem is as follows:

$$\max_{\nu \in R^1} \max_{x \in R^n} \rho\,\nu,$$

subject to

$$Ax + b\nu \leq 0.$$

We know that the penalty function method is applicable to general constrained problems with equality and inequality constraints. Also the starting design point can be arbitrary. Here several penalty functions can be defined. The most popular one is called the quadratic loss function defined as

$$\Theta(x, \nu, \varepsilon) = -\varepsilon\rho\nu + \frac{1}{2}\|(Ax + b\nu)_+\|^2, \quad (10)$$

where Θ is piecewise quadratic, convex, and once differentiable function. By using the above penalty function, we can find the minimum-norm solution to the system (4). Now, we consider the following quadratic programming problem:

$$\min_{x \in R^n} \min_{\nu \in R^1} \Theta(x, \nu, \varepsilon) = -\varepsilon\rho\nu + \frac{1}{2}\|(Ax + b\nu)_+\|^2. \quad (11)$$

Suppose $[x^{*\top}, \nu^*]^\top \in \arg\min_{x,\nu} \Theta(x, \nu, \varepsilon)$. Then the partial derivatives at the optimal solution would be:

$$\Theta_x(x^*, \nu^*, \varepsilon) = A^\top(Ax^* + b\nu^*)_+ = 0_m, \quad (12)$$
$$\Theta_\nu(x^*, \nu^*, \varepsilon) = -\varepsilon\rho + b^\top(Ax^* + b\nu^*)_+ = 0. \quad (13)$$

Therefore,

$$\tilde{u}^* = u(x^*, \nu^*, \varepsilon) = \frac{(Ax^* + b\nu^*)_+}{\varepsilon}, \quad (14)$$

is the solution of (4). From (14) we get

$$\tilde{u}^* - \frac{(Ax^* + b\nu^*)}{\varepsilon} \geq 0_m, \quad \tilde{u}^{*\top}(\tilde{u}^* - \frac{(Ax^* + b\nu^*)}{\varepsilon}) = 0, \quad \tilde{u}^* \geq 0_m. \quad (15)$$

The conditions (12) (13) and (15) are the Kuhn-Tucker optimality conditions [6] of the following problem:

$$\min_{u \in U} \frac{1}{2} \|u\|^2, \quad \text{where} \quad U = \{u \in R^m : A^\top u = 0_n, \ b^\top u = \rho, \ u \geq 0_m\}. \quad (16)$$

By [10,18] there exists some positive $\bar{\varepsilon}$ such that if we choose $\varepsilon \in (0, \ \bar{\varepsilon})$, then $\tilde{u}^* = \frac{1}{\varepsilon}(Ax^* + \nu^* b)_+$ is the minimum-norm solution of (4). Therefore, we have proved:

Theorem 1. *Let* $\Theta(x, \nu, \varepsilon) = -\varepsilon \rho \nu + \frac{1}{2}\|(Ax + b\nu)_+\|^2$ *and suppose* $[x^{*\top}, \nu^*]^\top \in \arg\min_{x, \nu} \Theta(x, \nu, \varepsilon)$. *Then, there exits some positive* $\bar{\varepsilon}$ *such that for all* $\varepsilon \in (0, \ \bar{\varepsilon})$, *vector* $\tilde{u}^* = \frac{1}{\varepsilon}(Ax^* + \nu^* b)_+$ *is the solution of the quadratic problem (16).*

Remark 1. We should note that we prefer to solve the system (1) instead of the system (2) because, in polyhedrons, the number of variables of the system (2) is much greater than the number of variables of the system (1). In addition, with our strategy, we solve not only the system (2) but also find its minimum-norm solution.

The next two theorems suggest a simpler algorithm for finding a family of separating hyperplanes. First, one solves an unconstrained minimization problem in a space of lower dimension and calculates the minimum-norm solution to system (4). Then, one constructs Γ using (8) and (9).

The approach of Eremin is to find an arbitrary solution to the consistent system (4), where the number of unknowns is m. Since we have $n < m$, the approaches suggested by Theorem 2 and Theorem 3 are preferable.

Theorem 2. *Let* $[x^{*\top}, \nu^*]^\top \in \arg\min_{x, \nu} \Theta(x, \nu, \varepsilon)$, $\tilde{u}_1^* = \frac{1}{\varepsilon}(A_1 x^* + \nu^* b_1)_+$, *and* $\tilde{u}_2^* = \frac{1}{\varepsilon}(A_2 x^* + \nu^* b_2)_+$ *then:*

1. $A_1^\top \tilde{u}_1^* = -A_2^\top \tilde{u}_2^* \neq 0$ *and* $c^\top x = \alpha \rho$, *for every* $\alpha \in [0,1]$, *separates two polyhedrons* X_1 *and* X_2 *where* $c = A_1^\top \tilde{u}_1^*$.
2. $\nu^* > 0$ *and if* $z^* = (\frac{-x^*}{\nu^*})$, *then* $z^* \in X_1^c \cap X_2^c$.

Proof. To prove (1), we suppose the converse, i.e. $A_1^\top \tilde{u}_1^* = 0$. Then $[x^{*\top}, \nu^*]^\top$ is a solution of the convex quadratic problem $\min_{x, \nu} \frac{1}{2}\|(A_1 x + b_1 \nu)_+\|^2$ and since $X_1 \neq \emptyset$, this implies that $\tilde{u}_1^* = (A_1 x^* + \nu^* b_1)_+ = 0$ and $b_1^\top \tilde{u}_1^* = 0$. Therefore, the system

$$A_2^\top u_2 = 0_n, \quad b_2^\top u_2 = \rho, \quad u_2 \geq 0_{m2} \quad (17)$$

has a solution and its alternative system $A_2 x \geq b_2$ is inconsistent which is a contradiction. Now, $c^\top x = \alpha \rho$ for every $\alpha \in [0,1]$, separates two polyhedra X_1 and X_2 since $\tilde{u}_1^*$ and $\tilde{u}_2^*$ satisfies (4).

To prove (2), first note that from (13) and (14), we have

$$\rho = b^\top \tilde{u}^*. \quad (18)$$

From (13) and the optimal solution of (10), we obtain

$$\Theta = -\varepsilon\rho\nu^* + \frac{1}{2}\varepsilon^2\|u^*\|^2, \tag{19}$$

by substituting (18) into (19), we have

$$\Theta = -\varepsilon b^\top \tilde{u}^*\nu^* + \frac{1}{2}\varepsilon^2\|u^*\|^2.$$

Now we add $\pm\frac{1}{2}\|b\nu^*\|^2$ to the above equation and get the following function:

$$\Theta = \frac{1}{2}\|\varepsilon u^* - b\nu^*\|^2 - \frac{1}{2}\|b\nu^*\|^2.$$

The minimum of this function will happen when: $\varepsilon u^* = b\nu^*$, by multiplying the both side of this equality in $\tilde{u}^{*\top}$, we have $\frac{\rho\nu^*}{\varepsilon} = \|\tilde{u}^*\|^2$ which implies that $\nu^* > 0$ and from here we obtain that $(Az^* - b)_+ = \frac{\varepsilon}{\nu^*}\tilde{u}^*$. This means $(A_1 z^* - b_1)_+ \neq 0$ and $(A_2 z^* - b_2)_+ \neq 0$ and therefore, $z^* \notin X_1 \cup X_2$. $\qquad\square$

The following theorem is proved in [11], but our proof essentially differs and is more straightforward than that mentioned in [11]. Let us $A \in R^{m\times n}$ be a given matrix and $b \in R^m$ a vector such that $X = \{x \in R^n : Ax \geq b\} = \emptyset$ and X^* be the solution set of the following problem:

$$\min_x \frac{1}{2}\|(b - Ax)_+\|^2. \tag{20}$$

Then we have:

Theorem 3. *If A_s and $A_{s'}$ be two arbitrary submatrices of A and, respectively b_s and $b_{s'}$ two subvectors of b such that,*

$$A = \begin{bmatrix} A_s \\ A_{s'} \end{bmatrix}, \quad b = \begin{bmatrix} b_s \\ b_{s'} \end{bmatrix},$$
$$X_s = \{x \in R^n : A_s x \geq b_s\} \neq \emptyset,$$
$$X_{s'} = \{x \in R^n : A_{s'} x \geq b_{s'}\} \neq \emptyset.$$

Then

$$X^* \subseteq \Gamma_s(\alpha),$$

where $\Gamma_s(\alpha) = \{x \in R^n : u_s^\top A_s x - b_s^\top u_s + \alpha\rho = 0\}$, $\alpha \in [0,1]$ is arbitrary, and $u = [u_s; u_s']$ is the solution of (16).

Proof. Let x^* be an optimal solution of the problem (20).
This problem, is the dual of the following optimization problem :

$$\max_u \ b^\top u - \frac{1}{2}\|u\|^2 \ \text{subject to} \ A^\top u = 0, \ u \geq 0. \tag{21}$$

According to the Kuhn-Tucker optimality conditions, we obtain that, $z^* = (b - Ax^*)_+$ is the optimal solution of (21) [11]. Now we consider the following quadratic problem:

$$\min_{u \in U} \frac{1}{2}\|u\|^2, \quad \text{where} \quad U = \{u \in R^m : A^\top u = 0_n, \ b^\top u = \|z^*\|^2, \ u \geq 0_m\}. \quad (22)$$

It is obvious that the optimal solution of (22) is also z^* and if $z_s^* = (b_s - A_s x^*)_+$ then, by defining $\alpha = \|z_s^*\|^2/\|z^*\|^2$ in $\Gamma_s(\alpha)$, we have

$$\Gamma_s(\alpha) = \{x \in R^n : z_s^{*\top} A_s x - b_s^\top z_s^* + \|z_s^*\|^2 = 0\}. \quad (23)$$

The hyperplane (23) separates two polyhedra X_s and $X_{s'}$ and $x^* \in \Gamma_s(\alpha)$. This completes the proof.   □

Algorithm 1. Generalized Newton Method with the Armijo Rule

Input: Choose any $p_0 = [x_0^\top, \nu_0]^\top \in R^{n+1}$ and let $\epsilon > 0$ be an error tolerance
    and $i = 0$.
    while $\|\nabla\Theta(p_i)\|_\infty \geq \epsilon$ **do**
        Choose $\alpha_i = \max\{1, \frac{1}{2}, \frac{1}{4}, \ldots\}$ *such that*
        $\Theta(p_i) - \Theta(p_i + \alpha_i d_i) \geq -\alpha_i \mu \nabla\Theta(p_i)d_i$,
        where $d_i = -\nabla^2\Theta(p_i)^{-1}\nabla\Theta(p_i)$, $s > 0$ is a constant, $\delta \in (0,1)$ and $\mu \in$
        $(0,1)$.
        $p_{i+1} = p_i + \alpha_i d_i$
        $i = i + 1$.
    end while

---

3   Algorithm

In this section, we describe an Algorithm to solve the unconstrained optimization problem (11). In this problem, our objective function is $\Theta(x, \nu, \varepsilon) = -\varepsilon\rho\nu + \frac{1}{2}\|(Ax + b\nu)_+\|^2$. This function is piecewise quadratic, convex, and differentiable, but it is not twice differentiable.

Suppose $x, y \in R^n$, and $\nu, u \in R$. Then for gradient of $\Theta(x, \nu, \varepsilon)$ we have

$$\|\nabla\Theta(x, \nu, \varepsilon) - \nabla\Theta(y, u, \varepsilon)\| \leq \|B\|\|B^\top\|(\|x - y\|^2 + \|\nu - u\|^2)^{\frac{1}{2}}$$

where $B = \begin{bmatrix} A & 0 \\ 0 & b \end{bmatrix}$.

This means $\nabla\Theta$ is globally Lipschitz continuous with constant $K = \|B\|\|B^\top\|$. Thus, for this function the generalized Hessian exists and is defined by the $2m \times 2m$ symmetric positive semidefinite matrix [7, 14, 18, 21, 25, 27].

$$\nabla^2\Theta(x, \nu, \varepsilon) = BD(z)B^\top,$$

where $D(z)$ denotes an $(n+1) \times (n+1)$ diagonal matrix with i-diagonal element z_i equals to 1 if $(Ax + b\nu)_i > 0$ and, equal to 0 otherwise.

Therefore we can use the generalized Newton method for solving this problem and to obtain global termination we should use a line-search algorithm [23]. In Algorithm 1, we apply the generalized Newton method with a line-search based on the Armijo rule [2].

In this algorithm, the generalized Hessian may be singular, thus we use a modified Newton direction Cholesky factorizations as the following:

$$M^\top M = (\nabla^2 \Theta(p_k) + \gamma I_m), \quad d_k = -(M^\top M)^{-1} \nabla \Theta(p_k),$$

where M is an upper triangular matrix, γ is a small positive number and I_m is the identity matrix of order m.

Now we introduce the following iterative process:

$$p^{k+1} = p^k + \alpha_k d_k,$$

and if

$$p^* = [x^{*\top}, \nu^*]^\top = \arg\min_{x,\ \nu}\{\Theta(x, \nu, \varepsilon) = -\varepsilon\rho\nu + \frac{1}{2}\|(Ax + b\nu)_+\|^2,$$

then, $\tilde{u}^* = \frac{1}{\varepsilon}(Ax^* + \nu^*b)_+$ is the solution of the quadratic problem (16). The proof of the finite global convergence of this algorithm is given in [2,21].

4   Numerical Results

We begin this section by providing a numerical example that demonstrates the correctness of the theory presented in the preceding sections. We then present some numerical results on various randomly generated problems to show the ability and efficiency of the proposed algorithm.

All the experiments were conducted by a computer with these specifications: Windows 11 Home with 16-GB RAM AMD Ryzen 7 5800H 3.20GHz with MAT-LAB R2019a.

Here, we illustrate the results of Sects. 2 and 3 with the following example.

Example 1. Suppose X_s and $X_{s'}$ are two polyhedra given by a system of linear inequalities. Let also A_s and $A_{s'}$ be the related matrices, b_s and $b_{s'}$ the related vectors as follows:

$$A_s = \begin{bmatrix} -2 & -2 & -5 \\ 0 & -3 & 6 \\ -2 & -3 & 5 \end{bmatrix}, \ A_{s'} = \begin{bmatrix} 3 & 6 & 0 \\ -5 & 5 & -4 \\ 4 & 0 & 6 \\ -6 & 2 & 5 \\ -7 & -2 & 3 \end{bmatrix}, \ b_s = \begin{bmatrix} -34 \\ 3 \\ -12 \end{bmatrix}, \ bs' = \begin{bmatrix} 23 \\ -55 \\ 51 \\ -17 \\ -44 \end{bmatrix}.$$

For this example we see that

$$X_s = \{x \in R^n : A_s x \geq b_s\} \neq \emptyset,$$
$$X_{s'} = \{x \in R^n : A_{s'} x \geq b_{s'}\} \neq \emptyset.$$

By solving the problem (11) and using of the Theorem 2 we obtain that

$$x^* = \begin{bmatrix} -9.9313 \\ -0.8710 \\ -6.3091 \end{bmatrix}, \quad \tilde{u}^* = \begin{bmatrix} \tilde{u}_s^* \\ \tilde{u}_{s'}^* \end{bmatrix} = \begin{bmatrix} 1 \\ 0 \\ 0 \\ 0.2581 \\ 0 \\ 0.6452 \\ 0.2258 \\ 0 \end{bmatrix}, \quad \nu^* = 1.5338.$$

We also see that X_s and $X_{s'}$ are two disjoint polyhedra, since $\|u^*\| = 1.2385$. By computing $c = A_s^\top \tilde{u}_s^* = \begin{bmatrix} -2, -2, 5 \end{bmatrix}^\top$ and putting $\gamma = 34$, we find the separating hyperplane $c^\top x - \gamma = 0$. We plot this example in Fig. 1 and this figure shows that we could separate two mentioned polyhedra correctly.

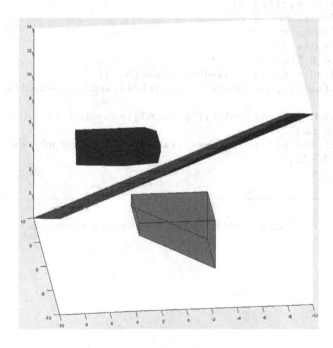

Fig. 1. Separation two polyhedra

By writing a program to generate a random matrix A for a given m, n and d (*density*) and then using our proposed method, we can learn more about the algorithm's computational ability. The following MATLAB code (Fig. 2) is how we came up with the random problems.

Computational results for the test problems are given in the following Table that are based on the penalty function (Theorem 2). Table 1 reports the following information for each test problem:

m, n, d a random $m \times n$ sparse matrix A, with uniformly distributed nonzero entries and density d ($0 \le d \le 1$).

$\tilde{u}^* = \frac{1}{\varepsilon}(Ax^* + \nu^* b)_+$ the solution of the quadratic problem (16).

f indicates $\|A^\top \tilde{u}^*\|_\infty$.

x_s^* the solution of the problem $\min_x \frac{1}{2}\|(b_s - A_s x_s^*)_+\|^2$.

$x_{s'}^*$ the solution of the problem $\min_x \frac{1}{2}\|(b_{s'} - A_s x_{s'}^*)_+\|^2$.

```
% Generate random insolvable system Ax >= b; and solvable
systems A1x >= b1, A2x >= b2;
% Input: m,n,d (density);
% Output: A, b, A1, b1, A2, b2;
pl=inline((abs(x)+x)/2);
m=input('enter m=' );
n=input('enter n= ');
d=input('enter d=' );
m1=m-round(0.5*m);
A1=sprand(m1,n,d);A1= (A1-0.5*spones(A1));
x=spdiags(ones(n,1)-sign(x),0,n,n)*10*(rand(n,1)-rand(n,1));
m2=m-m1;
b1=A1*x-spdiags((rand(m1,1)),0,m1,m1))*ones(m1,1);
u=randperm(m2);
A2=A1(u,:); b2=b1(u)- spdiags((rand(m2,1)),0,m2,m2)*ones(m2,1);
A=(5.)*[A1;-A2];
b=[b1;-b2];
rho=b'u;
```

Fig. 2. Code generation for random systems.

Table 1. Separating two polyhedra X_s and $X_{s'}$

m, n, d	f	ρ	$\|(b_s - A_s x_s^*)_+\|_\infty$	$\|(b_{s'} - A_{s'} x_{s'}^*)_+\|_\infty$
100,90,1	5.5843e-09	6.6781	4.0539e-10	1.2501e-14
200,150,1	1.1806e-008	1.1747e+001	9.6142e-010	8.3114e-010
400,300,1	8.8863e-008	2.9715e+001	1.6663e-009	1.8356e-009
500,300,.1	1.2571e-013	4.0651e+001	6.8556e-011	7.6328e-011
700,600,0.1	3.5599e-013	6.1905e+001	5.1154e-011	4.4498e-011
850,700,0.1	3.4905e-013	7.3053e+001	5.3766e-011	1.5113e-010
1000,800,0.1	3.6573e-013	7.9620e+001	3.0925e-010	1.2002e-010
1200,1000,0.1	7.9953e-013	9.9921e+001	5.0549e-011	3.5756e-010
1500,1200,0.1	8.6886e-013	1.3227e+002	1.2325e-010	2.2253e-010
1800,1500,0.1	1.0802e-012	1.5005e+002	2.0003e-010	2.9962e-010
2000,1900,0.01	6.4948e-014	1.6370e+002	1.1914e-011	2.1584e-011
2500,2000,0.01	7.3164e-014	2.0902e+002	1.7203e-011	3.6915e-011
3000,2500,0.01	9.5479e-014	2.3541e+002	5.3289e-011	2.7744e-011
4000,3000,0.01	1.5038e-013	3.3336e+002	3.1782e-011	5.9421e-011
5000,4000,0.01	2.5380e-013	4.0755e+002	5.8658e-011	1.2718e-010
7000,6000,0.001	7.6156e-011	5.7671e+002	4.8485e-011	6.1937e-012
10000,8000,0.001	5.1955e-013	8.3709e+002	1.4789e-011	5.5633e-011

The results in Table 1 show that $f = \|A^\top \tilde{u}^*\|_\infty$ is almost zero, ρ is positive, and also $\|(b_s - A_s x_s^*)_+\|_\infty$ and $\|(b_{s'} - A_{s'} x_{s'}^*)_+\|_\infty$ are almost zeros for all instances. Then the system (2) is feasible and the system (1) is infeasible, it means we could separate two polyhedrons successively.

5   Conclusion

In this paper, we have studied an algorithm to obtain a family of parallel hyperplanes that separate two disjoint nonempty polyhedrons. Our proposed algorithm is based on the penalty function (10) (Theorem 2). Also, we have proposed a fast generalized Newton method with the Armijo rule to solve our related unconstrained optimization problem. The numerical results illustrate our proposed method's effectiveness and performance, even for large-scale problems. In addition, there is another method for separating polyhedrons which is s based on the minimization problem (20) (Theorem 3) [11]. In this paper, we presented a new proof of this theorem. We should note that the approaches suggested by Theorems 2 and 3 are preferable to Eremin's approach. For future work, we can use kernels to separate two not-intersecting polyhedra that are not linearly separable.

Acknowledgments. H. Moosaei was supported by the Czech Science Foundation Grant 22-19353S. M. Hladík was supported by the Czech Science Foundation Grant P403-22-11117S.

Conflicts of Interest. The authors declare that they have no conflict of interest.

References

1. Arabasadi, Z., Alizadehsani, R., Roshanzamir, M., Moosaei, H., Yarifard, A.A.: Computer aided decision making for heart disease detection using hybrid neural network-Genetic algorithm. Comput. Methods Programs Biomed. **141**, 19–26 (2017)
2. Armijo, L.: Minimization of functions having lipschitz continuous first partial derivatives. Pac. J. Math. **16**(1), 1–3 (1966)
3. Astorino, A., Gaudioso, M.: Polyhedral separability through successive LP. J. Optim. theory appl. **112**(2), 265–293 (2002)
4. Astorino, A., Francesco, M.D., Gaudioso, M., Gorgone, E., Manca, B.: Polyhedral separation via difference of convex (DC) programming. Soft. Comput. **25**(19), 12605–12613 (2021). https://doi.org/10.1007/s00500-021-05758-6
5. Astorino, A., Fuduli, A.: Support vector machine polyhedral separability in semisupervised learning. J. Optim. Theory Appl. **164**(3), 1039–1050 (2015)
6. Bazaraa, M.S., Sherali, H.D., Shetty, C.M.: Nonlinear Programming: theory and algorithms. John Wiley Sons (2013)
7. Bazikar, F., Ketabchi, S., Moosaei, H.: DC programming and DCA for parametric-margin ν-support vector machine. Appl. Intell. **50**(6), 1763–1774 (2020)
8. Broyden, C.: On theorems of the alternative. Optim. methods softw. **16**(1–4), 101–111 (2001)
9. Cai, Y.D., Ricardo, P.W., Jen, C.H., Chou, K.C.: Application of SVM to predict membrane protein types. J. Theor. Biol. **226**(4), 373–376 (2004)
10. Eremin, I.I.: Theory Linear Optim. VSP, Utrecht (2002)
11. Evtushenko, Y.G., Golikov, A.I., Ketabchi, S.: Numerical methods for separating two polyhedra. In: Large-Scale Nonlinear Optimization, pp. 95–113. Springer (2006)
12. Grygarová, L.: A calculation of all separating hyperplanes of two convex polytopes. Optimization **41**(1), 57–69 (1997). https://doi.org/10.1080/02331939708844325
13. Grygarová, L.: On a calculation of an arbitrary separating hyperplane of convex polyhedral sets. Optimization **43**(2), 93–112 (1998). https://doi.org/10.1080/02331939808844377
14. Hiriart-Urruty, J.B., Strodiot, J.J., Nguyen, V.H.: Generalized hessian matrix and second-order optimality conditions for problems with $C^{1,1}$ data. Appl. Math. Optimi. **11**(1), 43–56 (1984)
15. Hladík, M.: Separation of convex polyhedral sets with column parameters. Kybernetika **44**(1), 113–130 (2008)
16. Hladík, M.: On the separation of parametric convex polyhedral sets with application in MOLP. Appl. Math. **55**(4), 269–289 (2010)
17. Javadi, S.H., Moosaei, H., Ciuonzo, D.: Learning wireless sensor networks for source localization. Sensors **19**(3), 635 (2019)
18. Kanzow, C., Qi, H., Qi, L.: On the minimum norm solution of linear programs. J. Optim. Theory Appl. **116**(2), 333–345 (2003)

19. Ketabchi, S., Moosaei, H., Razzaghi, M., Pardalos, P.M.: An improvement on parametric ν-support vector algorithm for classification. Ann. Oper. Res. **276**(1–2), 155–168 (2019)
20. Kundakcioglu, O.E., Seref, O., Pardalos, P.M.: Multiple instance learning via margin maximization. Appl. Numer. Math. **60**(4), 358–369 (2010)
21. Mangasarian, O.: A Newton method for linear programming. J. Optim. Theory Appl. **121**(1), 1–18 (2004)
22. Mitchell, J.S., Suri, S.: Separation and approximation of polyhedral objects. Comput. Geom. **5**(2), 95–114 (1995)
23. Nocedal, J., Wright, S.: Numer. Optim. Springer, New York (2006)
24. Pardalos, P.M.: Complexity Numer. Optim. World Scientific, Singapore (1993)
25. Pardalos, P.M., Ketabchi, S., Moosaei, H.: Minimum norm solution to the positive semidefinite linear complementarity problem. Optimization **63**(3), 359–369 (2014)
26. Rudin, W.: Functional Analysis. McGraw-Hill, New York (1991)
27. Salahi, M., Ketabchi, S.: Correcting an inconsistent set of linear inequalities by the generalized Newton method. Optim. Methods Softw. **25**(3), 457–465 (2010)
28. Wang, X.Y., Wang, T., Bu, J.: Color image segmentation using pixel wise support vector machine classification. Pattern Recogn. **44**(4), 777–787 (2011)

Contributed Papers

(Continued on page

A Composite Index Method
for Optimization Benchmarking

Yulan Bai(✉) 📵 and Eli Olinick📵

Department of Operations Research and Engineering Management,
Southern Methodist University, Dallas, TX 75205, USA
{yulanb,olinick}@smu.edu

Abstract. We propose a multi-criteria Composite Index Method (CIM)
to compare the performance of alternative approaches to solving an optimization problem. The CIM is convenient in those situations when neither approach dominates the other when tested on different sizes of problem instances. The CIM takes problem instance size and multiple performance criteria into consideration within a weighting scheme to produce
a single number that measures the relative improvement of one alternative over the other. Different weights are given to each dimension based
on their relative importance as determined by the end user. We summarize the successful application of the CIM to an $\mathcal{NP}$-hard combinatorial
optimization problem known as the backhaul profit maximization problem (BPMP). Using the CIM we tested a series of eleven techniques
for improving solution time using CPLEX to solve two different BPMP
models proposed in the literature.

Keywords: Performance · Benchmarking · Testing · Metric ·
Timing · Index · Routing · Backhaul · Pickup · Dropoff

1 Introduction

Using solution time as the key performance measure is a long-standing standard
practice in the optimization literature. However, now that computing environments take advantage of multiple processors and multiple threads while supporting concurrent running of multiple CPU-intensive processes have become
commonplace, measuring solution time is no longer straight-forward. Furthermore, it is often the case that there is a "crossover point" in problem instance
size below which one approach is generally "faster" than another, but above
which the second approach is faster. In this situation the second approach would
usually be favored because the emphasis in the literature is on solution time
as a function of problem instance size. In this study, however, we consider the
practical question of making a recommendation to a user who frequently solves
problems that range in size around the crossover point, and propose a multi-
criteria framework for comparing competing solution approaches. We propose
a Composite Index Method (CIM) that considers several weighted performance

ⓒ Springer Nature Switzerland AG 2022
D. E. Simos et al. (Eds.): LION 2022, LNCS 13621, pp. 43–57, 2022.
https://doi.org/10.1007/978-3-031-24866-5_4

measure factors and calculates a single number (a composite index) to measure the relative performance of two competing solution approaches.

The CIM was developed to evaluate two proposed mixed integer programming (MIP) formulations of the backhaul profit maximization problem (BPMP), the node-arc and triples formulations, each of which can be enhanced with a variety of solution techniques (e.g., branching-rules and cutting planes). The results of our application of the CIM to the BPMP are discussed in [10]. In this paper we focus on the process of using the CIM to arrive at the final "candidate" models in [10]. This falls into the area of optimization benchmarking. Beiranvand et al. [5] provide a recent comprehensive review of the benchmarking literature for optimization problems. As far as we know, the first published study in optimization benchmarking was by Hoffman et al. [13], in which different methods were proposed for linear programming and different test instances were used to compare algorithms based on the measures of CPU time, number of iterations, and convergence rate. Another important early paper by Box [6] considered the importance of model size and the number of function evaluations during comparison. Later, many researchers explored optimization benchmarking in various applications such as unconstrained optimization, nonlinear least squares, global optimization, and derivative-free optimization. Crowder et al. [8] proposed standards and guidelines for benchmarking algorithms. According to [5], the Performance Profile proposed by Dolan and Moré [9] has become the "gold standard" for optimization benchmarking (over 4,000 citations so far).

Given a set of solution approaches to an optimization problem, the procedure for using the Performance Profile may be summarized as follows. First, a single performance measure is selected (usually the computing time). Second, all candidate solution approaches are applied to each of a set of problem instances and the best-performing approach for each of the tested instances is used as the benchmark for assessing the performance of all of the other candidate approaches on that particular problem instance. In the case of CPU time as the selected metric, the relative performance of a particular approach on a particular instance is measured by a performance ratio obtained by dividing the CPU time of that particular approach by the CPU time of the best-performing approach. For a particular solution approach, the Performance Profile plots the cumulative distribution function of the performance ratio, the percentage of instances for which the ratio is less than x, over the range $1 \leq x \leq \infty$.

In cases such as our BPMP study where the Performance Profiles of the candidate approaches intersect and cross each other, it may be unclear which approach is the best overall. Problem size is an important consideration in the BPMP use case; this makes the Performance Profile inappropriate since it treats each problem instance equally regardless of its size. Furthermore, the Performance Profile only uses a single performance measure, which makes it difficult to use when there are multiple performance criteria such as when comparing the trade-off between solution time and solution quality with heuristics, or executing a solution approach on a system with multiple processors and/or threads.

To the best of our knowledge, parallel computing has received much less attention in the literature on optimization benchmarking despite the fact that it is now widely used in applied optimization. Barr and Hickman did pioneering studies in this area [3,4] and proposed solutions to the challenges parallelization brings to benchmarking. However, they did not suggest using a single measure for easy comparison. Hence, the CIM is an initial step in closing a gap in the literature.

The rest of this paper is structured as follows. We propose our Composite Index Method for benchmarking in Sect. 2. We describe the BPMP and node-arc and triples formulations in Sect. 3. We illustrate the application of the CIM to the node-arc formulation in Sect. 4 and summarize our results from applying the CIM to the triples formulation in Sect. 5. We draw conclusions in Sect. 6.

2   Performance Evaluation Using Composite Index Method (CIM)

We use the CPLEX MIP solver [14] to solve the BPMP instances described in [2,10]. There are three kinds of "solution time" in the CPLEX output: "CPU time", "real time", and "ticks". CPU time is a measure of the total time used by CPLEX to find an optimal solution; it is the total time used by all threads. Real time (also called wall clock time) is the time that elapsed during the CPLEX run. Both measures can vary noticeably between runs with identical input on identical hardware. Therefore, we solve each problem instance three times in each experiment and report the average CPU and real time over the three runs. The tick metric, also called deterministic time, is a proprietary measure of computation effort based on counting the number of instructions executed by the CPLEX solver and therefore shows no variation between multiple runs with the same inputs on a given hardware configuration [7].

For each of the time measures described above, we use a speedup measure to compare the solution time of two solution approaches, approach 1 versus approach 2. Note that in the BPMP application described herein, a solution approach is essentially a MIP model for the BPMP implemented in AMPL [1] and solved with CPLEX. In general, a solution approach could be a combination of a MIP model and an optimization algorithm. Hereinafter, Speedup is defined as the ratio

$$\text{Speedup} = (\text{Model 1 solution time}) \div (\text{Model 2 solution time}).$$

If Speedup > 1, model 2 is solved Speedup times faster than model 1; if Speedup $= 1$, model 2 has the same solution time as model 1, and if Speedup < 1, model 1 is solved 1/Speedup times faster than model 2. Due to the fact that CPU and real time are not completely reproducible, we suggest that neither one should be the sole basis for comparing solution approaches. Typically, ticks and real time are positively correlated (as are ticks and CPU time), however there does not appear to be a fixed relationship between ticks and the two time measures. For this reason, we cannot use ticks as the single measure to compare two models either. Instead, we propose a weighted combination of all three time measures.

For a given problem size, n, and timing measure (CPU time, real time, or ticks), we calculate a composite index based on a weighted combination of the minimum, median, mean, and maximum speedups among a set of problem instances. Thus, we obtain three composite indices: C_n, R_n, T_n for CPU time, real time, and ticks, respectively. To calculate these indices we denote the minimum, mean, median, and maximum speedups in CPU time by C_{min}, C_{mea}, C_{med}, and C_{max}, respectively, and define R_{min}, R_{mea}, R_{med}, R_{max}, T_{min}, T_{mea}, T_{med}, and T_{max} as the corresponding speedups for real time and ticks. Additionally, we define ω_{min}, ω_{mean} ω_{med} and ω_{max} for weighting of the minimum, mean, median and maximum statistics. We also define $\bar{\omega}$ as the summation of ω_{min}, ω_{mean}, ω_{med} and ω_{max}. The relative weights for CPU, real time, and ticks are ω_c, ω_r, and ω_t, respectively. Using this notation, the three composite indices are calculated as follows:

$$C_n = (\omega_{min}C_{min} + \omega_{mea}C_{mea} + \omega_{med}C_{med} + \omega_{max}C_{max})/\bar{\omega}$$
$$R_n = (\omega_{min}R_{min} + \omega_{mea}R_{mea} + \omega_{med}R_{med} + \omega_{max}R_{max})/\bar{\omega}$$
$$T_n = (\omega_{min}T_{min} + \omega_{mea}T_{mea} + \omega_{med}T_{med} + \omega_{max}T_{max})/\bar{\omega}$$

Next, we calculate a composite index, I_n, for problem size n as a weighted combination of indices C_n, R_n, and T_n:

$$I_n = (\omega_c C_n + \omega_r R_n + \omega_t T_n)/(\omega_c + \omega_r + \omega_t).$$

Given a set of problem instance sizes, $\mathcal{S}$, and weight ω_s for each $s \in \mathcal{S}$, we calculate the grand composite index (GCI), which is the weighted sum of the composite indices for each problem size given by

$$\text{GCI} = (\sum_{s \in \mathcal{S}} \omega_s I_s)/\sum_{s \in \mathcal{S}} \omega_s.$$

If the grand composite index GCI > 1, we say that model 2 performs GCI times better than model 1; if GCI $= 1$, model 2 performs the same as model 1, and if GCI < 1, model 1 performs $1/$GCI times better than model 2.

In summary, the composite index method (CIM) seeks to find a single number, GCI, in a parallel computing environment, to decide which solution approach is better, through instance testing. To do so, we first need to decide the performance measures; usually more than one measure is needed. Second, multiple runs are needed to reduce the variance of the measures for the same instance by averaging the measure. Third, for a fixed problem size, multiple instances should be randomly sampled. Along with the mean measure over the different instances of the same problem size, we consider the minimum (min), median, and maximum (max) measure to diminish the effects of outliers. The consideration of min, mean, median and max, is inspired by the famous PERT concept of project management, in which pessimistic, optimistic, and most likely task-completion times are considered with different weights. Finally, comparisons are made over a range of problem sizes and weighted accordingly.

The steps described above are for comparing two solution approaches. In Sect. 4 we describe how we apply CIM iteratively to compare multiple solution approaches. We illustrate this iterative process by applying it to the BPMP in Sects. 4 and 5.

3   The Backhaul Profit Maximization Problem (BPMP)

The BPMP requires simultaneously solving two problems: (1) determining how to route an empty delivery vehicle back from its current location to its depot by a scheduled arrival time, and (2) selecting a profit-maximizing subset of delivery requests between various locations on the route subject to the vehicle's capacity. Figure 1 illustrates a BPMP instance and solution.

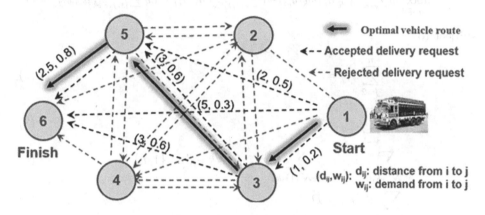

Fig. 1. BPMP example.

Figure 1 shows a network representation of the problem with an empty vehicle at a location represented by node 1. The vehicle weighs 1 ton and has a carrying capacity of $Q = 2$ tons of cargo. The vehicle needs to return to its depot, represented by node 6, within a fixed period of time. The vehicle's average traveling speed limits the route to node 6 to a maximum distance of 7 mi. The vehicle can make extra money by accepting delivery requests to pick up cargo at the locations represented by nodes 1 through 5, destined for locations represented by nodes 2 through 6 as long as it can get back to the depot on time. The tuple (d_{ij}, w_{ij}) indicates the distance (in miles) and the size of the delivery request (in tons) from node i to node j. The optimal solution indicated in Fig. 1 routes the vehicle on the path represented by the arc sequence (1, 3), (3, 5), (5, 6).

BPMP was first introduced by Dong et al. [10]. Yu and Dong [16] proposed a MIP formulation based on the traditional node-arc model of multicommodity flow. Dong [11] proposed an alternative MIP formulation of BPMP called the triples formulation. Thus, in the literature there are two kinds of BPMP MIP formulations: node-arc and triples. The purpose of our experimental study is to enhance both models as much as possible by applying candidate techniques, and compare their performance using the CIM method.

3.1 Node-arc Formulation

The following node-arc formulations is taken from [10]. The binary variable x_{ij} indicates whether or not the vehicle traverses arc (i,j), and binary variable y_{kl} indicates whether or not to accept request (k,l). Binary variable $z_{kl,ij}$ determines whether or not request (k,l) is performed via arc (i,j). Variable θ_{ij} represents the total flow (i.e., tons of cargo) transported on arc (i,j). Sequence variables $s_i \geq 0$, for $i = 1,\ldots,n$, track the relative order in which nodes are visited. The node-arc formulation for BPMP is

$$\max_{s,x,y,z,\theta} p\sum_{(k,l)\in\mathcal{R}} d_{kl}w_{kl}y_{kl} - c\sum_{(i,j)\in\mathcal{A}} d_{ij}\theta_{ij} - cv\sum_{(i,j)\in\mathcal{A}} d_{ij}x_{ij} \tag{1}$$

subject to

$$\sum_{j=2}^{n} x_{1j} = 1 \tag{2}$$

$$\sum_{i=1}^{n-1} x_{in} = 1 \tag{3}$$

$$\sum_{i\in\mathcal{N}\setminus\{k,n\}} x_{ik} = \sum_{j\in\mathcal{N}\setminus\{1,k\}} x_{kj} \qquad \forall k \in \mathcal{N}\setminus\{1,n\} \tag{4}$$

$$\sum_{(i,j)\in\mathcal{A}} d_{ij}x_{ij} \leq D \tag{5}$$

$$\sum_{i\in\mathcal{N}\setminus\{k,n\}} x_{ik} \leq 1 \qquad \forall k \in \mathcal{N}\setminus\{1\} \tag{6}$$

$$s_i - s_j + (n+1)x_{ij} \leq n \qquad \forall(i,j) \in \mathcal{A} \tag{7}$$

$$\sum_{(k,l)\in\mathcal{R}} z_{kl,ij} \leq Mx_{ij} \qquad \forall(i,j) \in \mathcal{A} \tag{8}$$

$$\sum_{j\in\mathcal{N}\setminus\{1,k\}} z_{kl,kj} = y_{kl} \qquad \forall(k,l) \in \mathcal{R} \tag{9}$$

$$\sum_{i\in\mathcal{N}\setminus\{l,n\}} z_{kl,il} = y_{kl} \qquad \forall(k,l) \in \mathcal{R} \tag{10}$$

$$\sum_{\{i\in\mathcal{N}:(i,h)\in\mathcal{A}\}} z_{kl,ih} = \sum_{\{j\in\mathcal{N}:(h,j)\in\mathcal{A}\}} z_{kl,hj} \qquad \forall(k,l)\in\mathcal{R}, h\in\mathcal{N}\setminus\{k,l\} \tag{11}$$

$$\theta_{ij} = \sum_{(k,l)\in\mathcal{R}} w_{kl}z_{kl,ij} \qquad (i,j)\in\mathcal{A} \tag{12}$$

$$\theta_{ij} \leq Q \qquad \forall(i,j)\in\mathcal{A} \tag{13}$$

The objective function is to maximize the total profit. Note that v represents the weight of the vehicle in tons. The vehicle's route is constrained to at most D miles by constraint (5). The *node-degree* (6), and *subtour elimination* constraints (7) ensure that the vehicle follows a simple path from node 1 to node n. The sequence variables determine the relative order in which nodes are visited by the vehicle. The logical connection between x_{ij} and $z_{kl,ij}$ is enforced by constraint set (8). Constraint sets (9) and (10) enforce the logical relationship between y_{kl} and $z_{kl,ij}$; Constraints (11) are flow-conservation constraints for intermediate nodes on the path the vehicle takes from node k to node l. The capacity limit is enforced by constraint set (13). We denote a solution to the node-arc formulation by a tuple of unsubscripted variables (x, y, z, s, θ).

3.2   Triples Formulation

The following description of the triples formulation of BPMP is adapted from [10]; it uses a compact formulation of multicommodity flow originally proposed by Matula [12,15] in which *triples variable* u_{ij}^k for node triple (i, j, k) represents the total flow on all paths from node i to node j with arc (i, k) as the first arc. In the triples formulation of BPMP, u_{ij}^k represents the tons of cargo that the vehicle carries from node i to node j on arc (i, k) and an unspecified path from node k to node j. In a feasible solution, the unspecified path turns out to be the route that the vehicle takes from node k to node j [10]. The triples formulation of the BPMP replaces the z variables of the node-arc formulation with triples variables. The multicommodity flow constraints (9)–(12) are replaced with the following set of *triples constraints* that relate the triples variables to arc flows:

$$\theta_{ij} = w_{ij}y_{ij} + \sum_{(i,k,j)\in\mathcal{T}} u_{ik}^j + \sum_{(k,j,i)\in\mathcal{T}} u_{kj}^i - \sum_{(i,j,k)\in\mathcal{T}} u_{ij}^k \quad \forall (i,j) \in \mathcal{A} \quad (14)$$

$$\theta_{ij} \leq Qx_{ij} \quad (i,j) \in \mathcal{A}. \quad (15)$$

The following constraints are imposed in order to force arc (i, k) to be on the vehicle's route if variable u_{ij}^k is positive:

$$u_{ij}^k \leq Qx_{ik} \quad \forall (i,j,k) \in \mathcal{T} \quad (16)$$

These constraints provide a logical linkage between the u variables and the x variables, and replace constraint set (8) of the node-arc formulation. For a detailed explanation of the triples formulation, interested readers are referred to [10].

4   Node-Arc Summary

In this section we use the GCI to evaluate the efficacy of various techniques (cuts, branching rules, etc.) designed to improve CPLEX's performance using the node-arc model given in Sect. 3. These techniques were selected and informally ranked by effectiveness from a larger set of candidates after preliminary experiments that we performed prior to developing the CIM. Before applying the first of the techniques, we established an "incumbent" enhanced node-arc formulation by determining a tight Big-M value for the x-z linking constraint set (8). We then applied the techniques sequentially according to the ranking from our preliminary experiments. Using "model 1" to refer to an incumbent solution approach and "model 2" to refer to the application of a particular technique to model 1. If GCI >1, we recommend adopting the technique, and making the resulting model the new incumbent. We also say, for convenience, that the model 2 is "GCI times faster" than model 1. If GCI ≤ 1, we recommend not adopting the technique.

4.1   Computing Environment and Weight Selection

The computations were performed on Dell R730 computers each with Dual 12 Core 2.6 GHz Intel Xeon processors and 380 GB RAM. The formulations were implemented in AMPL 10.00 and solved with CPLEX 12.6.0.0. We used the default settings for AMPL and CPLEX except where specified.

In our experience, practitioners solving real-world problems are much more concerned about real time as a performance measure than CPU time, and are often unaware of the tick measure. For our purposes, however, the reproducibility of the tick metric is quite important. Therefore, we used the following weights for each type of time speedup: $\omega_C = 6, \omega_R = 8$, and $\omega_T = 8$. Thus, real time and ticks were equally important and more important than CPU time by a factor of $1\frac{1}{3}$. For the ten instances of the same size problem, in reference to the PERT technique, we gave the largest weight to the median, the second largest to the mean, and the least weight to the min and max. For the problem sizes, we gave 30-node problems the largest weight, 20-node problems the second largest, and 10-node problems the least. The specific weights are listed in Table 1.

Table 1. Node-arc weights

CPU: $\omega_c = 6$	Ticks: $\omega_t = 8$	Real Time: $\omega_t = 8$
Median: $\omega_{med} = 40$	Mean: $\omega_{mea} = 10$	Min & Max: $\omega_{min} = \omega_{max} = 0.5$
10-node: $\omega_{10} = 1$	20-node: $\omega_{20} = 10$	30-node: $\omega_{30} = 12$

4.2   Initial Incumbent Formulation

We were only able to solve 10-node and 20-node instances with the incumbent solution approach (i.e., the original node-arc model). The model was solved three times for each problem instance. With no techniques applied, the median average real time for the 10-node instances was 1.15 s, and the median average real time for the 20-node instances was about 20 min. The results for $n = 20$ are shown in Table 2. We summarize the results for $n = 10$ using four-tuples listing the averages of the min, mean, median, and max values for CPU time, real time, and ticks. The CPU seconds, real time seconds, and ticks tuples are (2.28, 4.17, 4.23, 6.45), (0.50, 1.15, 0.99, 2.56), and (120.06, 332.90, 272.06, 645.50), respectively. That is, the maximum average CPU time for the 10-node instances was 6.45 s, the maximum average real time was 2.56 s, and the maximum average number of ticks was 645.50. Hereinafter, CPU and real time are reported in seconds.

Table 2. Test results of original Node-Arc model for $n = 20$.

Instance	CPU time				Real time				Ticks
	Run 1	Run 2	Run 3	Ave.	Run 1	Run 2	Run 3	Ave.	
1	12,654	12,310	12,172	12,379	1,937	1,896	1,734	1,856	1,244,880
2	1,745	1,757	1,716	1,739	355	372	344	357	272,703
3	2,457	2,544	2,393	2,465	497	536	467	500	382,350
4	24,954	26,053	27,313	26,106	1,891	2,044	1,943	1,960	1,211,197
5	2,964	3,036	2,981	2,994	632	681	628	647	583,158
6	32,635	35,586	34,566	34,262	1,720	1,859	1,778	1,786	1,005,079
7	2,760	2,770	2,615	2,715	502	512	482	499	422,516
8	5,873	5,888	5,738	5,833	845	857	806	836	674,853
9	28,620	32,115	28,746	29,827	2,007	2,226	1,967	2,067	1,318,085
10	17,096	17,745	18,202	17,681	1,571	1,749	1,606	1,642	1,091,225
Min	1,745	1,757	1,716	1,739	355	372	344	357	272,703
Mean	13,176	13,980	13,644	13,600	1,196	1,273	1,175	1,215	820,605
Median	9,263	9,099	8,955	9,106	1,208	1,303	1,206	1,239	839,966
Max	32,635	35,586	34,566	34,262	2,007	2,226	1,967	2,067	1,318,085

4.3   Technique 1: Conditional Arc Flow

The original node-arc model [16] uses constraint set (13), $\theta_{ij} \leq Q$, to ensure that the total amount of flow, θ_{ij}, on arc (i, j) is less than or equal to the vehicle capacity, Q. Notice that if the vehicle does not travel on arc (i, j), there should be no flow on the arc (i.e., if $x_{ij} = 0$, then $\theta_{ij} = 0$). If the vehicle does travel on arc (i, j), the maximum flow on the arc is Q (i.e., if $x_{ij} = 1$, then $\theta_{ij} \leq Q$). Therefore (13) can be replaced by the following constraint set which we call conditional arc-flow

$$\theta_{ij} \leq Qx_{ij} \qquad \forall(i, j) \in \mathcal{A}. \tag{17}$$

Yu and Dong [16] were unable to solve 30-node instances with the original node-arc model. We had a similar experience in our preliminary tests. Therefore, we tested this technique only on 10-, 20-, and 30-node instances. Table 3 gives detailed test results of three runs for the 20-node instances after applying the technique. The CPU time, real time, and ticks tuples for the 10-node instances are (0.90, 2.85, 2.91, 4.82) (0.26, 0.66, 0.61, 0.96), and (74.61, 237.37, 241.69,

384.78) respectively. The complete speedup summary is given in Table 4. Table 5 lists the composite indices and GCI. Speedups in bold are greater than 1.

Conclusion: After applying technique 1, conditional arc-flow, the GCI of speedups was 8.24, which means, on average, the model with conditional arc-flow was solved 8.24 times faster than the original model. Therefore, we adopted technique 1, replacing constraint set (13) with the conditional arc-flow constraints (17). Furthermore, after applying conditional arc-flow constraints we were able to solve 30-node instances. The average real times ranged from 4,421 s (1.23 h) to 17,829 s (4.95 h) with a mean and median of 8,546 s (2.37 h) and 7,635 s (2.12 h), respectively [2].

Table 3. Test Results of Incremental Effect of Conditional Arc-Flow for $n = 20$

Instance	CPU time				Real time				Ticks
	Run 1	Run 2	Run 3	Ave.	Run 1	Run 2	Run 3	Ave.	
1	1,306	1,258	1,266	1,276	96	100	97	98	54,992
2	720	701	702	708	63	63	64	63	37,316
3	1,974	1,689	1,729	1,797	143	146	134	141	81,335
4	4,945	3,848	3,836	4,210	358	362	318	346	202,725
5	450	409	408	422	94	94	92	93	70,847
6	2,152	1,752	1,764	1,890	196	200	182	193	119,363
7	1,243	1,105	1,120	1,156	81	81	77	80	39,778
8	1,189	1,080	1,010	1,093	74	76	68	73	35,113
9	673	630	620	641	127	130	125	128	96,582
10	811	759	740	770	123	125	119	122	85,867
Min	450	409	408	422	63	63	64	63	35,113
Mean	1,546	1,323	1,320	1,396	136	138	128	134	82,392
Median	1,216	1,093	1,065	1,125	110	112	108	110	76,091
Max	4,945	3,848	3,836	4,210	358	362	318	346	202,725

4.4   Technique 2: Relax Node-Degree Constraints

Yu and Dong [16] used the node-degree cuts (6) to ensure that the vehicle visits each location at most once. But, the MTZ subtour elimination constraints (7) also ensure that the vehicle visits each node at most once in an integer solution. Therefore, we can relax (drop) the node-degree constraints without losing validity of the integer model (the node-degree cuts can be violated in solutions to the LP relaxations). Table 6 gives the composite indices and GCI for this technique. Speedups greater than 1 are in bold.

Table 4. Summary of Incremental Effect of Conditional Arc Flow Constraints.

	Speedup		
	Ave. CPU Time	Ticks	Ave. Real Time
$n = 10$			
Min	0.95	0.91	**1.10**
Mean	**1.68**	**1.42**	**1.74**
Median	**1.46**	**1.32**	**1.71**
Max	**3.17**	**2.61**	**2.67**
$n = 20$			
Min	**1.37**	**4.70**	**3.55**
Mean	**12.21**	**11.35**	**9.74**
Median	**6.65**	**9.52**	**8.10**
Max	**46.52**	**22.64**	19.01

Table 5. Composite Indices and GCI for Technique 1

n	C_n (CPU)	T_n (Ticks)	R_n (Real Time)	I_n	GCI
10	**1.52**	**1.35**	**1.72**	**1.53**	**8.24**
20	**8.08**	**9.96**	**8.48**	**8.91**	

Table 6. Composite Indices and GCI for Technique 2

n	C_n (CPU)	T_n (Ticks)	R_n (Real Time)	I_n	GCI
10	0.73	**1.17**	0.90	**0.95**	**1.28**
20	**1.02**	**1.25**	**1.22**	**1.17**	
30	**1.43**	**1.28**	**1.48**	**1.40**	

Conclusion: After applying technique 2, relax node-degree constraints, the CGI of speedups was 1.28, which means, on average, the model relaxing the node-degree constraints was solved 1.28 times faster than the incumbent model. Therefore, we adopted technique 2 and dropped constraint set (6) from the incumbent.

4.5   Technique 3: Single-Node Demand Cuts

The single-node demand cuts state that the total weight of the delivery requests accepted from node i, or into node j, is at most the vehicle capacity, Q.

$$\sum_{j \in V \setminus \{1,i\}} w_{ij} y_{ij} \leq Q \ \forall i \in V \setminus \{n\} \tag{18}$$

$$\sum_{i \in V \setminus \{j,n\}} w_{ij} y_{ij} \leq Q \ \forall j \in V \setminus \{1\} \tag{19}$$

The above are valid inequalities that are satisfied by any feasible solution because the vehicle cannot simultaneously hold cargoes with total weight more than its capacity. This condition is not necessarily enforced by solutions to the LP relaxation because of the fractional y values. Table 7 gives the results from applying single-node demand cuts to the incumbent node-arc model.

Table 7. Composite Indices and GCI Technique 3

n	C_n (CPU)	T_n (Ticks)	R_n (Real Time)	I_n	GCI
10	**1.30**	0.96	**1.21**	**1.15**	**0.94**
20	0.86	0.91	0.88	0.89	
30	**1.01**	0.95	0.94	0.96	

Conclusion: After applying technique 3, single-node demand cuts, the GCI of speedups was 0.94, which means that solving the incumbent model was faster. Therefore, we did not adopt technique 3.

4.6 Best Node-Arc Model

In total, we tested a series of nine techniques for improving solution time using CPLEX to solve the node-arc formulation of BPMP. Four of the techniques were adopted resulting in the Best Node-Arc Model [2]. Table 8 summarizes the speedup of the Best Node-Arc Model compared to the original model proposed by Yu and Dong [16]. The CPU time, real time, and ticks tuples for $n = 30$ are (20,454, 55,459, 39,485, 140,354) (1,594, 3,198, 2,564, 6,166), and (812,789, 1,192,684, 1,147,441, 1,966,885), respectively. The CPU time, real time, and ticks tuples for $n = 40$ are (334,025, 6,643,337, 1,213,615, 52,502,367), (18,413, 329,773, 56,428, 2,652,518), and (6,601,682, 62,036,444, 15,873,253, 463,811,772), respectively.

Table 8. Best Node-Arc Model vs. Original Node-Arc Model

	Speedup					
	$n = 10$			$n = 20$		
	CPU	Ticks	Real time	CPU	Ticks	Real time
Min	0.58	**1.26**	**1.29**	**2.17**	**12.38**	**7.76**
Mean	**1.24**	**1.88**	**1.81**	**26.24**	**25.61**	**20.78**
Median	0.89	**1.9**	**1.63**	**17.97**	**24.58**	**19.56**
Max	**3.68**	**2.86**	**2.81**	**111.29**	**42.77**	**43.98**

5 Triples Summary

In [2] we described in detail how we applied the CIM to the triples formulation of the BPMP proposed by Dong [11]. We tested two techniques that are specific to the triples formulation, and six of the nine techniques we tested for the node-arc model. Due to the fact that preliminary studies showed that we could solve larger problem instances with the triples formulation than with the node-arc formulation, we used the following instance-size weights: $\omega_{10} = 6$, $\omega_{20} = 10$, $\omega_{30} = 13$, $\omega_{40} = 14$, and $\omega_{50} = 16$. The other weights were the same as those used for the node-arc formulation. Tables 9 and 10 summarize the speedup of the Best Triples Model compared to the original Triples model. Using the Best Triples Model, we were able to solve 50-node instances with a maximum average real time of 5,016 s (1.4 h) [2,10]. Tables 11 and 12 compare the final triples and node-arc formulations. We did not calculate the GGI for Tables 11 and 12, but it is clearly larger than 1. Thus, the Best Triples Model is the solution approach recommended by the CIM.

Table 9. Best Triples vs. Original Triples: $n = 10$, $n = 20$, and $n = 30$

	Speedup								
	$n = 10$			$n = 20$			$n = 30$		
	CPU	Ticks	Real	CPU	Ticks	Real	CPU	Ticks	Real
Min	**1.01**	0.49	0.44	0.80	0.74	0.84	0.71	0.69	0.60
Mean	**1.51**	**1.11**	**1.88**	**1.95**	**1.66**	**1.69**	**6.49**	**3.13**	**4.48**
Median	**1.48**	**1.02**	**1.87**	**1.53**	**1.17**	**1.28**	**6.43**	**3.18**	**4.50**
Max	**2.76**	**2.28**	**3.20**	**6.20**	**6.14**	**5.43**	**12.03**	**4.70**	**7.89**

Table 10. Best Triples vs. Original Triples: $n = 40$ and $n = 50$

	Speedup					
	$n = 40$			$n = 50$		
	CPU	Ticks	Real	CPU	Ticks	Real
Min	**1.49**	**1.11**	**1.31**	**3.91**	**2.00**	**4.62**
Mean	**7.97**	**3.17**	**6.59**	**10.34**	**3.65**	**9.07**
Median	**7.74**	**3.40**	**7.37**	**8.53**	**3.54**	**9.24**
Max	**14.87**	**5.30**	**12.27**	**24.11**	**5.82**	**17.17**

Table 11. Best Triples vs. Best Node Arc: $n = 10$ and $n = 20$

| | Speedup | | | | | |
| | $n = 10$ | | | $n = 20$ | | |
	CPU	Ticks	Real	CPU	Ticks	Real
Min	0.39	1.47	0.86	6.34	8.89	7.98
Mean	2.38	4.47	2.31	62.43	24.02	21.46
Median	2.4	4.08	2.22	68.97	25.39	20.01
Max	4.50	8.38	4.00	137.06	42.75	46.20

Table 12. Best Triples vs. Best Node Arc: $n = 30$ and $n = 40$

| | Speedup | | | | | |
| | $n = 30$ | | | $n = 40$ | | |
	CPU	Ticks	Real	CPU	Ticks	Real
Min	34.33	16.06	30.96	44.46	34.79	56.61
Mean	210.55	52.26	101.59	1,943.68	319.00	1,050.75
Median	153.62	44.40	106.01	327.41	96.47	225.86
Max	872.08	137.89	212.53	11,929.94	1,417.70	4,275.04

6 Conclusions

We have shown that the Composite Index Method (CIM) fills a gap in the area of optimization benchmarking. Calculating a single index, GCI, makes it much easier to select the best solution approach among multiple candidates. By applying CIM to the backhaul profit maximization problem (BPMP), we demonstrate the step-by-step details of the framework of CIM in a parallel computing environment. Although we focused on solution-time measures for finding a provably optimal solution, the CIM can be easily adapted to consider other dimensions of concern such as memory usage and solution quality (for heuristics). Furthermore, individual users can use their own weighting scheme to emphasize their personal preferences for making trade-offs between performance measures.

Although the Best Triples Model of the BPMP that we identified by our iterative application of the CIM is a significant improvement over the initial node-arc formulation proposed in the literature, it is possible that we could have discovered an even better model by testing the model enhancements (techniques) in a different order. Thus, an important question for future research is how to determine the order in which alternative solution approaches are compared using the CIM when the approaches are not mutually exclusive.

The successful utilization of the CIM lies in the wise selection of performance measures and weights in each dimension of concern. An illustrative study applying the CIM to well known optimization problems for different use cases is planned for the future.

References

1. AMPL: AMPL Version 10.6.16. AMPL Optimization LLC (2009)
2. Bai, Y., Olinick, E.V.: An empirical study of mixed integer programming formulations of the backhaul profit maximization problem (2019). https://scholar.smu.edu/engineering_management_research/1/. Accessed 20 Feb 2020
3. Barr, R.S., Hickman, B.L.: Reporting computational experiments with parallel algorithms: issues, measures, and experts' opinions. ORSA J. Comput. 5(1), 2–18 (1993)
4. Barr, R.S., Hickman, B.L.: Parallel simplex for large pure network problems: computational testing and sources of speedup. Oper. Res. 42(1), 65–80 (1994)
5. Beiranvand, V., Hare, W., Lucet, Y.: Best practices for comparing optimization algorithms. Optim. Eng. 18(4), 815–848 (2017). https://doi.org/10.1007/s11081-017-9366-1
6. Box, M.: A comparison of several current optimization methods, and the use of transformations in constrained problems. Comput. J. 9(1), 67–77 (1966)
7. Carle, M.A.: Deterministic behavior of CPLEX: ticks or seconds? (2019). https://tinyurl.com/kvy6jbbc. Accessed 14 Feb 2022
8. Crowder, H., Dembo, R.S., Mulvey, J.M.: On reporting computational experiments with mathematical software. ACM Trans. Math. Softw. (TOMS) 5(2), 193–203 (1979)
9. Dolan, E.D., Moré, J.J.: Benchmarking optimization software with performance profiles. Math. Program. 91(2), 201–213 (2002)
10. Dong, A., Bai, Y., Olinick, E.V., Yu, A.J.: The backhaul profit maximization problem: optimization models and solution procedures. INFORMS J. Optim. (2022). https://pubsonline.informs.org/doi/10.1287/ijoo.2022.0071
11. Dong, Y.: The Stochastic Inventory Routing Problem. Ph.D. thesis, Southern Methodist University (2015). https://search.proquest.com/docview/1757808242
12. Dong, Y., Olinick, E.V., Jason Kratz, T., Matula, D.W.: A compact linear programming formulation of the maximum concurrent flow problem. Networks 65(1), 68–87 (2015). https://doi.org/10.1002/net.21583, http://dx.doi.org/10.1002/net.21583
13. Hoffman, A., Mannos, M., Sokolowsky, D., Wiegmann, N.: Computational experience in solving linear programs. J. Soc. Ind. Appl. Math. 1(1), 17–33 (1953)
14. IBM. https://vdocuments.mx/ibm-ilog-cplex-user-manual-126.html. Accessed 14 Feb 2022
15. Matula: A new formulation of the maximum concurrent flow problem a proof of the maximum-concurrent-flow/max-elongation duality theorem (1986). https://s2.smu.edu/~matula/MCFP86.pdf
16. Yu, J., Dong, Y.: Maximizing profit for vehicle routing under time and weight constraints. Int. J. Prod. Econ. 145(2), 573–583 (2013)

Optimal Energy Management of Microgrid Using Multi-objective Optimisation Approach

Yahia Amoura[1,4,5](✉), Ana I. Pereira[1,2], José Lima[1,3], Ângela Ferreira[1],
and Fouad Boukli-Hacene[4]

[1] Research Centre in Digitalization and Intelligent Robotics (CeDRI), Instituto
Politécnico de Bragança, Bragança, Portugal
yahia@ipb.pt
[2] ALGORITMI Center, University of Minho, Braga, Portugal
[3] INESC TEC - INESC Technology and Science, Porto, Portugal
[4] Higher School in Applied Sciences of Tlemcen, Tlemcen, Algeria
[5] University of Laguna, Tenerife, Spain

Abstract. The use of several distributed generators as well as the energy storage system in a local microgrid require an energy management system to maximize system efficiency, by managing generation and loads. The main purpose of this work is to find the optimal set-points of distributed generators and storage devices of a microgrid, minimizing simultaneously the energy costs and the greenhouse gas emissions. A multi-objective approach called Pareto-search Algorithm based on direct multi-search is proposed to ensure optimal management of the microgrid. According to the non-dominated resulting points, several scenarios are proposed and compared. The effectiveness of the algorithm is validated, giving a compromised choice between two criteria: energy cost and GHG emissions.

Keywords: Microgrid · Power management · Energy management system · Multi-objective optimisation · Pareto-search algorithm

1 Introduction

An upcoming fossil fuel shortage is estimated in the coming years, on the other hand, problems related to global warming due to the increase in greenhouse gases (GHG) emissions have affected the world especially after the major peak in 2018 due to the massive use of fossil resources in power generation. About 64% of the world's production was based on oil and gas, however, 33.1 tons of CO_2 was released into the atmosphere [1]. According to the U.S. space agency (NASA) the average temperature of the earth surface has increased by 1 °C compared to the average of the 19th century [2]. As a result of this environmental conflict, a challenge was made by european union member states in the framework for action on climate and energy for the period 2021–2030. Three objectives have been set: reducing greenhouse gas emissions by at least 40%, increasing the contribution

© Springer Nature Switzerland AG 2022
D. E. Simos et al. (Eds.): LION 2022, LNCS 13621, pp. 58–76, 2022.
https://doi.org/10.1007/978-3-031-24866-5_5

of renewable energies to at least 32% and improving energy efficiency by at least 32.5% [3].

Nowadays there is an upward tendency for using small power systems, able to bring the energy production near to the consumption. In this type of system, the most important sources are renewable based (e.g., photovoltaic panels, wind turbines, etc.), due to their low environmental impact, in combination with diesel generators in order to obtain the necessary mix able to assure the balance between production and consumption. These small power-producing networks called microgrids need a distributed and autonomous power generation control [4]. Nevertheless, the dispatch problem is transversal to all power systems [5,23], in particular in the autonomous isolated microgrids with limited power sources.

A microgrid is based on the interconnection of small modular generation (micro-turbines, fuel cells, photovoltaic, among others), combined with storage devices (flywheels, energy capacitors or electrochemical batteries) and loads, some of them controllable, at low voltage distribution systems [6]. The operation of micro-sources in the network is complex but it can provide distinct benefits to the overall system performance if it is managed and coordinated efficiently [7,27].

The use of microgrids has become an attractive option for power utility companies since they can help to improve the power quality and power supply flexibility. Also, they can provide spinning reserves and reduce transmission and distribution costs. Moreover, they can be used to feed the customers in the event of an outage in the main grid [8,24].

Following this interest in microgrids, several works have been performed to ensure optimal management. Researchers in [9] proposed a genetic algorithm (GA) approach to solve the problem of electric power dispatch using a model that describes the load demand and environmental requirements. In [10] a multi-team particle swarm (MTPSO) algorithm is proposed to solve the microgrid schedule problem. The algorithm is based on swarm information to update the velocity (position) with faster and more stable convergence, the simulation results show that the proposed algorithm gives a better global search ability than the classic PSO. Real-time PSO-based energy management of a stand-alone hybrid wind, micro-turbine, and energy storage system is presented in [11], with the results being compared to sequential quadratic programming (SQP). The computation results show the reliability of the proposed PSO for energy management strategy in hybrid systems. However, due to the pollutants emission of fossil fuel generators, the economic objectives are not sufficient for optimal operation of the microgrids. Therefore, to achieve the best solutions, environmental and economic objectives must be considered simultaneously. Many researchers have considered both cost and gas emissions to schedule the output power of distributed generators in the microgrids. In [12] authors have converted the gas emissions objective to a constraint and have solved the problem as a single objective, but to find the Pareto optimal solutions, this method is not efficient. An improved modified bacterial foraging optimisation (MBFO) algorithm is proposed in [13] to solve the multi-objective problem for expert energy management of a microgrid considering wind energy uncertainty in such a way that the total operating costs and the net emissions are simultaneously minimized. Authors in [14] present an expert multi-objective adaptive modified particle swarm optimisation (AMPSO) algorithm for optimal operation of a typical MG with renewable energy sources to solve the multi-operation management problem in the microgrid, the numerical results indicate that the proposed method demonstrates superior performances and shows dynamic stability and excellent convergence.

The work proposed in this article consists in developing an energy management system dedicated to the scheduling of the distributed generators and the energy storage system of the microgrid considering the simultaneous optimisation of the economic and environmental criteria. The Pareto-search algorithm based on the direct multi-search method is proposed as an optimisation approach in the energy management system. The results allow to have a set of solutions called non-dominated solutions or optimal Pareto solutions. The Pareto solutions represent the compromise between the two criteria to be optimised: costs and GHG emissions. Finally, the obtained scenarios are analyzed and compared in order to have multiple scheduling choices while respecting the economic and environmental constraints.

The remaining of the paper is organized as follows: Sect. 2 presents the architecture of the proposed microgrid. In Sect. 3 the storage system is modelled. Section 4 formulates the multi-objective optimisation problem together with the related constraints and explains the concept of multi-objective optimisation. the Pareto-search Algorithm is presented in Sect. 5. Section 6 deals with the analysis and discussion of the results obtained after the implementation of the energy management system based on the Pareto-search Algorithm. Finally, Sect. 7 concludes the study and point out some further studies.

2   Microgrid Description

The proposed microgrid comprises two renewable sources: photovoltaic (PV) and wind-turbine (WT) additionally, it has a micro-turbine (MT), and an energy storage system (ESS). The microgrid can be explored connected to the main grid, which will act as a buffer if needed, or it can be explored off-grid, when internal resources are enough to satisfy the demand or even in case of a malfunction or failure of the grid. The connection is ensured through a transformer and common coupling point (PCC) as indicated in Fig. 1.

For a reliable operation process considering the economic and environmental constraints of the proposed management system, renewable energy sources can provide energy to loads and/or charge the battery. Excess energy, after satisfying local demands, can be fed into the main grid, reducing the total operating energy costs and GHG emissions from conventional generation, or it can be exchanged with other microgrids.

Regarding the energy storage system, it is assumed an exploitation mode able to contribute to its lifespan, avoiding deep discharges and reducing the number of charges-discharges cycles. Additionally, this work considers that the exchange of energy from the storage system to the main grid is not allowed.

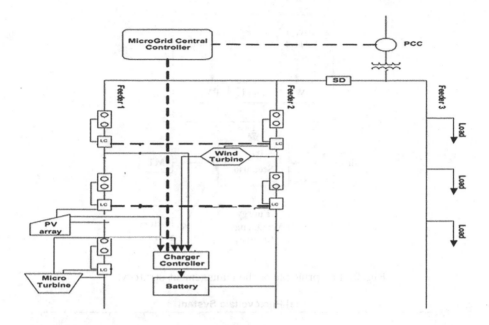

Fig. 1. The architecture of the microgrid

The energy management system will ensure the optimal control of the sources according to the dynamic market prices in a time span of 24 h. The load and power sources profiles of the microgrid proposed above are the same as the ones previously considered in [13]. The maximum power that can be produced by the photovoltaic panels is 10 kW and the maximum power of the wind turbine generator is 20 kW. To reduce the number of startup/shutdown, consequently, the maintenance requirements, the micro-turbine can operate in a power range from 6 kW to 30 kW. The maximum power exchanged with the main grid is limited to 90 kW. The energy storage system is designed to assure the load for a maximum time period of 1 h. Under this hypothesis, the total capacity of the energy storage system is $E^{max} = 180$ kWh, and it is considered an initial situation given by $E(1) = 52$ kWh. Figure 2 illustrates the principle scheme for the operation of the microgrid energy management system.

Figure 3 (a) and (b) show the maximal hourly power delivered by the renewable generators for a time span of 24 h. Figure 4 presents the variation of the hourly consumption of the microgrid under the same period of time.

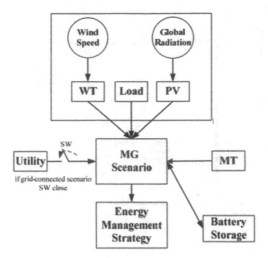

Fig. 2. The principle of the management strategy.

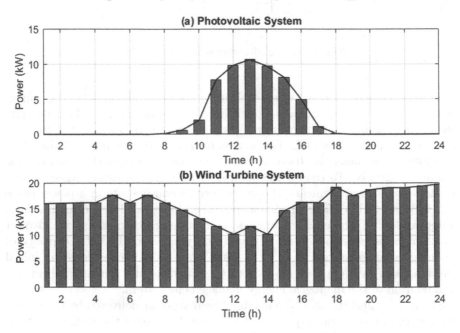

Fig. 3. The daily power profile from (a) PV system, (b) WT system.

3   Modeling of the Energy Storage System

The energy storage system has an important role in a microgrid exploitation because it allows the flexibility needed to assure the balance between the production and consumption, in the presence of variations of either loads or inter-

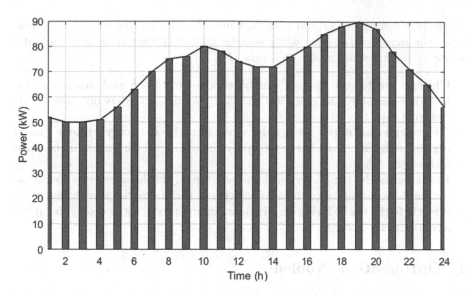

Fig. 4. The daily load profile.

mittent sources. Taking into account the microgrid storage requirements, the most appropriate storage form is the electrochemical battery [15]. This ESS is chosen for its long-term storage capacity and for the ease of bidirectional flow and fast power response, allowing a good frequency adjustment source in a microgrid to provide real-time dynamic balance. It has a positive significance for improving the power quality of the microgrid and ensuring stable operation [10]. In order to have reliable modeling of this ESS, several parameters have to be taken into account such as the nominal capacity and the rate of charge/discharge, this later is used to limit the deep discharge of the battery leading to a higher lifespan. Therefore, the battery usage is delimited by their minimum and maximum capacity allowed, respectively E^{min} and E^{max} with:

$$E^{min} \leq E(t) \leq E^{max} \tag{1}$$

The energy available in the battery is an important technical parameter to provide data support for the microgrid management, for instance, the quantity of energy at time $t + 1$ is related to the value at time t, and the charge and discharge energy of the battery can be expressed as follows [16]:

$$\begin{cases} E(t+1) = E(t) - \Delta_t P_c(t) \eta_c, & \text{charging mode} \\ E(t+1) = E(t) - \frac{\Delta_t P_d(t)}{\eta_d}, & \text{discharging mode} \end{cases} \tag{2}$$

where $P_c(t)$ and $P_d(t)$ are the charging and discharging power of the battery at time t, $E(t)$ and Δ_t are, respectively, the stored energy at time t and the interval of time considered and, finally, η_c and η_d are the charging and discharging efficiency, respectively.

Battery control is a crucial issue that must be taken into account when managing the microgrid, so the energy storage system (ESS) can only be operated as one of the following modes at a time [17]:

- **Charge mode**: the battery can be charged from the grid, micro-turbine, and/or renewable energies with an energy quantity that is not beyond the charging rate.
- **Discharge mode**: the battery delivers a quantity of energy without exceeding the limit rate of discharge to supply the microgrid consumers if the prices of Kwh are high.
- **Inactive mode**: the battery will not perform any of the above two operations (charge/discharge), since the grid utility and the microgrid provide electricity directly to the loads during certain hours in order to consider economic and/or environmental features.

4 Optimisation Problem

The problem of optimal scheduling of distributed microgrid generators and storage system is defined as a problem of allocating optimal power generation set points, in such a way that the operating cost and the net emission of pollutants from conventional sources in the microgrid are minimized simultaneously while satisfying all constraints imposed by the energy management system (EMS). The mathematical model of the problem can be presented in the following sections.

4.1 Objective Functions

Objective 1: Operating Cost Minimization. The definition of the operating cost function depends on several parameters, mainly the architecture of the microgrid. The cost of the distributed resources and the storage system is considered dynamic throughout the day, also the cost of selling/buying energy supplied by or injected into the grid varies during the day, being the main objective of the cost function is to satisfy the load demand during the day in the most economical way. So, in each hour t, for a time span of 24 h of operation, the objective function can be expressed as follows [18]:

$$C(t) = \sum_{i=1}^{N_g} U_i(t) P_{DGi}(t) B_{DGi}(t) + \sum_{j=1}^{N_s} U_j(t) P_{ESSj}(t) B_{ESSj}(t) + P_{Grid}(t) B_{Grid}(t)$$

$$(3)$$

where N_g and N_s are the total number of generators and storage devices, respectively. $B_{DGi}(t)$ and $B_{ESSj}(t)$ represent the bids of ith DG unit and j^{th} storage device at hour t. $P_{Grid}(t)$ is the active power which is bought (sold) from (to) the utility grid at hour t and $B_{Grid}(t)$ is the bid of the utility grid at hour t. $U_i(t)$ and $U_j(t)$ are the operation mode of the ith generator and the jth storage device (ON or OFF), respectively. Table 1 present distributed energy resources, storage and grid bids.

The optimisation model of the first objective function can be written as follows:

$$f_1 = \sum_{t=1}^{T} \min C(t) \tag{4}$$

The optimisation model will lead to find P_{DGi}, P_{ESSj} and P_{Grid}, $i.e$, the optimum set points of distributed generators, energy storage system and main grid respectively that ensures a low total energy price in each hour t.

Table 1. The hourly unit prices of the distributed generators, storage system and main grid of the proposed microgrid (Euro/kWh) [13]

Time (h)	PV	WT	MT	ESS	GRID
01:00	0	0.021	0.0823	0.1192	0.033
02:00	0	0.017	0.0823	0.1192	0.027
03:00	0	0.0125	0.0831	0.1269	0.020
04:00	0	0.011	0.0831	0.1346	0.017
05:00	0	0.051	0.0838	0.1423	0.017
06:00	0	0.085	0.0838	0.15	0.029
07:00	0	0.091	0.0846	0.1577	0.033
08:00	0.0646	0.110	0.0854	0.1608	0.054
09:00	0.0654	0.140	0.0862	0.1662	0.215
10:00	0.0662	0.143	0.0862	0.1677	0.572
11:00	0.0669	0.150	0.0892	0.1731	0.572
12:00	0.0677	0.155	0.09	0.1769	0.572
13:00	0.0662	0.137	0.0885	0.1692	0.215
14:00	0.0654	0.135	0.0885	0.16	0.572
15:00	0.0646	0.132	0.0885	0.1538	0.286
16:00	0.0638	0.114	0.09	0.15	0.279
17:00	0.0654	0.110	0.0908	0.1523	0.086
18:00	0.0662	0.0925	0.0915	0.15	0.059
19:00	0	0.091	0.0908	0.1462	0.050
20:00	0	0.083	0.0885	0.1462	0.061
21:00	0	0.033	0.0862	0.1431	0.181
22:00	0	0.025	0.0846	0.1385	0.077
23:00	0	0.021	0.0838	0.1346	0.043
24:00	0	0.017	0.0831	0.1269	0.037

Objective 2: GHG Emissions Minimization. The environmental footprint from atmospheric pollutants is considered the second objective. Emissions include the polluting gases responsible for the greenhouse gas effect such as

nitrogen oxides (NO_x), sulfur dioxide (SO_2), and carbon dioxide (CO_2). Table 2 presents the emission factors as defined in [13].

Table 2. Pollutants emission factors [13]

EF	Micro-turbine (Kg/MWh)	Grid (Kg/MWh)
CO_2	724	922
NO_X	0.2	2.295
SO_2	0.00136	3.583

The mathematical formulation of the second objective can be described as follows [13]:

$$EM(t) = \sum_{i=1}^{N_g} U_i(t)P_{DGi}(t)EF_{DGi}(t) + P_{Grid}(t)EF_{Grid}(t) \qquad (5)$$

where $EF_{DGi}(t)$ and $EF_{Grid}(t)$ are GHG emissions factors describing the amount of pollutants emission in kg/MWh for each distributed generator and utility grid at hour t, respectively.

The optimisation model of the second objective function can be written as follows:

$$f_2 = \sum_{t=1}^{T} \min EM(t) \qquad (6)$$

The optimisation model will lead to find P_{DGi} and P_{Grid}, i.e, the optimum set points of distributed generators and main grid, respectively, that ensures a low total emission amount in each hour t.

4.2 Constraints Functions

Power Balance Constraint. Total demand (including storage) and transmission losses must be covered by the total power generation. The active power balance, in terms of frequency stability, is the precondition for a stable operation. The losses in transmission are considered numerically small, being ignored in this article. The condition of the power balance assumes the following form:

$$\sum_{i=1}^{N_g} P_{DGi}(t) + \sum_{j=1}^{N_s} P_{ESSj}(t) + P_{Grid}(t) = P_L(t) \qquad (7)$$

being $P_L(t)$ the total electrical load demand at hour t. Moreover, the power of the energy storage system $P_{ESSj}(t)$ can be positive in case of discharging or negative in the case of charging.

Electrical Limits of Generators Constraint. The microgrid distributed generators must not operate beyond their limits, and the energy exchanged between the microgrid and the main grid is also limited. Each DG and main grid's active power output is limited by the lower and upper limits, as follows:

$$P_{DGi}^{min}(t) \leq P_{DGi}(t) \leq P_{DGi}^{max}(t) \tag{8}$$

$$P_{SDj}^{min}(t) \leq P_{ESSj}(t) \leq P_{ESSj}^{max}(t) \tag{9}$$

$$P_g^{min}(t) \leq P_{Grid}(t) \leq P_{Grid}^{max}(t) \tag{10}$$

where $P^{min}(t)$ and $P^{max}(t)$ are the minimum and the maximum powers of the distributed generator (DG), energy storage system (ESS) and the grid (Grid) at the time t, respectively.

Storage System Limits Constraint. The battery must maintain within the limits of its capacity and limited by a maximum rate (charging/discharging) that must not be exceeded.

$$E^{min}(t) \leq E(t) \leq E^{max}(t) \tag{11}$$

$$\begin{cases} -P_c(t)\eta_c \leq P_c^{max} & \text{charging mode,} \quad P_c(t) < 0 \\ \dfrac{P_d(t)}{\eta_d} \leq P_d^{max} & \text{discharging mode,} \quad P_d(t) > 0 \end{cases} \tag{12}$$

where $E_{min}(t)$ and $E_{max}(t)$ are the minimum and maximum energy levels of the battery, P_c^{max} and P_d^{max} are the maximum rate of charge/discharge of the battery that be must respected in each operation.

4.3 The Multi-objective Optimisation Problem

Many real optimisation problems require the simultaneous optimisation of different and often conflicting objectives, characterized by the term of multi-objective optimisation. The solution to these multi-criteria optimisation problems is not a unique optimal point, but a set of solutions called the non-dominated, indifferent, or Pareto-optimal solutions, corresponding to the best possible compromise, since, one particular solution is not the best with regard to all the objectives. Generally, in a multi-objective optimisation problem, different objective functions must be simultaneously optimised taking into account a set of equality and inequality constraints, as follows [19]:

$$\min F = \{f_1, f_2\} \tag{13}$$

where F is a vector composed by the two objective functions (cost & emissions) defined on Sect. 4.1. The minimization problem defined on (13) is subject to the constraints defined in previous Sect. 4.2.

Assuming that the two solutions to the multi-objective problem are x_1 and x_2, the two solutions may have one of two properties: one dominates the other, or none dominates the other. In a minimization problem, without loss of generality, a solution x_1 dominates x_2 if both following two conditions are satisfied [18]:

$$\exists i \in 1, 2 : f_i(x_1) < f_i(x_2) \tag{14}$$

The non-dominated set of solutions is referred as the optimal set called the Pareto front. On the other hand, to establish the acceptability of each solution to be included in the non-dominated solutions repository, the concept of Pareto dominance is used.

5   Pareto-Search Algorithm

The Pareto search algorithm is a direct multiple search algorithm that uses pattern searches on a set of points to iteratively search for non-dominated points [20]. It is based on the search/polling model of direct directional search methods and uses the concept of pareto dominance to keep a list of non-dominated points by satisfying all linear/nonlinear bounds and constraints at each iteration [21]. The pattern search is intended to find the best correspondence $i.e$, the solution with the smallest error value in the multidimensional possibility analysis space. The Pareto search algorithm employs a number of intermediary and tolerance variables in its search mechanism. [21].

At each iteration, the algorithm is structured along a search step and a probe step, which are important considerations for achieving convergence results. The searching step is used to improve the performance of the algorithm. The polling step performs a local search around one of the non-dominated points chosen by the search step, which constitutes an iteration point or interrogation center. In both steps, search and polling, a provisional list is first generated, which keeps all the points of the actual iteration list and all the estimated points around this step. This list is then filtered by removing all dominated points and retaining the non-dominated points. A trial list L_{trial} is then retrieved from the filtered list of non-dominated points and must eventually contain all non-dominated points that are part of the considered iteration list in the preceding iteration [21,22]. The steps of the Pareto search algorithm are explained as follows.

1. **Initialization** To generate the starting set of points, Pareto-search algorithm will produce a set of random points that satisfies the bounds of the problem.
2. **Poll to Find Better Points**
   The Pareto search algorithm interrogates the points of the iterates, with the interrogated points inheriting the associated mesh size of the point in the iterates. The algorithm uses a query that keeps the feasibility relative to the limits and all linear constraints. If the model has non-linear constraints, the Pareto search computes the feasibility of each interrogated point and keeps the unfeasible points score separate from the feasible points value. The score of a feasible point is the vector of values of the objective function of this point

while the score of an infeasible point is the sum of the nonlinear infeasibilities. The Pareto search algorithm interrogates each point by iterations. If the interrogated points result in at least one non-dominated point compared to the existing (original) point, the interrogation is considered as successful. Otherwise, the algorithm continues to interrogate until it reaches an undominated point or there are no more points in the model.

3. **Stopping Conditions**

For three or less objective functions, the Pareto search algorithm uses volume and spread as stopping criteria. For four or more, the Pareto search algorithm employs distance and spread as stopping parameters.

6   Numerical Results and Discussion

The energy management system (EMS) proposed in this work consists in the scheduling of the microgrid production sources by taking into account the simultaneous minimization of both cost and GHG emission criteria through the Pareto-search optimisation algorithm.

Table 3 presents the set of non-dominanted solutions obtained by the implementation of the optimisation algorithm in the energy management system. These results represent the best trade-off between the two targets under minimisation.

Table 3. The non-dominanted solutions obtained by Pareto-search Algorithm

Scenarios	Total Energy Cost (euro)	Total Emissions (kg)
01	161.0118	1.2795×10^3
02	159.9786	1.2809×10^3
03	159.7697	1.2823×10^3
04	158,5841	$1,2872 \times 10^3$
05	157.5630	1.2984×10^3
06	156,7201	1.3120×10^3
07	155,8556	$1,3239 \times 10^3$
08	155.3164	1.3387×10^3
09	154.9797	1.3524×10^3

The non-dominated points are classified in Pareto front as shown in Fig. 5. All these points represent several scheduling scenarios for the distributed generators of the microgrid, the energy storage system and energy exchanged between the main grid and the microgrid.

According to the trade-off obtained from the non-dominated points, two cases are highlighted to illustrate the energy management process of the energy storage

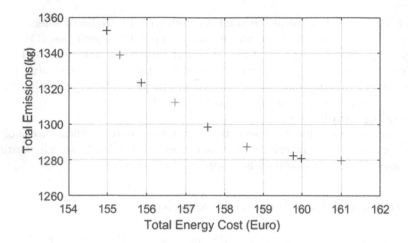

Fig. 5. Pareto front.

system, the best environmental trade-off and the best economic trade-off, scenarios 1 and 9, respectively. Figure 6 show the two power profiles of the energy storage system of the microgrid corresponding to those scenarios. It is possible to verify that, for both cases, the energy storage system is mainly used to compensate the lack of energy during peak hours.

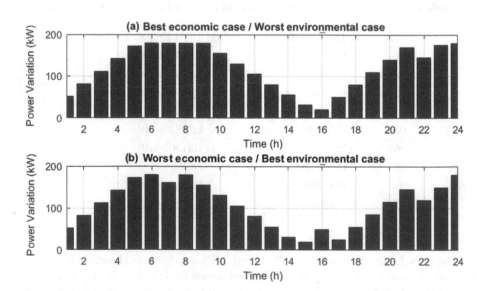

Fig. 6. Storage system power variation considering (a) scenario 1 and (b) scenario 9.

Figure 7 illustrates the daily power exchange of the energy storage system with the microgrid. The areas below the zero axes represent the energy during the charging process while the remaining areas represent the energy delivered to the microgrid. The null values, between 6 and 8 am from the best economic scenario (Fig. 7(a)), indicate the inactive mode of the energy storage system, which translates that the energy of the storage system has reached its maximum limit E^{max}, and therefore the energy storage system stop charging.

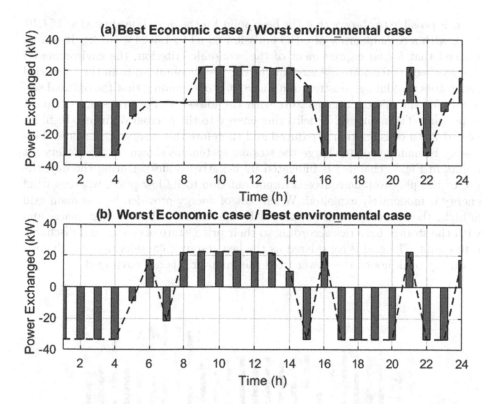

Fig. 7. Power exchange of the batteries with the microgrid during the day considering (a) scenario 1 and (b) scenario 9.

Based on the analysis of the non-dominated points, the discussion is divided into two cases, the first one mainly characterize the economic criterion, while the second one is related to the environmental criterion, discussed hereinafter.

6.1 Economic Criterion

Table 4 characterizes the classification of the prices according to three states: best, average, and worst, identifying scenarios one, five, and nine, respectively.

Table 4. Comparison of results considering the economic criterion

Scenarios	Total Energy Cost (euro)
The worst	161.01
The average	157.56
The best	154.97

It is possible to observe that the best point for the price is evaluated at $154, 97$ euro, with a total quantity of GHG emission equal to 1.3524×10^3 kg. It can be noticed that for an improvement of the economic criterion, the environmental one has been deteriorated. According to the results obtained from the microgrid generators scheduling, illustrated in Fig. 8, it is outstanding that the optimal set-points for the microgrid generators with the lowest energy prices are the most important. The main grid is delivering energy to the microgrid during the night period when consumption is reduced and therefore the energy price is low. This energy is mainly used to charge the storage system, as shown by the battery set points in Fig. 8 (charging is indicated by negative values). During the day, the use of the photovoltaic source is important due to its low price, whereas wind energy is moderately exploited. When cost of energy provided by the main grid is high, the consumption is supported by the micro-turbine in first place, and with the storage batteries according to their price, state of charge and discharge rate limits. The grid is considered as the last resource considering its high cost, ie during peak hours, the power from the main grid is not envisaged.

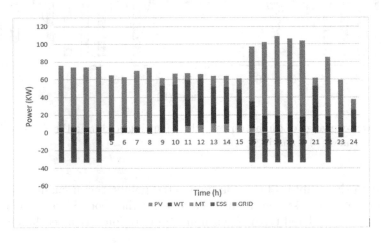

Fig. 8. Hourly dispatching set points of generators considering the best economical solution of multi-objective situation.

However, the first scenario takes mainly into account the economic criterion by favoring the cheapest sources and considering the fact that the environmental criterion will not be much affected since it is a simultaneous optimisation of two objectives.

6.2 Environmental Criterion

Table 5. Comparison of results considering the environmental criterion

Scenarios	Total Emissions (Kg)
The worst	1.3524×10^3
The average	1.2984×10^3
The best	1.2795×10^3

Table 5 characterizes the classification of emissions according to three states best, average and worst case, identifying scenarios nine, five, and one, respectively.

The best point for emissions is evaluated at 1.2795×10^3 kg with a total energy price equal to $161,01$ euro. It can be noticed that for an improvement of the environmental criterion by 5.39%, the economic criterion is deteriorated by 3.75%. According to the results presented in Fig. 9, the hourly set points from renewable sources (wind turbines and photovoltaic) are the most important. The photovoltaic source is fully exploited during the day due to its encouraging price, and being non-polluting also the wind source is considerably exploited to reduce the use of conventional sources responsible for greenhouse gas emissions (GHG). On the other hand, the use of conventional sources is classified according to the emission factor, the lack of energy is compensated by the micro-turbine due to its reduced emission factor compared to that of the main grid, for this reason, their set-points are important comparing with the previous case which takes into account much more the economic criteria. Furthermore, the main grid is less interrogated since it is considered a strong emission source. The purpose of battery discharging is to compensate the lack of energy and limit the energy exchange from the main grid to the microgrid in order to reduce greenhouse gas emissions responsible for global warming. The second case will take into consideration the scheduling of the microgrid production sources while favoring the environmental aspect without affecting the economic aspect illustrated by the total cost of energy.

The performance of the Energy Management System (EMS) based on the Pareto-search Algorithm is demonstrated by the non-dominant points obtained which represent trade-off cases between cost and emissions, allowing the achievement of several scenarios and offering several choices to the grid operator for the scheduling of the microgrid generators taking as reference the points located in the Pareto front.

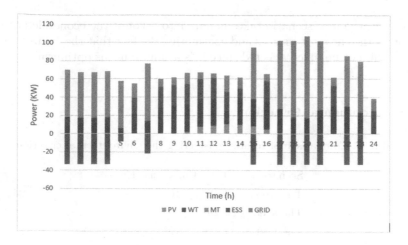

Fig. 9. Hourly dispatching set-points of generators considering the best environmental solution of multi-objective situation.

7 Conclusions and Future Work

In this paper, an energy management system based on a multi-objective optimisation approach has been proposed to solve the problem of optimal energy management in microgrids. Both economic and environmental aspects were simultaneously considered and optimised through the Pareto-search Algorithm. The results present a set of non-dominated solutions placed on a Pareto front, allowing the achieving of several microgrid scheduling scenarios. The proposed methodology provides a set of effective Pareto-optimal solutions respecting the technical-economic and environmental considerations of the problem under study and offering to the microgrid operator a variety of options for selecting an appropriate energy allocation scenario based on environmental or economic considerations. The wind turbine represents one of the permanent producers of the microgrid, however, this latter faces several obstacles, mainly the fluctuating effect. A wind speed forecasting model to predict the available capacity of wind energy production in the microgrid is important to improve the reliability of the system, to do that, a forecasting model based on the Artificial intelligence of the Neural Network (ANN) is proposed as future work. In the same context, another algorithm based on artificial intelligence will be proposed to ensure the demand scheduling of a smart city under economic and environmental considerations to further optimise the management of the microgrid and increase its efficiency.

References

1. Escrivani, G.R., Luna, A.S., Torres, A.R.: Operating parameters for bio-oil production in biomass pyrolysis: a review. J. Anal. Appl. Pyrol. **129**, 134–149 (2018)

2. Vittal, H., Oldrich, R., Yannis, M.: Increased future occurrences of the exceptional 2018–2019 Central European drought under global warming. Sci. Rep. Nature Publisher Group **10**(1)(2020)
3. Tsikalakis, A.G., Hatziargyriou, N.D.: Centralized control for optimizing microgrids operation. In: 2011 IEEE Power and Energy Society General Meeting. IEEE (2011)
4. Morais, H., Kádár, P., Faria, P., Vale, Z.A., Khodr, H.M.: Optimal scheduling of a renewable micro-grid in an isolated load area using mixed-integer linear programming. Renew. Energy **35**(1), 151–156 (2010)
5. Shahidehpour, S.M., Tong, S.K.: An overview of power generation scheduling in the optimal operation of a large scale power system. Electr. Mach. Power Syst. **19**(6), 731–762 (1991)
6. Jiayi, H., Chuanwen, J., Rong, X.: A review on distributed energy resources and MicroGrid. Renew. Sustain. Energy Rev. **12**(9), 2472–2483 (2008)
7. Voumvoulakis, E., Skotinos, I., Tsouchakinas, A.: Transient analysis of microgrids in Grid-connected and islanded mode of Operation. Fuel **2**, 30 (2004)
8. Lund, P.D., et al.: Review of energy system flexibility measures to enable high levels of variable renewable electricity. Renew. Sustain. Energy Rev. **45**, 785–807 (2015)
9. Mohamed, F.A., Koivo, H.N.: Online management genetic algorithms of microgrid for residential application. Energy Convers. Manage. **64**, 562–568 (2012)
10. Liu, Z., Chen, C., Yuan, J.: Hybrid energy scheduling in a renewable micro grid. Appl. Sci. **5**(3), 516–531 (2015)
11. Kani, S.A.P., Colson, C.M., Nehrir, H., Wang, C.: Real-time energy management of a stand-alone hybrid wind-microturbine energy system using particle swarm optimization. IEEE Trans. Sustain. Energy **3**, 193–201 (2010)
12. Granelli, G., Montagna, M., Pasini, G., Marannino, P.: Emission constrained dynamic dispatch. Electric Power Sys. **24**, 56–64 (1992)
13. Motevasel, M., Seif, A.R.: Expert energy management of a micro-grid considering wind energy uncertainty. Energy Convers. Manage. **83**, 58–72 (2014)
14. Moghaddam, A.A., Seif, A., Niknam, T., Pahlavani, M.R.A.: Multi-objective operation management of a renewable mg (micro-grid) with back-up micro-turbine/fuel cell/battery hybrid power source. Energy **36**, 6490–6507 (2011)
15. Chen, C., Cai, D., Hu, L.: Smart energy management system for optimal microgrid economic operation. IET Renew. Power Gener. **5**, 258–267 (2011)
16. Chen, J., Wang, C., Zhao, B., Zhang, X.: Economic operation optimization of a stand-alone microgrid system considering characteristics of energy storage system. Electric Power Syst. **6**, 25–31 (2012)
17. Hossain, M.A., Pota, H.R., Squartini, S., Abdou, A.F.: Modified PSO algorithm for real-time energy management in grid-connected microgrids. Renew. Energy **136**, 746–757 (2018)
18. Moghaddam, A.A., et al.: Multi-objective operation management of a renewable MG (micro-grid) with back-up micro-turbine/fuel cell/battery hybrid power source. Energy **36**(11), 6490–6507 (2011)
19. Lin, C., Gen, M.: Multi-criteria human resource allocation for solving multistage combinatorial optimization problems using multiobjective hybrid genetic algorithm. Expert Syst. Appl. **34**, 2480–2490 (2008)
20. Audet, C., Dennis, E.: Analysis of generalized pattern searches. SIAM J. Optim. **13**, 889–903 (2002)

21. Chang, P., Chen, S., Liu, C.: Sub-population genetic algorithm with mining gene structures for multiobjective flowshop scheduling problems. Expert Syst. Appl. **33**, 762–771 (2007)
22. Audet, C., Dennis, E.: Mesh adaptive direct search algorithms for constrained optimization. SIAM J. Optim. **17**, 188–217 (2006)
23. Amoura, Y., Pereira, A.I., Lima, J., Ferreira, A., Boukli-hacene, F.: Optimal energy management of a microgrid system. In: Symposium of Applied Science for Young Researchers SASYR (2021)
24. Amoura, Y., Pereira, A.I., Lima, J.: A short term wind speed forecasting model using artificial neural network and adaptive neuro-fuzzy inference system models. In: International Conference on Sustainable Energy for Smart Cities (2021)
25. Amoura, Y., Ferreira, A., Lima, J., Pereira, A.I.: Optimal sizing of a hybrid energy system based on renewable energy using evolutionary optimization algorithms. In: International Conference on Optimization, Learning Algorithms and Applications (2021)
26. Amoura, Y., Pereira, A.I, Lima, J.: Optimization methods for energy management in a microgrid system considering wind uncertainty data. In: Proceedings of International Conference on Communication and Computational Technologies (2021)
27. Amoura, Y., Pereira, A.I., Lima, J., Ferreira, A., Boukli-hacene, F., Kerboua, A.: Smart Microgrid Management: a Hybrid Optimisation Approach, Preprint on Energy Sustainability and Society

A Stochastic Alternating Balance
k-Means Algorithm for Fair Clustering

Suyun Liu$^{(\boxtimes)}$ and Luis Nunes Vicente

Department of Industrial and Systems Engineering, Lehigh University, Bethlehem,
PA 18015, USA
sul217@lehigh.edu, lnv@lehigh.edu

Abstract. In the application of data clustering to human-centric
decision-making systems, such as loan applications and advertisement
recommendations, the clustering outcome might discriminate against
people across different demographic groups, leading to unfairness. A natural conflict occurs between the cost of clustering (in terms of distance
to cluster centers) and the balance representation of all demographic
groups across the clusters, leading to a bi-objective optimization problem
that is nonconvex and nonsmooth. To determine the complete trade-off
between these two competing goals, we design a novel stochastic alternating balance fair k-means (SAfairKM) algorithm, which consists of alternating classical mini-batch k-means updates and group swap updates.
The number of k-means updates and the number of swap updates essentially parameterize the weight put on optimizing each objective function.
Our numerical experiments show that the proposed SAfairKM algorithm
is robust and computationally efficient in constructing well-spread and
high-quality Pareto fronts both on synthetic and real datasets.

Keywords: k-means clustering · Unsupervised machine learning ·
Data mining · Fairness · Bi-objective optimization · Pareto front

1 Introduction

Clustering is a fundamental task in data mining and unsupervised machine learning with the goal of partitioning data points into clusters, in such a way that
data points in one cluster are very similar and data points in different clusters are
quite distinct [16]. It has become a core technique in a huge amount of application fields such as feature engineering, information retrieval, image segmentation,
targeted marketing, recommendation systems, and urban planning. Data clustering problems take on many different forms, including partitioning clustering like
k-means and k-median, hierarchical clustering, spectral clustering, among many
others [8, 16]. Given the increasing impact of automated decision-making systems

L. N. Vicente—Support for this author was partially provided by the Centre for
Mathematics of the University of Coimbra under grant FCT/MCTES UIDB/MAT/
00324/2020.

© Springer Nature Switzerland AG 2022
D. E. Simos et al. (Eds.): LION 2022, LNCS 13621, pp. 77–92, 2022.
https://doi.org/10.1007/978-3-031-24866-5_6

in our society, there is a growing concern about algorithmic unfairness, which in the case of clustering may result in discrimination against minority groups. For instance, females may receive proportionally fewer job recommendations with high salary [13] due to their under-representation in the cluster of high salary recommendations. Such demographic features like gender and race are called *sensitive* or *protected* features, which we wish to be fair with respect to.

Related Work. An extensive literature work studying algorithmic fairness has been focused on developing universal fairness definitions and designing fair algorithms for supervised machine learning problems. Among the broadest representative fairness notions proposed for classification and regression tasks are *disparate impact* [6] (also called *demographic parity* [10]), *equalized odds* [19], and individual fairness [15], based on which the fairness notions in clustering were proposed accordingly. There are a number of classes of fairness definitions proposed and investigated for the clustering task [1,11,12,18,23,27]. The most widely used fairness notion is called *balance*. It was proposed by [12], and it has been extended in several subsequent works [7,20,30]. As a counterpart of the disparate impact concept in fair supervised machine learning, balance essentially aims at ensuring that the representation of protected groups in each cluster preserves the global proportion of each protected group.

Depending on the stage of clustering in which the fairness requirements are imposed, the prior works on fair clustering are categorized into three families, namely pre-processing, in-processing, and post-processing. A large body of the literature work [5,12,20,30] falls into the pre-processing category. The whole dataset is first decomposed into small subsets named *fairlets*, where the desired balance can be guaranteed. Any resulting solution from classical clustering algorithms using the set of fairlets will then be fair. Chierichetti et al. [12] focused on the case of two demographic groups and formulated explicit combinatorial problems (such as perfect matching and minimum cost flow problems) to decompose the dataset into minimal fair sets defining the fairlets. Their theoretical analysis gave strong guarantees on the quality of the fair clustering solutions for k-center and k-median problems. Following that line of work, Backurs et al. [5] embedded the whole dataset into a hierarchical structure tree and improved the time complexity of the fairlet decomposition step from quadratic to nearly linear time (in the dataset size). Schmidt et al. [30] introduced the notion of fair *coresets* and proposed an efficient streaming fair clustering algorithm for k-means. They introduced a near-linear time algorithm to construct coresets that helps reduce the input data size and hence speeds up any fair clustering algorithm. Huang et al. [20] further boosted the efficiency of coresets construction and made a generalization to multiple non-disjoint demographic groups for both k-means and k-median.

On the contrary, post-processing clustering methods [3,7,22,29] modify the resulting clusters from classical clustering algorithms to improve fairness. For example, Bera et al. [7] proposed a fair re-assignment problem as a linear relaxation of an integer programming model given the clustering results from any vanilla k-means, k-median, or k-center algorithms. They showed how to derive a

($\rho+2$)-approximation fair clustering algorithm from any ρ-approximation vanilla clustering algorithm within a theoretical bound of fairness constraints violation. Moreover, their framework works for datasets with multiple and potentially overlapping demographic groups. Lastly, in-processing methods incorporate the fairness constraints into the clustering process [2,24,32]. Our approach falls into this category and allows for the determination of the trade-offs between clustering costs and fairness. To our knowledge, the only such other in-processing approach is the one of Ziko et al. [32], where the clustering balance is approximately measured by the KL-divergence and imposed as a penalty term in the fair clustering objective function. The penalty coefficient is then used to control the trade-offs between clustering cost/fairness.

Our Contribution. The partitioning clustering model, also referred to as the center-based clustering model, consists of selecting a certain number K of centers and assigning data points to their closest centers. In this paper, we will focus on the well-known k-means model, and we will introduce a novel fair clustering algorithm using the balance measure. The main challenge of the fair clustering task comes from the violation of the assignment routine, which then indicates that a data point is no longer necessarily assigned to its closest cluster. The higher the balance level one wants to achieve, the more clustering cost is added to the final clustering. Hence, there exists a natural conflict between the fairness level, when measured in terms of balance, and the classical k-means clustering objective.

We explicitly formulate the trade-offs between the k-means clustering cost and the fairness as a bi-objective optimization problem, where both objectives are written as nonconvex and nonsmooth functions of binary assignment variables defining point assignments in the clustering model (see (2) further below). Our goal is to construct an informative approximation of the Pareto front for the proposed bi-objective fair k-means clustering problem, without exploring exhaustively the binary nature of the assignment variables. The most widely used method in solving general bi-objective optimization problems is the so-called weighted-sum method [17]. There, one considers a set of single objective problems, formed by convex linear combinations of the two functions, and (a portion of) the Pareto front might be approximated by solving the corresponding weighted-sum problems. However, this methodology has no rigorous guarantees due to the nonconvexity of both objective functions. Also, the non-smoothness of the fairness objective poses an additional difficulty to the weighted-sum method, as one function will be smooth and the other one no. Moreover, even ignoring the nonconvexity and non-smoothness issues, the two objectives, namely the clustering cost and the clustering balance, can have significantly different magnitudes. One can hardly preselect a good set of weights corresponding to decision-makers' preferences to capture a well-spread Pareto front.

Therefore, we were motivated to design a novel stochastic alternating balance fair k-means (SAfairKM) algorithm, inspired from the classical mini-batch k-means algorithm, which essentially consists of alternatively taking pure mini-batch k-means updates and swap-based balance improvement updates. In fact, the number of k-means updates (denoted by n_a) and the number of swap updates (denoted by n_b) play a role similar to the weights in the weighted-sum method, parameterizing the efforts of optimizing each objective. In the pure mini-batch k-means updates, we focus on minimizing the clustering cost. A mini-batch of points is randomly drawn and assigned to their closest clusters, after which the set of centers are updated using mini-batch stochastic gradient descent. In the swap-based balance improvement steps, we aim at increasing the overall clustering balance. For this purpose, we propose a simple swap routine that is guaranteed to increase the overall clustering balance by swapping data points between the minimum balance cluster and a target well-balanced cluster. Similarly to the k-means updates, the set of centers are updated using the batch of data points selected to swap. While the k-means updates reproduce the stochastic gradient descent directions for the clustering cost function, the swap updates can be seen as taking steps along some increasing directions for the clustering balance objective (not necessarily the best ascent direction).

We have evaluated the performance of the proposed SAfairKM algorithm using both synthetic datasets and real datasets. To endow SAfairKM with the capability of constructing a Pareto front in a single run, we use a list of nondominated points updated at every iteration. The list is randomly generated at the beginning of the process. At every iteration, and for every point in the current list, we apply SAfairKM for all considered pairs of (n_a, n_b). For each pair (n_a, n_b), one does n_a k-means updates and n_b swap updates. At the end of each iteration, we remove from the list all dominated points (those for each there exists another one with higher clustering cost and lower clustering balance). Such a simple mechanism is also beneficial for excluding bad local optima, considering that the two objectives are nonconvex. We will present the full trade-offs between the two conflicting objectives for four synthetic datasets and two real datasets. A numerical comparison with the fair k-means algorithm proposed in [7,32] further confirms the robustness and efficiency of the proposed algorithm in constructing informative and high-quality trade-offs.

2   The Mini-Batch k-means Algorithm

In the classical k-means problem, one aims to choose K centers (representatives) and to assign a set of points to their closest centers. The k-means objective is the sum of the minimum (squared Euclidean) distance of all points to their corresponding centers. Given a set of N points $P = \{x_p\}_{p=1}^N$, where x_p is the non-sensitive feature vector, the goal of clustering is to assign N points to K clusters identified by K centroids $C = [c_1, \ldots, c_K]^\top$. Let $[M]$ denote the set of positive integers up to M for any $M \in \mathbb{N}$. The k-means clustering problem is formulated as the minimization of a nonsmooth function of the set of centroids:

$$\min \ f_1^{KM}(C) \ = \ \frac{1}{2} \sum_{p=1}^N \min_{k \in [K]} \|x_p - c_k\|^2 \tag{1}$$

Since each data point is assigned to the closest cluster, the K cluster centroids are implicitly dependent on the point assignments. Let $s_{p,k} \in \{0,1\}$ be an assignment variable who takes the value 1 if point x_p is assigned to cluster k, and 0 otherwise. For simplicity, we denote s_k, $k \in [K]$, as an N-dimensional assignment vector for cluster k, and $s_p, p \in [N]$, as a K-dimensional assignment vector for point x_p. Let $X \in \mathbb{R}^{N \times d}$ be the data matrix stacking N data points of dimension d and $e_N \in \mathbb{R}^N$ be an all-ones vector. Then one can compute each centroid using $c_k = X^\top s_k / e_N^\top s_k$.

In practice, Lloyd's heuristic algorithm [26], also known as the standard batch k-means algorithm, is the simplest and most popular k-means clustering algorithm, and converges to a local minimum but without worst-case guarantees [21,31]. The main idea of Lloyd's heuristic is to keep updating the K cluster centroids and assigning the full batch of points to their closest centroids.

In the standard batch k-means algorithm, one can compute the full gradient of the objective function (1) with respect to k-th center by $\nabla_{c_k} f_1^{KM}(C) = \sum_{x_p \in \mathcal{C}_k} (c_k - x_p)$, where $\mathcal{C}_k, k \in [K]$, is the set of points assigned to cluster k. Whenever there exists a tie, namely a point that has the same distance to more than one cluster, one can randomly assign the point to any of such clusters. A full batch gradient descent algorithm would iteratively update the centroids by $c_k^{t+1} - c_k^t = \alpha_k^t \sum_{x_p \in \mathcal{C}_k} (x_p - c_k), \forall k \in [K]$, where $\alpha_k^t > 0$ is the step size. Let N_k^t be the number of points in cluster k at iteration t. It is known that the full batch k-means algorithm with $\alpha_k^t = 1/N_k^t$ converges to a local minimum as fast as Newton's method, with a superlinear rate [9].

The standard batch k-means algorithm is proved to be slow for large datasets. Bottou and Bengio [9] proposed an online stochastic gradient descent (SGD) variant that takes a gradient descent step using one sample at a time. Given a new data point x_p to be assigned, a stochastic gradient descent step would look like $c_k^{t+1} = c_k^t + \alpha_k^t (x_p - c_k^t)$ if x_p is assigned to cluster k. While the SGD variant is computationally cheap for large datasets, it finds solutions of lower quality than the batch algorithm due to the stochasticity. The mini-batch version of the k-means algorithm uses a mini-batch sampling to lower stochastic noise and, in the meanwhile, speed up the convergence. The detailed mini-batch k-means is given in Algorithm 1.

3   A New Stochastic Alternating Balance Fair k-means Method

3.1   The Bi-objective Balance k-means Formulation

Balance [12] is the most widely used fairness measure in the literature of fair clustering. Consider J disjoint demographic groups. Let V_j represent the set of points in demographic group $j \in [J]$. Then, $v_{p,j}$ takes the value 1 if point $x_p \in V_j$. We denote v_j as an N-dimensional indicator vector for the demographic group $j \in [J]$. The balance of cluster k is formally defined as $b_k = \min_{j \neq j'} v_j^\top s_k / v_{j'}^\top s_k \leq 1, \forall k \in [K]$, which calculates the minimum ratio among different pairs of protected groups. The overall clustering balance is the

---

Algorithm 1. Mini-batch k-means algorithm

1: **Input:** The set of points P and an integer K.
2: **Output:** The set of centers $C = \{c_1, \ldots, c_K\}$.
3: Randomly select K points as initial centers.
4: **for** $t = 0, 1, 2, \ldots$ **do**
5:     Randomly sample a batch of points B_t.
6:     **for** $k = 1, \ldots, K$ **do**
7:         Identify the set of points $B_t^k \subseteq B_t$ whose closest center is c_k.
8:         $N_k = N_k + |B_t^k|$.
9:         $c_k = c_k + \frac{1}{N_k} \sum_{x_p \in B_t^k} (x_p - c_k)$.

---

minimum balance over all clusters, i.e., $b = \min_{k=1}^K b_k$. The higher the overall balance, the fairer the clustering.

By the definition of cluster balance given above, the balance function can be easily computed only using the assignment variables. The k-means objective (1) can be rewritten as a function of the assignment variables as well. Hence, one can directly formulate the inherent trade-off between clustering cost and balance as a bi-objective optimization problem, i.e.,

$$\min \quad (f_1(s), -f_2(s)) \quad \text{s.t.} \quad \sum_{k=1}^K s_{p,k} = 1, \forall p \in [N], \quad s \in \{0,1\}^{N \times K}, \quad (2)$$

where s is the binary-valued assignment matrix with column vectors $s_k, k \in [K]$, and row vectors $s_p, p \in [N]$, and

$$f_1(s) = \frac{1}{N} \sum_{k=1}^K \sum_{p=1}^N s_{p,k} \|x_p - c_k\|^2, \quad \text{with } c_k = \frac{X^\top s_k}{e_N^\top s_k} = \frac{\sum_{p=1}^N x_p s_{p,k}}{\sum_{p=1}^N s_{p,k}},$$

$$f_2(s) = \min_{k \in [K]} \min_{\substack{j \neq j' \\ j,j' \in [J]}} \frac{v_j^\top s_k}{v_{j'}^\top s_k}$$

The two constraints in (2) ensure that one point can only be assigned to one cluster. Note that both objectives are nonconvex functions of the binary assignment variables.

3.2   The Stochastic Alternating Balance Fair k-means Method

We propose a novel stochastic alternating balance fair k-means clustering algorithm to compute a nondominated solution on the Pareto front. We will use a simple but effective alternating update mechanism, which consists of improving *either* the clustering objective *or* the overall balance, by iteratively updating cluster centers and assignment variables. Specifically, every iteration of the proposed algorithm contains two sets of updates, namely pure k-means updates and pure swap-based balance improvement steps. The pure k-means updates were introduced in Sect. 2, and will consist of taking a certain number of stochastic k-means steps. In the balance improvement steps, a certain batch of points is selected and swapped between the minimum balanced cluster and a target well-balanced cluster.

Balance Improvement Steps. At the current iteration, let C_l be the cluster with the minimum balance. Then C_l is the bottleneck cluster that defines the overall clustering balance. Without loss of generality, we assume that $b_l = |C_l \cap V_1|/|C_l \cap V_2|$, which then implies that the pair of demographic groups (V_1, V_2) forms a key to improve the balance of cluster C_l, as well as the overall clustering balance. In terms of the assignment variables, we have

$$b = b_l = \frac{v_1^\top s_l}{v_2^\top s_l} = \frac{\sum_{p=1}^{N} v_{p,1} s_{p,l}}{\sum_{p=1}^{N} v_{p,2} s_{p,l}} \tag{3}$$

One way to determine a target well-balanced cluster C_h is to select it as the one with the maximum ratio between V_1 and V_2, i.e.,

$$h \in \mathrm{argmax}_{k \in [K]} \left\{ v_1^\top s_k / v_2^\top s_k, v_2^\top s_k / v_1^\top s_k \right\} \tag{4}$$

Another way to determine such a target cluster is to select a cluster C_h that is closest to C_l, i.e.,

$$h \in \mathrm{argmin}_{k \in [K], k \neq l} \| c_k - c_l \| \tag{5}$$

We call the target cluster computed by (4) a *global* target and the one selected by (5) a *local* target. Using a global target cluster makes the swap updates more efficient and stable in the sense that the target cluster is only changed when the minimum balanced cluster changes. Instead, swapping according to the local target leads to less increase in clustering costs.

To improve the overall balance, one swaps a point in cluster C_l belonging to V_2 with a point in cluster C_h belonging to V_1. Each of these swap updates will guarantee an increase in the overall balance. The detailed stochastic alternating balance fair k-means clustering algorithm is given in Algorithm 2. At each iteration, we alternate between taking k-means updates using a drawn batch of points (denote the batch size by n_a) and "swap" updates using another drawn batch of points (denote the batch size by n_b). The generation of the two batches is independent. The choice of n_a and n_b influences the nondominated point obtained at the end, in terms of the weight put into each objective.

Instead of randomly selecting points to swap in line 11 of Algorithm 2, in our experiments we have used a more accurate swap strategy by increasing the batch size. Basically, we randomly sample a batch of points from $C_l \cap V_2$ (resp. $C_h \cap V_1$) and select x_p (resp. x'_p) as the one closest to C_h (resp. C_l). The batch size could be increased as the algorithm proceeds. Our numerical experiments show that the combination of local target clusters and the increasingly accurate swap strategy result in better numerical performance.

Algorithm 2. Stochastic alternating balance fair k-means clustering
(SAfairKM) algorithm

1: **Input:** The set of points P, an integer K, and parameters n_a, n_b.
2: **Output:** The set of clustering labels $\Delta = \{\delta_1, \ldots, \delta_N\}$, where $\delta_p \in [K]$.
3: Randomly initialize labels $\{\delta_1, \ldots, \delta_N\}$ and a set of counters $\{N_1, \ldots, N_K\}$. Compute k-means centers $\{c_1, \ldots, c_K\}$ and balances $\{b_1, \ldots, b_K\}$ for all clusters.
4: **for** $t = 1, 2, \ldots$ **do**
5:    Randomly sample a batch of n_a points $B_t \subseteq P$ without replacement.
6:    **for** $x_p \in B_t$ **do**
7:        Decrease the counter $N_{\delta_p} = N_{\delta_p} - 1$ for the previous clustering label.
8:        Identify its closest center index i_p. Update clustering label $\delta_p = i_p$.
9:        Increase the counter $N_{\delta_p} = N_{\delta_p} + 1$ and center $c_{\delta_p} = c_{\delta_p} + \frac{1}{N_{\delta_p}}(x_p - c_{\delta_p})$.
10:    **for** $r = 1, 2, \ldots, n_b$ **do**
11:        Identify $\mathcal{C}_l$, $\mathcal{C}_h$, and the pair of demographic groups (V_1, V_2) according to (3) and (5).
12:        Randomly select points $x_p \in \mathcal{C}_l \cap V_2$ and $x_{p'} \in \mathcal{C}_h \cap V_1$.
13:        Swap points: set $\delta_p = h$ and $\delta_{p'} = l$.
14:        Update centers $c_l = c_l + \frac{1}{N_l}(x_{p'} - c_l)$ and $c_h = c_h + \frac{1}{N_h}(x_p - c_h)$.
15:        Update balance for clusters $\mathcal{C}_l$ and $\mathcal{C}_h$.

One could have converted the bi-objective optimization problem (2) into a weighted-sum function using the weights associated with the decision-maker's preference. However, optimizing such a weighted-sum function hardly reflects the desired trade-off due to significantly different magnitudes of the two objectives. Moreover, the existing k-means algorithm frameworks, including Lloyd's heuristic algorithm, are not capable of directly handling the weighted-sum objective function. In our proposed SAfairKM algorithm, the pair (n_a, n_b) plays a role similar to the weights in the weighted-sum method.

4   Numerical Experiments

4.1   Pareto Front SAfairKM Algorithm

In our implementation[1], to obtain a well-spread Pareto front, we frame the SAfairKM algorithm into a Pareto front version using a list updating mechanism. See Algorithm 3 for a detailed description. In the initialization phase, we specify a sequence of pairs of the number of k-means updates and swap updates $\mathcal{W} = \{(n_a, n_b) : n_a + n_b = n_{\text{total}}, n_a, n_b \in \mathbb{N}_0\}$, and we generate a list of random initial clustering labels $\mathcal{L}_0$. Then we run Algorithm 2 for a certain number of iterations ($q = 1$ in our experiments) parallelly for each label in the current list

---

[1] Our implementation code is available at https://github.com/sul217/SAfairKM. All the experiments were conducted on a MacBook Pro Intel Core i5 processor.

$\mathcal{L}_t$, resulting in a new list of clustering labels $\mathcal{L}_{t+1}$. At the end of each iteration, the list is cleaned up by removing all the dominated points from $\mathcal{L}_{t+1}$. Using this algorithm, the list of nondominated points is refined towards the true real Pareto front. The process can be terminated when either the number of nondominated points is greater than a certain budget (1500 in our experiments) or when the total number of iterations exceeds a certain limit (depending on the size of the dataset).

Algorithm 3. Pareto-Front SAfairKM Algorithm

1: Generate a list of starting labels $\mathcal{L}_0$. Select parameter $q \in \mathbb{N}$ and a sequence of
   pairs $\mathcal{W} = \{(n_a, n_b) : n_a + n_b = n_{\text{total}}, n_a, n_b \in \mathbb{N}_0\}$.
2: **for** $t = 0, 1, \ldots$ **do**
3:     Set $\mathcal{L}_{t+1} = \mathcal{L}_t$.
4:     **for** each clustering label Δ in the list $\mathcal{L}_{k+1}$ **do**
5:         **for** $(n_a, n_b) \in \mathcal{W}$ **do**
6:             Apply q iterations of Algorithm 2 starting from Δ using the parameters
   (n_a, n_b).
7:             Add the final output label to the list $\mathcal{L}_{t+1}$.
8:     Remove all the dominated points from $\mathcal{L}_{t+1}$: **for** each label Δ in the list $\mathcal{L}_{t+1}$
   do
9:         If $\exists\ \Delta' \in \mathcal{L}_{t+1}$ such that $f_1(\Delta') < f_1(\Delta)$ and $f_2(\Delta') > f_2(\Delta)$ hold, remove Δ.

To the best of our knowledge, the only approach in the literature providing a mechanism of controlling trade-offs between the two conflicting objectives was suggested by [32] and briefly described in Appendix A. Their approach (here called VfairKM) consists of solving (6) for different penalty coefficients μ, resulting in a set of solutions from which we then remove dominated solutions to obtain an approximated Pareto front. To ensure a fair comparison, we select a set of penalty coefficients evenly from 0 to an upper bound μ_{max}, which is determined by pre-experiments such that the corresponding fairness error is less than 0.01 or no longer possibly decreased when further increasing its value. In some cases, we found that VfairKM is not able to produce a fairer clustering outcome when the penalty coefficient is greater than μ_{max} due to numerical instability.

In addition, we compare the Pareto fronts computed by SAfairKM with the fair k-means solution obtained from the postprocessing fair assignment approach proposed in [7] (marked as FairAssign). In [7], the fairest clustering solution is computed by first using a standard clustering algorithm and then applying the so-called fair assignment procedure. Such a fair assignment procedure consists of solving a linear programming relaxation of an integer programming problem, followed by an iterative rounding procedure to satisfy the bound constraints $\beta_j \leq |\mathcal{C}_k \cap V_j|/|\mathcal{C}_k| \leq \gamma_j, \forall j \in [J], k \in [K]$, where $\beta_j \in [0, 1]$ and $\gamma_j \in [0, 1]$ are lower and upper fairness bounds respectively. In our case, however, and in order to get the fairest solution, we set $\beta_j = \gamma_j = |V_j|/N$ which is exactly the proportion of demographic group j in the input dataset. Finally, we also present as benchmarks the k-means solutions obtained by both the state-of-the-art Lloyd's algorithm (denoted as VanillaKM) and the mini-batch k-means algorithm (denoted as MinibatchKM). Both VanillaKM and MinibatchKM were equipped with the well-known k-means++ initialization [4].

4.2 Numerical Results

Trade-Offs for Synthetic Datasets. We randomly generated four synthetic datasets from Gaussian distributions, and their demographic compositions are given in Fig. 2 of Appendix B. Each synthetic dataset has 400 data points in the $\mathbb{R}^2$ space and two demographic groups ($J = 2$) marked by black/circle and purple/triangle.

Using the list update mechanism (described by Algorithm 3), we are able to obtain a well-spread Pareto front with comparable quality for each of the synthetic datasets. Recall that we are minimizing the clustering cost and maximizing the clustering balance. The closer the Pareto front is to the upper left corner, the higher its quality. In particular, Fig. 1 (a) gives the approximated Pareto front for the Syn_unequal_ds2 dataset with $K = 2$, which confirms the natural conflict between the clustering cost and the clustering balance. One can see that the VfairKM algorithm is not able to output any trade-off information as it always finds the fairest solution regardless of the value of μ. Due to the special composition of this dataset, the Pareto front generated by SAfairKM is disconnected (the point around $(1.25, 0.35)$ is both VfairKM and SAfairKM). Results for the other three synthetic datasets are given by Figs. 3-6 in Appendix B. For all the synthetic datasets, the left end point on the Pareto front given by SAfairKM is consistent with the solution of VanillaKM. On the right end of the Pareto fronts, the fair solution given by FairAssign is dominated by the fairest solution identified by our approach.

Trade-Offs for Real Datasets. Two real datasets *Adult* [25] and *Bank* [28] are taken from the UCI machine learning repository [14]. The *Adult* dataset contains 32,561 samples. Each instance is characterized by 12 nonsensitive features (including age, education, hours-per-week, capital-gain, and capital-loss, etc.). For the clustering purpose, only five numerical features among the 12 features are kept. The demographic proportion of the *Adult* dataset is $[0.67, 0.33]$ in terms of gender ($J = 2$), which corresponds to a dataset balance of 0.49. The *Bank* dataset contains 41,108 data samples. Six nonsensitive numerical features (age, duration, number of contacts performed, consumer price index, number of employees, and daily indicator) are selected for the clustering task. Its demographic composition in terms of marital status ($J = 3$) is $[0.11, 0.28, 0.61]$, and hence the best clustering balance one can achieve is 0.185.

For the purpose of a faster comparison, we randomly select a subsample of size 5000 from the original datasets and set the number of clusters to $K = 10$. The resulting solutions from the five algorithms are given in Fig. 1 (b)-(c). For both datasets, SAfairKM is able to produce more spread-out Pareto fronts which capture a larger range of balance, and hence provide more complete trade-offs between the two conflicting goals. In terms of Pareto front quality (meaning dominance of one over the other), SAfairKM also performs better than VfairKM. In fact, we can see from Fig. 1 (b)-(c) that the Pareto fronts generated by SAfairKM dominate most of the solutions given by VfairKM and FairAssign. Also, the left end of the Pareto front generated by SAfairKM is much closer to the solution given by VanillaKM than VfairKM. The Pareto fronts corresponding to $K = 5$ are also given in Fig. 7 of Appendix B. Overall, SAfairKM results in a Pareto front of higher spread and slightly lower quality than VfairKM for the Adult dataset, while the Pareto front output from SAfairKM has better spread and higher quality for the Bank dataset.

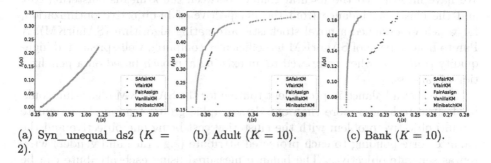

(a) Syn_unequal_ds2 ($K =$ 2).

(b) Adult ($K = 10$).

(c) Bank ($K = 10$).

Fig. 1. Pareto fronts: SAfairKM: 400 iterations for Syn_unequal_ds2, 2500 iterations for Adult, and 8000 iterations for Bank, 30 starting labels, and 4 pairs of (n_a, n_b); VfairKM: $\mu_{\max} = 0$ for Syn_unequal_ds2, $\mu_{\max} = 3260$ for Adult, and $\mu_{\max} = 2440$ for Bank.

Performance in Terms of Spread and Quality of Pareto Fronts. SAfairKM is able to generate more spread-out and higher-quality Pareto fronts regardless of the data distribution (see the trade-off results for the four synthetic datasets). The robustness partially comes from the list update mechanism which establishes a connection among parallel runs starting from different initial points and pairs (n_a, n_b), and thus helps escape from bad local optima.

Table 1. Average CPU times per nondominated solution.

Dataset	SAfairKM	VfairKM	Dataset	SAfairKM	VfairKM
Syn_equal_ds1	0.80	1.06	Adult ($K = 10$)	18.52	40.43
Syn_unequal_ds1	0.81	0.98	Bank ($K = 10$)	59.12	76.29
Syn_equal_ds2	0.70	1.29	Adult ($K = 5$)	14.88	15.08
Syn_unequal_ds2	0.80	0.10	Bank ($K = 5$)	11.97	50.31

Performance in Terms of Computational Time. Since the two algorithms (SAfairKM and VfairKM) generally produce Pareto fronts of different cardinalities, we evaluate their computational efforts by the average CPU time spent per computed nondominated solution (see Table 1). Our algorithm was shown to be clearly more computationally efficient than VfairKM.

5   Concluding Remarks

We have investigated the natural conflict between the k-means clustering cost and the clustering balance from the perspective of bi-objective optimization, for which we designed a novel stochastic alternating algorithm (SAfairKM). A Pareto front version of SAfairKM has efficiently computed well-spread and high-quality trade-offs, when compared to an existing approach based on a penalization of fairness.

Note that a balance improvement routine for the SAfairKM algorithm could be derived to handle more than one demographic group. One might formulate a multi-objective problem with the clustering cost being one objective and the balance corresponding to each protected attribute (e.g., race and gender) written as separate objectives. The balance measured using each attribute can be improved via alternating swap updates with respect to each balance objective.

A   Description of an Existing Approach for Comparison

The authors in [32] considered the fairness error computed by the Kullback-Leibler (KL)-divergence, and added it as a penalized term to the classical clustering objective. When using the k-means clustering cost, the resulting problem takes the form:

$$\min \ f_1(s) + \mu \sum_{k=1}^{N} \mathcal{D}_{KL}(U \| \mathbb{P}_k) \quad \text{s.t.} \quad \sum_{k=1}^{K} s_{p,k} = 1, \forall p \in [N], \tag{6}$$

where $\mathcal{D}_{KL}$ is the KL divergence between the desired demographic proportion $U = [u_j, j \in [J]]$ (usually specified by the demographic composition of the whole dataset) and the marginal probability $\mathbb{P}_k = [\mathbb{P}(j|k) = s_k^\top v_j / e_N^\top s_k, j \in [J]]$. The penalty coefficient μ associated with the fairness error is the tool to control the trade-offs between the clustering cost and the clustering balance. To solve problem (6) for a fixed $\mu \geq 0$, the authors in [32] have developed an optimization scheme based on a concave-convex decomposition of the fairness term.

B   More Numerical Results

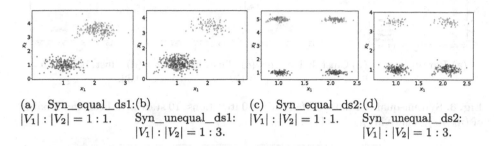

(a) Syn_equal_ds1:(b)
$|V_1| : |V_2| = 1 : 1$.

Syn_unequal_ds1:
$|V_1| : |V_2| = 1 : 3$.

(c) Syn_equal_ds2:(d)
$|V_1| : |V_2| = 1 : 1$.

Syn_unequal_ds2:
$|V_1| : |V_2| = 1 : 3$.

Fig. 2. Demographic composition of four synthetic datasets.

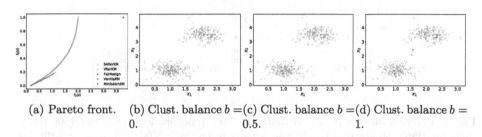

(a) Pareto front.   (b) Clust. balance $b =$(c) Clust. balance $b =$(d) Clust. balance $b =$
0.   0.5.   1.

Fig. 3. Syn_equal_ds1 data: SAfairKM: 400 iterations, 10 starting labels, and 3 pairs of (n_a, n_b); VfairKM: $\mu_{max} = 202$.

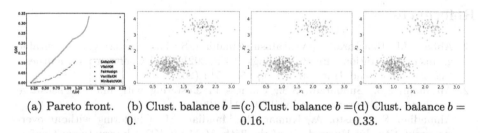

(a) Pareto front.   (b) Clust. balance $b =$(c) Clust. balance $b =$(d) Clust. balance $b =$
0.   0.16.   0.33.

Fig. 4. Syn_unequal_ds1 data: SAfairKM: 400 iterations, 10 starting labels, and 3 pairs of (n_a, n_b); VfairKM: $\mu_{max} = 223$.

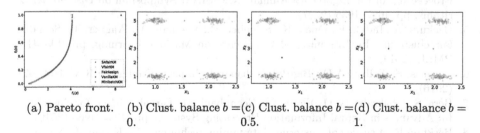

(a) Pareto front.   (b) Clust. balance $b =$(c) Clust. balance $b =$(d) Clust. balance $b =$
0.   0.5.   1.

Fig. 5. Syn_equal_ds2 data: SAfairKM: 400 iterations, 10 starting labels, and 3 pairs of (n_a, n_b); VfairKM: $\mu_{max} = 60$.

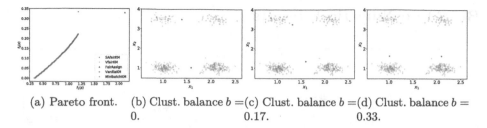

(a) Pareto front.   (b) Clust. balance b =(c) Clust. balance b =(d) Clust. balance b =
                          0.   0.17.   0.33.

Fig. 6. Syn_unequal_ds2 data: SAfairKM: 400 iterations, 10 starting labels, and 3 pairs of (n_a, n_b); VfairKM: $\mu_{\max} = 0$.

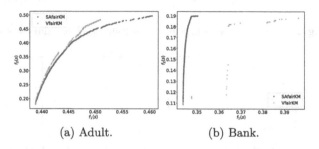

(a) Adult.   (b) Bank.

Fig. 7. Pareto fronts for $K = 5$: SAfairKM: 2500 iterations for Adult and 1500 iterations for Bank, 30 starting labels, and 4 pairs of (n_a, n_b); VfairKM: $\mu_{\max} = 6190$ for Adult and $\mu_{\max} = 4790$ for Bank.

References

1. Abbasi, M., Bhaskara, A., Venkatasubramanian, S.: Fair clustering via equitable group representations. In: Proceedings of the 2021 ACM Conference on Fairness, Accountability, and Transparency, pp. 504–514 (2021)
2. Abraham, S.S., Sundaram, S.S.: Fairness in clustering with multiple sensitive attributes. arXiv preprint arXiv:1910.05113 (2019)
3. Ahmadian, S., Epasto, A., Kumar, R., Mahdian, M.: Clustering without over-representation. In: Proceedings of the 25th ACM SIGKDD International Conference on Knowledge Discovery and Data Mining, pp. 267–275 (2019)
4. Arthur, D., Vassilvitskii, S.: k-means++ the advantages of careful seeding. In: Proceedings of the Eighteenth Annual ACM-SIAM Symposium on Discrete Algorithms, pp. 1027–1035 (2007)
5. Backurs, A., Indyk, P., Onak, K., Schieber, B., Vakilian, A., Wagner, T.: Scalable fair clustering. In: International Conference on Machine Learning, pp. 405–413. PMLR (2019)
6. Barocas, S., Selbst, A.D.: Big data's disparate impact. Calif. Law Rev. **104**, 671 (2016)
7. Bera, S., Chakrabarty, D., Flores, N., Negahbani, M.: Fair algorithms for clustering. In: Advances in Neural Information Processing Systems, pp. 4954–4965 (2019)
8. Berkhin, P.: A survey of clustering data mining techniques. In: Kogan, J., Nicholas, C., Teboulle, M. (eds.) Grouping Multidimensional Data. Springer, Berlin, Heidelberg (2006). https://doi.org/10.1007/3-540-28349-8_2

9. Bottou, L., Bengio, Y.: Convergence properties of the k-means algorithms. In: Advances in Neural Information Processing Systems, pp. 585–592 (1995)
10. Calders, T., Kamiran, F., Pechenizkiy, M.: Building classifiers with independency constraints. In: 2009 IEEE International Conference on Data Mining Workshops, pp. 13–18. IEEE (2009)
11. Chen, X., Fain, B., Lyu, L., Munagala, K.: Proportionally fair clustering. In: International Conference on Machine Learning, pp. 1032–1041 (2019)
12. Chierichetti, F., Kuma, R., Lattanzi, S., Vassilvitskii, S.: Fair clustering through fairlets. In: Advances in Neural Information Processing Systems, pp. 5029–5037 (2017)
13. Datta, A., Tschantz, M.C., Datta, A.: Automated experiments on ad privacy settings: a tale of opacity, choice, and discrimination. Proc. Priv. Enhancing Technol. **2015**, 92–112 (2015)
14. Dua, D., Graff, C.: UCI Machine Learning Repository (2017). http://archive.ics.uci.edu/ml
15. Dwork, C., Hardt, M., Pitassi, T., Reingold, O., Zemel, R.: Fairness through awareness. In: Proceedings of the 3rd Innovations in Theoretical Computer Science Conference, pp. 214–226. ACM (2012)
16. Gan, G., Ma, C., Wu, J.: Data Clustering: Theory, Algorithms, and Applications. SIAM, Philadelphia (2020)
17. Gass, S., Saaty, T.: The computational algorithm for the parametric objective function. Nav. Res. Logist. Q. **2**, 39–45 (1955)
18. Ghadiri, M., Samadi, S., Vempala, S.: Socially fair k-means clustering. In: Proceedings of the 2021 ACM Conference on Fairness, Accountability, and Transparency, pp. 438–448 (2021)
19. Hardt, M., Price, E., Srebro, N.: Equality of opportunity in supervised learning. In: Advances in Neural Information Processing Systems, pp. 3315–3323 (2016)
20. Huang, L., Jiang, S., Vishnoi, N.: Coresets for clustering with fairness constraints. In: Advances in Neural Information Processing Systems, pp. 7589–7600 (2019)
21. Kanungo, T., Mount, D.M., Netanyahu, N.S., Piatko, C.D., Silverman, R., Wu, A.Y.: A local search approximation algorithm for k-means clustering. Comput. Geom. **28**, 89–112 (2004)
22. Kleindessner, M., Awasthi, P., Morgenstern, J.: Fair k-center clustering for data summarization. In: International Conference on Machine Learning, pp. 3448–3457. PMLR (2019)
23. Kleindessner, M., Awasthi, P., Morgenstern, J.: A notion of individual fairness for clustering. arXiv preprint arXiv:2006.04960 (2020)
24. Kleindessner, M., Samadi, S., Awasthi, P., Morgenstern, J.: Guarantees for spectral clustering with fairness constraints. In: International Conference on Machine Learning, pp. 3458–3467. PMLR (2019)
25. Kohavi, R.: Scaling up the accuracy of Naive-Bayes classifiers: a decision-tree hybrid. In: Proceedings of the Second International Conference on Knowledge Discovery and Data Mining, pp. 202–207. KDD1996, AAAI Press (1996)
26. Lloyd, S.: Least squares quantization in PCM. IEEE Trans. Inf. Theory **28**, 129–137 (1982)
27. Mahabadi, S., Vakilian, A.: Individual fairness for k-clustering. In: Proceedings of the 37th International Conference on Machine Learning. Proceedings of Machine Learning Research, vol. 119, pp. 6586–6596. PMLR, Virtual (13–18 Jul 2020)
28. Moro, S., Cortez, P., Rita, P.: A data-driven approach to predict the success of bank telemarketing. Decis. Support Syst. **62**, 22–31 (2014)

29. Rösner, C., Schmidt, M.: Privacy preserving clustering with constraints. In: 45th International Colloquium on Automata, Languages, and Programming. Schloss Dagstuhl-Leibniz-Zentrum fuer Informatik (2018)
30. Schmidt, M., Schwiegelshohn, C., Sohler, C.: Fair coresets and streaming algorithms for fair k-means. In: Bampis, E., Megow, N. (eds.) WAOA 2019. LNCS, vol. 11926, pp. 232–251. Springer, Cham (2020). https://doi.org/10.1007/978-3-030-39479-0_16
31. Selim, S.Z., Ismail, M.A.: k-means-type algorithms: a generalized convergence theorem and characterization of local optimality. IEEE Trans. Pattern Anal. Mach. Intell. **PAMI-6**(1), 81–87 (1984)
32. Ziko, I.M., Granger, E., Yuan, J., Ayed, I.B.: Variational fair clustering. In: Proceedings of the AAAI Conference on Artificial Intelligence. vol. 35, pp. 11202–11209 (2021)

Binary Black Widow Optimization Algorithm for Feature Selection Problems

Ahmed Al-Saedi[✉] and Abdul-Rahman Mawlood-Yunis

Physics and Computer Science Department, 75 University Ave W, Waterloo,
ON N2L 3C5, Canada
alsa0290@mylaurier.ca, amawloodyunis@wlu.ca

Abstract. In this research work, we study the ability of a nature-inspired algorithm called the Black Widow Optimization (BWO) algorithm to solve feature selection (FS) problems. We use the BWO as a base algorithm and propose a new algorithm called the Binary Black Widow Optimization (BBWO) algorithm to solve FS problems. The evaluation method used in the algorithm is the wrapper method, designed to keep a degree of balance between two objectives: (i) minimize the number of selected features, (ii) maintain a high level of accuracy. We use the k-nearest-neighbor (KNN) machine learning algorithm in the learning stage to evaluate the accuracy of the solutions generated by the BBWO. This study has two main contributions: (a) applying the BBWO algorithm to solve FS problems efficiently, and (b) test results. The performance of the BBWO is tested on twenty-eight UCI benchmark datasets and the test results were compared with six well-known FS algorithms (namely, the BPSO, BMVO, BGWO, BMFO, BWOA, and BBAT algorithms). The test results show that the BBWO is as good as, or even better in some cases than the FS algorithms compared against. The obtained results can be used as new a benchmark and provide new insights about existing FS solutions.

Keywords: Black widow optimization algorithm · Classification · Data mining · Feature selection · Metaheuristic algorithms

1 Introduction

Feature selection (FS) algorithms are used to determine the best subset of instructive features while preserving a high level of classification accuracy in portraying the original dataset features. They are used as a pre-processing stage in many machine learning algorithms and data mining applications. For example, in data mining, FS is used as a pre-processing stage to remove redundant and inconsequential features, as well as to determine a final set of features that cast the greatest degree of light on the original data [1, 2].

FS has been studied using classical approaches such as random search, complete search, breadth search, and depth search [1]. However, even though these methods ensure the optimal solution for small datasets, their render is impractical for large datasets

D. E. Simos et al. (Eds.): LION 2022, LNCS 13621, pp. 93–107, 2022.
https://doi.org/10.1007/978-3-031-24866-5_7

because of the enormous amount of computational power required and the excessive amount of time taken up [1, 2].

In the last few years, metaheuristic algorithms (MAs) have been considered to be the ideal and most reliable optimization algorithms for solving FS problems, particularly in cases involving the challenges presented by high-dimensional problems. Researchers employ MAs as FS algorithms because of their ability and the outstanding results that have been attained. Examples of such works include Simulated Annealing (SA) [3], Ant Colony Optimization (ACO) [4], Particle Swarm Optimization (PSO) [5], Genetic Algorithm (GA) [6], Whale Optimization Algorithm (WOA) [7], Mine Blast Algorithm (MBA) [20], Ant Lion Optimizer (ALO) [6], Grey Wolf Optimizer (GWO) [8], and Bat Algorithm (BA) [9].

Metaheuristic algorithms are the most appropriate alternative method for addressing the limitations of lengthy, far-reaching searches that entail high computational cost [17]. But each dataset has a different number of features, and no single method is the most appropriate for solving FS problems, i.e., one can still find room for improvements in the results. These shortcomings urge researchers to find new means to overcome the limitation of the current FS algorithms.

In this research work, we investigated the ability of a nature-inspired algorithm called the Black Widow Optimization algorithm (BWO) to solve feature selection problems efficiently. The algorithm is a population metaheuristic algorithm recently proposed with the intention of optimizing engineering design.

The BWO process is inspired, in essence, by the singular mating behavior exhibited by black widow spiders, a process that includes an exclusive stage called cannibalism [10]. It follows Darwin's natural selection theory, which is defined as generational descent accompanied by modification, species being subtly adjusted over time, and new species arising as a result. The BWO approach is designed to deliver rapid convergence and to avoid local optima, and it is, therefore, particularly appropriate for solving several kinds of optimization problems that involve several local optima. This is because BWO maintains equilibrium between the exploration and exploitation stages [10].

We build upon the BWO base algorithm and propose a Binary Black Widow Optimization (BBWO) algorithm to solve FS problems. The FS evaluation method used in the BBWO is the wrapper method, designed to keep a degree of balance between two objectives: (i) minimize the number of selected features, (ii) maintain a high level of accuracy. To achieve this, we have used the k-nearest-neighbor (KNN) algorithm in the learning stage with the intent to evaluate the accuracy of the solutions generated by the BBWO.

The main contributions of this study are (a) applying the BBWO algorithm to solve FS problems efficiently, (b) test results. We tested the performance of our proposed algorithm on twenty-eight University of California Irvine (UCI) benchmark datasets that include low, medium, and high dimensional datasets. The obtained results can be used as a new benchmark and provide new insights about existing FS solutions.

The rest of this paper is organized as follows. In section two, we describe the BBWO algorithm, in Sect. 3, we present experimental results, and in section four, we conclude the work and identify future works to improve the performance of the BBWO algorithm.

2 BBWO Algorithm for FS

In this section, we describe the Binary Black Widow Optimization (BBWO) algorithm for feature selection problems. The algorithm is made of multiple steps. These steps are listed and described below.

1) Solutions Representation
In the BWO, a possible solution to every problem is envisioned in terms of the attributes of the black widow spider. To solve the optimization problem, the structure is viewed as an array in a N_{var} dimension. A widow is an array describing the solution of the problem, and it can be defined as:

$$w = \left[x_1, , x_3, \ldots, x_{N_{var}} \right]$$

In the BBWO, we use *binary* representation to represent a population of solutions (N_{pop}). Each solution represents a single widow and is shown by a one-dimensional vector. The length of the vector varies in accordance with the feature number of the original dataset. For example, if S features are contained in the dataset, this means that the solution length is S. The cell value in the vector is indicated by a '1' or a '0'. The value '1' indicates that the corresponding feature is selected, whereas '0' indicates that the feature is not selected.

Length S

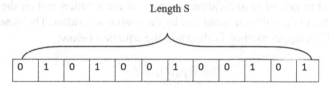

Fig. 1. A solution representation.

Since the BBWO operates on a population of solutions, the population is represented by an array, where each row represents one candidate solution (Fig. 1). Assuming that the number of features is N_f. And the population size is $\left| N_{pop} \right|$, the array size will be $N_f \times \left| N_{pop} \right|$.

2) Initialization
The population of solutions offered by the BBWO for the FS problems is randomly generated by assigning to each cell of the solution a value of either "0" or "1". The process begins by initializing the *population-size* and the *number of features*. The algorithm then arbitrarily assigns either '0' or '1' by looping through each solution in the population. This process is repeated until all solutions in the population have been initialized. The population generation procedure is presented in Algorithm 1.

Algorithm 1 Population Generation procedure

Set population size, $|N_{pop}|$

Set the number of features, N_f

$X_j^i = \theta$

For i=1 to $|N_{pop}|$ do

  For j=1 to N_f do

  X_j^i = select a randomly number either "0" or "1"

  Add X_j^i to $|N_{pop}|$

  End for

End for

---

3) Fitness Function and Evaluation Method

FS can be considered as a multi-objective optimization problem where two contradictory objectives are to be achieved: a minimal number of selected features and the highest classification accuracy. The smaller the number of features in the solution and the higher the classification accuracy, the better the solution is.

Each solution is evaluated according to the proposed fitness function, which depends on the classifier to get the classification accuracy of the solution and on the number of selected features in the solution generated by the search algorithm. The fitness function based on the FS wrapper method is shown in the equation below:

$$f = \alpha \gamma_R(D) + \beta \frac{|R|}{|C|} \tag{1}$$

where $\gamma_R(D)$ represents the classification error rate of a given classifier, $|R|$ is the cardinality of the selected subset, $|C|$ is the total number of the original features in the dataset, and α, β are two weight parameters corresponding to the importance of classification quality and subset length, $\alpha \in [0,1]$ and $\beta = (1 - \alpha)$.

After initializing the population of solutions, we assign to each solution (widow) a fitness value, which represents the quality of the solution. The FS wrapper method uses the classification performance (accuracy) of a classifier to evaluate the solutions. In particular, we use the KNN classifier as a learning stage algorithm to assess the accuracy of the solutions generated by the BBWO. KNN is a supervised ML algorithm used for both classification and regression problems. It calculates the distance between the testing sample and the samples in the training dataset based on specific metrics like Euclidean distance, then sorts the calculated distances in ascending order and picks the first k neighbours. The final step is to predict the response based on neighbours' voting, where each neighbour votes for its class feature, then takes the majority vote as the prediction. The fitness and evaluation methods followed in [6, 8, 11–14] are similar to the described approach.

4) Transformation Function

The positions of the search agents generated from the standard BWO are continuous values. This cannot, therefore, be directly applied to our problem because it contradicts the binary nature of FS selection or non-selection (0 or 1). The sigmoidal function in Eq. (2) and (3), which is considered a form of the transformation function, is used in our proposed method as a part of the reproduction process to convert any continuous value to a binary equivalent. The performance of the transformation function has been investigated and adopted in other works, e.g., [6, 8, 11].

$$z_{s_w} = \frac{1}{1 + e^{-z_w}} \tag{2}$$

$$z_{binary} = \begin{cases} 0, \text{ if } rand < z_{s_w} \\ 1, \text{ if } rand \geq z_{s_w} \end{cases} \tag{3}$$

The z_{s_w} in Eq. (2) is a continuous value (feature) in the search agent for the S-shaped function, specifically in the solution w at dimension d ($w = 1, \ldots, d$). The rand variable in Eq. (3) is a random number drawn from the uniform distribution $\in [0, 1]$. Lastly, the z_{binary} value can be 0 or 1 in accordance with the value of a *rand* in comparison with the values of z_{s_w}, where e is a mathematical constant known as Euler's number.

5) Reproduction Process

To bring forth the new generation, the procreation process begins, and parents (in pairs) are selected randomly to perform the procreating steps by mating. An array known as Alpha should also be generated to complete further reproduction. Offspring c_1 and c_2 will be produced by taking α with the following equation in which w_1 and w_2 are parents [10].

$$\begin{cases} c_1 = \alpha \times w_1 + (1 - \alpha) \times w_2 \\ c_2 = \alpha \times w_2 + (1 - \alpha) \times w_1 \end{cases} \tag{4}$$

This process is repeated for all pairs, where no repetition of randomly selected parents should take place. Lastly, the children and maternal parents are added to an array and sorted in accordance with their fitness value. Figure 2 is an assuming example of the procreate process for a child, $y1$.

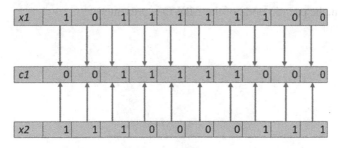

Fig. 2. An assuming example of the procreate process.

6) Cannibalism Process

Cannibalism can be classified into three kinds: sexual cannibalism where the husband gets eaten by the female black widow during or after mating, sibling-cannibalism where the weaker sibling spiders get eaten by stronger siblings, and the last kind where the mother gets eaten by her strongest child [10]. The proposed method (BBWO) determines the weak or strong spiders by calculating and evaluating their fitness values. Therefore, the best solutions (surviving spiders) from the reproduction process will be selected and stored in a variable pop2.

7) Mutation Process

The procedure of mutations begins by randomly selecting a number of solutions (widows) from the *pop1* population which will be mutated individually. Two cells from each selected solution (widow) are randomly exchanged, and the new mutation solutions will be kept in *pop3*. Figure 3 shows an example of the mutation structure for an individual solution.

Fig. 3. Mutation structures.

8) New Population Generation

The new population can finally be generated as a combination of *pop2* and *pop3*, which will then be evaluated to return the optimal solution (W^*) of values bearing the N dimension. The BWO algorithm contains some of the parameters with which exceptional results can be achieved. These involve the cannibalism rate, the procreation rate (P_r) and the mutation rate (M_r). The proposed BBWO algorithm determines the cannibalism rate in accordance with the fitness values Eq. (1), where the same parameters (P_r and M_r rates) of the standard BWO have been used.

2.1   The Pseudo-code of the Proposed Method (BBWO)

In this section, the pseudo-code of the BBWO is presented.

Algorithm 2 Pseudo-Code of BBWO Algorithm

Set the parameters value:

population size (N_{pop}) = 20; number of iterations $(maxIteration)$ = 10; number of features (N_f) = dimension size; procreate rate (P_r) = 0.6; mutation rate (M_r) = 0.4

\# Initialization process

Generating the initial population of solutions randomly $(N_{pop} \times N_f)$. Each solution represents one widow, which is indicated in one-dimension vector $1 \times N_f$

Calculate the fitness value for each solution using Eq. (1)

Evaluate all solutions in the population based on their fitness value and save them in *pop1*

Set the best solution in the population as W^*

based on P_r calculate the number of reproductions N_r

based on M_r calculate the number of mutations N_m

Define I=0

while $I < maxIteration$ **do**

#Procreate and cannibalism processes

for i = 1 to $(N_r/2)$ **do**

Randomly select two solutions w_1, w_2 as parents from *pop1*

Generate two children c_1, c_2 using Eq. (4)

Transformation c_1, c_2 to binary nature using Eq. (1 and 2)

Calculate the fitness value of c_1, c_2 using Eq. (1)

Destroy the father w_1 or w_2 based on their fitness value (cannibalism process)

Remove one of the two children c_1 or c_2 based on their fitness value (sibling cannibalism)

Save the remaining solutions in *pop2*

end for

#Mutation process

for i = 1 to N_m **do**

Randomly select a solution from *pop1*

Apply the mutation process on the selected solution

Save the result (the new solution) in *pop3*

end for

#Update the population

Update the population = *pop2+pop3*

Evaluate all solutions in the population using Eq. (1)

Update W^* if there is a better solution

$I= I+1$

end while

returning the best solution W^*

---

3 Experiments

In this section, we evaluate the performance of the BBWO against well-known heuristic FS algorithms. Specifically, we compare the performance of our BBWO algorithm against the BPSO [14], BMVO [16, 17], BGWO [8, 11], BMFO [18], BWOA [7], and BBAT [19] algorithms.

3.1 Implementation Setup

Python programming is used to implement the proposed BBWO algorithm, and the work was carried out via a Windows 10, 64-bit operating system, Core i5 processor, operating at 1.8 GHz and with 8 GB of RAM. A wrapper approach based on a KNN classifier (where $K = 5$ [7, 20]) is used to evaluate the fitness value of the selected feature subsets generated by the BBWO. Twenty-eight well known datasets from the University of California Irvine (UCI) machine learning repository [7, 8, 20, 21] have been used to investigate the performance and the strength of the BBOW. The dataset is randomly split into 80% for the training set and 20% for the test set.

3.2 Evaluation Criteria and Parameters Setting

The performance evaluation is based on the two criteria: classification accuracy and features selected. A calculation of the classification accuracy and the features selected were carried out by taking the average accuracy and the average number of features selected for the optimum solution of independent runs. To ensure an impartial comparison and a correct evaluation between our proposed method and other FS algorithms, we reimplemented the six FS algorithms (BPSO, BMVO, BGWO, BMFO, BWOA, BBAT) using the same parameter values (population-size, iterations, runs, K (KNN classifier), α, β) as illustrated in Table 1, and the same transformation function as explained in Sect. 2. The description of the data used in the testing is shown in Table 2.

Table 1. Parameters values

Parameters name	Value	Parameters name	Value
Population-size	20	pr	0.6
No. of iterations	10	mr	0.4
Number of independent runs	20	α	0.99
K (*KNN* classifier)	5	β	0.01
Dimension-size	No. of features		

Table 2. Dataset description

No.	Datasets	Features	Objects	Classes	Domain
1	Breastcancer	9	699	2	Medical
2	BreastEW	30	569	2	Medical
3	CongressEW	16	435	2	Politics
4	Exactly	13	1000	2	Medical
5	Exactly2	13	1000	2	Medical
6	HeartEW	13	270	5	Medical
7	IonosphereEW	34	351	2	Electronic
8	Lymphography	18	148	4	Medical
9	M-of-n	13	1000	2	Medical
10	PenglungEW	325	73	2	Medical
11	SonarEW	60	208	2	Medical
12	SpectEW	22	267	2	Medical
13	Tic-tac-toe	9	958	2	Game
14	Vote	16	300	2	Politics
15	WaveformEW	40	5000	3	Physical
16	Zoo	16	101	7	Artificial
17	Colon	2000	62	2	Medical
18	Parkinsons	22	195	2	Medical
19	Lungcancer	21	226	2	Medical
20	Leukemia	7129	72	2	Medical
21	Dermatology	34	366	6	Medical
22	Semeion	256	1593	10	Handwriting
23	Satellite	36	5100	2	Physical
24	Spambase	57	4601	2	Computer
25	Segment	19	2310	7	Images
26	Credit	20	1000	2	Business
27	KrvskpEW	36	3196	2	Game
28	Plants-100	64	1599	100	Agriculture

3.3 Experiment Results and Discussion of BBWO

The classification accuracy and number of features selected for the BBWO algorithm is presented in In Table 3.

Table 3. Experiment results for the BBWO

Datasets name	Number of features	Classification accuracy	Feature selected
Breastcancer	9	0.97	3.00
BreastEW	30	0.94	12.25
CongressEW	16	0.95	4.60
Exactly	13	0.91	3.75
Exactly2	13	0.77	3.65
HeartEW	13	0.84	3.80
IonosphereEW	34	0.88	13.75
Lymphography	18	0.85	6.80
M-of-n	13	0.95	7.00
PenglungEW	325	0.90	150.75
SonarEW	60	0.86	24.40
SpectEW	22	0.81	8.50
Tic-tac-toe	9	0.80	3.80
Vote	16	0.93	4.05
WaveformEW	40	0.88	20.60
Zoo	16	0.92	5.05
Parkinsons	22	0.89	7.70
Lungcancer	21	0.90	8.40
Colon	2000	0.87	959.20
Leukemia	7129	0.86	3531.90
Dermatology	34	0.97	15.70
Semeion	256	0.93	130.00
Satellite	36	0.99	12.20
Spambase	57	0.92	28.60
Segment	19	0.96	8.70
Credit	20	0.79	7.60
KrvskpEW	36	0.95	19.00
Plants-100	64	0.80	33.50

The BBWO results compared to the six FS algorithms (BPSO, BMVO, BGWO, BMFO, BWOA, BBAT), based on the classification accuracy (maximizing), and the number of features selected (minimizing) are shown in Table 4 and Table 5 respectively.

Table 4. Comparison BBWO with all algorithms based on the classification accuracy

Datasets name	BBWO	BPSO	BMVO	BGWO	BMFO	BWOA	BBAT
Breastcancer	**0.97**	0.96	**0.97**	0.96	**0.97**	**0.97**	0.96
BreastEW	0.94	0.94	0.94	**0.95**	0.94	0.93	0.94
CongressEW	**0.95**	0.92	**0.95**	**0.95**	**0.95**	**0.95**	0.94
Exactly	**0.91**	0.76	0.89	0.74	0.90	**0.91**	0.73
Exactly2	**0.77**	**0.77**	0.76	0.75	0.76	0.74	0.74
HeartEW	0.84	0.81	**0.85**	0.84	**0.85**	**0.85**	0.82
IonosphereEW	**0.88**	0.86	**0.88**	**0.88**	**0.88**	**0.88**	**0.88**
Lymphography	**0.85**	0.82	0.84	0.82	**0.85**	0.83	0.81
M-of-n	0.95	0.83	**0.99**	0.88	0.98	0.98	0.81
PenglungEW	**0.90**	0.87	0.89	0.89	0.89	0.88	0.88
SonarEW	0.86	0.86	**0.87**	0.86	0.86	**0.87**	0.86
SpectEW	0.81	0.81	0.81	**0.82**	**0.82**	0.81	0.81
Tic-tac-toe	0.80	0.74	0.81	0.78	**0.82**	0.81	0.76
Vote	0.93	0.91	**0.94**	**0.94**	**0.94**	**0.94**	0.93
WaveformEW	**0.88**	0.86	**0.88**	0.87	**0.88**	**0.88**	0.83
Zoo	**0.92**	0.89	0.89	0.88	0.88	0.90	0.89
Parkinsons	**0.90**	0.88	0.89	0.86	0.89	0.89	0.88
Lungcancer	0.90	0.88	**0.91**	0.90	**0.91**	**0.91**	0.90
Colon	0.87	0.86	**0.89**	0.87	0.87	0.87	0.87
Leukemia	**0.86**	0.83	**0.86**	0.85	**0.86**	**0.86**	0.85
Dermatology	**0.97**	0.89	0.96	0.95	**0.97**	**0.97**	0.92
Semeion	**0.93**	0.92	**0.93**	0.92	**0.93**	**0.93**	0.92
Satellite	**0.99**	**0.99**	**0.99**	**0.99**	**0.99**	**0.99**	**0.99**
Spambase	**0.93**	0.88	**0.93**	0.92	**0.93**	**0.93**	0.89
Segment	**0.96**	0.94	**0.96**	**0.96**	**0.96**	**0.96**	0.94
Credit	**0.79**	0.76	**0.79**	0.77	0.78	**0.79**	0.78
KrvskpEW	0.95	0.90	0.96	0.95	**0.97**	**0.97**	0.87
Plants-100	**0.80**	0.78	0.79	0.78	**0.80**	0.79	0.77
Average	0.8932	0.8614	0.8935	0.8760	0.8939	0.8925	0.8632
Rank	3	7	2	5	1	4	6

From the best results, which are showed in Tables 4 and 5, we can conclude that the BBWO produced competitive results in terms of the average total classification accuracy for all datasets in comparison with other the six FS algorithms. This result is summarized

Table 5. Comparison BBWO with all algorithms based on the feature selected

Datasets name	BBWO	BPSO	BMVO	BGWO	BMFO	BWOA	BBAT
Breastcancer	**3.00**	3.40	4.55	5.15	4.35	4.60	3.45
BreastEW	12.25	11.40	**10.95**	13.55	13.20	12.12	13.15
CongressEW	4.60	5.20	4.25	5.95	5.40	**4.20**	5.55
Exactly	**3.75**	5.30	7.25	7.05	7.05	6.65	5.80
Exactly2	3.65	3.95	2.33	5.40	3.15	**2.10**	3.50
HeartEW	3.80	4.25	3.45	4.15	3.70	**3.45**	4.65
IonosphereEW	13.75	14.35	13.35	15.90	15.00	**12.55**	16.55
Lymphography	6.80	7.60	6.66	7.56	7.35	**6.35**	7.55
M-of-n	7.00	**5.50**	7.22	8.00	6.75	7.35	6.15
PenglungEW	151.75	154.80	152.35	155.20	152.45	**146.35**	156.85
SonarEW	24.40	27.05	25.55	28.05	28.95	**23.85**	27.00
SpectEW	8.50	8.95	8.22	10.15	**8.20**	8.75	9.85
Tic-tac-toe	**3.80**	4.20	4.55	4.55	4.41	4.05	4.28
Vote	**4.05**	5.05	4.95	6.35	5.85	4.50	6.15
WaveformEW	20.60	22.00	22.15	21.45	21.35	**19.45**	20.25
Zoo	**5.05**	5.59	6.35	6.65	6.13	5.75	6.50
Parkinsons	**7.70**	8.00	8.45	9.15	9.10	8.20	9.25
Lungcancer	**7.00**	7.25	8.66	9.35	8.90	8.05	8.95
Colon	980.22	961.65	963.25	965.55	962.15	**943.55**	963.35
Leukemia	3531.90	3555.82	3571.85	3535.85	3534.55	**3511.35**	3513.50
Dermatology	**14.70**	16.85	15.95	16.70	16.60	16.45	16.50
Semeion	130.00	131.85	127.00	128.6	131.60	126.80	126.70
Satellite	**9.20**	13.01	10.55	12.40	11.40	10.10	12.45
Spambase	28.60	29.77	26.55	30.50	**26.25**	26.50	27.25
Segment	**7.70**	8.72	9.25	9.90	9.95	8.90	9.60
Credit	**7.22**	8.41	7.95	8.50	8.30	7.72	8.75
KrvskpEW	19.00	19.73	19.81	21.30	18.56	**17.92**	18.45
Plants-100	**32.50**	35.55	33.15	33.80	33.32	34.15	35.50
Average	180.44	181.61	181.66	181.66	180.85	178.27	180.26
Rank	3	5	6	6	4	1	2

in Fig. 4. Moreover, The BBWO shows impressive results in term of minimizing the features selected. This performance is shown in Fig. 5. The rank of the BBWO is third out of seven, based on the average total features selected of all datasets in comparison with the other FS algorithms.

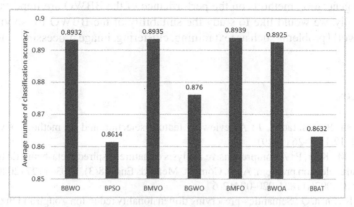

Fig. 4. Average number of classification accuracy of all algorithms (Maximizing)

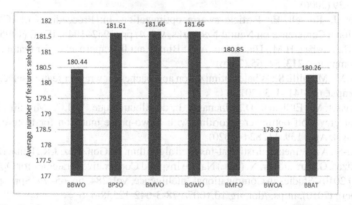

Fig. 5. Average number of features selected of all algorithms (Minimizing)

4 Conclusion and Future Works

In this research work, we investigated the ability of a nature-inspired algorithm called the Black Widow Optimization algorithm (BWO) to solve feature selection problems. We used the essence of the BWO to create an algorithm called the Binary Black Widow Algorithm (BBWO) and applied it to solve feature selection problems. We described the BBWO concepts and algorithm steps and showed how it can be applied to solve feature selection problems. The performance of the BBWO was tested on twenty-eight UCI benchmark datasets and the test results were compared with six well-known FS algorithms (BPSO, BMVO, BGWO, BMFO, BWOA, BBAT). The results showed that the BBWO is as good as, or even better in some cases, than the other FS algorithms.

In future works, we would like to combine the BBWO with other metaheuristics algorithms, for example, the hill-climbing algorithm, to further improve the performance of the BBWO. The impact of the parameter adaptation schemes and different population generation methods on the performance of the BBWO are important future works. Finally, we would like to study the suitability of the BBWO for solving many other real-world problems such as text mining, clustering, image processing, and routing problems.

References

1. Venkatesh, B., Anuradha, J.: A review of feature selection and its methods. Cybern. Inf. Technol. **19**(1), 3–26 (2019)
2. Sharma, M., Kaur, P.: A comprehensive analysis of nature-inspired meta-heuristic techniques for feature selection problem. Arch. Comput. Methods Eng. **28**(3), 1103–1127 (2020). https://doi.org/10.1007/s11831-020-09412-6
3. Jensen, R., Shen, Q.: Semantics-preserving dimensionality reduction: rough and fuzzy-rough-based approaches. IEEE Trans. Knowl. Data Eng. **16**(12), 1457–1471 (2004)
4. Dorigo, M., Birattari, M., Stutzle, T.: Ant colony optimization. IEEE Comput. Intell. Mag. **1**(4), 28–39 (2006)
5. Kennedy, J., Russell, E.: Particle swarm optimization. In: Proceedings of ICNN 1995-International Conference on Neural Networks, vol. 4, pp. 1942–1948. IEEE (1995)
6. Emary, E., Zawbaa, H.M., Hassanien, A.E.: Binary ant lion approaches for feature selection. Neurocomputing. **213**, 54–65 (2016)
7. Mafarja, M., Mirjalili, S.: Whale optimization approaches for wrapper feature selection. Appl. Soft Comput. **62**, 441–453 (2018)
8. AbdelBasset, M., ElShahat, D., Elhenawy, I., de Albuquerque, V.H.C., Mirjalili, S.: A new fusion of grey wolf optimizer algorithm with a two-phase mutation for feature selection. Expert Syst. App. **139**, 112824 (2020)
9. Yang, X.-S.: A new metaheuristic bat-inspired algorithm. In: González, J.R., Pelta, D.A., Cruz, C., Terrazas, G., Krasnogor, N. (eds.) Nature Inspired Cooperative Strategies for Optimization (NICSO 2010). Studies in Computational Intelligence, vol. 284, pp. 65–74. Springer, Berlin, Heidelberg (2010). https://doi.org/10.1007/978-3-642-12538-6_6
10. Hayyolalam, V., Kazem, A.A.P.: Black widow optimization algorithm: a novel meta-heuristic approach for solving engineering optimization problems. Eng. Appl. Artif. Intell. **87**, 103249 (2020)
11. Emary, E., Zawbaa, H.M., Hassanien, A.E.: Binary grey wolf optimization approaches for feature selection. Neurocomputing. **172**, 371–381 (2016)
12. Taradeh, M., et al.: An evolutionary gravitational search-based feature selection. Inf. Sci. **497**, 219–239 (2019)
13. Mafarja, M., Jaber, I., Ahmed, S., Thaher, T.: Whale optimisation algorithm for high-dimensional small-instance feature selection. Int. J. Parallel, Emerg. Distrib. Syst. **36**, 1–17 (2019)
14. Shaban, W.M., Rabie, A.H., Saleh, A.I., Abo-Elsoud, M.A.: A new COVID-19 patients detection strategy (CPDS) based on hybrid feature selection and enhanced KNN classifier. Knowl. Based Syst. **205**, 106270 (2020)

15. Mafarja, M., Jarrar, R., Ahmad, S., Abusnaina, A.A.: Feature selection using binary particle swarm optimization with time varying inertia weight strategies. In: Proceedings of the 2nd International Conference on Future Networks and Distributed Systems, pp. 1–9 (2018)
16. Mirjalili, S., Mirjalili, S.M., Hatamlou, A.: Multi-verse optimizer: a nature-inspired algorithm for global optimization. Neural Comput. App. **27**(2), 495–513 (2015). https://doi.org/10.1007/s00521-015-1870-7
17. Al-Madi, N., Faris, H., Mirjalili, S.: Binary multi-verse optimization algorithm for global optimization and discrete problems. Int. J. Mach. Learn. Cybern. **10**(12), 3445–3465 (2019). https://doi.org/10.1007/s13042-019-00931-8
18. Zawbaa, H.M., Emary, E., Parv, B., Sharawi, M.: Feature selection approach based on moth-flame optimization algorithm. In: 2016 IEEE Congress on Evolutionary Computation (CEC), pp. 4612–4617. IEEE (2016)
19. Mirjalili, S., Mirjalili, S.M., Yang, X.-S.: Binary bat algorithm. Neural Comput. Appl. **25**(3–4), 663–681 (2013). https://doi.org/10.1007/s00521-013-1525-5
20. Alweshah, M., Alkhalaileh, S., Albashish, D., Mafarja, M., Bsoul, Q., Dorgham, O.: A hybrid mine blast algorithm for feature selection problems. Soft. Comput. **25**(1), 517–534 (2020). https://doi.org/10.1007/s00500-020-05164-4
21. https://archive.ics.uci.edu/ml/datasets.php

Learning to Solve a Stochastic Orienteering Problem with Time Windows

Fynn Schmitt-Ulms[1], André Hottung[2(✉)], Meinolf Sellmann[3],
and Kevin Tierney[2]

[1] McGill University, Montreal, Canada
fynn.schmitt-ulms@mcgill.ca
[2] Decision and Operation Technologies Group, Bielefeld University,
Bielefeld, Germany
{andre.hottung,kevin.tierney}@uni-bielefeld.de
[3] InsideOpt, Dover, DE, USA

Abstract. Reinforcement learning (RL) has seen increasing success at solving a variety of combinatorial optimization problems. These techniques have generally been applied to deterministic optimization problems with few side constraints, such as the traveling salesperson problem (TSP) or capacitated vehicle routing problem (CVRP). With this in mind, the recent IJCAI AI for TSP competition challenged participants to apply RL to a difficult routing problem involving optimization under uncertainty and time windows. We present the winning submission to the challenge, which uses the policy optimization with multiple optima (POMO) approach combined with efficient active search and Monte Carlo roll-outs. We present experimental results showing that our proposed approach outperforms the second place approach by 1.7%. Furthermore, our computational results suggest that solving more realistic routing problems may not be as difficult as previously thought.

Keywords: Learning to optimize · Stochastic optimization · Deep reinforcement learning · Orienteering problem

1 Introduction

Deep reinforcement learning (DRL) approaches represent an exciting new research avenue in artificial intelligence (AI) and operations research (OR) for automatically creating heuristics to solve combinatorial optimization (CO) problems. These approaches are attractive as they can solve CO problems with little domain knowledge by iteratively building a solution through a construction process. A central goal of these approaches is to make optimization technology more accessible to audiences without expertise in operations research and perhaps even limited problem domain knowledge. Since the approach "learns" a heuristic on its own, the overall process of solving the CO problem is transformed to a data science task, rather than an OR task.

© Springer Nature Switzerland AG 2022
D. E. Simos et al. (Eds.): LION 2022, LNCS 13621, pp. 108–122, 2022.
https://doi.org/10.1007/978-3-031-24866-5_8

RL has seen particular success in solving deterministic routing problems with few side constraints [18], even if concerns about generalization to very large problem sizes remain [15]. The recent IJCAI AI for TSP competition [4] aimed to expand the horizon of RL techniques in CO to stochastic problems with side constraints. The competition posed a stochastic orienteering problem with time windows to be solved with RL, thus presenting a difficult problem with novel components not yet solved with RL techniques in the literature. Given a set of customers, each providing a reward if they are visited, and a time window in which they can be visited, the objective of the time-dependent orienteering problem with stochastic weights and time windows (TDOP[1]) is to construct a tour from a depot through a subset of the customers that maximizes the total rewards and returns to the depot before a maximum time is reached. A penalty is incurred for tours that exceed the maximum time, or that arrive at customer nodes after their time windows have closed. Travel times are subject to uncertainty, while the stay duration at customers is instant, provided the tour arrives after the start of the time window. However, tours that arrive at a customer before the time window are forced to wait until the start of the window.

The competition posed two tracks: supervised learning and reinforcement learning. In the supervised learning track, a complete solution to an instance of the TDOP is provided before the tour is traveled, meaning no recourse actions can be taken if delays are incurred while carrying out the tour. In the reinforcement learning track, the goal is to learn a policy for picking the next node to visit on a TDOP tour. Recourse is allowed, thus at each node visited in the solution, the future nodes to be visited can be adjusted according to how much time is left in the time horizon.

We propose a solution procedure using the policy optimization with multiple optima (POMO) method [19] and efficient active search (EAS) [12] that won first place in the reinforcement learning track[2]. POMO is a DRL approach that exploits symmetries in the solution space of CO problems. EAS is an RL-based search method originally designed for deterministic problems.

POMO and EAS have been successfully applied to deterministic problems. In contrast to problems considered in earlier work, the TDOP is a stochastic problem in which the travel times between customers are only revealed during solution construction. This means that a solution needs to be generated online (taking into account the already realized travel times at each decision step) rather than offline. Furthermore, the TDOP is heavily constrained and there are no obvious symmetries in the solution space that POMO can exploit. Our contribution is as follows. (1) We adapt POMO to stochastic problems without symmetries in the solution space. (2) We use EAS as a method for fine-tuning a given RL policy, and further extend EAS to use entropy regularization to improve exploration. (3) We use Monte Carlo rollouts for the final online solution

---

[1] We note that the TDOP is abbreviated as the TD-OPSWTW in some works.

[2] Although not the focus of our research, our approach can also generate complete solutions using the expected travel time for the supervised learning track, and these tie the winning team's solutions and generate them in less computation time.

generation. We discuss our entry into the competition and show how the addition of EAS and Monte Carlo rollouts yields a novel, competition-winning technique.

2   Related Work

Models for automatically learning to solve routing problems have advanced significantly since [10] solved small TSP instances with a Hopfield network. The pointer network architecture [23] in particular has allowed deep networks to make high-quality recommendations for choosing the next step in a construction process. Further insights, such as using actor-critic RL [2], attention [18], multiple rollouts [19], and simulation-guided beam search [7], have further closed the gap to state-of-the-art, handcrafted routing algorithms.

While learned models are good at generating solutions to routing problems, search is still required to find high quality solutions comparable to those found with handcrafted heuristics. The neural network's output distribution is sampled in [18], while [16] use a beam search with guidance from the neural network. The work of [13] proposes an improvement method that integrates a neural network into the repair operator of a large neighborhood search. Other iterative improvement methods for routing problems are proposed in [6,8,20,26]. Generative machine learning models can also be used to create a searchable latent space as in [11], in which a generated latent space is searched with an unconstrained continuous optimizer for good solutions. The DPDP approach from [17] uses a dynamic programming algorithm that is guided by a "heatmap" of suggestions generated by the neural network from [16]. In NeuroLKH [27] a neural network is integrated into the well-known LKH algorithm [9], thus relying on LKH to perform its search. Finally, [12] introduces EAS to adjust a subset of (model) parameters at test time to better solve a given problem instance.

In contrast to deterministic routing problems, stochastic routing problems have seen little attention in the ML literature. DRL is used in [14] to solve a dynamic vehicle routing problem with stochastic customers and time-windows. In this VRP variant, new customers (requests) can arise at any time during the solution execution, resulting in already planned routes to be adjusted online. A similar VRP variant with stochastic customers is solved by [21] via DRL, but this variant also considers vehicles with limited capacity. A multiagent RL approach is proposed in [5] to solve the dynamic CVRP with stochastic travel times and stochastic customers with time windows. Finally, [1] consider the problem of routing a single electric vehicle with a reliable charge, taking into account stochastic energy consumption and dynamic customer requests. They propose an RL method that learns a policy aiming to minimize the risk of battery depletion by planning charging stops.

3   Background

We provide background information modeling routing problems using RL. As our approach uses a transformer architecture, we focus on this in the following. We note, however, that a number of options exist for RL for routing and combinatorial optimization problems in general (see [3]). We begin with a general description of modeling routing problems in a sequential decision process, followed by the POMO method and, finally, how to apply EAS.

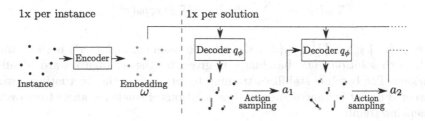

Fig. 1. POMO solution construction, from [12].

Solving Routing Problems with RL. Most ML-based approaches for solving routing problems formulate the solution construction as a sequential decision making problem. Starting from a start node, an actor decides, at each decision step $t \in \{0, ..., T\}$, which node should be visited next. The actor p_θ is usually a deep neural network with weights θ that outputs a probability value for each possible action in the given state. The starting state s_0 describes the given problem instance (e.g., the positions of the customers and the depot) and the state s_{t+1} is obtained by applying the action a_t chosen at step t to the state s_t. Once a complete solution $\pi = (a_0, ..., a_T)$ that satisfies all constraints of the problem is constructed, the objective function value of the solution can be computed (e.g., the tour length for the TSP). For the training of the network, most existing approaches use the REINFORCE algorithm [24] which adjusts the network weights based on the objective function value of the complete solution.

POMO. One state-of-the-art RL approach for sequential solution construction is the POMO approach [19]. POMO is an end-to-end approach that exploits symmetries in the solution space of combinatorial optimization problems to enforce exploration during the training phase. POMO uses a transformer-based network p_θ [22] that consists of an encoder and decoder. The encoder learns to generate an internal representation ω of a given problem instance (i.e., an embedding) and the decoder q_ϕ learns to construct a solution based on this embedding sequentially. Note that the weights ϕ of the decoder are a subset of all model weights θ. Figure 1 shows the embedding generation and the autoregressive solution construction for a TSP instance. At each decoding step, an action is sampled according to the output distribution by the decoder.

Note, that once an instance embedding has been generated by the encoder, the decoder can be used to generate multiple solutions for that instance. POMO exploits this fact by sampling multiple solutions for each problem instance. Furthermore, POMO uses symmetries in the solution space to enforce exploration. For example, for the TSP with n cities, POMO constructs n solutions $\{\pi^1, ..., \pi^n\}$ from the same embeddings. By starting each solution construction process from a different starting city, POMO ensures that each of the solutions is unique.

During training, POMO uses gradient ascent to update the neural network weights based on a set of sampled solutions $\{\pi^1, ..., \pi^n\}$ with the objective to maximize the expected reward

$$\nabla_\theta J(\theta) \approx \frac{1}{n} \sum_{i=1}^{n} (R(\pi^i) - b) \nabla_\theta \log p_\theta(\pi^i \mid s_0^i) \tag{1}$$

where $p_\theta(\pi^i \mid s_0^i) \equiv \prod_{t=0}^{T} p_\theta(a_t^i \mid s_t^i)$, b is the baseline, and $R(\pi^i)$ is the reward of the i-th solution. The baseline b is given by the average reward of all n solutions. The baseline stabilizes training by preventing the fluctuating reward between different instances and multiple solutions of single instances from overly influencing training.

EAS. Given a model trained as described above, we now focus on the search phase in which we are given an instance we have never seen before and need to solve it. Active search [2] performs an extensive search for high-quality solutions to a single test instance by iteratively solving the instance and adjusting the model parameters at the end of each iteration. In other words, active search performs a search for solutions by fine-tuning a given model towards a single instance using reinforcement learning. Over the course of the search/training, the model performance on the single test usually improves, and high-quality solutions are found. The best found solution at the end of the search/training is returned as the final solution. After solving the instance, the adjusted parameters are discarded and the model is returned to its original state, as they likely will not generalize to any other instances. While active search finds high-quality solutions, it is very slow because all instances have to be solved sequentially and a full update of the model parameters must be performed.

A recent extension to active search, called EAS [12], only updates a subset of (model) parameters during the search This significantly reduces the runtime and GPU memory requirements, since most model parameters are not updated and many model operations can hence be performed on a batch of test instances in parallel. For example, an extra layer can be added for each instance and only the parameters of this layer are updated during search. In this work, we use an EAS variant that adjusts the embeddings generated by the encoder, since both this and the extra layer versions of EAS show similar performance in [12].

4   The TDOP

In the TDOP, we are given a graph, $G = (V, E)$, with nodes V representing customers and a single depot, and edges E between all nodes. Each node i is assigned a location (x_i, y_i) in a Euclidean plane, a time window $(\underline{w}_i, \bar{w}_i)$, as well as a reward r_i. The goal of the problem is to construct a tour starting and ending at the depot that maximizes the reward earned by visiting customers (nodes). If a node is visited, it must be visited during its time window. If a node is visited early, the model is forced to wait until the beginning of the time window. If a node is visited late, a penalty p is incurred. The travel time between nodes, $\hat{t}_{ij}$,

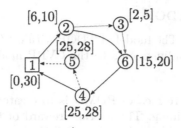

Fig. 2. An instance and solution (black, solid line) to the TDOP with time windows shown in square brackets. Alternate solutions are given by the red, dotted line and blue, dashed line. (Color figure online)

is stochastic, but bounded above by the euclidean distance between each pair of nodes. The visit duration at each node is instantaneous. The maximum travel time is given by D, and the penalty $p|V|$ is added to the objective function if it is exceeded. Note that in this version of the orienteering problem, no costs are incurred traveling between nodes. Thus, given a variable $x_i \in \{0,1\}$ that is set to 1 iff the route visits node $i \in V$, and a variable $\delta_i \in \{0,1\}$ that is equal to 1 iff the time window at node $i \in V$ is violated, the objective function can be formulated as $\max \sum_{i \in V} (r_i x_i - p'_i \delta_i)$, with p'_i defined as the penalty p for all nodes i that are not the depot, and $p|V|$ for the depot.

Figure 2 shows a TDOP instance with time windows for each node. For the purpose of illustration, assume the reward at every node is one. One possible solution is given by the black, solid line; assume the stochastic travel times between nodes are realized such that none of the time windows visited along the tour are violated. The tour thus incurs no penalty. The blue, dashed line shows an alternative end of the tour that would earn an extra reward. If we arrive at node 4 early enough, visiting node 5 may still be viable within its time window. This decision would be made on the fly as the tour is carried out. The red, dotted line, however, shows an alternative solution that is guaranteed to suffer a penalty regardless of the realization of the travel times, as the time window ends before the previous node's time window begins.

5   Solving the TDOP

Our solution approach for the TDOP consists of three steps. First, we use the POMO approach to learn a problem size-specific policy. To this end, we slightly adjust the POMO model to TDOP, e.g., by changing the structure of the input that the model accepts. In the next step, we use EAS to enhance the learned policy. EAS has been originally proposed as a search method for non-stochastic problems. We adjust EAS to the stochastic search setting and deploy it with the objective to find a fine-tuned policy to a given test instance. In the third step, we use Monte Carlo rollouts to construct the final solutions using the policies fine-tuned via EAS. In the following, we describe each of these three steps in more detail.

5.1   POMO for the TDOP

Modeling the TDOP. The model of TDOP within POMO is straightforward and only requires a few adjustments to the CVRP model presented by [19] to output decent solutions.

POMO Rollouts. The core idea of POMO is to create n diverse solutions for each instance during training. The average reward of the n solutions for one instance can then be used as a baseline. POMO uses problem-specific mechanisms to enforce diversity among the solution rollouts for each instance. For the TSP, POMO starts each of the rollouts from a different starting city. For the CVRP, POMO visits a different customer first for each rollout. For both problems, the quality of the different rollouts for an instance can be directly compared, since the diversity enforcing mechanism only marginally (or not at all) influences the quality of the solutions that can be found. For the TDOP it is not possible to use a similar diversity enforcing mechanism, because the solution quality heavily depends on the selected start node. Instead, we leverage the stochastic nature of the TDOP and sample the travel times independent for each rollout. This encourages diversity among the different rollouts and hence increases exploration. We still use the average reward of all rollouts as the shared baseline. Future work could investigate if an individual baseline for each rollout taking into account the sampled travel times can improve performance.

Model Input. POMO accepts a vector representing each node of the problem instance. For example, for the TSP, POMO is given a vector consisting of the x and y-coordinates of each node. For the TDOP, we provide the vector $(x_i, y_i, r_i, \underline{w}_i, \bar{w}_i)$ for each node i. Note that we scale $\underline{w}_i$ and $\bar{w}_i$ based on the maximum travel time D.

Decoder Context. During decoding, POMO constructs a solution autoregressively. Starting from an empty solution it sequentially decides which customer should be visited next. To choose a customer, the POMO model has to consider the context of the decision (e.g., the currently visited customer). This context is provided to the model in the form of a vector. For example, for the TSP the POMO model is given the embedding of the current node, the embedding of the start node and the global graph embedding as context. For the TDOP, we additionally provide the decoder with the current time at which the last node in the partial solution is visited as input. Note that we scale current time based on the maximum travel time D.

Masking Schema. The output space of POMO potentially includes actions that are infeasible, such as visiting a previously visited node, in addition to actions that are clearly bad, such as visiting a node with a clearly violated time window. A simple mechanism for avoiding such actions is *masking*, which sets the probability of an undesired action in the output space to 0. We apply masking in both the train and test phases. Masking the previously visited nodes is sufficient to ensure that the solution created is always feasible.

We also use masking in the context of the time windows. Each state s_t is associated with the current time at which the last node in the partial solution is visited. We mask any node with an end time window $\bar{w}_i$ that is less than the current time, as these nodes can clearly no longer be visited without incurring a large penalty[3]. It is still possible to get hit with a penalty due to the stochastic travel times, but we let the model choose its own risk/reward trade-off. Furthermore, we forbid actions that correspond to traveling to a node i where $\underline{w}_i > D$ and actions that involve traveling to a customer that has already been visited.

5.2 EAS for Stochastic Problems

EAS has originally been proposed as a search method for deterministic problems. In that setting, EAS fine-tunes a given model to a single test instance via reinforcement learning. The best solution observed during this fine-tuning process is returned as the final solution. In the stochastic setting considered in this work, we can not generate multiple solutions and pick the best one, because the solution quality depends on the realized travel times, and these are not known in advance. Hence, we do not use EAS to search for a solution, but as a tool to generate a fine-tuned, robust policy for each test instance. These instance-specific policies can then be used to generate the final solutions at test time.

In [12], three different variants of EAS are proposed. In this work, we use the EAS-Emb variant that updates the embeddings generated by the POMO encoder throughout the search. Note that we consider these instance embeddings a part of the policy. The overall EAS process then works as follows. For each single test instance, we first use the POMO encoder to generate a corresponding embedding ω. We then fine-tune ω in an iterative process by repeatedly sampling a set of solutions $\{\pi^1, ..., \pi^n\}$ using the POMO decoder $q_\phi(\pi \mid \omega)$ and adjusting the embeddings ω via gradient ascent using the gradient

$$\nabla_\omega \mathcal{J}_1(\omega) \approx \frac{1}{n} \sum_{i=1}^{n} \left[(R(\pi^i) - b)\nabla_{\hat{\omega}} \log q_\phi(\pi^i \mid \omega) \right] \tag{2}$$

where $q_\phi(\pi^i \mid \omega) \equiv \prod_{t-0}^{T} q_\phi(a_t^i \mid s_t^i, \omega)$, and b is the POMO baseline. Note that in contrast to (1), here we adjust the embedding ω rather than the model parameters θ.

Entropy Regularization. As previously discussed, we are not able to force diverse rollouts as POMO does on the TSP and VRP, and instead sample n solutions independently. This hampers the exploration during the search, and we noticed that EAS often quickly converges towards a single solution. We propose to use entropy regularization [25] to increase the exploration during the search. Entropy regularization aims to increase the entropy of the model output (i.e., the distribution over all possible actions) and penalizes assigning a very high probability

---

[3] We assume the penalty is large enough $(p > r_i)$ such that, in the version of the problem with recourse, we should always avoid the late arrival penalty at nodes.

values to a single action. We use entropy regularization by considering a second gradient during the search that is defined as

$$\nabla_\omega \mathcal{J}_H(\omega) \approx -\frac{1}{n} \sum_{i=1}^{n} \sum_{t=0}^{T} \sum_{a \in A(s_t^i)} \left[\nabla_{\hat{\omega}} q_\phi(a \mid s_t^i, \omega) \log q_\phi(a \mid s_t^i, \omega) \right] \qquad (3)$$

where $A(s_t^i)$ is the set of all actions that can be applied to the state s_t^i.

The gradient of the overall objective $\mathcal{J}_2$ is defined as

$$\nabla_{\hat{\omega}} \mathcal{J}_2(\omega) = \nabla_\omega \mathcal{J}_1(\omega) + \beta \cdot \nabla_\omega \mathcal{J}_H(\omega) \qquad (4)$$

where β is a hyperparameter that defines the regularization strength.

Since choosing a value for β is not trivial, we perform EAS with m different values (see Sect. 6 for details). To do this, we create m copies of the initial embedding ω prior to the EAS search and then fine-tune each embedding completely separately (but in parallel) using different β values. This allows us to effectively fill the available GPU memory. After the search, we evaluate each of the m fine-tuned embeddings on a separate validation set, and we discard all but the best performing embedding. We note that β could also be tuned in a hyperparameter tuning phase.

5.3 Solution Construction Using Monte Carlo Rollouts

For the final solution construction, we use the fine-tuned policies and Monte Carlo rollouts to determine each node to visit. We solve the instances one at a time, but exploit the batching capability of the GPU in our method. Since we reveal the true travel time of each arc as we solve a TDOP instance, we can only solve it for "real" a single time. Once we commit to a move, the move is performed and the real travel time is revealed.

Monte Carlo Rollouts. Monte Carlo rollouts are a well-known mechanism for examining the quality of an action in a sequential decision process. In our case, we could of course just rely on the argmax action of the EAS-trained policy, but this is akin to assuming our model never makes mistakes. Naturally, our model is not always correct. We thus roll out, or complete, the solution from the top five model-recommended nodes using a simulation. The expected value of each of the top actions is then computed and we select the one with the highest expected value. Thus, the Monte Carlo rollouts are a recourse mechanism. If the decision maker is risk averse, one could also use an alternative criteria to the expected value, such as the conditional value-at-risk. However, the penalty in the competition is so high that our learned policy avoids risk as much as possible.

Figure 3 shows an example application of the Monte Carlo rollouts. From the depot, POMO is queried and provides a probability distribution over all unmasked nodes. As discussed, the Monte Carlo rollouts are only computed for the top actions, in this case the top 5 ranked descending. We thus receive an expected value for each action and see that b has the best value, even though the

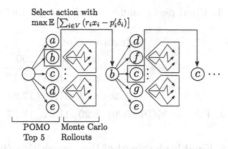

Select action with
$$\max \mathbb{E}\left[\sum_{i \in V}(r_i x_i - p'_i \delta_i)\right]$$

POMO   Monte Carlo
Top 5   Rollouts

Fig. 3. Illustration of the POMO Monte Carlo procedure.

network ranked it below a. We select b, realize the stochastic travel time, and query POMO again, yielding the second set of nodes. While some are the same as in the first iteration, some are new, which corresponds to the fact that we are now at a different node and some actions may no longer be feasible or could be too risky to carry out. The rollouts are again performed on the top nodes and c is selected. The process continues until the route is complete.

6 Computational Results

We evaluate our approach using the competition environment. We answer the following research questions:

(RQ1) How does our approach perform on the competition test set? Can EAS and Monte Carlo rollouts improve the performance?

(RQ2) Does entropy regularization improve the performance of EAS and what is the search trajectory of EAS?

(RQ3) How often do the actions selected by the Monte Carlo rollouts diverge from greedy action selection, and in what cases?

Dataset. We use the competition dataset to evaluate our approach. The dataset consists of four different sizes of instances, with 25, 50, 100 and 200 nodes each, respectively. Each size category contains 250 instances. Nodes are assigned x, y coordinates in a Euclidean plane according to a uniform distribution. Time windows are generated according to a nearest neighbor procedure described in [4] in more detail. The rewards at the nodes are generated according to their Euclidean distance to the depot at $(0, 0)$. The penalty p for missing a time window is 1. The instance generator is available at the competition's GitHub page: https://github.com/paulorocosta/ai-for-tsp-competition.

Setup. For POMO, we train 4 separate models to solve instances with 20, 50, 100, and 200 nodes, respectively. During training, we generate training instances on the fly using the competition instance generator. We train until we achieve full convergence on a single Nvidia V100 GPU. This takes between five hours

Table 1. Final performance on the test set instances

Method	Instance size				Avg.
	20	50	100	200	
POMO	5.27	8.03	11.24	17.17	10.43
POMO & EAS	5.36	8.14	11.49	17.68	10.67
POMO & EAS & MCR	5.40	8.20	11.62	17.87	10.77

Table 2. Final leaderboard of the competition (top 3).

Team	Average Reward
RISE up (ours)	10.77341
Ratel	10.58859
ML for TSP	10.39341

(for the 20 node model) up to a full day (for the 100 and 200 node model). For EAS, we perform 1500 iterations per instance with $m = 120$ and set β to values in the range $[0, 3]$. For the Monte Carlo rollouts, we perform 600 rollouts for each possible action at each decision step. On average, the EAS and MC computation takes between 7 min for 20 node instances and 48 min for 200 node instances.

6.1   RQ1: Test Set Performance

We evaluate our approach on the competition dataset and report results after each step in Table 1. Across all different instance sizes, EAS can improve the performance of POMO by 2.3% and the Monte Carlo rollouts (MCR) improve the performance of POMO and EAS by an additional 0.9%. The relative improvement of EAS is significantly higher on larger instances (e.g., 1.7% on instances of size 20 and 3.0% on instances of size 200). In contrast, for Monte Carlo rollouts the offered improvement is only slightly higher on larger instances (e.g., 0.8% on instances of size 20 and 1.1% on instances of size 200).

In Table 2 we report the final results for the reinforcement learning track of the competition. Our approach significantly outperforms the second best approach (by 1.75%) and the third best approach (by 3.66%). We note that we do not know the computational costs associated with the approaches of the other competition participants. However, our approach would have also won the competition without the expensive Monte Carlo rollout phase.

6.2   RQ2: EAS

EAS is examined in [12] for deterministic problems. There is no guarantee that using EAS on a stochastic problem like the TDOP will result in good performance. Thus, we examine how the reward improves with each iteration of EAS. Furthermore, we evaluate if entropy regularization is able to improve the quality of the solutions seen during EAS.

Figure 4 shows the average best reward in each EAS iteration for the four different problem sizes. That is, we report the average over the best reward of each instance at each iteration of EAS. We show results for EAS with and without entropy regularization. Note that although the total percentage improvement is not that high, for the class of TDOP investigated, it is rather significant since most of the rewards are earned from a few nodes. Thus, the challenge is to maximize the reward of the remaining route. In all problem sizes, EAS quickly increases the performance, with the average best reward at iteration 250 already near the quality achieved after 1500 iterations. Also note that on size 200 instances, EAS has not converged by iteration 1500, although the remaining performance improvement is likely rather small.

We now briefly examine entropy regularization. As can be seen, it provides a small boost in the average best reward on all sizes of instances. The additional exploration provided by entropy regularization is beneficial to the overall search, although we note that on size $n = 200$ instances the default performance catches up given enough time. It is possible that a higher weight must be given to entropy regularization on this size of instance or that a reactive mechanism is necessary to properly balance the loss functions.

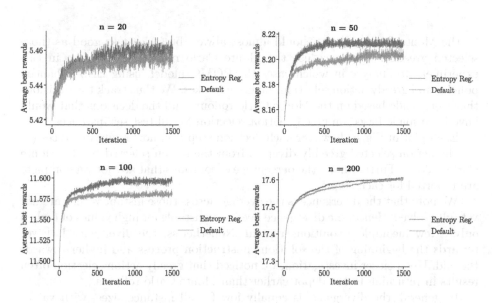

Fig. 4. Average of the best reward at each EAS iteration over all instances.

6.3   RQ3: Monte Carlo Rollouts

While Monte Carlo rollouts are computationally expensive, they significantly improve the performance over greedy action selection on the test dataset. If enough rollouts are performed for each action, we get an accurate estimate of the expected total reward for each action. Consequently, the actions selected

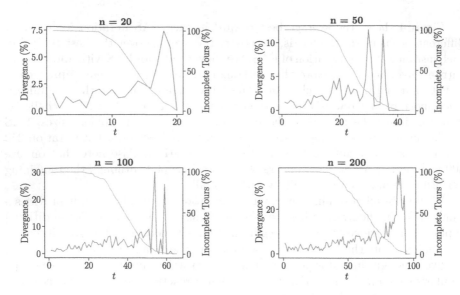

Fig. 5. Average of the best reward seen at each EAS iteration over all instances.

by the Monte Carlo rollouts should almost always be at least as good as those selected greedily. We hence use the Monte Carlo rollouts mechanism in this experiment to analyze in which cases the POMO model, using EAS enhanced policies and greedy action selection, makes mistakes. We thus track the decisions that were made based on the Monte Carlo rollouts and the decisions that would have been made based on greedy action selection for all test set instances.

Each plot in Fig. 5 shows for each decision step t on how many instances (in %) the action selected greedily diverges from the action selected by the Monte Carlo rollouts. Furthermore, the percentage of solutions that are not yet complete are reported for each step t.

We note that the divergence is the average across those instances that are not yet fully solved. Hence, the divergence becomes unstable for high values of t when only a few incomplete solutions remain. Nonetheless, the divergence is lower towards the beginning of the solution construction process and higher towards the end. Upon closer investigation, we noticed that greedy action selection often results in returning to the depot earlier than Monte Carlo rollouts.

In general, the divergence is equally low for all instance sizes, with values staying below 5% for the majority of the solution construction process in all cases. This is especially surprising for instances with 100 and 200 nodes where the number of possible (unmasked) actions is much higher at each decision step than for instances with fewer nodes.

7 Conclusion

We presented an RL approach for solving the TDOP that won first place in the RL track of the IJCAI AI4TSP competition in 2021. Our approach modifies the POMO method and extends it with EAS and Monte Carlo rollouts. First, we enable the POMO method to handle stochastic problems without symmetries in the solution space. Then we use EAS with entropy regularization to fine-tune POMO's learned policies. Finally, we use Monte Carlo rollouts to assist in solution construction. We show experimentally that each of these steps contributes towards the good performance of our method, and that entropy regularization can significantly improve the performance of EAS. In future work, we will examine the approach on different distributions of travel times and time windows.

Acknowledgments. Fynn Schmitt-Ulms was supported by the German Academic Exchange Service Research Internships in Science and Engineering (DAAD RISE) program. The computational experiments in this work have been performed using the Bielefeld GPU cluster. We thank the Bielefeld HPC.NRW team for their support.

References

1. Basso, R., Kulcsár, B., Sanchez-Diaz, I., Qu, X.: Dynamic stochastic electric vehicle routing with safe reinforcement learning. Transp. Res. Part E: Logistics Transp. Rev. **157**, 102496 (2022)
2. Bello, I., Pham, H., Le, Q.V., Norouzi, M., Bengio, S.: Neural Combinatorial Optimization with Reinforcement Learning. arXiv:1611.0 (2016)
3. Bengio, Y., Lodi, A., Prouvost, A.: Machine learning for combinatorial optimization: a methodological tour d'horizon. Eur. J. Oper. Res. **290**, 405–421 (2020)
4. Bliek, L., et al.: The first AI4TSP competition: learning to solve stochastic routing problems (2022). https://doi.org/10.48550/arXiv.2201.10453
5. Bono, G.: Deep multi-agent reinforcement learning for dynamic and stochastic vehicle routing problems. Ph.D. thesis, Université de Lyon (2020)
6. Chen, X., Tian, Y.: Learning to perform local rewriting for combinatorial optimization. In: Advances in Neural Information Processing Systems, pp. 6278–6289 (2019)
7. Choo, J., et al.: Simulation-guided beam search for neural combinatorial optimization (2022). https://doi.org/10.48550/arXiv.2207.06190
8. de O. da Costa, P.R., Rhuggenaath, J., Zhang, Y., Akcay, A.: Learning 2-opt heuristics for the traveling salesman problem via deep reinforcement learning. In: Asian Conference on Machine Learning, pp. 465–480. PMLR (2020)
9. Helsgaun, K.: An effective implementation of the Lin-Kernighan traveling salesman heuristic. Eur. J. Oper. Res. **126**, 106–130 (2000)
10. Hopfield, J.J.: Neural networks and physical systems with emergent collective computational abilities. Proc. Nat. Acad. Sci. U.S.A. **79**(8), 2554–2558 (1982)
11. Hottung, A., Bhandari, B., Tierney, K.: Learning a latent search space for routing problems using variational autoencoders. In: International Conference on Learning Representations (2021)
12. Hottung, A., Kwon, Y.D., Tierney, K.: Efficient active search for combinatorial optimization problems. In: International Conference on Learning Representations (2022)

13. Hottung, A., Tierney, K.: Neural large neighborhood search for the capacitated vehicle routing problem. In: European Conference on Artificial Intelligence, pp. 443–450 (2020)
14. Joe, W., Lau, H.C.: Deep reinforcement learning approach to solve dynamic vehicle routing problem with stochastic customers. Proc. Int. Conf. Autom. Plann. Sched. **30**, 394–402 (2020)
15. Joshi, C.K., Cappart, Q., Rousseau, L.M., Laurent, T.: Learning TSP requires rethinking generalization. In: Michel, L.D. (ed.) 27th International Conference on Principles and Practice of Constraint Programming (CP 2021). Leibniz International Proceedings in Informatics (LIPIcs), vol. 210, pp. 33:1–33:21. Schloss Dagstuhl - Leibniz-Zentrum für Informatik, Dagstuhl, Germany (2021)
16. Joshi, C.K., Laurent, T., Bresson, X.: An efficient graph convolutional network technique for the travelling salesman problem. arXiv:1906.01227 (2019)
17. Kool, W., van Hoof, H., Gromicho, J., Welling, M.: Deep policy dynamic programming for vehicle routing problems. In: International Conference on Integration of Constraint Programming, Artificial Intelligence, and Operations Research, pp. 190–213 (2022)
18. Kool, W., van Hoof, H., Welling, M.: Attention, learn to solve routing problems! International Conference on Learning Representations (2019). https://doi.org/10.48550/arXiv.1803.08475
19. Kwon, Y.D., Choo, J., Kim, B., Yoon, I., Gwon, Y., Min, S.: POMO: policy optimization with multiple optima for reinforcement learning. In: Larochelle, H., Ranzato, M., Hadsell, R., Balcan, M.F., Lin, H. (eds.) Advances in Neural Information Processing Systems. vol. 33, pp. 21188–21198. Curran Associates, Inc. (2020)
20. Li, S., Yan, Z., Wu, C.: Learning to delegate for large-scale vehicle routing. In: Advances in Neural Information Processing Systems. **34** (2021)
21. Sultana, N.N., Baniwal, V., Basumatary, A., Mittal, P., Ghosh, S., Khadilkar, H.: Fast approximate solutions using reinforcement learning for dynamic capacitated vehicle routing with time windows. arXiv preprint arXiv:2102.12088 (2021)
22. Vaswani, A., Shazeer, N., Parmar, N., Uszkoreit, J., Jones, L., Gomez, A.N., Kaiser, Ł., Polosukhin, I.: Attention is all you need. In: Guyon, I., Luxburg, U.V., Bengio, S., Wallach, H., Fergus, R., Vishwanathan, S., Garnett, R. (eds.) Advances in Neural Information Processing Systems. vol. 30. Curran Associates, Inc. (2017)
23. Vinyals, O., Fortunato, M., Jaitly, N.: Pointer networks. Adv. Neural Inf. Process. Syst. **28**, 2692–2700 (2015)
24. Williams, R.J.: Simple statistical gradient-following algorithms for connectionist reinforcement learning. Mach. Learn. **8**(3–4), 229–256 (1992)
25. Williams, R.J., Peng, J.: Function optimization using connectionist reinforcement learning algorithms. Connection Sci. **3**(3), 241–268 (1991)
26. Wu, Y., Song, W., Cao, Z., Zhang, J., Lim, A.: Learning improvement heuristics for solving routing problems. IEEE Trans. Neural Netw. Learn. Syst. **33**(9), 5057–5069 (2021)
27. Xin, L., Song, W., Cao, Z., Zhang, J.: NeuroLKH: combining deep learning model with lin-kernighan-helsgaun heuristic for solving the traveling salesman problem. In: Advances in Neural Information Processing Systems. **34** (2021)

ML-Based Approach for Accelerating Global Search Algorithm for Solving Multicriteria Problems

Konstantin Barkalov$^{(\boxtimes)}$ ⓘ, Vladimir Grishagin ⓘ, and Evgeny Kozinov ⓘ

Lobachevsky State University of Nizhni Novgorod, Nizhni Novgorod, Russia
{konstantin.barkalov,evgeny.kozinov}@itmm.unn.ru, vagris@unn.ru

Abstract. The paper considers a new approach to solving multicriterial optimization problems on the base of criteria scalarization, dimensionality reduction via Peano curves and efficient global search algorithm. The novelty of the approach consists in application of machine learning methods combined with utilizing a posteriori information for acceleration of the search. Effectiveness of the proposed approach has been demonstrated by means of solving a set of multiextremal multicriterial optimization problems.

Keywords: Multicriterial problems · Global optimization · Machine learning · Logistic regression

1 Introduction

Machine learning (ML) methods are powerful tools that are applied in many areas of research. In particular, the ML methods are successfully used for solving complicated problems of computational mathematics. One of such problem classes to which ML can be applied is the class of multicriterial optimization (MCO) models. In this case, ML is used, as a rule, in combination with metaheuristic algorithms [11,17,18] that in the case of multiextremal criteria lose in efficiency in comparison with deterministic methods [10,15].

Among the most qualitative deterministic methods for solving these problems the information-statistical global search algorithm [14,16] can be considered. This method proposed initially for solving scalar problems was successfully extended to MCO models [4,7].

The paper reflects results of a new research direction connected with utilizing ML methods for acceleration of the global search algorithm in the case of its application to MCO problems. The proposed approach is based on building a number of separating planes in the criteria space to segregate the Pareto set and accelerate the process of its construction. The efficiency of the approach is estimated experimentally compared to several other MCO methods.

This study was supported by the Russian Science Foundation, project No 21-11-00204.

2   Problem Statement

The MCO problem to be considered is formulated as follows:

$$f(y) = (f_1(y), f_2(y), \ldots, f_s(y)) \to \min, \; y \in D, \tag{1}$$

$$D = \left\{ y \in R^N : a_i \leq y_i \leq b_i, \; 1 \leq i \leq N, \; a, b \in R \right\}. \tag{2}$$

where $f_i(y)$, $1 \leq i \leq s$, are criteria, $y = (y_1, y_2, \ldots, y_N)$ is a vector of independent variables, N is the dimension of the problem. The criteria $f_i(y)$, $1 \leq i \leq s$, are supposed to be multiextremal, to be given as a "black-box" functions and to satisfy the Lipschitz condition

$$|f_i(y') - f_i(y'')| \leq L_i \|y' - y''\|, \; y', y'' \in D, \; 1 \leq i \leq s, \tag{3}$$

where $L_i > 0$, $1 \leq i \leq s$, are a priori unknown Lipschitz constants.

Any effective (Pareto-optimal) variant in which it is impossible to reduce the values of all criteria $f_i(y)$, $1 \leq i \leq s$, at once by choosing another option can be considered as a partial solution to the MCO problem. In practice, various scalarization techniques are often used to find effective solutions [2,8,12,13]. The present study uses the minimax convolution of partial criteria that possesses good theoretical properties and allows one to find effective points by solving the problem

$$\max_{1 \leq i \leq s} \{\lambda_i f_i(y)\} \to \min, \; y \in D, \tag{4}$$

where $\lambda_i \geq 0$, $1 \leq i \leq s$, $\sum_{i=1}^{s} \lambda_i = 1$, are numerical indicators (weights) of the significance of each criterion.

In general case, the entire set of Pareto-optimal variants is taken as a complete solution to the MCO problem. For the numerical building of an approximation of the Pareto set a number of scalar optimization problems (4) are solved with different coefficients $\lambda_i \geq 0$, $1 \leq i \leq s$, distributed uniformly.

3   General Computational Scheme

In the framework of the proposed approach the information-statistical theory of global search is used for solving the scalar optimization problems (4). For problems with several variables ($N > 1$) the dimensionality reduction on the base of Peano-type mappings [14,16] is applied. While solving a problem (4) the algorithm generates a sequence of trials where the term "trial" means evaluation of criteria at a point of the feasible domain D and computation of the convolution (4) at this point.

Coordinates of a new trial are chosen according to the following rules:

1. Obtain univariate images of all points of preceding trials mapped onto one-dimensional interval $[0, 1]$ by means of Peano curves.
2. Partition the segment $[0, 1]$ into subintervals according to the images of trials performed.

3. For each subinterval compute a numerical value called *characteristic* of this subinterval.
4. Select the subinterval with the highest characteristic.
5. Place a point of new trial within the subinterval with the highest characteristic.
6. Map the new trial point to the point $y \in D$ using Peano evolvent and compute the value of the function (4) at this point.

The algorithm completes execution when the required accuracy is achieved. After stopping the global search algorithm, an estimate of the Pareto set is built on the base of the accumulated information obtained in the course of optimization. If the quality of the Pareto set estimate is not sufficient, then new preferences (new coefficients λ) are set and the search process continues.

4   Approaches to Improving Search Efficiency

Two techniques are used to improve the search efficiency of the Pareto area assessment.

The first one is to jointly solve a series of global search problems from (4). The essence of the technique consists in the accumulation of search information during the optimization process and in reusing afterwards this information while solving a problem (4) with new weight coefficients λ. A detailed description of the technique can be found in [4, 7].

The second improvement is connected with a new method for calculating the characteristics of subintervals formed during univariate optimization on the interval [0, 1].

Let $R(i)$ be the characteristic of the i-th subinterval. This characteristic is supposed to consist of two parts:

$$R(i) = R_{ags}(i) + \alpha R_{PS}(i). \tag{5}$$

The term $R_{ags}(i)$ allows you to select a subinterval oriented at finding the global minimum of the current optimization task (4), while $R_{PS}(i)$ influences the selection of a subinterval to improve the evaluation of the Pareto area. The coefficient α from (5) allows adjusting the contribution of each of parts.

$R_{ags}(i)$ is calculated in accordance with the expression from the global search algorithm [4, 7, 14, 16].

Calculation of $R_{PS}(i)$ is based on machine learning methods. When solving the first global search problem (4), the value of $R_{PS}(i)$ is assumed to be zero. To solve each subsequent scalar problem (4), $R_{PS}(i)$ is calculated as follows.

In the course of Pareto set assessment all trial points are partitioned into two classes. This bipartition is based on belonging to the Pareto set. For these classes, a separating hyperplane is constructed in the domain of criteria values using the logistic regression [19]. Examples of separating planes are shown in Fig. 1. During the choice of a new trial point, the distances of all obtained criteria values from

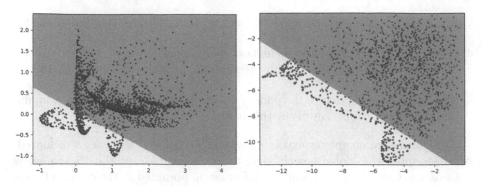

Fig. 1. Separating hyperplanes in the criteria space

the separating plane are calculated and the value of $R_{PS}(i)$ is built depending on these distances.

Hereinafter the algorithm taking into account distances to separating hyperplane and accumulated information will be called ML_MGSA.

5   Results of Computational Experiments

Computational testing was carried out on the "Lobachevsky" supercomputer of Nizhny Novgorod State University in the environment of the Globalizer software system [6].

The first series of experiments was performed to compare the algorithm ML_MGSA with a number of widely known multiobjective optimization algorithms using a test two-criterion problem [3]

$$f_1(y) = (y_1 - 1)y_2^2 + 1, \ f_2(y) = y_2, \ 0 \le y_1, y_2 \le 1. \tag{6}$$

In these experiments, a numerical approximation of the Pareto set was built. The quality of approximation is estimated by such indicators as the hypervolume index (HV) and distribution uniformity index (DU) [3,20]. In the framework of the experiment 5 algorithms of multicriterial optimization were compared: the Monte-Carlo (MC) method, the genetic algorithm SEMO [1], the Non-uniform coverage (NUC) method [3], the bi-objective Lipschitz optimization (BLO) method [20] and algorithm ML_MGSA, proposed in this article.

ML_MGSA solved 25 problems (4) with different convolution coefficients λ, distributed uniformly. The results of the conducted experiments are shown in Table 1 where ML_MGSA with $\alpha = 0$ corresponds to the version without machine learning. As it follows from the results presented in Table 1, ML_MGSA demonstrates the better quality compared to other methods.

In the second group of experiments 100 bi-criteria MCO problems have been solved. As the criteria, the multiextremal functions

Table 1. Efficiency of the multicriterial algorithms compared

Algorithm	MC	SEMO	NUC	BLO	ML_MGSA ($\alpha = 0$)	ML_MGSA ($\alpha = 1.5$)
Number of trials	500	500	515	498	302	269
Number of Pareto points found	67	104	29	68	92	79
HV index (the more the better)	0.300	0.312	0.306	0.308	0.312	0.312
DU index (the less the better)	1.277	1.116	0.210	0.175	0.101	0.103

$$f(y) = -(AB + CD)^{1/2}$$
$$AB = \left(\sum_{i=1}^{7} \sum_{j=1}^{7} [A_{ij} a_{ij}(y_1, y_2) + B_{ij} b_{ij}(y_1, y_2)] \right)^2$$
$$CD = \left(\sum_{i=1}^{7} \sum_{j=1}^{7} [C_{ij} a_{ij}(y_1, y_2) - D_{ij} b_{ij}(y_1, y_2)] \right)^2 \tag{7}$$
$$a_{ij}(y_1, y_2) = \sin(\pi i y_1) \sin(\pi j y_2), b_{ij}(y_1, y_2) = \cos(\pi i y_1) \cos(\pi j y_2),$$

$0 \leq y_1, y_2 \leq 1$, have been taken where parameters $-1 \leq A_{ij}, B_{ij}, C_{ij}, D_{ij} \leq 1$ are the independent random numbers distributed uniformly [5,9].

In order to estimate efficiency of machine learning procedure in ML_MGSA, this method tested in 2 variants: with ($\alpha > 0$) and without ($\alpha = 0$) use of separating planes in the criteria space. Each multicriterial problem was reduced to 50 scalar problems (4) with different weight coefficients λ, distributed uniformly. Other parameters of the algorithm were the same. The averaged results can be found in Table 2. Figure 2 additionally shows the average value of HV index depending on the number of trials.

Table 2. Efficiency of the multicriterial algorithms compared

	ML_MGSA ($\alpha = 0.0$)	ML_MGSA ($\alpha = 0.01$)
Average number of iterations	1902.4	658.6
Average value of DU index	1.32	1.55
Average value of HV index	92.1	91.9
Reducing the number of iterations	1	2.9

The results show that embedding the machine learning into the global search algorithm leads to a significant reduction of the number of trials (almost 3 times) with maintaining close values of indicators HV and DU.

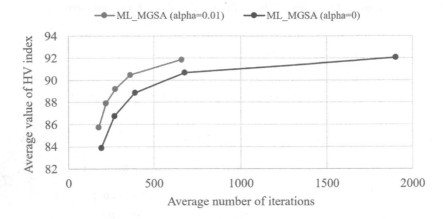

Fig. 2. Efficiency of the multicriterial algorithms compared

References

1. Bleuler, S., Laumanns, M., Thiele, L., Zitzler, E.: PISA — a platform and programming language independent interface for search algorithms. In: Fonseca, C.M., Fleming, P.J., Zitzler, E., Thiele, L., Deb, K. (eds.) EMO 2003. LNCS, vol. 2632, pp. 494–508. Springer, Heidelberg (2003). https://doi.org/10.1007/3-540-36970-8_35
2. Ehrgott, M.: Multicriteria Optimization. Springer, Berlin, Heidelberg (2005)
3. Evtushenko, Y.G., Posypkin, M.A.: A deterministic algorithm for global multiobjective optimization. Optim. Methods Softw. **29**(5), 1005–1019 (2014)
4. Gegrel, V., Kozinov, E.: Accelerating multicriterial optimization by the intensive exploitation of accumulated search data. AIP Conf. Proc. **1776**, 090003 (2016)
5. Gergel, V., Grishagin, V., Israfilov, R.: Adaptive dimensionality reduction in multiobjective optimization with multiextremal criteria. Lecture Notes in Computer Science **11331**, 129–140 (2019)
6. Gergel, V.P., Barkalov, K.A., Sysoyev, A.V.: A novel supercomputer software system for solving time-consuming global optimization problems. Numer. Algebra Control Optimi. **8**(1), 47–62 (2018)
7. Gergel, V.P., Kozinov, E.A.: Efficient multicriterial optimization based on intensive reuse of search information. J. Glob. Optim. **71**(1), 73–90 (2018)
8. Gergel, V., Kozinov, E.: Parallel computations for various scalarization schemes in multicriteria optimization problems. In: Wyrzykowski, R., Deelman, E., Dongarra, J., Karczewski, K. (eds.) PPAM 2019. LNCS, vol. 12043, pp. 174–184. Springer, Cham (2020). https://doi.org/10.1007/978-3-030-43229-4_16
9. Grishagin, V., Israfilov, R.: Multidimensional constrained global optimization in domains with computable boundaries. CEUR Workshop Proc. **1513**, 75–84 (2015)
10. Kvasov, D.E., Mukhametzhanov, M.S.: Metaheuristic vs. deterministic global optimization algorithms: the univariate case. Appl. Math. Comput. **318**, 245–259 (2018)
11. Ma, X., et al.: MOEA/D with opposition-based learning for multiobjective optimization problem. Neurocomputing **146**, 48–64 (2014)

12. Marler, R., Arora, J.: Survey of multi-objective optimization methods for engineering. Struct. Multi. Optim. **26**(6), 369–395 (2004)
13. Pardalos, P.M., Žilinskas, A., Žilinskas, J.: Non-Convex Multi-Objective Optimization. SOIA, vol. 123. Springer, Cham (2017). https://doi.org/10.1007/978-3-319-61007-8
14. Sergeyev, Y.D., Strongin, R.G., Lera, D.: Introduction to Global Optimization Exploiting Space-Filling Curves. Springer Briefs in Optimization, Springer, New York (2013)
15. Sergeyev, Y.D., Kvasov, D.E., Mukhametzhanov, M.S.: On the efficiency of nature-inspired metaheuristics in expensive global optimization with limited budget. Sci. Rep. **8**(1), 435 (2018)
16. Strongin, R.G., Sergeyev, Y.D.: Global Optimization with Non-Convex Constraints. Sequential and parallel algorithms. Kluwer Academic Publishers, Dordrecht (2000)
17. Subraveti, S., Li, Z., Prasad, V., Rajendran, A.: Machine learning-based multiobjective optimization of pressure swing adsorption. Ind. Eng. Chem. Res. **58**(44), 20412–20422 (2019)
18. Talbi, E.G.: Hybrid metaheuristics for multi-objective optimization. J. Algorithms Comput. Technol. **9**(1), 41–63 (2015)
19. Yu, H.F., Huang, F.L., Lin, C.J.: Dual coordinate descent methods for logistic regression and maximum entropy models. Mach. Learn. **85**(1), 41–75 (2011)
20. Žilinskas, A., Žilinskas, J.: Adaptation of a one-step worst-case optimal univariate algorithm of bi-objective lipschitz optimization to multidimensional problems. Commun. Non-linear Sci. Numer. Simulat. **21**(1–3), 89–98 (2015)

The Skewed Kruskal Algorithm

Ermanno Righini and Giovanni Righini[✉][ID]

University of Milan, Milan, Italy
ermanno.righini@studenti.unimi.it, giovanni.righini@unimi.it

Abstract. Kruskal algorithm is one of the most efficient algorithms to compute a minimum spanning tree (MST) of a given weighted unoriented and connected graph. The edge list is sorted and edges are iteratively selected from it until a MST is found. To improve its performance, edge sorting can be interleaved with edge selection, so that only the relevant part of the edge list is actually sorted. Filtering techniques can also be used so that the size of some parts of the edge list is reduced before sorting and selection. Here we examine a further idea, i.e. to produce an unbalanced initial partition, with the aim of early guessing which part of the edge list actually needs being considered for filtering, sorting and edge selection.

Keywords: Minimum spanning tree · Kruskal algorithm

1 Kruskal Algorithm

The minimum spanning tree problem consists of finding a the minimum weight spanning tree (MST) of an unoriented connected weighted graph $G = (V, E)$ with $|V| = n$ vertices and $|E| = m$ edges. The most used and successful algorithms to solve this problem are Prim algorithm [6] and Kruskal algorithm [3].

In Kruskal algorithm the list of edges is sorted by non-decreasing weight and a forest is initialized where each vertex represents a tree on its own. Then starting from the smallest edge, all edges that do not close cycles are selected and inserted, thus connecting trees to form larger trees until the forest is made by a unique component which is a MST.

---

procedure Kruskal(IN: V, E, OUT: T)
  $T \leftarrow \emptyset$
  Sort(E)   ▷ Edge sorting (by non-decreasing weight)
  for $(x, y) \in E$ **do**
    if Find(x) = Find(y) **then**   ▷ Edge selection
      Union(x, y)
      $T := T \cup \{(x, y)\}$

Edge selection implies two main operations: Find detects the tree to which a vertex belongs; Union merges two trees into a single tree. An edge is selected to

© Springer Nature Switzerland AG 2022
D. E. Simos et al. (Eds.): LION 2022, LNCS 13621, pp. 130–135, 2022.
https://doi.org/10.1007/978-3-031-24866-5_10

be inserted in the MST if and only if its endpoints belong to different trees of the current forest. For an efficient implementation of the algorithm a Union-Find data structure can be used [7].

The time complexity of this implementation is $O(m \log m)$, owing to edge sorting. The second part of the algorithm, after sorting, can be implemented with worst-case time complexity $O(m + n \log n)$.

2 Improvements

Two main improvements to Kruskal algorithm consist of delaying or avoiding the need of sorting unnecessary edges.

2.1 QuickKruskal Algorithm

The QuickKruskal algorithm variation [5] interleaves edge sorting with edge selection. It performs edge sorting as it is done in QuickSort, i.e. partitioning the edge list in two lists by comparing all edges with a suitably chosen pivot element; in this way the edges in the second list remain unsorted if the MST is found within the edges of the first list. This procedure is recursively applied to both lists.

---

procedure QuickKruskal(IN: V, E. OUT: T)
  if $|E| < \theta$ **then return** Kruskal(V, E, T)
  $p \leftarrow$ SelectPivot(E)
  $(E_1, E_2) \leftarrow$ Partition(E, p)
  QuickKruskal(V, E_1, T)
  if $|T| < n - 1$ **then**
    QuickKruskal(V, E_2, T)

With this strategy Kruskal algorithm becomes much faster for random graphs with randomly generated weights, where the last edge contained in the MST is expected to be found within the $\frac{1}{2} n \log n$ smallest edges [2].

On the other hand, QuickKruskal is not very robust: if the input graph has at least one heavy edge, then the entire edge list up to that edge needs to be sorted and the number of operations to be executed falls back to the standard Kruskal algorithm, possibly with a small overhead due to the comparisons with the pivot elements.

For the same reason, i.e. to avoid this overhead, when the size of a list is below a suitably tuned threshold θ, then the list is completely pre-sorted as in the classing Kruskal algorithm. The experimental results presented in Sect. 3 were obtained with $\theta = 1000$.

As shown in [5], the *average* time complexity of this implementation is $O(m + n \log^2 n)$ when G is a random graph with randomly generated weights.

2.2 FilterKruskal Algorithm

In the FilterKruskal algorithm, presented in [4], each edge list is recursively partitioned into two parts as in QuickKruskal algorithm, but the second list is filtered before being sorted: the list is scanned and the edges whose endpoints belong to the same component in the current forest are deleted. This simple modification leads to better performance since, unless the graph is very sparse, most of the edges do not belong to the MST and they are filtered before being sorted.

---

procedure FilterKruskal(IN: V, E. OUT: T)
    if $|E| < \theta$ **then return** Kruskal(V, E, T)
    $p \leftarrow$ SelectPivot(E)
    $(E_1, E_2) \leftarrow$ Partition(E, p)
    FilterKruskal(V, E_1, T)
    if $|T| < n - 1$ **then**
        for $[i, j] \in E_2$ **do**
            if Find(i) = Find(j) **then**
                $E_2 \leftarrow E_2 \backslash \{[i, j]\}$ ▷ Filtering
        FilterKruskal(V, E_2, T)

As shown in [4], the *average* time complexity of the algorithm for random graphs with random edge weights is $O(m + n \log n \log \frac{m}{n})$.

2.3 SkewedFilterKruskal Algorithm

To further improve the FilterKruskal algorithm, we try to guess the size of the edge list containing the MST at the first level of the recursion. Following [2], the MST of a random graph with random edges is expected to be contained in the $\frac{1}{2} n \log n$ smallest edges. More generally, in most graphs the MST is contained in a relatively small part of the edge list if the graph is dense enough.

To exploit this observation, the pivot element in the first recursion level of FilterKruskal is chosen to produce a skewed partitioning, so that E_1 is likely to have size $kn \log n$ where k is a constant slightly larger than $1/2$. In all other recursion levels the algorithm tries to achieve a balanced partition, as in Quick-Kruskal and FilterKruskal.

In order to achieve balanced or skewed partitions, pivot elements must be suitably chosen. For this purpose we tested an s-samples policy for different values of s. The policy consists of selecting s samples at random from the edge list and then to select a pivot among them in order to partition their sorted list according to the desired proportion between the two parts.

To achieve balanced partitions in FilterKruskal, the 1-sample policy is obviously the fastest, but it allows for large variability in the size of the partitions and hence in the computing time. In particular, in the first recursion level a bad selection of the pivot (a too large value) can worsen the performance significantly. Better results are achieved by increasing s (keeping it odd) and selecting the median of the samples as the pivot.

To achieve an unbalanced partition in SkewedFilterKruskal, it is necessary to be more precise with pivot selection. Therefore we opted for a $\sqrt{n}$-samples policy. Clearly, we do not use the median value but the one in position $\lceil \frac{\sqrt{n}\log n}{2m} \rceil$ in the sorted list of the samples. This policy produces accurate unbalanced partitions although it is definitely slower than those that are effective for balanced partitions. An idea we successfully tested to mitigate this performance degradation is to pick a sample of size $\sqrt{n}$ only once at the start of the procedure and to reuse the same samples at every recursion level. When a range must be partitioned and there are no samples in it, then the algorithm turns to the 1-sample policy.

3   Computational Results

In our tests, we considered three types of random graphs:

- Random graphs with random edge weights. In these graphs every edge has the same probability to belong to E and edge weights are randomly generated with uniform probability distribution in a predefined interval.
- Random graphs with a long edge. These graphs are generated in the same way as the previous type but the longest edge in E belongs to the MST. These rather artificial graphs are used to test the algorithms when they are forced to scan the entire edge list.
- Random 2D Euclidean graphs. In these graphs, vertices are points in a 2D plane and edge weights are Euclidean distances. For each vertex, the edges linking it to its k nearest neighbors are added to the graph. Graphs in this class tend to exhibit a more clustered structure.

In out tests we compared the following algorithms:

- Prim algorithm [6] implemented with pairing heaps [1];
- the basic implementation of Kruskal algorithm [3];
- the FilterKruskal algorithm [4];
- the SkewedFilterKruskal algorithm.

Our tests were done on graphs with $n = 60000$ and several values of density. All tests were performed on a Intel Core i5-6600 CPU, 3.9 GHz. The source code as well as instances and solutions are available from the authors upon request.

Plots in Fig. 1 show the comparison between the computing time taken by above algorithms for graphs of the three classes with different density. The computing time does not take into account the time needed to create the data-structures, which is significant for Prim algorithm. The best performances for all graph types were achieved by SkewedFilterKruskal, even in the case of graphs with a long edge, where the procedure is forced to read and filter all edges, as well as with Euclidean random graphs, where the presence of clusters introduces heavier edges into the MST.

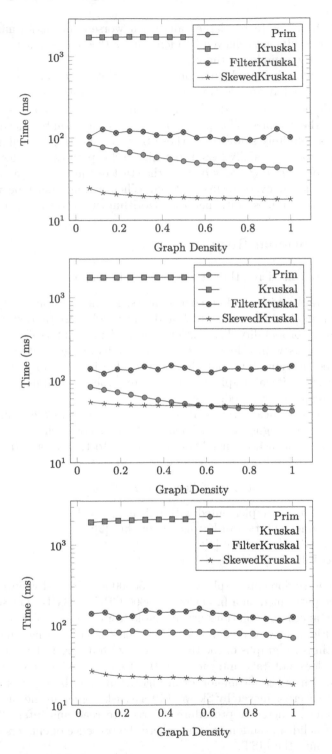

Fig. 1. Comparison of different algorithms on different graph types. From top to bottom: graphs with random weights, graphs with a long edge, random Euclidean graphs.

4   Conclusions and Open Questions

These experiments show that, in spite of the remarkable improvements achieved in the last years compared with the basic version of Kruskal algorithm, there is still room for further enhancements.

In particular, the idea of producing an unbalanced partition since the very beginning of the FilterKruskal algorithm (SkewedFilterKruskal) would benefit from further investigation to answer the following questions:

- Is it possible to use learning techniques to guess the size of the edge list containing all the edges of the MST?
- Assuming the desired size of the initial partition is known, how can one optimally tune an s-samples policy, i.e. the trade-off between the number of samples used (and sorted) and the accuracy in the actual resulting partition?

References

1. Fredman, M.L., Sedgewick, R., Sleator, D.D., Tarjan, R.E.: The pairing heap: a new form of self-adjusting heap. Algorithmica **1**(1–4), 111–129 (1986)
2. Janson, S., Knuth, D.E., Łuczak, T., Pittel, B.: The birth of the giant component. Random Struct. Algorithms **4**(3), 233–358 (1993)
3. Kruskal, J.B.: On the shortest spanning subtree of a graph and the traveling salesman problem. Proc. Am. Math. Soc. **7**(1), 48–50 (1956)
4. Osipov, V., Sanders, P., Singler, J.: The filter-kruskal minimum spanning tree algorithm. In: 2009 Proceedings of the Eleventh Workshop on Algorithm Engineering and Experiments (ALENEX), pp. 52–61. SIAM (2009)
5. Paredes, R., Navarro, G.: Optimal incremental sorting. In: 2006 Proceedings of the Eighth Workshop on Algorithm Engineering and Experiments (ALENEX), pp. 171–182. SIAM (2006)
6. Prim, R.C.: Shortest connection networks and some generalizations. Bell Syst. Tech. J. **36**(6), 1389–1401 (1957)
7. Tarjan, R.E.: A class of algorithms which require nonlinear time to maintain disjoint sets. J. Comput. Syst. Sci. **18**(2), 110–127 (1979)

Bounds for Sparse Solutions of K-SVCR Multi-class Classification Model

Hossein Moosaei[1,2]($\boxtimes$)(iD) and Milan Hladík[2](iD)

[1] Department of Informatics, Faculty of Science, Jan Evangelista Purkyně
University, Ústí nad Labem, Czech Republic
hmoosaei@gmail.com, hossein.moosaei@ujep.cz
[2] Department of Applied Mathematics, Faculty of Mathematics and Physics, Charles
University, Prague, Czech Republic
hladik@kam.mff.cuni.cz

Abstract. The support vector classification-regression machine for k-class classification (K-SVCR) is a novel multi-class classification approach based on the "1-versus-1-versus-rest" structure. In this work, we suggested an efficient model by proposing the p-norm ($0 < p < 1$) instead of the 2-norm for the regularization term in the objective function of K-SVCR that can be used for feature selection. We derived lower bounds for the absolute value of nonzero entries in every local optimal solution of the p-norm based model. Also, we provided upper bounds for the number of nonzero components of the optimal solutions. We explored the link between solution sparsity, regularization parameters, and the p-choice.

Keywords: K-SVCR · Multi-class classification · First-order optimally condition · Sparsity · p-norm

1 Introduction

Support vector machines (SVM) were developed by Vapnik and his colleagues for binary classification problems [5,10]. They were employed in a variety of applications, including facial recognition [11], heart disease detection [3], energies prediction [1,23,26], Raman spectroscopy [14], biomedicine [22], and many others. The concept behind this approach is to identify the largest margin between two hyperplanes, which leads to solving a constraint quadratic programming problem (QPP). Many versions, modifications, and applications of SVM and other binary classification methods have been presented during the last few decades [4,12,13,15,19,24].

The task of categorizing things into several classes is known as multi-class classification. It is not limited to any number of classes, unlike binary classification. Multi-class classification problems occur in many real-world applications. Although a problem with multi-class categories can be transformed into a series of binary classification problems by "1-versus-rest" [6] and "1-versus-1" [16] methods, studies such as [25] have shown that some of these strategies are frequently ineffective when applied directly to the multi-class problems.

© Springer Nature Switzerland AG 2022
D. E. Simos et al. (Eds.): LION 2022, LNCS 13621, pp. 136–144, 2022.
https://doi.org/10.1007/978-3-031-24866-5_11

In [2], a new and effective multi-class classification technique based on "1-versus-1-versus-rest" structure, named support vector classification regression for k-class classification (K-SVCR), was presented by Angulo et al. for k-class classification problems. This approach builds $\frac{1}{2}k(k-1)$ classifiers such that each of them is trained with all of the training data, eliminating the risk of information loss and class imbalance issues. As a result, the K-SVCR outperforms SVM methods for multi-class classification tasks. As a consequence of its superior predicting outcomes, the K-SVCR has gotten much interest [18,20].

As a fundamental challenge in classification, feature selection removes irrelevant data characteristics to enhance the algorithm's performance. K-SVCR has been a promising tool in machine learning, but it does not directly obtain the feature importance. Identifying a collection of traits that contribute most to categorization is another critical challenge in classification. For the K-SVCR problem, most papers focused on solving this problem and using all the features. In recent years, p-norm ($0 < p < 1$) has been widely used in optimization. The main idea is to use p-norm to obtain sparse solutions for the classification problem [17,21]. The final classifier uses all features when the 2-norm is used in K-SVCR. Naturally, employing the p-norm ($0 < p < 1$) in the regularization term should provide a sparser solution than using the 2-norm and even 1-norm, as proven by various investigations [8,9]. This paper suggests a lower bound of the solution to identify zero and nonzero entries in every local solution of the K-SVCR model. This lower bound clearly highlights the link between the sparsity of the solution and the choice of the regularization parameters and the value of p. The theorem may be used to assist in the selection of appropriate model parameters and norms in K-SVCR. It may also be used to distinguish between zero and nonzero elements in an optimum numerical solution. In addition, we derived upper bounds for the number of nonzero components of the optimal solutions.

The rest of this paper is organized as follows: Sect. 2 briefly describes K-SVCR. Section 3 presents absolute lower bounds for nonzero entries in the local solution of K-SVCR and upper bounds for the number of nonzero components, and concluding remarks are given in Sect. 4.

Notations. If f is a real-valued function defined on the n-dimensional real space R^n, the gradient of f with respect to x is denoted by $\nabla_x f$, which is a column vector in R^n. By A^T we mean the transpose of a matrix A. For two vectors x and y in the n-dimensional real space, $x^T y$ denotes the scalar product. For $x \in R^n$, $\|x\|$ denotes 2-norm, $\|x\|_0$ denotes the 0-norm of x, which is the number of nonzero entries of x, and a general p-norm is defined as $\|x\|_p = (\sum_{i=1}^{p} |x_i|^p)^{1/p}$. A column vector of ones of arbitrary dimension is indicated by e. The positive part of a real r is denoted by $r_+ = \max(r, 0)$; for vectors, we use it entrywise. For $A \in R^{m \times n}$ and $B \in R^{n \times l}$, the kernel $k(A, B)$ is an arbitrary function which maps $R^{m \times n} \times R^{n \times l}$ into $R^{m \times l}$. In particular, if x and y are column vectors in R^n and $A \in R^{m \times n}$, then $k(x^T, y)$ is a real number, $k(x^T, A^T)$ is a row vector in R^m, and $k(A, A^T)$ is an $m \times m$ matrix.

2   K-support Vector Classification Regression

K-SVCR, which is a new method of multi-class classification with ternary outputs $\{-1, 0, +1\}$, proposed in [2]. This method introduces the support vector classification-regression machine for k-class classification. This new machine evaluates all the training data into a "1-versus-1-versus-rest" structure during the decomposing phase using a mixed classification and regression support vector machine (SVM). Figure 1 from [20] illustrates the K-SVCR method graphically.

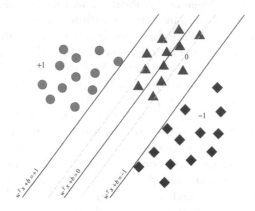

Fig. 1. Geometric representation of K-SVCR method.

K-SVCR can be formulated as a convex quadratic programming problem as follows:

$$\min_{w,b,\zeta_1,\zeta_2,\phi,\phi^*} \quad \frac{1}{2}\|w\|^2 + c_1(e^T\zeta_1 + e^T\zeta_2) + c_2 e^T(\phi + \phi^*) \tag{1}$$

$$\text{subject to} \quad Aw + eb \geqslant e - \zeta_1,$$
$$- (Bw + eb) \geqslant e - \zeta_2,$$
$$- \delta e - \phi^* \leqslant Cw + eb \leqslant \delta e + \phi,$$
$$\zeta_1, \zeta_2, \phi, \phi^* \geqslant 0.$$

where $c_1 > 0$ and $c_2 > 0$ are the regularization parameters, and ζ_1, ζ_2, ϕ and ϕ^* are positive slack variables. To avoid overlapping, the positive parameter δ must be lower than 1. We can define the nonlinear K-SVCR as follows:

$$\min_{w,b,\zeta_1,\zeta_2,\phi,\phi^*} \quad \frac{1}{2}\|w\|^2 + c_1(e^T\zeta_1 + e^T\zeta_2) + c_2 e^T(\phi + \phi^*) \tag{2}$$

$$\text{subject to} \quad k(A, D^T)w + eb \geqslant e - \zeta_1,$$
$$- (k(B, D^T)w + eb) \geqslant e - \zeta_2,$$
$$- \delta e - \phi^* \leqslant k(C, D^T)w + eb \leqslant \delta e + \phi,$$
$$\zeta_1, \zeta_2, \phi, \phi^* \geqslant 0,$$

where $k(\cdot, \cdot)$ is an appropriate kernel function and $D = [A^T \ B^T \ C^T]^T$.

3   Lower Bound for Nonzero Components of Solutions

This subsection contains lower bounds on the absolute values of the optimum solution's non-zero components. We establish such lower and upper boundaries that any component of the optimal solution that fits inside the bounds must be 0. We will demonstrate a connection between the penalty parameters and p and the sparsity of the solution.

At first, we can slightly improve the primal problem of K-SVCR (1) as (3), which uses the square of 2-norm of slack variables ζ_1, ζ_2, ϕ and ϕ^* instead of 1-norm slack variables in the objective function and we also improved the regularization term $\frac{1}{2}\|w\|^2$ to general form $\frac{c}{2}(\|w\|^2 + b^2)$. Then, the following minimization problem can be considered.

$$\min_{w, b, \zeta_1, \zeta_2, \phi, \phi^*} \frac{c}{2}(\|w\|^2 + b^2) + c_1(\|\zeta_1\|^2 + \|\zeta_2\|^2) + c_2(\|\phi\|^2 + \|\phi^*\|^2) \quad (3)$$
$$\text{subject to } Aw + eb \geqslant e - \zeta_1,$$
$$-(Bw + eb) \geqslant e - \zeta_2,$$
$$Cw + eb \leqslant \delta e + \phi,$$
$$-Cw - eb \leqslant \delta e + \phi^*,$$
$$\zeta_1, \zeta_2, \phi, \phi^* \geqslant 0.$$

where $c > 0$, $c_1 > 0$ and $c_2 > 0$ are the regularization parameters, and ζ_1, ζ_2, ϕ and ϕ^* are positive slack variables. Indeed, we want to discover the sparest solution of the problem (3), which implies we should minimize the number of nonzero components as well as objective functions. To reach this goal, we substitute $\|w\|^2$ with $\|w\|_0$ in the above problem to find the following hard problem:

$$\min_{w, b, \zeta_1, \zeta_2, \phi, \phi^*} c_1(\|\zeta_1\|^2 + \|\zeta_2\|^2) + c_2(\|\phi\|^2 + \|\phi^*\|^2) + \frac{c}{2}(\|w\|_0 + b^2) \quad (4)$$
$$\text{subject to } \zeta_1 \geqslant e - (Aw + eb),$$
$$\zeta_2 \geqslant e + (Bw + eb),$$
$$Cw + eb - \delta e \leqslant \phi,$$
$$-Cw - eb - \delta e \leqslant \phi^*,$$
$$\zeta_1, \zeta_2, \phi, \phi^* \geqslant 0.$$

For the optimal solution, we have $\zeta_1 = (e - (Aw + eb))_+$ and similarly for variables ζ_2, ϕ, and ϕ^*. Thus, problem (4) is equivalent to the following unconstrained optimization problem:

$$\min_{w, b} c_1\|(e - (Aw + eb))_+\|^2 + c_1\|(e + Bw + eb)_+\|^2$$
$$+ c_2\|(-\delta e - Cw - eb)_+\|^2 + c_2\|(-\delta e + Cw + eb)_+\|^2 + \frac{c}{2}(\|w\|_0 + b^2). \quad (5)$$

The $\|x\|_0$ is naturally linked to the $\|x\|_p$ with $0 < p < 1$, because we have the following relation [7]:

$$\|x\|_0 = \lim_{p \to 0} \|x\|_p^p = \lim_{p \to 0} \sum_{i=1}^{n} |x_i|^p. \tag{6}$$

Thus, in the objective functions of problems (4) and also (5), the term $\|w_1\|_0$ can be approximated by the p-norm, $\|w\|_p^p$ with $0 < p < 1$. Therefore instead of problem (5) we have the following problem:

$$\min_{w,b} \Phi(w,b) = \min_{w,b} c_1 \|(e - (Aw + eb))_+\|^2 + c_1 \|(e + Bw + eb)_+\|^2$$

$$+ c_2 \|(-\delta e - Cw - eb)_+\|^2 + c_2 \|(-\delta e + Cw + eb)_+\|^2 + \frac{c}{2}(\|w\|_p^p + b^2). \tag{7}$$

Now, we try to find lower bounds on the absolute values of non-zero components of the optimal solution. Indeed, we want to find such lower and upper boundaries that each component of the optimal solution that lies within the bounds must be 0.

Now we introduce the main theorem of this paper as follows:

Theorem 1. *Let (w^*, b^*) is a optimal solution of problem (7) and (w_0, b_0) be an arbitrary point. For any $i \in \{1, 2, \ldots, n\}$, if $w_i^* \in (-I, I)$, then $w_i^* = 0$, where*

$$I = \left[\frac{\frac{c}{2} p}{2(c_1 \|A\| + c_1 \|B\| + 2c_2 \|C\|)\sqrt{\Phi(w_0, b_0)}} \right]^{\frac{1}{1-p}}.$$

Proof. Let $k = \|w^*\|_0$. Assume, without loss of generality, that the optimal solution has the form $w^* = (w_1^*, w_2^*, \ldots, w_k^*, 0, \ldots, 0)^T$. Let $Z^* = (w_1^*, w_2^*, \ldots, w_k^*)^T$ be nonzero components of the optimal solution. Now, consider the optimization problem

$$\min_{Z,b} F(Z,b) = c_1 \|(e - (\tilde{A}Z + eb))_+\|^2 + c_1 \|(e + \tilde{B}Z + eb)_+\|^2 \tag{8}$$

$$+ c_2 \|(-\delta e - \tilde{C}Z - eb)_+\|^2 + c_2 \|(-\delta e + \tilde{C}Z + eb)_+\|^2 + \frac{c}{2}(\|Z\|_p^p + b_1^2).$$

In the above problem, $\tilde{A}$, $\tilde{B}$ and $\tilde{C}$ represent the positive class, the negative class and the zero class in the new training set, respectively.

It is obvious that (Z^*, b^*) is a local optimal solution of (8). So, the first-order necessary condition is satisfied at (Z^*, b^*). Since all components of Z^* are nonzero, the function $F(Z, b)$ is differentiable at (Z^*, b^*) and we have

$$0 = -2c_1 \tilde{A}^T (e - (\tilde{A}Z^* + eb^*))_+ + 2c_1 \tilde{B}^T (e + \tilde{B}Z^* + eb^*)_+ \tag{9}$$

$$- 2c_2 \tilde{C}^T (-\delta e - \tilde{C}Z^* - eb^*)_+ + 2c_2 \tilde{C}^T (-\delta e + \tilde{C}Z^* + eb^*)_+ + \frac{c}{2}(\nabla_Z^* \|Z^*\|_p^p),$$

In view of $\nabla_Z \|Z^*\|_p^p = p \operatorname{diag}(|Z^*|)^{p-1} \operatorname{sgn}(Z^*)$, we derive

$$\frac{c}{2} p \|(\operatorname{diag}(|Z^*|)^{p-1} \operatorname{sgn}(Z^*))\| \tag{10}$$

$$= \|2c_1 \tilde{A}^T (e - (\tilde{A}Z^* + eb^*))_+ - 2c_1 \tilde{B}^T (e + \tilde{B}Z^* + eb^*)_+$$
$$+ 2c_2 \tilde{C}^T (-\delta e - \tilde{C}Z^* - eb^*)_+ - 2c_2 \tilde{C}^T (-\delta e + \tilde{C}Z^* + eb^*)_+\|.$$

Thus

$$\frac{c}{2} p \| \operatorname{diag}(|Z^*|)^{p-1} \| \tag{11}$$

$$\leqslant 2c_1 \|\tilde{A}^T\| \cdot \|(e - (\tilde{A}Z^* + eb^*))_+\| + 2c_1 \|\tilde{B}^T\| \cdot \|(e + \tilde{B}Z^* + eb^*)_+\|$$
$$+ 2c_2 \|\tilde{C}^T\| \cdot \|(-\delta e - \tilde{C}Z^* - eb^*)_+\| + 2c_2 \|\tilde{C}^T\| \cdot \|(-\delta e + \tilde{C}Z^* + eb^*)_+\|$$
$$\leqslant (2c_1 \|A^T\| + 2c_1 \|B^T\| + 4c_2 \|C^T\|) \sqrt{F(w_0, b_0)}.$$

Then we obtain

$$\frac{c}{2} p \min_{1 \leqslant i \leqslant k} |Z_i^*|^{p-1} \leqslant \frac{c}{2} p \| \operatorname{diag}(|Z^*|)^{p-1} \|$$

$$\leqslant 2(c_1 \|A\| + c_1 \|B\| + 2c_2 \|C\|) \sqrt{F(w_0, b_0)}.$$

Finally, we can conclude that:

$$|Z_i^*| \geqslant \min_{1 \leqslant i \leqslant k} |Z_i^*| \geqslant \left(\frac{\frac{c}{2} p}{2(c_1 \|A\| + c_1 \|B\| + 2c_2 \|C\|) \sqrt{F(w_0, b_0)}} \right)^{\frac{1}{1-p}} \tag{12}$$

Therefore, given any local optimal solution (w^*, b^*) of problem (8), we have that $w_i^* \in (-I, I)$ implies $w_i^* = 0$, for $i = 1, \ldots, n$. □

The above theorem has shown the choice of penalty parameters and p and the sparsity of the solution have a link.

Now we find an upper bound for the number of nonzero components of the optimal solution. This indicates at least how many features can be removed.

Corollary 1. *Suppose that* (w^*, b^*) *is a optimal solution of problem (7) and* (w_0, b_0) *be an arbitrary point. Then we have:*

$$\|w^*\|_0 \leqslant \min \left\{ n, \frac{2\Phi(w_0, b_0)}{cI^p} \right\}.$$

Proof. Let $k = \|w^*\|_0$. Assume, without loss of generality, that the optimal solution has the form $w^* = (w_1^*, w_2^*, \ldots, w_k^*, 0, \ldots, 0)^T$. Let $Z^* = (w_1^*, w_2^*, \ldots, w_k^*)^T$ be the nonzero components of the optimal solution. It is obvious that

$$\frac{c}{2} k \min_{1 \leqslant i \leqslant k} |Z_i^*|^p \leqslant \frac{c}{2} \|Z^*\|_p^p = \frac{c}{2} \|w^*\|_p^p \leqslant \Phi(w^*, b^*) \leqslant \Phi(w_0, b_0).$$

One the one hand, from Theorem 1, we know

$$\min_{1 \leqslant i \leqslant k} |Z_i^*| \geqslant I.$$

On the other hand, we have:

$$k = \|w^*\|_0 \leqslant n.$$

This completes the proof.

The above corollary demonstrates that, naturally, the maximum number of nonzero components of the optimal solution depends on parameters and the number of features (n).

The nonlinear form of the problem (5) is equivalent to the following optimization problem.

$$\min_{w,b} \ c_1 \|(e - (k(A, D^T)w + eb))_+\|^2 \tag{13}$$

$$+ c_1 \|(e + k(B, D^T)w + eb)_+\|^2 + c_2 \|(-\delta e - k(C, D^T)w - eb)_+\|^2$$

$$+ c_2 \|(-\delta e + k(C, D^T)w + eb)_+\|^2 + \frac{c}{2}(\|w\|_0 + b^2).$$

Then by using p-norm we have:

$$\min_{w,b} \ \Psi(w, b) = \min_{w,b} \ c_1 \|(e - (k(A, D^T)w + eb))_+\|^2 \tag{14}$$

$$+ c_1 \|(e + k(B, D^T)w + eb)_+\|^2 + c_2 \|(-\delta e - k(C, D^T)w - eb)_+\|^2$$

$$+ c_2 \|(-\delta e + k(C, D^T)w + eb)_+\|^2 + \frac{c}{2}(\|w\|_p^p + b^2).$$

The following theorem allows us to determine the non-zero components of any optimum solution of the nonlinear K-SVCR. Its proof is quite similar to the proof of Theorem 1, so we omitted it from here.

Theorem 2. *Let (w^*, b^*) is a optimal solution of problem (14) and (w_0, b_0) be an arbitrary point. For any $i \in \{1, 2, \ldots, n\}$, if $w_i^* \in (-L, L)$, then $w_i^* = 0$, where*

$$L = \left[\frac{\frac{c}{2}p}{(2c_1 \|k(A, D^T)\| + 2c_1 \|k(B, D^T)\| + 4c_2 \|k(C, D^T)\|) \sqrt{\Psi(w_0, b_0)}} \right]^{\frac{1}{1-p}}.$$

The following corollary provides an upper bound for the number of nonzero components of the optimal solution for nonlinear K-SVCR. It can be proved similarly to Corollary 1.

Corollary 2. *Suppose (w^*, b^*) is a optimal solution of problem (14) and (w_0, b_0) be an arbitrary point. Then we have:*

$$\|w^*\|_0 \leqslant \min\left\{ N, \frac{2\Psi(w_0, b_0)}{cL^p} \right\},$$

where N is the number of features in the nonlinear space.

4   Conclusion

An innovative multi-class technique is the support vector classification-regression machine for k-class classification (K-SVCR). K-SVCR evaluates all training data into the "1-versus-a-versus-rest" structure with ternary outputs $\{-1, 0, +1\}$. We suggested an efficient model with feature selection duty by improving the K-SVCR model by using the p-norm instead of 2-norm in the original problem.

By using the first-order necessary condition for a the local optimal solution, a lower bound for the absolute value of nonzero entries for any local optimal solution of the new model is investigated. Furthermore, the lower bounds indicate the link between the sparsity of the solution and the regularization parameters and norm selection. As future work, an implementation of the proposed method can be considered to obtain sparse solutions for K-SVCR numerically.

Acknowledgments. The authors were supported by the Czech Science Foundation Grant P403-22-11117S. H. Moosaei was also supported by the Center for Foundations of Modern Computer Science (Charles Univ. project UNCE/SCI/004).

References

1. Ahmad, A.S., et al.: A review on applications of ANN and SVM for building electrical energy consumption forecasting. Renew. Sustain. Energy Rev. **33**, 102–109 (2014)
2. Angulo, C., Català, A.: K-SVCR. a multi-class support vector machine. In: López de Mántaras, R., Plaza, E. (eds.) ECML 2000. LNCS (LNAI), vol. 1810, pp. 31–38. Springer, Heidelberg (2000). https://doi.org/10.1007/3-540-45164-1_4
3. Arabasadi, Z., Alizadehsani, R., Roshanzamir, M., Moosaei, H., Yarifard, A.A.: Computer aided decision making for heart disease detection using hybrid neural network-genetic algorithm. Comput. Methods Programs Biomed. **141**, 19–26 (2017)
4. Bazikar, F., Ketabchi, S., Moosaei, H.: DC programming and DCA for parametric-margin ν-support vector machine. Appl. Intell. **50**(6), 1763–1774 (2020)
5. Boser, B.E., Guyon, I.M., Vapnik, V.N.: A training algorithm for optimal margin classifiers. In: Haussler, D. (ed.) Proceedings of the Fifth Annual Workshop on Computational Learning Theory, pp. 144–152. COLT 1992, ACM, New York (1992)
6. Bottou, L., et al.: Comparison of classifier methods: a case study in handwritten digit recognition. In: Proceedings of the 12th IAPR International Conference on Pattern Recognition, Vol. 3-Conference C: Signal Processing (Cat. No. 94CH3440-5). vol. 2, pp. 77–82. IEEE (1994)
7. Bruckstein, A.M., Donoho, D.L., Elad, M.: From sparse solutions of systems of equations to sparse modeling of signals and images. SIAM Rev. **51**(1), 34–81 (2009)
8. Chartrand, R.: Exact reconstruction of sparse signals via nonconvex minimization. IEEE Signal Process. Lett. **14**(10), 707–710 (2007)
9. Chartrand, R.: Nonconvex regularization for shape preservation. In: 2007 IEEE International Conference on Image Processing. vol. 1, pp. 293–297. IEEE (2007)
10. Cortes, C., Vapnik, V.: Support-vector networks. Mach. Learn. **20**(3), 273–297 (1995)

11. Déniz, O., Castrillon, M., Hernández, M.: Face recognition using independent component analysis and support vector machines. Pattern Recogn. Lett. **24**(13), 2153–2157 (2003)
12. Ding, S., Shi, S., Jia, W.: Research on fingerprint classification based on twin support vector machine. IET Image Proc. **14**(2), 231–235 (2019)
13. Ding, S., Zhang, N., Zhang, X., Wu, F.: Twin support vector machine: theory, algorithm and applications. Neural Comput. Appl. **28**(11), 3119–3130 (2017)
14. Fenn, M.B., Xanthopoulos, P., Pyrgiotakis, G., Grobmyer, S.R., Pardalos, P.M., Hench, L.L.: Raman spectroscopy for clinical oncology. Adv. Opt. Technol. **2011**, 213783 (2011)
15. Ketabchi, S., Moosaei, H., Razzaghi, M., Pardalos, P.M.: An improvement on parametric ν -support vector algorithm for classification. Ann. Oper. Res. **276**(1–2), 155–168 (2019)
16. Kreßel, U.H.G.: Pairwise Classification and Support Vector Machines, pp. 255–268. MIT Press, Cambridge (1999)
17. Li, G., Yang, L., Wu, Z., Wu, C.: DC programming for sparse proximal support vector machines. Inf. Sci. **547**, 187–201 (2021)
18. Ma, J., Zhou, S., Chen, L., Wang, W., Zhang, Z.: A sparse robust model for large scale multi-class classification based on K-SVCR. Pattern Recogn. Lett. **117**, 16–23 (2019)
19. Moosaei, H., Ketabchi, S., Razzaghi, M., Tanveer, M.: Generalized twin support vector machines. Neural Process. Lett. **53**(2), 1–20 (2021)
20. Moosaei, H., Hladík, M.: Least squares approach to K-SVCR multi-class classification with its applications. Ann. Math. Artif. Intell. **90**, 873–892 (2022). https://doi.org/10.1007/s10472-021-09747-1
21. Moosaei, H., Mousavi, A., Hladík, M., Gao, Z.: Sparse universum quadratic surface support vector machine models for binary classification. arXiv preprint arXiv:2104.01331 (2021)
22. Pardalos, P.M., Boginski, V.L., Vazacopoulos, A.: Data Mining in Biomedicine, Springer Optimization and Its Applications, vol. 7. Springer (2007). https://doi.org/10.1007/978-0-387-69319-4
23. Shao, M., Wang, X., Bu, Z., Chen, X., Wang, Y.: Prediction of energy consumption in hotel buildings via support vector machines. Sustain. Cities Soc. **57**(6), 102128 (2020)
24. Trafalis, T.B., Ince, H.: Support vector machine for regression and applications to financial forecasting. In: Proceedings of the IEEE-INNS-ENNS International Joint Conference on Neural Networks. IJCNN 2000. Neural Computing: New Challenges and Perspectives for the New Millennium. vol. 6, pp. 348–353. IEEE (2000)
25. Xu, Y., Guo, R., Wang, L.: A twin multi-class classification support vector machine. Cogn. Comput. **5**(4), 580–588 (2013)
26. Zhao, H.x., Magoulès, F.: A review on the prediction of building energy consumption. Renew. Sustain. Energy Rev. **16**(6), 3586–3592 (2012)

Integer Linear Programming in Solving an Optimization Problem at the Mixing Department of the Metallurgical Production

Damir N. Gainanov[1,2], Dmitriy A. Berenov[1], Egor A. Nikolaev[2], and Varvara A. Rasskazova[2(✉)]

[1] Ural Federal University, Ekaterinburg, Russia
berenov@dc.ru
[2] Moscow Aviation Institute (National Research University), Moscow, Russia
varvara.rasskazova@mail.ru

Abstract. The paper is devoted to investigate of the optimization problem on the mixing department processes at the metallurgical production. The problem is to unload transportation ladles into the iron ladles in such a way, to provide a timely and continues exchange between domain department and the mixing one, as well as between mixing department and converter shop-floor. This stage of technological chain plays the most important role for timely delivery of iron ladles to the converter shop-floor and for execution of the production plan in general.

To solve the problem under consideration there proposed an integer linear programming model, which takes into account all technological restrictions on the mixing department processes. There constructed a special set of variables, which allowed one to formalize both a complex system of constraints an objective function.

To demonstrate an effectiveness and powerful of the proposed approach, there were carried out a computational experiment using real-world data on the mixing department processes at the metallurgical production.

Keywords: Integer linear programming · Mathematical modeling · Metallurgical production · Mixing department

1 Introduction

Integer linear programming provides a wide class of methods and techniques for solving different problems on planning and management, including transportation and production problems. In [1,2] there were proposed integer linear programming models for solving several applied problems on railway planning. In [3] there also considered a transportation problem in frame of extreme flows in networks. An approaches to solving scheduling problems, including railway scheduling problems and production chains management, could be found in [4,5].

© Springer Nature Switzerland AG 2022
D. E. Simos et al. (Eds.): LION 2022, LNCS 13621, pp. 145–161, 2022.
https://doi.org/10.1007/978-3-031-24866-5_12

For several planning problems regarding production, integer linear models were proposed in [6]. When solving an applied problem, one need to pay the most attention to various technologies features of the problem under consideration. Indeed, an applied problems are often characterized by a complex and hard system of constraints, and that is why various models are investigated for each unique problem separately. In this paper we propose an integer linear programming model for solving an applied problem on optimization of mixing department processes at the metallurgical production.

In [7–10] one could find a full-scale overview on methods and applications of the integer linear programming. In [7, 8] there were discussed with details a classic statements and foundation methods on solving linear programs, including integer linear programs and Boolean ones. In [9, 10] there were investigated various methods of integer linear modeling to solve problems on planning, decision-making and management in applied areas. The present paper also propose a scalable approach, which could be continued to solve optimization problems arising both in mixing department and others at the metallurgical production.

It is well known, that integer linear programming is one of the classical $\mathcal{NP}$-hard problems, and that is why there continues developing a modern and powerful methods for solving this problem fast and effectively. In [11, 12] there presented a full-scale overview on modern methods and algorithms for solving integer linear programming problem. Due to existing of the various solvers integrated into the software of the most languages, we do not discuss ways for solving integer linear programs in the present paper. But the most attention we pay for developing a full-scale and effectively integer linear programming model to solve an applied optimization problem under consideration.

In [13–15] there proposed a various methods for solving applied problems on planning and management at the metallurgical production. In [13] there proposed an effective methodology for solving decision-making problem on management of the technological chains at the metallurgical production. A robust approach to solve problems on steel production management were also proposed in [16–18]. In [14,15] there were developed an effectively algorithms for predicting the quality of the final production at the rolling mills. However, there is a rather poor covering on problems related with dispatching of shop-floors at the metallurgical production, and in this paper one of such problems is investigated. It is need to be mentioned, that the problem under consideration serve as a preliminary stage of the production chain. Indeed, at the mixing department there preparing an iron ladles for further processing at the converter shop-floor. And when there occurs various interruptions in timely delivery of the iron ladles, it lead to connected interruptions in processing of the production plan in general.

2   Statement of the Problem

Let us consider the set of ladles which needs to be loaded by cast iron (iron ladles). We denote such a set by C, and mean that for each element of the set C there be given a set of parameters of the form

$$C = \{c_i \colon i, id_i, l\_nom_i, p_i, w\_min_i, w\_max_i, grade_i, Si_i, Mn_i, S_i, P_i\},$$

where i is an order number, and id_i is the unique identifier of the ladle, and l_nom_i is an absolutely number of the ladle, and p_i is a place at the mixing department, where corresponding iron ladle was standed for loading, and w_min_i and w_max_i are allowed weight for the ladle in tonnes of iron, and $grade_i$ is the steel grade with respect to production order, and Si_i, Mn_i, S_i, P_i correspond to chemical requirements on silicium, marganec, sulfur and phosphor, with respect to production order also.

Let there be also given a set of ladles with cast iron, which came from the domain department by railway trains (transportation ladles). These ladles are available for loading iron ladles of the corresponding set C, and each transportation ladle is characterized by the following set of parameters:

$$T = \{t_i \colon tr_i, path_i, nom_i, w_i, T\_Si_i, T\_Mn_i, T\_S_i, T\_P_i, cr_i\},$$

where tr_i is the number of train, which delivered a corresponding transportation ladle to the mixing department, and $path_i$ is the number of railway track, where corresponding train was fixed, and nom_i is a number of transportation ladle, and w_i is an amount of transportation ladle in tonnes of iron, and T_Si_i, T_Mn_i, T_S_i, and T_P_i are chemical features of the transportation ladle with regards to silicium, marganec, sulfur and phosphor respective, and cr_i is a list of cranes, which could take up the transportation ladle from the railway track for delivering to the place, where some iron ladle stands.

One need to construct a plan for loading each iron ladle from the set C, using transportation ladles from the set T, with respect to following conditions:

1. each iron ladle from the set C could use not more than 4 transportation ladles from the set T,
2. each transportation ladle from the set T could be used for loading not more than 2 iron ladles from the set C,
3. not more then 2 sequentially arriving trains could be used in parallel for loading of iron ladles from the set C; that is if there be used a train with number i, then either $(i-1)$-th or $(i+1)$-th train could be also used,
4. each transportation ladle from the set T needs to become either empty, or not used, or partitional used (and stay at the crane) due to loading all ladels from the set C; at the same time there could be stay the only one partitional used transportation ladle from the set T,
5. the weight of each loaded iron ladle i from the set C needs to be not less than corresponding w_min_i, and also not more than w_max_i.

We introduce a parameter of $step\_c$ to allow one a partially loading of iron ladles from the set C. Similarly, we introduce a parameter of $step\_t$ to allow one a partially unloading of transportation ladles from the set T. It is clear, that the values of $step\_c$ and $step\_t$ are closely connected with a dimensional of computational model. And if these values will be rather big, then the model could become not-representative. So, using these parameters, one need to achieve a balance between both a powerful and dimensional of the model under construction.

Let us suppose, that there is known how many iron ladles from C one could to load using fixed 2 trains with transportation ladles from T. In general case this is a separate problem, but it does not matter for goals of the further explanation. We denote such a number of iron ladles from C by k, and let j_1, j_2 be corresponding trains with transportation ladles from T. Thus, there be given

$$C' = \{c_{i_1}, c_{i_2}, ..., c_{i_k}\}, C' \subseteq C,$$

and

$$T' = \{t_{i_1}, t_{i_2}, ..., t_{i_l}\}, T' \subseteq T,$$

where $tr_{i_j} = j_1$ or $tr_{i_j} = j_2$ for all $t_{i_j} \in T'$.

So, for each iteration of the solving procedure there be given a set of iron ladles, which need to be loaded using the set of transportation ladles. In manufacturing terms, iron ladles contain three hundred tonnes of iron, and each transportation ladle contains from eighty to one hundred tonnes of iron. That is why each iron ladle could be loaded using three or four transportation ladles. But in case of using four transportation ladles the last one will be unloaded not fully, and needs to be unloaded into the nearest iron ladle arriving to the mixing department.

It needs to be also mentioned, that iron ladles from the set C', as well as transportation ladles from the set T', could be either fully loaded, or partially loaded in case of partially loading or unloading at the previous iteration. But all ladles from the set T' and belonging to the train, which is new arriving from the domain department, need to be necessary fully loaded. And this condition provide a sequential unloading of transportation ladles, as well as periodical and timely exchange of transportation ladles between domain and mixing departments corresponding.

3   Variable Set

Let us consider the set of C' and the set of T'. Each variable of the model under construction will corresponds to the possible unloading of the transportation ladle from the set T' into some iron ladle from the set of C'. Thus, for each variable of the model there will be defined a set of parameters of the form:

$$x: x\_load, x\_crane, x\_c, x\_t, x\_path, x\_w, coeef,$$

where x_load is an indeficator of the loading, and x_crane is a subset of cranes, which could deliver a choosen transportation ladle to the place with corresponding iron ladle, and x_c is a number of iron ladle, and x_t is a number of transportation ladle, and x_path is a railway track, where corresponding transportation ladle was fixed, and x_w is an amount of iron which is unloading from transportation ladle into corresponding iron ladle, and $coeff$ is a coefficient of the objective function.

We state $x = 1$ if the transportation ladle x_t is unloading into the iron ladle x_c with the amount x_w tonnes, and $x = 0$ otherwise.

We will call as *true* variables such ones, which correspond to real unloading from transportation ladle into iron one. And for each true variable we introduce a special connected variable corresponds to symmetric unloading of the same transportation ladle. So, for the fixed iron ladle c_i, transportation ladle t_j, and the value of a parameter $step_t$, a subset of variables, including both true variables and corresponding connected variables for symmetric unloading, will take the following form:

$$x_k:$$
$$x\_load_k = id_i, x\_crane_k = \text{``}p_i \cap cr_j\text{''}, x\_c_k = l\_nom_i,$$
$$x\_t_k = t_j, x\_path_k = path_j, x\_w_k = w_j - k \cdot step\_t, coeff_k,$$
$$x_{k+1}:$$
$$x\_load_{k+1} = id_i, x\_crane_{k+1} = \text{``}p_i \cap cr_j\text{''}, x\_c_{k+1} = l\_nom_i,$$
$$x\_t_{k+1} = t_j, x\_path_{k+1} = path_j, x\_w_{k+1} = w_j - k \cdot step\_t, coeff_{k+1},$$

where $k = \left[0, \lceil \frac{w_j}{step\_t} \rceil \right]$.

To provide a regular ladle exchange between domain department and the mixing one, we also introduce a *fake* variables with respect to possible surpluses on the transportation ladles. A parameter of the type x_c corresponds to iron ladle from the set C' and needs to be deleted for such variables. It is clear that for each fake variable there needs to be created a corresponding symmetric variable also. And this way will allow one to describe any partially unloading with respect to the value of the parameter $step\_t$.

It is need to be mentioned, that if two trains with transportation ladles are fixed with arrival time, than fake variables could not be introduced for the first arrived train due to condition on sequential unloading. So, let us suppose that there fixed some transportation ladle t_j: $path_j = j_2$, where j_2 is a number of train which arrived at the mixing department later, than another train with number j_1. Then a subset of fake variables for this ladle will take the form:

$$x_k:$$
$$x\_load_k = \text{`` -- ''}, x\_crane_k = \text{`` -- ''}, x\_c_k = \text{`` -- ''},$$
$$x\_t_k = t_j, x\_path_k = j_2, x\_w_k = w_j - k \cdot step\_t, coeff_k,$$
$$x_{k+1}:$$
$$x\_load_{k+1} = \text{`` -- ''}, x\_crane_{k+1} = \text{`` -- ''}, x\_c_{k+1} = \text{`` -- ''},$$
$$x\_t_{k+1} = t_j, x\_path_{k+1} = j_2, x\_w_{k+1} = k \cdot step\_t, coeff_{k+1}.$$

The above procedure for generating of fake variables is directed on possible partially unloading of transportation ladles. But the same conditions could also occur for any iron ladle, when there are not available transportation ladles for loading this iron ladle. In other words, this case take place when available number of transportation ladles is not enough for loading all given iron ladles. With respect to these cases, we also introduce an additional group of fake variables with lost parameters of the type x_t, x_path and x_w corresponding. So, for the fixed iron ladle $c_i \in C'$ and the value of a parameter $step\_c$, a subset of fake variables will take the form:

$$x_k:$$
$$x\_load_k = id_i, x\_crane_k = " - ", x\_c_k = l\_nom_i,$$
$$x\_t_k = " - ", x\_path_k = " - ", x\_w_k = w\_max_i - k \cdot step\_c, coeff_k.$$

Such fake variables will correspond to a partially loading of the iron ladles, and it is not necessary to generate a connected symmetric variables.

4   Coefficients

The most targets of the service on optimal iron mixing are the following:

1. decreasing of the number of iron ladles loads using four transportation ladles;
2. decreasing of the number of iron ladles loads, which need for a long-time improvement of chemical features of the iron using special machines.

The first target is due to the following. When there used four transportation ladles for loading some iron ladle, then there occurs a significant delay for ladles exchange between domain department and mixing one. Indeed, four fully loaded transportation ladles could not be unloaded into the same iron ladle. That is why the last not fully unloaded transportation ladle will stay at the mixing department, and wait for the next iron ladle available for loading. But the train, which delivered this transportation ladle to the mixing department, could not be returned to the domain department until all ladles become empty. And thus, there occurs a delay in timely delivery of transportation ladles from the domain department at the mixing one. As a corollary there occurs a significant delay for delivery iron ladles to the convertor department, and further delay in manufacturing plan running. So, the decreasing of the number of iron ladles loads using four transportation ladles provide a priority fully unloading for each transportation ladle, and leads to the timely trains returning back to the domain department.

The second target above is also directed to provide a timely iron ladles delivery to the converter department. But the highest priority is connected with the first mentioned condition, which regards to using three transportation ladles for loading each iron ladle. And thus, the second condition could be serve as a constraint of the model when the first condition is holds, and the minimum number of loads using four transportation ladles is already achieved.

Indeed, if even some iron ladle needs to be improved with respect to chemical features, then a timely exchange of transportation ladles increase a possibility of mixing department for preparing an iron ladle with given chemical requirements.

In the present paper we do not investigate the second target above, and stay with details on the first one. To minimize a number of iron ladles loads using four transportation ladles, we will consider the following structure of the coefficients.

It is clear, that each fully unloading of the transportation ladle has a higher priority instead of partially unloading of the same transportation ladle. Then for each true variables, that is

$$x_k: x\_load_k = id_i, x\_crane_k,$$
$$x\_c_k, x\_t_k = t_j, x\_path_k, x\_w_k = 0 \text{ or } x\_w_k = w_j, coeff_k,$$

one need to state the minimum value of the coefficient (in case of problem on minimum).

At the same time, transportation ladles from the first arrived train have the higher priority for unloading to the current iron ladle. That is if j_1 and j_2 be a fixed numbers of trains, sequentially arrived to the mixing department, and the train j_1 be the first arrived, then the holds the following:

$$coeff_k < coeff_{k'},$$

where

$$x_k: x\_load_k = id_i, x\_crane_k, x\_c_k, x\_t_k, x\_path_k = j_1, x\_w_k, coeff_k,$$
$$x_{k'}: x\_load_{k'} = id_i, x\_crane_{k'}, x\_c_{k'}, x\_t_{k'}, x\_path_{k'} = j_2, x\_w_{k'}, coeff_{k'}.$$

As for partially unloading is allowed for the last arrived train only, then coefficients of the corresponding variables will have the same form with x' above. That is

$$coeff_k < coeff_{k'},$$

where

$$x_k: x\_load_k = id_i, x\_crane_k, x\_c_k, x\_t_k, x\_path_k = j_1, x\_w_k, coeff_k,$$
$$x_{k'}: x\_load_{k'} = \text{``} - \text{''}, x\_crane_{k'}, x\_c_{k'}, x\_t_{k'}, x\_path_{k'} = j_2, x\_w_{k'}, coeff_{k'}.$$

Sometimes further, for the more compact presentation we will drop such a parameters of variables, which are not significant in the frame of current reasoning. Thus, for true variables the following hold:

$$coeff_{k_1} < coeff_{k_2} <$$
$$< coeff_{k_3} = coeff_{k_4} < coeff_{k_5} = coeff_{k_6},$$

where

$$x_{k_1}: x\_load_{k_1} = id_i, x\_t_{k_1} = t_j,$$
$$x\_path_{k_1} = j_1, x\_w_{k_1} = 0 \text{ or } x\_w_{k_1} = w_j, coeff_{k_1},$$

and

$$x_{k_2}: x\_load_{k_2} = id_i, x\_t_{k_2} = t_j,$$
$$x\_path_{k_2} = j_1, x\_w_{k_2} \neq 0 \text{ and } x\_w_{k_2} \neq w_j, coeff_{k_2},$$

and

$$x_{k_3}: x\_load_{k_3} = id_i, x\_t_{k_3} = t_{j'},$$
$$x\_path_{k_3} = j_2, x\_w_{k_3} = 0 \text{ or } x\_w_{k_3} = w_{j'}, coeff_{k_3},$$

and

$$x_{k_4}: x\_load_{k_4} = \text{"} - \text{"}, x\_t_{k_4} = t_{j'},$$
$$x\_path_{k_4} = j_2, x\_w_{k_4} = 0 \text{ or } x\_w_{k_4} = w_{j'}, coeff_{k_4},$$

and

$$x_{k_5}: x\_load_{k_4} = id_i, x\_t_{k_5} = t_{j'},$$
$$x\_path_{k_5} = j_2, x\_w_{k_5} \neq 0 \text{ and } x\_w_{k_5} \neq w_{j'}, coeff_{k_5},$$

and

$$x_{k_6}: x\_load_{k_6} = \text{"} - \text{"}, x\_t_{k_6} = t_{j'},$$
$$x\_path_{k_6} = j_2, x\_w_{k_6} \neq 0 \text{ and } x\_w_{k_6} \neq w_{j'}, coeff_{k_6},$$

To simplification, the above coefficients of the type k_1, k_2, k_3, k_4, k_5, k_6 could be described using arbitrary monotone increasing function. For example, it could be a linear function of the form $1, 2, 3, 3, 4, 4$.

Let us consider the subset of variables, which correspond to the partially loaded iron ladles now. That is the case, when given set of transportation ladles is not enough for loading all given iron ladles. As it was already mentioned, such variables are characterized by the set of parameters of the following form:

$$x_k: x\_load_k = id_i, x\_crane_k = \text{"} - \text{"}, x\_c_k = l\_nom_i, x\_t_k = \text{"} - \text{"},$$
$$x\_path_k = \text{"} - \text{"}, x\_w_k, coeff_k.$$

It is clear, that in such a case there is no question on the number of transportation ladles using for the load. But one need to guarantee the volume of the iron ladle only. With respect to this condition, there is no reasons to differ priorities among possible partially leavings of the iron ladle, because of the most priority way for loading this ladled will be included into solution due to coefficients for transportation ladles. Then it will be enough to introduce an arbitrary coefficients for such fake variables, with simply difference from the corresponding fake variables for transportation ladles. That is,

$$coeff_{k_1} < coeff_{k_2} < coeff_{k_3} = coeff_{k_4} < coeff_{k_5} =$$
$$= coeff_{k_6} < coeff_{k_7} < coeff_{k_8},$$

where coefficients of the type k_1-k_6 were described above, and coefficients of the type k_7 and k_8 correspond to fully and partially loads, and looks as following:

$$x_{k_7}: x\_load_{k_7} = id_i, x\_w_{k_7} = x\_max_i, coeff_{k_7},$$
$$x_{k_8}: x\_load_{k_8} = id_i, x\_w_{k_8} \neq x\_max_i, coeff_{k_8}.$$

Note, that difference between coefficients of the type k_7 and k_8 were introduced for presentation purposes only, and is not loaded with significant ideology sense.

It should be also mentioned, that in terms of objective, the transfer from minimization problem to the maximization one could be achieved by simply circulation of coefficients as following:

$$coeff_{k_1} > coeff_{k_2} > coeff_{k_3} = coeff_{k_4} > coeff_{k_5} =$$
$$= coeff_{k_6} > coeff_{k_7} > coeff_{k_8}.$$

But in terms of applied features of the problem, such a relation will be still directed to the minimization of the number of iron ladles loads using four transportation ladles.

5 Constraints

As we do not consider a group of constraints on chemical features of the iron ladles, we will stay on discussion of numerical constraints only.

At first it should be mentioned, that no one transportation ladle could be unloaded into more, then two different iron ladles. In case of fully unloading, it will be the only one iron ladle, and in case of partially unloading the leaved part needs to be unloaded into the nearest available iron ladle. Thus there occur a constraint of the following form:

$$\sum_k \{x_k : x\_t_k = t_j\} = 2 \text{ for all } t_j \in T'. \tag{1}$$

And also

$$\sum_k \{x_k \cdot x\_w_k : x\_t_k = t_j\} = w_j \text{ for all } t_j \in T'. \tag{2}$$

Note, that namely for constraints of the form (2) there were generated symmetric variables in functional space. That is for example, if transportation ladle t_j will be fully unloaded into some iron ladle c_i, then a solution will contain both the following variables:

$$x_{k_1} = 1 : x\_c_{k_1} = c_i, x\_t_{k_1} = t_j, x\_w_{k_1} = w_j,$$
$$x_{k_2} = 1 : x\_c_{k_2} = c_i, x\_t_{k_2} = t_j, x\_w_{k_2} = 0.$$

Let us consider an iteration $nom_i t$ such, that for some transportation ladle $t_j \in T'$ there holds the following:

$$\sum_k \{x_k : x\_t_k = t_j\} = 2,$$

$$\sum_k \{x_k \cdot x\_w_k : x\_t_k = t_j\} = w_j.$$

And let there also holds $x\_load_{k'} = \text{"} - \text{"}$ for some $x_{k'} = 1$ included into the sum above.

The described case means, that during the current iteration $nom_i t$ a transportation ladle t_j was not unloaded in full. Thus, this transportation ladle t_j needs to transfer into a functional space for the next iteration as an element of the train with number j_1. That is, this transportation ladle becomes to the next iteration with a highest priority for unloading. Then a tuple of constraints of the form (1)-(2) for this transportation ladle will take the form:

$$\sum_k \{x_k : x\_t_k = t_j\} = 1, \tag{3}$$

$$\sum_k \{x_k \cdot x\_w_k : x\_t_k = t_j\} = w_j - \sum_{k,nom\_it} \{x\_w_k : x_k = 1\}. \tag{4}$$

Let us now consider a group of constraints on the number of transportation ladles using for loading iron ladles, as well as constraints on the integrated volume of the iron ladles. Due to the maximum number of transportation ladles using for each iron ladle is equal to four, then

$$\sum_k \{x_k : x\_load_k = id_i\} \leqslant 4 \text{ for all } c_i \in C'. \tag{5}$$

And also with respect to integrated volume of iron ladle, there needs to hold the following:

$$\sum_k \{x_k \cdot x\_w_k : x\_load_k = id_i\} \geqslant w\_min_i \text{ for all } c_i \in C', \tag{6}$$

$$\sum_k \{x_k \cdot x\_w_k : x\_load_k = id_i\} \leqslant w\_max_i \text{ for all } c_i \in C'. \tag{7}$$

Similarly to the notation of the (3) and (4), there occur a specific modification of constraints of the form (5) and (6), when a process become from the one iteration to another. That is for some iron ladle $c_i \in C'$ there needs to hold the following:

$$\sum_k \{x_k : x\_load_k = id_i\} \leqslant 4 -$$

$$- \sum_{k,nom\_it} \{x_k : x\_load_k = id_i, x\_w_k \neq 0\}, \tag{8}$$

and

$$\sum_k \{x_k \cdot x\_w_k : x\_load_k = id_i\} \geqslant w\_min_i -$$

$$- \sum_{k,nom\_it} \{x_k \cdot x\_w_k : x\_load_k = id_i, x\_t_k \neq \text{``} - \text{''}\}, \tag{9}$$

and the constraint of the form (7) is not changed.

It should be also mentioned, that above comments regarding constraints modification of the form (3), (4), and (8), (9), plays a significant role in the code realization of the model only. But in general case for the purposes of the present paper it does not matter, and we will restrict our description by the fixed iteration of the solving process. So, to get a correct solution one need to satisfied the constraints of the form (1), (2), and (5)-(7) at the each iteration.

6 Objective

As it was already noted, the present paper consider an objective case directed on decreasing of the number of iron ladles loades using four transportation ladles. Reminder, that a problem under investigation is also characterized with a condition on sequential unloading of trains, which delivery transportation ladles to the mixing department. Due to these reasons, a solving procedure is an iteration process, and there fixed two trains with transportation ladles on each step. These trains are with numbers j_1 and j_2 respectively, where train with number j_1 arrives to the mixing department early, than a train with number j_2. That is why we state, that a train with number j_1 needs to be unloaded with a higher priority, than a train with number j_2. And also we state, that fake variables on partially unloading of transportation ladles are allowed for the train with number j_2 only.

So, let us consider an arbitrary iteration nom_it, and the sets of C' and T' of iron and transportation ladles corresponding. A functional space for this iteration nom_it will take the form:

$$x_k: x\_load_k = id_i,$$
$$x\_crane_k = \text{``}p_i \cap cr_j\text{''},$$
$$x\_c_k = l\_nom_i,$$
$$x\_t_k = t_j,$$
$$x\_path_k = j_1 \text{ or } x\_path_k = j_2,$$
$$x\_w_k = k \cdot step\_t \text{ or } x\_w_k = w_j - k \cdot step\_t,$$

$$coeff_k = \begin{cases} 1, \text{ if } x\_path_k = j_1 \text{ and } x\_w_k = 0 \text{ or } x\_w_k = w_j, \\ 2, \text{ if } x\_path_k = j_1 \text{ and } x\_w_k \neq 0 \text{ and } x\_w_k \neq w_j, \\ 3, \text{ if } x\_path_k = j_2 \text{ and } x\_w_k = 0 \text{ or } x\_w_k = w_j, \\ 4, \text{ if } x\_path_k = j_2 \text{ and } x\_w_k \neq 0 \text{ and } x\_w_k \neq w_j, \end{cases}$$

or

$$x_k : x\_load_k = \text{`` -- ''},$$
$$x\_crane_k = \text{`` -- ''},$$
$$x\_c_k = \text{`` -- ''},$$
$$x\_t_k = t_j,$$
$$x\_path_k = j_2,$$
$$x\_w_k = k \cdot step\_t \text{ or } x\_w_k = w_j - k \cdot step\_t,$$
$$coeff_k = \begin{cases} 3, & \text{if } x\_w_k = 0 \text{ or } x\_w_k = w_j, \\ 4, & \text{if } x\_w_k \neq 0 \text{ and } x\_w_k \neq w_j, \end{cases}$$

or

$$x_k : x\_load_k = id_i,$$
$$x\_crane_k = \text{`` -- ''},$$
$$x\_c_k = l\_nom_i,$$
$$x\_t_k = \text{`` -- ''},$$
$$x\_path_k = \text{`` -- ''},$$
$$x\_w_k = w\_max_i - k \cdot step\_c,$$
$$coeff_k = \begin{cases} 5, & \text{if } x\_w_k = w\_max_i, \\ 6, & \text{otherwise.} \end{cases}$$

We state, that $x_k = 1$ if and only if, a transportation ladle t_j is unloading to the iron ladle c_i with number l_nom_i in the amount of x_w_k tons, and $x_k = 0$ otherwise. Then the integer linear program under construction will take the following form:

$$\sum_k x_k \cdot coeff_k \longrightarrow \min \tag{10}$$

with constraints

$$\begin{cases} \sum_k \{x_k : x\_t_k = t_j\} = 2 \text{ for all } t_j \in T', \\ \sum_k \{x_k \cdot x\_w_k : x\_t_k = t_j\} = w_j \text{ for all } t_j \in T', \\ \sum_k \{x_k : x\_load_k = id_i\} \leqslant 4 \text{ for all } c_i \in C', \\ \sum_k \{x_k \cdot x\_w_k : x\_load_k = id_i\} \geqslant w\_min_i \text{ for all } c_i \in C', \\ \sum_k \{x_k \cdot x\_w_k : x\_load_k = id_i\} \leqslant w\_max_i \text{ for all } c_i \in C', \\ x_k \in \{0, 1\}. \end{cases} \tag{11}$$

Thus, an optimization problem on decreasing the number of iron ladles loads using four transportation ladles is reduced to solving a sequence of integer linear programs of the form (10), (11), using at the each step a correction procedure of the (3), (4), (8) and (9).

7   Computational Results

A numerical experiment was carried out, using real world data on the mixing department processes at the Novolipetsk Metallurgical Enterprise at 2020.

A parameter of $step\_t$ was stated as equal to 10 (tonnes), which is rather significant restriction on optimization possibilities. Indeed, this parameter is closely connected with the dimension of a functional space, and if the value become less, than a set of variables increase significantly. And due to these reasons there could occur another problem, with another global optimal solution. But even with the value of a parameter of $step\_t$ is chosen to be rather big, the obtained results demonstrate a powerful of the proposed approach with respect to the purposes on decreasing the number of iron ladles loads using four transportation ladles.

Computational results are described in the tables below, and show corresponding values of observed parameters for each month of a base period. In Tables 1, 2, 3, 4, 5, 6, 7, 8 and 9 there used the following notations:

- "Day" is a calendar day, for which a calculations were carried out,
- "Iron ladles" is a number of iron ladles, for which there were fixed a full load (sometimes it was not possible to get full-scale data, and that is why this number could be rather small),
- "Avg, transportation ladles (day)" is an average number of transportation ladles, which was recommended for optimal loading of each iron ladle in the day under consideration; for example, in Table 1 at the April, 16-th there was obtained a solution for loading 23 iron ladels using 3.5 transportation ladles for each load,
- "Avg, transportation ladles (month)" is an average number of transportation ladles, which was recommended for optimal loading of each ladle in the month under consideration (among all iron ladles).

In Table 10 there shown a percent relations for loads, which use three and four transportation ladles respectively in obtained solution. It should be mentioned, that values of 40 % and 50 % for loads, which use three transportation ladles were obtained in May, June, July and August, due to small base samples, and in all other cases an obtained percent increasing significantly.

Historical data under consideration in general is characterized by relation of 40 % and 60 % for loads using three and four transportation ladles corresponding. And in the Table 11 there shown, that obtained solution provide an opposite relation and significantly better result of 60 % for the number of loads, which use three transportation ladles.

Thus, one can conclude a powerful and effectiveness of the proposed integer linear programming approach, to minimize number of loads using four transportation ladles.

Table 1. Computational results (2020, April)

Day	16	19	21	22	27
Iron ladles	23	23	3	11	21
Avg, transportation ladles (day)	3.5	3.6	3	3.1	3.7
Avg, transportation ladles (month)	3.38				

Table 2. Computational results (2020, May)

Day	2	6	7	8	9	11	13	16	17	18	19	21
Iron ladles	24	4	21	19	14	24	20	21	38	26	24	2
Avg, transportation ladles (day)	3.6	3.2	3.5	3.4	3.5	3.2	3.2	3.5	3.7	3.6	3.8	3.5
Day	22	24	25	27	28							
Iron ladles	13	4	3	17	17							
Avg, transportation ladles (day)	3.3	3.2	4	3.7	3.7							
Avg, transportation ladles (month)	3.53											

Table 3. Computational results (2020, June)

Day	4	8	9	10	28
Iron ladles	42	5	2	10	17
Avg, transportation ladles (day)	3.5	3.8	3	3.8	3.7
Avg, transportation ladles (month)	3.56				

Table 4. Computational results (2020, July)

Day	4	19	20	24	27	31
Iron ladles	8	9	9	11	14	13
Avg, transportation ladles (day)	3.8	3.5	3.7	3.7	3.8	3.6
Avg, transportation ladles (month)	3.68					

Table 5. Computational results (2020, August)

Day	2	5	6	7	11	15	16	18	19	20
Iron ladles	18	19	19	18	13	17	19	19	14	18
Avg, transportation ladles (day)	3.7	3.6	3.7	3.7	3.5	3.6	3.6	3.7	3.7	3.5
Avg, transportation ladles (month)	3.63									

Table 6. Computational results (2020, September)

Day	1	2	3	4	6	7	10	11	12	15	17	18
Iron ladles	18	19	19	15	22	33	13	15	36	18	38	20
Avg, transportation ladles (day)	3.5	3.6	3.3	3.5	3.6	3.5	3.3	3.1	3.6	3.6	3.5	3.4
Day	20	21	22	23	26	27	30					
Iron ladles	19	17	17	28	2	15	10					
Avg, transportation ladles (day)	3.3	3.5	3	3.3	4	3.6	10					
Avg, transportation ladles (month)	3.47											

Table 7. Computational results (2020, October)

Day	2	3	4	9	10	12	13	14	15	16	19	20
Iron ladles	13	36	18	21	14	4	16	15	15	28	12	16
Avg, transportation ladles (day)	3	3.5	3.6	3.3	3.7	3.7	3.6	3.4	3.4	3.7	3.1	3.5
Day	22	23	24	25	26	28	29	30	31			
Iron ladles	3	28	38	28	10	22	31	32	13			
Avg, transportation ladles (day)	3	3.6	3.4	3.4	3.6	3.4	3.3	3.5	3.5			
Avg, transportation ladles (month)	3.43											

Table 8. Computational results (2020, November)

Day	1	3	4	5	6	8	9	10	11	16	17	19
Iron ladles	12	34	28	28	18	18	18	18	16	16	11	19
Avg, transportation ladles (day)	3.2	3.2	3.2	3.3	3.3	3.2	3.3	3.2	3.6	3.5	3.1	3.4
Day	20	21	22	24	25	26	29					
Iron ladles	12	18	18	17	19	18	4					
Avg, transportation ladles (day)	3.6	3.6	3.6	3.5	3.4	3.5	3					
Avg, transportation ladles (month)	3.35											

Table 9. Computational results (2020, December)

Day	1	3	5	6	7	8	10	11	12	13	15	16
Iron ladles	15	15	25	23	19	17	25	24	21	26	4	21
Avg, transportation ladles (day)	3.6	3.5	3.4	3.7	3.4	3.4	3.5	3.2	3.6	3.5	3.3	3.5
Day	19	22	23	24	26	27	28	29				
Iron ladles	12	20	26	21	28	26	21	27				
Avg, transportation ladles (day)	3.5	3.2	3.3	3.3	3.3	3.1	3.2	3.4				
Avg, transportation ladles (month)	3.39											

Table 10. Computational results

Month	Three transportation ladles	Four transportation ladles
April	60%	40%
May	50%	50%
June	50%	50%
July	40%	60%
August	40%	60%
September	60%	40%
October	60%	40%
November	70%	30%
December	70%	30%

Table 11. Computational results

	Avg, fact	Avg, optimal
Number of loads using three transportation ladles	40%	60%
Number of loads using four transportation ladles	60%	40%

8   Conclusion

The paper investigates an optimization problem on decreasing a number of iron ladles loads, which use four transportation ladles at the mixing department of the metallurgical production. Such a problem arises due to requirements on regular and timely return of transportation ladles back to the domain department, as well as on delivery of iron ladles to the converter shop-floor. The present paper does not consider the second condition above, which is the most related with chemical requirements on iron ladles. But we only directed to satisfy a numerical conditions on using transportation ladles. And this condition have a highest priority, and that is why it was considered in the present paper in details.

The proposed integer linear programming model takes into account all technological restrictions and rules regarding availability of transportation ladles. There constructed a full-scale mathematical model, where were taken into account both numerical constraints and restrictions on iron amount, from unloading out of transportation ladles until loading into the iron ladle.

A computational experiment was carried out using real-world data on the mixing department processes at the Novolipetsk Metallurgical Enterprise. This experiment shows an effectiveness and powerful of the proposed approach, even using rather big step for a generating process of the functional space.

Acknowledgement. The paper was prepared with support of the Russian Scientific Fund. Project No̱ 23-21-00293 "Optimization of discrete systems in problems of logistic using integer linear programming approach and graph theory methods".

References

1. Lazarev, A.A., Musatova, E.G.: Integer statements of the problem of forming railway trains and schedules of their movement. Manage. Large Syst. **38**, 161–169 (2012)
2. Gainanov, D.N., Ignatov, A.N., Naumov, A.V., Rasskazova, V.A.: On track procession assignment problem at the railway network sections. Autom. Remote. Control. **81**(6), 967–977 (2020)
3. Hu, T.: Integer programming and threads in networks. Mir, Moscow (1974)
4. Ryan, D. M., Foster, B. A.: An integer programming approach to scheduling. In: Wren, A. (eds.) Computer Scheduling of Public Transport Urban Passenger Vehicle and Crew Scheduling. Amsterdam: North-Holland, pp. 269–280 (1981)
5. Wagner, H.M.: An integer linear-programming model for machine scheduling. Nav. Res. Logist. Quart. **6**(2), 131–140 (1959)
6. Pochet, Y., Wolsey, L. A.: Production planning by mixed integer programming. In: Mikosh, T. V., Resnick, S. I., Robinson, S. M. (eds.) Springer Series in Operations Research & Financial Engineering (2006). https://doi.org/10.1007/0-387-33477-7
7. Shevchenko, V.N., Zolotykh, N.Y.: Linear and integer linear programming. Nizhny Novgorod State University named after N. I. Lobachevsky, Nizhny Novgorod (2004)
8. Schraver, A.: Theory of linear and integer programming. Mir, Moscow (1991)
9. Segal, I.K., Ivanova, A.P.: Introduction to applied discrete programming: models and computational algorithms. FIZMATLIT, Moscow (2007)
10. Appa, G. M., Pitsoulis, L. S., Paul, W. H.: Handbook on modeling for discrete optimization. Springer Series, International Series in Operations Research & Management Science, vol. 88, XXII (2006)
11. Wolsey, L.A.: Integer programming. John Wiley & Sons, NJ (2020)
12. Hu, T.C., Kahng, A.B.: Linear and integer programming made easy. Springer (2016). https://doi.org/10.1007/978-3-319-24001-5
13. Kabulova, E.G.: Intelligent management of multi-stage systems of metallurgical production. Model. Optim. Inf. Technol. **7**(24), 341–351 (2018)
14. Gitman, M.B., Trusov, P.V., Fedoseev, S.A.: On optimization of metal forming with adaptable characteristics. J. Appl. Math. Comput. **7**(2), 387–396 (2020)
15. Gainanov, D.N., Berenov, D.A.: Algorithm for predicting the quality of the product of metallurgical production. In: CEUR Workshop Proceedings, vol. 1987, 194–200 (2017)
16. Qiu, Y., Wang, L., Xu, X., Fang, X., Pardalos, P.M.: Scheduling a realistic hybrid flow shop with stage skipping and adjustable processing time in steel plants. Appl. Soft Comput. **64**, 536–549 (2018)
17. Kong, M., Pei, J., Xu, J., Liu, X., Pardalos, P.M.: A robust optimization approach for integrated steel production and batch delivery scheduling with uncertain rolling times and deterioration effect. Int. J. Prod. Res. **58**(17), 5132–5154 (2020). https://doi.org/10.1080/00207543.2019.1693659
18. Long, J., Sun, Z., Pardalos, P.M., Bai, Y., Zhang, S., Li, C.A.: robust dynamic scheduling approach based on release time series forecasting for the steelmaking continuous casting production. Appl. Soft Comput. **92**, 106271 (2020). https://www.sciencedirect.com/science/article/pii/S1568494620302118

Realtime Gray-Box Algorithm Configuration

Dimitri Weiss[✉] and Kevin Tierney

Decision and Operation Technologies Group, Bielefeld University, Bielefeld, Germany
{dimitri.weiss,kevin.tierney}@uni-bielefeld.de

Abstract. A solver's runtime and the quality of the solutions it generates are strongly influenced by its parameter settings. Finding good parameter configurations is a formidable challenge, even for fixed problem instance distributions. However, when the instance distribution can change over time, a once effective configuration may no longer provide adequate performance. Realtime algorithm configuration (RAC) offers assistance in finding high-quality configurations for such distributions by automatically adjusting the configurations it recommends based on instances seen so far. Existing RAC methods treat the solver to be configured as a black box, meaning the solver is given a configuration as input, and it outputs either a solution or runtime as an objective function for the configurator. However, analyzing intermediate output from the solver can enable configurators to avoid wasting time on poorly performing configurations. To this end, we propose a gray-box approach that utilizes intermediate output during evaluation and implement it within the RAC method CPPL. We apply cost sensitive machine learning with pairwise comparisons to determine whether ongoing evaluations can be terminated to free resources. We compare our realtime gray-box configurator to a black-box equivalent on several experimental settings and show that our approach reduces the total solving time in several scenarios.

Keywords: Algorithm configuration · Cost sensitive learning · SAT · MILP

1 Introduction

Changing a solver's parameters can have a drastic impact on its performance. The right parameters can lead to runtime reductions of multiple orders of magnitude or significantly improved solution quality, as demonstrated in [3,17]. However, finding good parameters is time-consuming and a challenge for both solver developers and users due to large parameter spaces and long evaluation times of the solvers. Automated algorithm configuration (AC) has been proposed to assist algorithm designers and users in this daunting task. Approaches like SMAC [16], GGA [2] or irace [21], for offline configuration, as well as realtime configuration methods like [10,12], automatically identify high-quality parameter configurations. These approaches have proven themselves highly capable in recent years on a wide range of problems [18].

© Springer Nature Switzerland AG 2022
D. E. Simos et al. (Eds.): LION 2022, LNCS 13621, pp. 162–177, 2022.
https://doi.org/10.1007/978-3-031-24866-5_13

Offline AC provides a high-quality configuration after an offline search on a problem instance set. A core assumption of this approach is that the instance distribution does not change over time, not even when the training phase is over and the configuration is used in practice. Thus, offline AC only requires a sufficiently large and representative problem instance set and enough time for the configuration process to be completed. However, in many real-world settings, instance distributions change over time in unpredictable ways, for example if a company grows and encounters larger instances or different types of customers. Furthermore, there may be insufficient instances to train configurators offline in advance of seeing new instances. Thus, specialized AC methods are required for this setting to provide high-quality configurations to decision makers.

Realtime AC considers a stream of problem instances to be solved in which the configuration used to solve them can be adjusted on an instance-by-instance basis. It is assumed that the instances are subject to changes in statistical properties, i.e. *concept drift*, which is known to be a challenge in realtime data streams [27]. Recently proposed RAC methods like RAC through tournaments (ReACT) [12], RAC through tournament rankings (ReACTR) [11] and Contextual Preselection with Plackett-Luce (CPPL) [10] are designed to handle this setting. They propose configurations that may be good at solving a given instance and run the configurations in parallel in a tournament format. The methods adjust learned models based on feedback from each instance solved. In this way, they dynamically respond to changing instance distributions and ensure the stream of instances is solved quickly.

AC methods generally consider the solver being configured (henceforth the *target algorithm*) to be a *black box*, meaning the target algorithm accepts parameters and a problem instance as input and outputs a solution, with no view of the internals of the approach. However, under this assumption the AC method must wait until the target algorithm has finished executing, even though it may be able to gain insights about the quality of a configuration by monitoring its output. For further detail, see e.g., [22]. An alternative to the black box is a *white box*, which assumes full information about a solver's behavior. For example, the method would know all implementation details of a branch-and-bound procedure for a particular optimization problem. This would require customizing the configurator for each solver it encounters, which is impractical both in terms of development time and the lack of access to the solver internals in many cases. Thus, we seek a middle ground in which the configurator can monitor the status of a target algorithm and gain actionable insights from it.

Most solvers provide intermediate output that can be aggregated into a time series of feature vectors that describe the current status of the solution process. We consider such target algorithms to be a *gray box*, and utilize this information to support an AC process. This concept has been in the focus of research in the Bayesian optimization framework as well, see e.g. [4]. In [26], such information is used to adapt parameters during runtime dynamically for specially instrumented target algorithms. Thus, in this paper, we propose a realtime, gray-box algorithm configuration approach. We extend the RAC approach CPPL [10] to gather and utilize information provided by the target algorithm during runtime to guide the configuration process. The contributions in this work are as follows.

1. We implement the first realtime configurator that assumes the target algorithm is a gray box.
2. We develop an approach based on cost sensitive learning to detect and eliminate underperforming configurations in a RAC environment.
3. We empirically assess the performance gains of viewing the target algorithm as a gray box.

This paper is organized as follows. First, we provide an overview of automated AC and related work in Sect. 2. Our gray-box extension is detailed in Sect. 3. We experimentally evaluate the gray-box mechanism in Sect. 4, and we conclude and discuss promising future work in Sect. 5.

2   Automated AC

We provide an overview of related work and the AC methods extended in this article. We first discuss offline AC, in which a single configuration must be chosen to work well on a static set of instances. Then, we introduce RAC, in which the instance distribution is allowed to change, and we suggest new configurations for each instance to be solved. For a full overview of the area of AC, see [25].

2.1   Offline AC

We first consider offline AC, in which the goal is to find a high-quality configuration for a given solver *offline*, i.e., in advance of using the configuration to solve unseen instances. To this end, we adopt the notation of [17]. In this setting, we are given a set of problem instances $\Pi \subseteq \hat{\Pi}$ and a parameterized algorithm $\mathbb{A}$. The objective of the offline AC problem is to find a configuration θ in the space of all possible parameter settings Θ of $\mathbb{A}$ that optimizes a performance metric. The performance metric $m : \Pi \times \Theta \to \mathbb{R}$ describes, e.g., the runtime $\mathbb{A}$ requires to achieve a specified solution quality (*runtime configuration*), or the solution quality found within a given time limit l (*quality configuration*). Thus, the goal in the case of runtime configuration is to minimize $\sum_{\pi \in \hat{\Pi}} m(\pi, \theta)$, whereas for quality configuration we maximize the previous sum. Note that in the context of this work, we focus solely on runtime minimization.

Offline AC has been tackled successfully by a variety of approaches that employ search paradigms including evolutionary algorithms, local search, and model-based learning. The CALIBRA method [1] is one of the first general AC methods and employs a fractional factorial design to generate initial configurations, followed by a local search to refine them. ParamILS [17] also uses a local search, but embeds a single-change neighborhood within an iterative local search. The method F-race [8] empirically evaluates a finite set of configuration candidates with a racing mechanism that discards low-quality configurations using a Friedman test. ParamILS, CALIBRA and F-race require a discretized search space.

More recent AC methods relax the requirement of a discretized search space and thus support continuous, discrete and categorical parameters. The genetic algorithm-based GGA [3] uses a racing mechanism and a crossover operator designed to capture parameter dependencies to generate new configurations. GGA is extended in [2] to include a specialized random forest surrogate to assist in generating new configurations. The configurator SMAC [16] is based on sequential model-based optimization, meaning it uses a Bayesian optimization paradigm for suggesting high-quality configurations. The irace configurator [21] samples configurations based on a distribution and evaluates them by racing. After each race, it updates the sampling distribution based on the best performing configurations. This procedure is iterated to enhance the configuration qualities until a given threshold is reached. Finally, the method GPS [23] exploits the structure of parameter space landscapes with the assumption of weak interaction of parameters and uni-modal response to value changes in the performance of the algorithm.

2.2   RAC

In contrast to the offline AC setting, in which the problem instance set is assumed to be given in advance, the RAC setting assumes the problem instance set is given as a sequence of problem instances that must be solved without a training phase. Thus, at each time i, a corresponding instance π_i must be solved. RAC represents real-world settings in which, for example, instances must be solved on a daily basis to plan vehicle tours, employee schedules, etc. In such cases, instances will likely change over time, either in terms of their size or in terms of their structure (e.g., more customers of a certain type or changing time windows for deliveries). Thus, the underlying distribution of the problem instance set $\hat{\Pi}$ is not fixed, meaning learning a single offline configuration in advance would be insufficient. On the single instance Π_i, the goal of RAC is to find a configuration $\theta(\pi_i)$ optimizing $m(\pi_i, \theta(\pi_i))$ over the set of possible configurations $\Theta(\Pi_i)$, only using the information gained from solving instances $\pi_1, \ldots, \pi_{i-1}$.

On a sequence of instances, the goal of a possible RAC setting is to minimize their solution time, as depicted in Fig. 1. Configurations are evaluated in tournaments to compute the evaluation metric. The basic assumption here is that there is a machine available with multiple cores to run several copies of the solver in parallel, thus taking advantage of the ever-increasing parallel capabilities of modern CPUs. A tournament involves running the solver in parallel with different configurations θ_j chosen from a pool of configurations. The best performing configuration $\theta^*_{\pi_i}$ is the configuration to first lead to a solution of the problem instance π_i. Once it is finished, the tournament can be stopped, as we have a solution to the instance. The configuration runtime is used to update the scores of the configurations in the pool. If any scores reach a specified threshold, the corresponding configuration is removed from the pool. Furthermore, the scores are used to decide which configurations should be run on the next instance.

The concept drift present in the instances of the RAC setting requires a significantly different approach than in the offline setting. For RAC, we require mechanisms to adjust the best configuration over time and discount past knowledge gained on instances that may no longer be relevant. Furthermore, the total

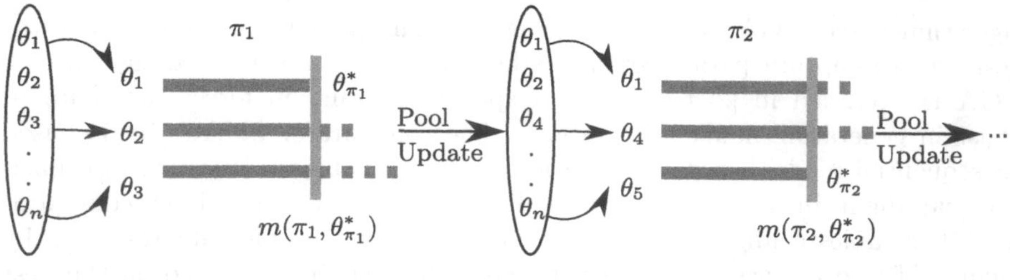

$$\pi_1 \qquad\qquad \pi_2$$

$$\theta_1,\theta_2,\theta_3,\cdot,\cdot,\theta_n \quad \theta_1\ \theta_2\ \theta_3 \qquad \theta^*_{\pi_1}$$

$$m(\pi_1,\theta^*_{\pi_1})$$

Pool Update

$$\theta_1,\theta_2,\theta_4,\cdot,\cdot,\theta_n \quad \theta_1\ \theta_4\ \theta_5 \qquad \theta^*_{\pi_2}$$

$$m(\pi_2,\theta^*_{\pi_2})$$

Pool Update ...

Fig. 1. RAC in tournaments.

resources available are assumed to be significantly lower than in the offline AC setting, since we cannot expect users of RAC methods to have hundreds of cores available to commit to solving their problem. Two types of RAC methods can be identified in the literature. The first, based on the work of [12], uses statistical tests to manage a pool of configurations that can be used to solve instances. The second employs a bandit approach [10] to select configurations from its pool to solve each instance. The information gained from solving the instance is provided back to the bandit approach.

RAC Through Tournaments (ReACT). The first RAC method introduced is ReACT in [12]. A pool of configurations is randomly initialized that includes the default configuration of the target algorithm. For the first problem instance, configurations are chosen randomly to run in the tournament and include the default configuration to avoid a bootstrapping phase where the algorithm might perform worse than the default. A score keeping component follows the configurations' performance. For each new instance, configurations are sampled from the pool weighted by how well they performed on past instances. Configurations are removed from the pool if they lose in too many tournaments. New configurations are generated randomly.

The ReACT approach is extended with the TrueSkill ranking system [15] in [11] for managing the pool of configurations. This approach, called ReACTR, uses TrueSkill to provide a score and confidence value for each configuration. TruSkill estimates the quality of configurations in comparison to each other without them needing to be evaluated in the same tournament. ReACTR greedily selects a fixed number of configurations from the pool to run in a tournament for a given instance based on their TrueSkill rankings. A fixed portion of the tournament is filled with random configurations to further explore the parameter space. New configurations are generated by genetic crossover of the configurations ranked highest by TrueSkill. These are then mutated to further encourage exploration of high-quality configurations. Randomly generated configurations replace low-quality configurations with a given probability. Configurations are removed from the pool if an empirically defined TrueSkill threshold is exceeded.

Contextual Preselection with Plackett-Luce (CPPL). CPPL [10] is currently a state-of-the-art RAC method based on contextual bandits. It harnesses a Plackett-Luce model for online preselection with context information [9]. The preselection bandit approach solves a sequential online decision problem in which every decision yields feedback. Furthermore, a context $\mathbf{X_i} = (\mathbf{x_{i,1}} \ldots \mathbf{x_{i,n}})$ for the problem instance i provides information about each of the n available configurations. Thus, the context consists of the parameter values of a configuration and problem instance features. Given an instance i to be solved, CPPL decides the subset of configurations in the pool that are most likely to solve the instance. The chosen subset of configurations is evaluated by running the parameterized solvers in parallel in a tournament. The feedback for the decision is the information about performance of the configurations that have been evaluated.

In addition to determining each tournament, the preselection bandit model is used to decide if a configuration in the pool should be replaced by a new one. To this end, all pairs of configurations in the pool are examined. If a configuration is dominated by any other configuration, it is discarded. Configuration θ_j is dominated by configuration θ_k, with $\theta_j \neq \theta_k$, if the inequality $\hat{v}_{i,k} - c_{i,k} \geq \hat{v}_{i,j} + c_{i,j}$ holds true for $\theta_j \neq \theta_k$, $\hat{v}$ is the weight of a configuration and c is a confidence in the learned weight. The learned model also guides the generation process of new configurations. To this end, configurations are generated by a crossover mechanism with mutations or random generation with a small probability. These configurations are then assessed by the model and the best ranked configurations are used to replace the discarded configurations in the pool, akin to the genetic engineering procedure in [2]. Note that the context vector provided to the learned model always includes the context of the problem instance in the next time step.

3   Gray-box Method

We propose a gray-box approach to improve the RAC process of the chosen framework CPPL. In this approach, output data of the solver during runtime is gathered to help improve the RAC process. Specifically, we learn a cost sensitive model that is used to terminate underperforming configurations. Termination of configurations frees resources for the evaluation of other configurations in the pool.

3.1   Identifying Underperforming Configurations

In the proposed approach, target algorithm output data is gathered for each configuration j running on an instance i and condensed into a feature vector $\boldsymbol{f}_{j,t}$ at preset time points t with n features. Note that for notational convenience, we do not index this vector with the current instance. This data indicates the current state of the target algorithm and is gathered from direct output of the solver. Feature vectors are gathered from each configuration running in the tournament if the algorithm provided output until t. Algorithms may not provide output in

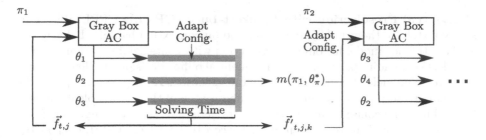

Fig. 2. Adapting configurations during tournaments based on solver output data.

preprocessing phases or if no progress was made in between time points. The output can be used to classify configurations.

As illustrated in Fig. 2, the runtime output data of the target algorithm is monitored during the tournament and fed back to the gray-box algorithm configurator. The algorithm configurator classifies pairs of configurations, if enough meaningful output data is available. Configurations in the tournament can be terminated based on the classifications. If configurations are removed from the tournament, resources are freed. All data gathered during the tournament is processed in between solving problem instances by the gray-box algorithm configurator.

Given $\boldsymbol{f}_{j,t}$ and the (censored) runtime of each configuration, it is possible to learn a model to predict whether a given configuration will win the tournament or not. However, this is a practically impossible task in isolation, i.e., to know whether a configuration will win (or lose) a tournament, we need to know about the other configurations in our feature vector. Thus, we propose a preference learning approach using pairwise comparisons. This enables us to assess configurations in relation to the other configurations in the tournament. To this end, we compute a feature vector for the pairwise comparison of configurations j and k, $\boldsymbol{f}'_{t,j,k} = (\boldsymbol{f}_{j,t}; \boldsymbol{f}_{k,t}; \boldsymbol{f}_{j,t} - \boldsymbol{f}_{k,t})$, which concatenates the vectors of each configuration and their difference at time t. The model predicts whether j outperforms k, i.e. $m(\pi_i, \theta_j) < m(\pi_i, \theta_k)$, or the opposite case $m(\pi_i, \theta_j) > m(\pi_i, \theta_k)$, where k outperforms j.

After solving each instance, we update our model using the information gained in the previous tournament. We note that the model can also be updated at a fixed interval. The training data consists of the pairwise feature vectors gathered during previous tournaments, with the label set as indicated above. If both configurations j and k fail to solve an instance, we ignore all corresponding vectors $\boldsymbol{f}'_{t,j,k}$.

3.2   Applying Cost-Sensitive Classification

A drawback of the previously proposed mechanism for comparing configurations is that it does not consider the performance difference between configurations. Consider configurations A and B and C, with runtimes 4, 5 and 30 s, respectively. It is clearly worse for our model to suggest that C will finish before A than it is for the model to suggest that B will finish before A. We thus propose using cost-sensitive classification so that configurations with small differences in runtime receive a different label than those with large runtime differences.

Cost-sensitive learning requires an objective function value for each possible combination of ground truth and prediction of the model. We thus construct the following cost matrix at time t given configurations j and k:

$$C_{t,j,k} = \begin{pmatrix} C_{TN} & C_{FN} \\ C_{FP} & C_{TP} \end{pmatrix}.$$

The cost matrix entries are computed for every pairwise comparison. The indices TN, FN, FP and TP of the cost entries stand for true negative, false negative, false positive and true positive, respectively. The positive ground truth is the case of configuration j outperforming configuration k, and negative ground truth the opposite case.

We assign the following costs to C_{TN}, C_{FN}, C_{FP}, and C_{TP}:

$$C_{t,j,k} = \begin{pmatrix} 0 & \max\left(0, \frac{m(\pi_i,\theta_k)-m(\pi_i,\theta_j)}{l}\right) \\ \max\left(0, \frac{m(\pi_i,\theta_j)-m(\pi_i,\theta_k)}{l}\right) & 0 \end{pmatrix}$$

The intuition of the cost matrix is simple: the model is penalized proportionally to how "wrong" it is. Here, the difference in runtime is divided by the timeout l. The maximum term ensures there are no negative entries, although these would never be chosen anyway due to the combination of ground truth and prediction in such cases. The entries for true positives and negatives are naturally set to 0. We note that we did not experiment significantly with the cost matrix, and there may be more effective alternatives.

3.3 Terminating Underperforming Configurations

The cost sensitive pairwise comparison model allows us to rank configurations, but it does not directly tell us which ones ought to be terminated. We thus introduce a simple heuristic for eliminating the underperformers. The heuristic is designed conservatively, as we must avoid removing the tournament winner (which at this stage is hidden among the potential losers in the ongoing tournament). The heuristic only considers configurations for which there is output available. Empirically, there did not seem to be any useful information from a lack of information, i.e., just because a solver does not output anything does not necessarily mean it will not win the tournament (at least for the solvers we tested). Thus, we just ignore such configurations until they begin to output.

The heuristic works as follows. First, we rank the configurations by how often the cost sensitive model says they is better than a different configuration. Note that if there is a tie between all configurations, we do not terminate any configuration. However, if multiple configurations are tied for last place we kill all of them. We run this check as soon as there is enough output to make pairwise comparisons. If there is a meaningful difference in the ranking between configurations, we terminate all configurations with the worst ranking from the current tournament and do not terminate any configurations in this tournament again. In other words, our goal is to free up resources as soon as possible, but once that has been done, we do not want to risk terminating something else.

3.4   Utilizing Freed Resources

Once one or more configurations are removed from the current tournament, computational resources are free and can be reallocated. There are two options for using these resources: (1) we can run a different configuration from the pool or (2) we can run the next instance from the instance sequence. In option (1), the new configuration will start late compared to other configurations and will thus be at a clear disadvantage. It is unlikely that this new configuration will ever win unless configurations are stopped very early, which is often not possible. Thus, we select option (2), in which we start the next instance early.

In the proposed strategy, resources freed in the tournament on problem instance π_i are used to start the evaluation of configurations on the problem instance π_{i+1} early. We assume that problem instances in the sequence are available, and that there is no dependence between them (such as in rolling horizon planning). Starting a configuration from the pool with the next problem instance π_{i+1} before the current tournament π_i has been finished gives a runtime advantage to the configuration that starts early. Note that due to this advantage, in the following tournament we will not actually know whether a configuration that started later may have actually been the fastest. This could have an impact on CPPL and our cost-sensitive learning over time, since we could end up providing incorrect data to the models. Empirically, however, this does not seem to cause any issues.

4   Computational Experiments

We assess the performance of our proposed gray-box approach and assess the overall need of a gray box within the CPPL approach. We run all experiments with 16 Intel Xeon Gold Skylake 6148 cores running at 2.4 GHz. We answer the following research questions:

RQ1: Does CPPL always select the best configuration in the pool?
RQ2: Can underperforming configurations be identified based on gray-box data?
RQ3: Can a gray-box version of CPPL outperform a black-box version?

4.1   Dataset and Solver

We consider a set of three scenarios also used in [10]. Table 1 provides a summary. We use two SAT solvers, CaDiCaL [7] and Glucose [5]. Since CaDiCaL and Glucose are targeted to different types of SAT instances, we match each solver with a separate dataset of 1,000 instances. Thus, CaDiCaL is tasked with solving instances from the power-law random SAT generator [13], whereas Glucose

Table 1. The RAC scenarios we consider in this work.

Scenario	Solver	Problem	Instance set
1	CaDiCaL [7]	SAT	Power-law random SAT [13]
2	Glucose [5]	SAT	Modularity-based random SAT [14]
3	CPLEX [19]	MILP	Frequency assignment problem [24]

is assigned instances from the modularity-based random SAT instance generator [14]. In the third scenario, CPLEX is given 1,000 instances of the frequency assignment problem [24].

All instance sets are generated to include concept drift. The problem instance set in scenario 1 is generated with 10,000 variables, 93,000 clauses, 4 literals per clause, a power-law exponent of variables set to 18 and a power-law exponent of clauses changing in 10 equally distributed steps from 12.5 to 2.5. The second problem instance set is generated with 10,000 variables, 60,000 clauses, 4 literals per clause, 600 communities and gradually changing the modularity factor from 0.4 to 0.35 as in the scenarios before. The MILP instances in scenario 3 are generated setting the number of cells to 5, the variance of channel requirements per cell to 1.5, drawing the necessary distance between channels from a normal distribution and gradually changing the mean requirement of channels per cell from 8 to 18 in 10 equally distributed steps.

We use the first 50 instances in all scenarios as a validation set. These instances are used to bootstrap the bandit model in CPPL. The reason for this is that CPPL has a burn-in period in which it is exploring the space of configurations. This is significantly cheaper than running standard offline configuration, requiring under a second of computation time. We note that this step can be skipped in cases where no instances are available before the instance stream must be solved, or performed with randomly generated instances matching the initial instance distribution of the instance stream. We do not include these instances in our results and consider them to be a training set.

4.2   RQ1: Does CPPL Find the Best Configuration?

We run an experiment to examine the accuracy of the selection mechanism of CPPL for each scenario. CPPL is run with a time limit l of 300s, with tournaments of 16 configurations and a pool of 30 configurations at all times. As the sequence of problem instances is solved, we record the state of the configuration pool, the selection of configurations to run in the tournaments and the results of the tournaments. The results of the tournaments include the winning configuration as well as its runtime needed to solve the problem instance. For each problem instance, we run the solver with the 14 configurations that did not compete in the tournament, using the runtime of the winning configuration as the time limit. To reduce the expense of this experiment, the time limit l is set to the formerly recorded winning time. This way, we identify each configuration which would have won the tournament but was not selected.

Table 2. Underperformance of the CPPL selection from the pool.

Scenario	Time loss		Average number of
	Total (s)	%	better configs
1	1090	8.0	4.95
2	853	28.9	2.75
3	324	0.1	0.01

Table 2 shows the aggregated performance of CPPL over each of the scenarios. The time loss is the unrealized improvement in runtime over the whole instance set, i.e. if CPPL had always chosen the best configuration, this amount of time could have been saved. This time loss is stated in the total amount of seconds and in percent of the overall runtime needed for the respective scenario. Additionally, the average number of better performing configurations per problem instance is given for each scenario.

Overall, there is room for improvement in all scenarios. The message from these results is that CPPL is not perfect, and that an intervention in the tournament during runtime might improve performance. We note, however, that while these results motivate attempting to select new configurations out of the pool, as discussed the time potentially given to new configurations is not enough to outperform the already running configurations.

4.3   RQ2: Quality of Prediction Based on Gray-Box Data

We conduct experiments offline and during realtime configuration to examine whether the gray-box model can detect low-quality configurations. In the first experiment, we consider an offline setting and examine the prediction quality of the cost-sensitive machine learning method. The second experiment is performed in a realtime setting to examine the decisions of our termination heuristic. That is, the question is whether we erroneously eliminate high-quality configurations. In all experiments, we use the cost-sensitive classification method CostCla [6].

Offline Classification. We conduct an offline experiment to gauge whether it is possible to train a cost sensitive model to identify underperforming configurations. First, we initialize 160 configurations randomly and let them simultaneously run in 10 tournaments of 16 configurations each on the first 100 problem instances in each scenario, resulting in 1,000 tournaments. Each run of a configuration on an instance has a time limit of 300 s. We let every configuration run until it solves the instance or reaches the time limit, even if there is a winner in the tournament. The gray-box data from the solvers is recorded every 2 s. Note that the solvers do not all output at every time point, nor is the feature vector always complete, since there can be a lack of output for some features. Thus, we fill missing output with the last known feature vector and missing entries in the feature vector with zeros.

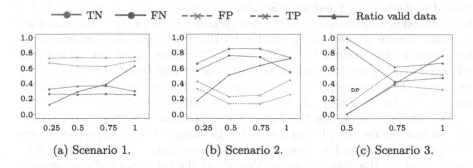

(a) Scenario 1.   (b) Scenario 2.   (c) Scenario 3.

Fig. 3. Pairwise comparison classification in offline setting. (Color figure online)

Figure 3 shows the performance of the model at time points, which are computed as a percentage of the average runtime of all scenarios. We run a five fold cross validation split by problem instances.

The blue line provides the amount of valid data at each time point, since sometimes solvers take some time before they first output anything. During this time, we simply have no information whether a configuration is good or bad. The other lines provide the labels of the confusion matrix. Given a pairwise comparison between configurations i and j, we consider the positive case to be when the runtime of i is less than the runtime of j. Note that the fluctuations in the classification accuracy are mainly due to the amount of data available (shown in blue).

We first note that the model accuracy is mostly stable over time, notwithstanding the randomness of not all data being available in early time points in scenarios 1 and 2. This is somewhat unexpected, as more information is available as solving progresses. This could indicate a weakness of the model, or just that the information we need to make a decision is available early. In scenario 3, the model classifies more pairwise comparisons as positive and less as negative with more available data. Since in this experiment all configured solvers run until the time limit, even if there is a winner in the tournament, more configurations begin to solve the instance. This makes the pairwise comparison challenging. However, when using our classification model for a real configuration task, all remaining configurations are terminated as soon as there is a winner in the tournament, thus making the data less ambiguous.

Overall, performance of the model in this case is somewhat mixed. While the model excels at determining which configuration is better in scenario 2, the performance drops off in scenario 1. However, this performance is good enough, as our main goal is not to get every single i,j pair ordered correctly, but to build a model capable of figuring out which configurations underperform. Hence, in the following, we adjust the way the model is used to more accurately reflect its use in a RAC setting.

Classification in a Realtime Setting. To gain insight about the effect of termination of configurations in the RAC setting, we implement the model and pairwise comparison mechanism within CPPL. Each time our heuristic would terminate a configuration, we log the state of the tournament, but do not actually carry out the termination. In all scenarios, we find that the winning configuration of the tournament is never terminated. Thus, our heuristic is sufficiently conservative to avoid choosing the best performing configuration.

We compute the average rankings of the winners and the other configurations in each tournament for all scenarios to find the range of rankings between them. Our goal is to assess how close the tournament winner is to the top of the tournament. In this way, we get an idea about the robustness of the approach. In scenario 1, on average 1.9 configurations are predicted to be better than the winning configuration. Compared to that, the average for all non-winning configurations is 3.3. Configurations predicted to be better than the winning configuration in scenario 2 are 4.5 on average. For all configurations not winning the tournament, 4.7 configurations are predicted to be better. In scenario 3, on average 14.6 configurations are predicted to be better than the winner and 15.9 configurations are predicted to be better than the losing configurations.

Overall, the results of this experiment are very promising. In all scenarios run in this experiment, the winning configuration never would have been terminated. This is due to the big enough range between the average numbers of predictions of winning or losing configurations to be better than the compared configuration such that it is possible to reliably distinguish the winner from the remaining configurations. Since this is the case, the termination of configurations does not increase the runtime over the sequence of problem instance.

4.4 RQ3: Black-Box vs. Gray-Box CPPL

We directly compare CPPL as a black-box and a gray-box approach. We again use a time limit of 300 s and tournaments of 16 configurations each. All scenarios are run three times each using the black-box version of CPPL and the gray-box implementation.

Figure 4 illustrates the direct comparison of the approaches as a rolling average of the runtime needed to complete an instance. The blue line depicts the performance of CPPL and the orange line the performance of the gray-box implementation. Note that the concept drift of the problem instances makes solving problem instances more difficult over the course of the problem instance sequence. In all scenarios, the gray-box outperforms the black-box version. In scenario 1, the gray-box outperforms the black-box by 37.1% of the overall runtime the black-box needed to solve the problem instance set. In scenario 2 the gray-box implementation needs 19.4% less of the total black-box runtime. The gray-box implementation outperforms the black-box in scenario 3 by 11.9%. While the concept drift affects the gray-box algorithm configurator in scenario 2 as well, it is better able to adapt and keep the average runtime low.

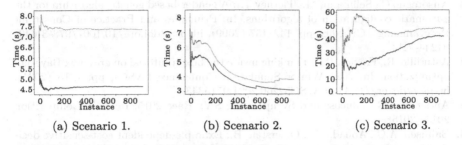

(a) Scenario 1.   (b) Scenario 2.   (c) Scenario 3.

Fig. 4. Direct comparison of the black-box (blue, dotted) and gray-box (orange, solid) implementations of CPPL. (Color figure online)

5 Conclusion and Future Work

We introduced a RAC method based on a gray-box view of the target algorithm. Our approach uses a dynamically learned cost-sensitive pairwise comparison mechanism combined with a simple heuristic to terminate poorly performing configurations early. We show that exploiting the intermediate output of the solver can lead to significant performance gains over ignoring it. This is especially promising, as the solvers we used were not specifically designed for gray-box configuration. On the contrary, a solver specifically designed for this setting could likely output information that is actionable earlier, leading to larger performance gains. For future work, we will investigate how to better bootstrap realtime configuration with offline configuration and how to improve the model performance of the cost sensitive classification in the gray box setting. Furthermore, the identification of underperforming configurations will be investigated with other methods, e.g. the bandit-based approach Hyperband [20].

Acknowledgements. The authors are supported in part by the funding program *Zentrales Innovationsprogramm Mittelstand (ZIM)* (Grant No. ZF4622601LF8) of the German Federal Ministry for Economic Affairs and Climate Action, and the project *Maschinelle Intelligenz für die Maschinelle Intelligenz für die Optimierung von Wertschöpfungsnetzwerken (MOVE)* (Grant No. 005-2001-0042) of the "it's OWL" funding of the Ministry of Economics, Innovation, Digitalization and Energy of the German state of North Rhine-Westphalia. The authors would also like to thank the Paderborn Center for Parallel Computation (PC²) for the use of the OCuLUS and Noctua clusters.

References

1. Adenso-Díaz, B., Laguna, M.: Fine-tuning of algorithms using fractional experimental designs and local search. Operat. Res. **54**, 99–114 (2006)
2. Ansótegui, C., Malitsky, Y., Samulowitz, H., Sellmann, M., Tierney, K.: Model-based genetic algorithms for algorithm configuration. In: International Joint Conferences on Artificial Intelligence Organization (IJCAI) (2015)

3. Ansótegui, C., Sellmann, M., Tierney, K.: A gender-based genetic algorithm for the automatic configuration of algorithms. In: Principles and Practice of Constraint Programming - CP 2009, pp. 142–157 (2009). https://doi.org/10.1007/978-3-642-04244-7_14

4. Astudillo, R., Frazier, P.I.: Thinking inside the box: a tutorial on grey-box Bayesian optimization. In: 2021 Winter Simulation Conference (WSC), pp. 1–15 (2021). https://doi.org/10.1109/WSC52266.2021.9715343

5. Audemard, G.: Glucose and Syrup in the SAT Race 2015. In: SAT Competition 2015 (2015)

6. Bahnsen, A.C., Aouada, D., Ottersten, B.: Example-dependent cost-sensitive decision trees. Expert Syst. Appl. **42**(19), 6609–6619 (2015). https://doi.org/10.1016/j.eswa.2015.04.042

7. Biere, A.: CaDiCaL at the SAT Race 2019. In: SAT Race 2019 - Solver and Benchmark Descriptions, p. 2 (2019)

8. Birattari, M., Stützle, T., Paquete, L., Varrentrapp, K.: A racing algorithm for configuring metaheuristics. In: Proceedings of the Genetic and Evolutionary Computation Conference (GECCO), pp. 11–18 (2002)

9. El Mesaoudi-Paul, A., Bengs, V., Hüllermeier, E.: Online Preselection with Context Information under the Plackett-Luce Model (2020)

10. El Mesaoudi-Paul, A., Weiß, D., Bengs, V., Hüllermeier, E., Tierney, K.: Pool-based realtime algorithm configuration: a preselection bandit approach. In: Kotsireas, I.S., Pardalos, P.M. (eds.) LION 2020. LNCS, vol. 12096, pp. 216–232. Springer, Cham (2020). https://doi.org/10.1007/978-3-030-53552-0_22

11. Fitzgerald, T., Malitsky, Y., O'Sullivan, B.: ReACTR: realtime algorithm configuration through tournament rankings. In: International Joint Conferences on Artificial Intelligence Organization (IJCAI), pp. 304–310 (2015)

12. Fitzgerald, T., Malitsky, Y., O'Sullivan, B.J., Tierney, K.: ReACT: real-time algorithm configuration through tournaments. In: Annual Symposium on Combinatorial Search (SoCS) (2014)

13. Friedrich, T., Krohmer, A., Rothenberger, R., Sutton, A.: Phase Transitions for scale-free SAT formulas. In: Association for the Advancement of Artificial IntelligenceSPONSORSHIP (AAAI), pp. 3893–3899 (2017)

14. Giráldez-Cru, J., Levy, J.: A modularity-based random SAT instances generator. In: International Joint Conferences on Artificial Intelligence Organization (IJCAI), pp. 1952–1958 (2015)

15. Guo, S., Sanner, S., Graepel, T., Buntine, W.L.: Score-based Bayesian skill learning. In: European conference on Machine Learning and Knowledge Discovery in Databases (ECMLPKDD), pp. 106–121 (2012). https://doi.org/10.1007/978-3-642-33460-3_12

16. Hutter, F., Hoos, H.H., Leyton-Brown, K.: Sequential model-based optimization for general algorithm configuration. In: Learning and Intelligent Optimization (LION), p. 507–523 (2011)

17. Hutter, F., Hoos, H.H., Leyton-Brown, K., Stützle, T.: ParamILS: an automatic algorithm configuration framework. J. Artif. Intell. Res. (JAIR), p. 267–306 (2009)

18. Hutter, F., et al.: Aclib: a benchmark library for algorithm configuration. In: International Conference on Learning and Intelligent Optimization (LION), pp. 36–40 (2014). https://doi.org/10.1007/978-3-319-09584-4_4

19. IBM: IBM ILOG CPLEX Optimization Studio: CPLEX User's Manual (2016). https://www.ibm.com/support/knowledgecenter/SSSA5P_12.7.0/ilog.odms.studio.help/pdf/usrcplex.pdf

20. Li, L., Jamieson, K.G., DeSalvo, G., Rostamizadeh, A., Talwalkar, A.: Efficient hyperparameter optimization and infinitely many armed bandits. CoRR abs/1603.06560 (2016). http://arxiv.org/abs/1603.06560
21. López-Ibáñez, M., Dubois-Lacoste, J., Stützle, T., Birattari, M.: The irace package: iterated racing for automatic algorithm configuration. Operat. Res. Perspect. pp. 43–58 (2016). https://doi.org/10.1016/j.orp.2016.09.002
22. Pardalos, P.M., Rasskazova, V., Vrahatis, M.N.: black box optimization, machine learning, and no-free lunch theorems. Springer International Publishing (2021). https://doi.org/10.1007/978-3-030-66515-9
23. Pushak, Y., Hoos, H.: Golden parameter search: exploiting structure to quickly configure parameters in parallel. In: Proceedings of the Genetic and Evolutionary Computation Conference (GECCO), pp. 245–253 (2020). https://doi.org/10.1145/3377930.3390211
24. Santos, H., Toffolo, T.: Python MIP: Modeling examples (2018–2019). Accessed 23 Jan 2020. https://engineering.purdue.edu/~mark/puthesis/faq/cite-url/
25. Schede, E., et al.: A survey of methods for automated algorithm configuration (2022)
26. Speck, D., Biedenkapp, A., Hutter, F., Mattmüller, R., Lindauer, M.: Learning heuristic selection with dynamic algorithm configuration. CoRR abs/2006.08246 (2020). https://arxiv.org/abs/2006.08246
27. Tsymbal, A.: The problem of concept drift: definitions and related work. Tech. rep., Department of Computer Science, Trinity College, Dublin (2004)

Dynamic Urban Solid Waste Management System for Smart Cities

Adriano S. Silva[1,2,3,4](✉) , Thadeu Brito[1] , Jose L. Diaz de Tuesta[2,5] ,
José Lima[1] , Ana I. Pereira[1] , Adrián M. T. Silva[3,4] ,
and Helder T. Gomes[2]

[1] Research Centre in Digitalization and Intelligent Robotics (CeDRI), Instituto
Politécnico de Bragança, 5300-253 Bragança, Portugal
{adriano.santossilva,brito,jllima,apereira}@ipb.pt
[2] Centro de Investigação de Montanha (CIMO). Instituto Politécnico de Bragança,
5300-253 Bragança, Portugal
[3] Laboratory of Separation and Reaction Engineering - Laboratory of Catalysis
and Materials (LSRE-LCM), Faculty of Engineering, University of Porto,
Rua Dr. Roberto Frias, 4200-465 Porto, Portugal
[4] Associate Laboratory in Chemical Engineering (ALiCE), Faculty of Engineering,
University of Porto, Rua Dr. Roberto Frias, 4200-465 Porto, Portugal
[5] Department of Chemical and Environmental Technology, ESCET,
Rey Juan Carlos University, Madrid, Spain

Abstract. Increasing population in cities combined with efforts to
obtain more sustainable living spaces will require a smarter Solid Waste
Management System (SWMS). A critical step in SWMS is the collection
of wastes, generally associated with expensive costs faced by companies
or municipalities in this sector. Some studies are being developed for the
optimization of waste collection routes, but few consider inland cities as
model regions. Here, the model region considered for the route optimization using Guided Local Search (GLS) algorithm was Bragança, a city
in the northeast region of Portugal. The algorithm used in this work is
available in open-source Google OR-tools. Results show that waste collection efficiency is affected by the upper limit of waste in dumpsters.
Additionally, it is demonstrated the importance of dynamic selection of
dumpsters. For instance, efficiency decreased 10.67% for the best upper
limit compared to the traditional collection in the regular selection of
dumpsters (levels only). However, an improvement of 50.45% compared
to traditional collection was observed using dynamic selection of dumpsters to be collected. In other words, collection cannot be improved only
by letting dumpsters reach 90% of waste level. In fact, strategies such as
the dynamic selection here presented, can play an important role to save
resources in a SWMS.

Keywords: Smart city · Waste management systems · Vehicle routing
problem

Supported by Fundação para Ciência e a Tecnologia and MIT Portugal Program.

D. E. Simos et al. (Eds.): LION 2022, LNCS 13621, pp. 178–190, 2022.
https://doi.org/10.1007/978-3-031-24866-5_14

1   Introduction

Cities worldwide occupy only 3% of the earth's land area, but they consume 75% of natural resources and produce 60–80% of global greenhouse emissions. Their impact on the environment will become even more expressive since, by 2050, most of the worldwide population (*ca.* 70%) will be concentrated in urban areas [21]. Cities with high population density will require a more innovative sustainable infrastructure to manage the generated wastes [6,25,26]. Solid Waste Management System (SWMS) is a complex system that deals with activities that include waste collection, transportation, handling, and proper disposal [17]. Solid waste generation is currently more than 4 billion tons per year, and management systems globally are not taking advantage of technological development to improve their services. . For urban solid waste collection, companies (or municipal authorities in some countries) deploy a fleet of trucks on predetermined routes to collect the waste from dumpsters. The collection and transportation represent 60–80% of the total SWMS cost [1,15].

The initial strategy adopted to address collection route problems for Bragança, a city located in the Northeast region of Portugal, will be reported in this work. The ongoing project's main goal is to implement a wireless sensors network to collect real-time information regarding the waste level and use these data for collection route optimization. However, before the implementation, a study about how the upper limit of waste in dumpsters (level of waste in which dumpsters will be considered for collection) affects collection efficiency and how dynamic selection of dumpsters to be collected can improve efficiency was necessary. Thus, a GLS algorithm was used to evaluate the effect of the upper limit on the overall performance. Furthermore, a dynamic method for selecting dumpsters to be collected was assessed. Daily waste levels were determined considering the demographic factor of each dumpster as starting approach.

The rest of the paper is organized as follows. Section 2 brings an updated overview of waste collection problems, some examples of waste collection in smart cities, and algorithms used to solve these problems. Section 3 presents the methodology employed, Sect. 4 summarizes the results, and Sect. 5 carries the main findings of the present work and future steps.

2   Related Literature

Literature used here was obtained by exploring the most relevant documents in this area. In brief, keywords "smart cities" and "waste management" were introduced in the Web of Science^{TM} (WoS) search engine and a small database of 552 documents published from 2015 to 2021 was obtained. Using biblioshiny (free tool from bibliometrix R package), "waste collection" cluster was identified and used to narrow documents of interest in this work. The final sub-cluster used to select documents was "vehicle routing problem", and the 29 documents found will be presented in the next sections.

2.1 Waste Collection in Smart Cities

Urban informatization increased significantly last years due to the rapid development and wide applications of Information and Communication Technologies (ICT) in urban areas to assemble future cities, or smart cities. The concept is mainly related to the smarter management of the large population content in cities [19,22,27]. For SWMS in smart cities, the effort has been directed towards solving the waste collection problem using ICT tools. For instance, several works are reporting the utilization of a wireless sensors network to acquire real-time data that helps the decision-making process of scheduling collection routes [24,28].

There are already some cities and countries worldwide leading digital transition in waste management systems with smart waste collection technologies. Brisbane, a town in Australia, signed a cooperation agreement with the local waste management company to install induction devices and develop a strategy to empty them according to remaining space in each bin. In Denmark, the country's largest telecommunications company invested in SmartBin technology to install sensors in waste bins and integrate data in the smart city digital platform through an agreement with Cisco. In Shangai, China, the company responsible for the waste management system created a dynamic management platform covering 16 districts and 82 streets. This platform improved waste collection quality through digitization, refinement, and visualization of waste classification [15].

2.2 Waste Collection Route Problems

Frequently, well-studied vehicle routing problems are used as starting approach for solving waste collection problems. Classic formulations currently used include (but are not limited to) Traveling Salesman Problem (TSP) and Vehicle Routing Problem (VRP).

TSP illustrates the case in which a business traveler needs to visit M customers from surrounding locations K just once, returning to the point of origin. In this problem, the time required to travel from location i to j is defined as t_{ij}, and the objective is to find the shortest path in which these locations should be visited. VRP represents an extending TSP. In this problem, the objective is to find the best route for multiple vehicles visiting multiple locations. Several works use VRP as initial formulation and introduce new constraints to the problem to reach real-case scenarios. From this attempt, more formulations arise, such as Capacitated Vehicle Routing Problem (CVRP), Vehicle Routing Problem With Time Windows (VRPTW), Multi-Depot Vehicle Routing Problem (MDVRP), the Site-Dependent Vehicle Routing Problem (SDVRP), and the Open Vehicle Routing Problem (OVRP) [4].

For solid waste collection, CVRP is frequently used. In CVRP, vehicles have a limited capacity and need to pick up, or deliver, items to customers with a given demand. The problem can be summarized as picking up, or delivering, the items for the minimum cost, never exceeding the vehicle's maximum capacity [13].

2.3   Algorithms for Route Optimization

Algorithms used to solve vehicle routing problem in waste collection range from conventional to meta-heuristics. The first is known to have an easier implementation but lacks in handling uncertain parameters and multi-objective functions so far. However, advances in mathematical programming are allowing the coupling of uncertain parameters with many real-world problems using mixed-integer linear programming (MILP), for example. Nevertheless, MILP has difficulties solving problems demanding large computational efforts and risk of high dimensionality, which can hinder their use to solve large dimension VRP problems. In spite of that, there are interesting works in the literature using this approach. Youseflloo et al. for example used MILP to optimize the total cost of collection routes considering also environmental aspects [30]. Other works used MILP to find the optimal routes for solid waste collection, as the case study approach by Erdinç et al. optimized waste collection routes in the city of Istanbul, Turkey [5].

Most of the algorithms implemented to solve routing problems in waste collection can be divided into bio-inspired algorithms. Each one is developed considering the problem approached, with specific constraints and objectives. Despite that, problems in this area generally focus on reducing the total cost associated with the routes, considering the capacity constraints of both trucks and depots. A great number of works have already explored the use of Ant Colony Optimization [9], Genetic Algorithms [12], Large Neighborhood Search [18], Tabu Search [23], Variable Neighborhood Search [10] among others. There is also a significant number of works in the literature using Geographic Information Systems (GIS) to deal with waste collection [29]. GIS-based algorithms are able to give a strong base to support truck routes, being considered a valuable tool for a better decision-making [11].

The increasing need to deal with environmental problems and achieve highly sustainable solutions has also collaborated to develop algorithms for multi-objective problems. Studies in this approach consider constraints and objectives directly associated to waste collection routes and socioeconomic aspects. For instance, the work presented by Yanfang et. al. used the NSGA-based algorithm to find the best locations for recycling plants considering the obnoxious effect [16]. Another study dealing with smart system for waste collection and recycling waste is presented in Xulong et. al. developing a hybrid algorithm based on whale optimization algorithm and genetic algorithm to solve a bi-objective mathematical programming model [15].

Search algorithms, such as Guided Local Search (GLS), have also proven to be a good solution to various combinatorial optimization problems, including VRP [2,3]. GLS is a meta-heuristic algorithm that operates building penalties during search procedure. The penalty factors are taken into consideration by the objective function when the search is too close to a previously visited local minima. With this strategy, the algorithm is able to migrate to neighborhood solutions and escape local searches [14].

3   Methodology

In brief, locations known to have a dumpster in the city of Bragança were selected to assess the waste collection efficiency considering different upper limit values for waste containers (regular). Daily waste oscillation was approached by analyzing the nearby of each dumpster. After that, a method for dynamic selection of dumpsters for collection routes was applied to evaluate how much this strategy can improve the collection efficiency (dynamic). The overall procedure is illustrated in Fig. 1.

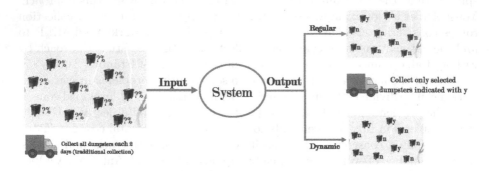

Fig. 1. General procedure approached.

3.1   Problem Assembly

The company in charge of solid waste management in the Northeast region of Portugal is Resíduos do Nordeste (see company website). The company is responsible for the waste management of 13 municipalities and around 134000 citizens, with a production of wastes estimated in 58000 tons/year (2019). In 2019, 7.9% of this amount represented selective waste collection, and the number of waste collected by this means increased 77.6% compared to 2014. Paper is the type of waste most discharged/found in dumpsters, representing 33.5% of the total waste collected in the selective collection (see more details in [20]).

This work will approach the selective collection of paper waste. The population of Bragança represents 27.11% of the collected waste that is selected for recycling. For this reason, the collection region considered for this study has 10 waste dumpsters of paper. The problem can be summarized as a CVRP, where the objective is to minimize total distance traveled for waste collection, reducing total cost and carbon emission. As constraints it is necessary to taking in account that:

– Trucks have limited capacity K_{max}.
– Each route begin and ends at central depot.

- Dumpsters must be served when predicted level of waste reach C_{max}.
- Waste must be transported to central depot when full capacity of truck is reached.
- Trucks only collect one type of waste (*i.e.* paper).

3.2 Level of Dumpsters

Initial levels $(L_{0,j})$ inside the chosen dumpsters $j = 1, ..., k$ were calculated using a uniform distribution probability. Oscillation of waste level was considered constant throughout the days, and calculated based on evaluation of 150 m around each dumpster using the software Google Earth. This evaluation made it possible to calculate a filling velocity that expresses the daily oscillation of waste in percentage for each dumpster [7]. The formula for calculating the filling velocity (fv) of dumpster j, the daily waste level (L), and amount of waste (Lc) in m^3 on the day i of the dumpster j are presented.

$$fv_j = \frac{FA_j}{TA}P_M \wedge L_{i,j} = L_{i-1,j} + fv_j \wedge Lc_{i,j} = L_{i,j}\frac{CD}{100} \tag{1}$$

In which fv_j is the filling velocity of the dumpster j, FA_j represents the filled area with buildings around dumpster j, TA is the total area around each dumpster, P_M is the maximum filling velocity. The $L_{i,j}$ and $Lc_{i,j}$ represents the waste level and amount in day i for dumpster j, respectively. $L_{i-1,j}$ represents the level of waste from the day before $i - 1$, and CD represents total volume of the dumpster. In Fig. 2 collection points are represented (left), as well as one example of how the filled area was determined (right).

Fig. 2. Chosen locations for waste collection (on left) and example of how filled area was determined (on right).

3.3 Dynamic Selection

In this work, GLS algorithm was used to solve CVRP in order to compare distances traveled for waste collection during TDa days. Dumpsters suitable for the collection were chosen based on a daily level. In regular selection of dumpster to be collected, when a given dumpster reached C_{max}, it should be considered for collection.

For the dynamic selection of dumpsters to be collected, upper level was considered as $C_{max} - 2fv_j$. Within this strategy, it is possible to previously collect dumpsters that would be full after 2 days. The parameter CM represents the average maximum capacity of the dumpsters. Algorithm 1 illustrates the complete procedure for dynamic waste collection.

Algorithm 1. Dynamic waste collection

1: $L_{0,j} \leftarrow rand(0, CM - 1)$
2: $fv = [fv_1, ..., fv_k]$
3: **for** i in *days* **do**
4:      **for** j in *dumpsters* **do**
5:          $L_{i,j} = L_{i-1,j} + fv_j$
6:          **if** $L_{i,j} \geq C_{max} - 2fv_j$ **then**
7:              $Lc_{i,j} = L_{i,j}\dfrac{CD}{100}$
8:              $L_{i,j} = 0$
9:              $col_{i,j} = j$
10:         **end if**
11:     **end for**
12: **end for**
13: **for** i in *days* **do**
14:     $DM_i \longleftarrow distance\_matrix(col_i)$
15:     $TD_i, load_i \longleftarrow GLS(DM_i, Lc_{i,j})$
16: **end for**

Dumpsters selected for collection were stored on the list $col_{i,j}$. At last, daily distance matrix DM_i should be calculated considering all dumpsters locations associated to the day i. Also, the amount of waste $Lc_{i,j}$ in m^3 was used as argument by the GLS algorithm $GLS()$. The algorithm returns the total distance traveled daily for collection TD_i and the total load $load_i$ for the day i.

4   Results and Discussion

Generally, authors consider the upper limit for collection (C_{max}) in 70–90%, but few explain the choice [8]. Here, GLS-based algorithm was compiled for values of 70, 80, and 90% for C_{max} to determine which is the best C_{max} value for both regular and dynamic selection method. For comparison purposes, a traditional collection (TC) simulation was also performed considering the best route for one collection of all dumpsters every two days.

A period of 30 days was approached in this study ($TDa = 30$), and collection was modeled with a small fleet (3 trucks). Truck's capacity is $K_{max} = 16m^3$, and total volume of dumpsters is $CD = 2.5m^3$, based on information obtained from

Resíduos do Nordeste. Average maximum capacity for determination of initial level was 80% ($CM = 80$).

4.1 Waste Level Throughout Days

For the determination of daily levels of solid waste, two parameters are important: initial level ($L_{0,j}$) and filling velocities (fv_j). The calculation of fv_j was performed by individual analysis nearby each dumpster and defining P_{max} as 20% per day. Results obtained for these parameters are expressed below in Table 1 together with latitude and longitude of each location. Central depot (location 0) is located in coordinates (41.37233, -7.14233).

Table 1. Definition of $L_{0,j}$, fv_j, and dumpsters location.

Dumpster	$L_{0,}$ (%)	fv (%/day)	Latitude	Longitude
1	43	6.85	41.79323	−6.76898
2	12	5.01	41.79971	−6.76808
3	73	11.80	41.80608	−6.75995
4	54	7.64	41.80049	−6.76400
5	38	5.74	41.79887	−6.77037
6	18	3.50	41.80399	−6.76464
7	76	15.81	41.80294	−6.76206
8	21	10.50	41.80282	−6.77679
9	35	12.28	41.80729	−6.77366
10	19	7.64	41.78781	−6.77335

Waste levels during studied days is directly related to the selection of dumpsters for collection since collected dumpsters have their level reset to 0. In this regard, situations of overfilling might happen depending on the strategy adopted for the waste collection Dumpsters overfilled should be avoided in real scenarios as this could lead to population unsatisfied with collection service and more serious problems with the accumulation of waste in manhole, leading to sewer flooding and disease spread. To study the influence of the chosen strategy on overfilling of dumpsters the percentage of collection days with overfilled dumpsters was determined. Results obtained using 70% and 80% did not showed significant differences, with only one dumpster overfilled (Dumpster 7) 6.67% and 16.67% of collection days using 70% and 80% as upper threshold of waste, and no overfilled dumpster in dynamic approach. Considering 90% as maximum level of waste overcome results with higher differences in regular approach, as shown in Fig. 3.

For dynamic selection using 90% as maximum waste level, no dumpster showed overfilling. On the other hand, most dumpsters were overfilled in regular selection. Higher overfilling is proportional to filling velocities since the

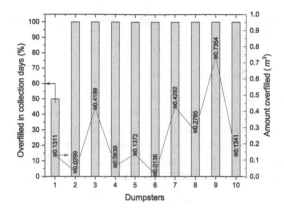

Fig. 3. Overfilling results for regular approach using maximum waste level of 90%.

most overfilled dumpster (9) is the one with highest fv, and the less overfilled dumpster (6) has the lower fv. The absence of overfilled dumpsters in dynamic approach is associated to the early collection performed in this approach, that considers the level after 2 days to select suitable dumpsters for collection.

4.2   Numerical Results Discussion

The analysis of separate results of total distance and load carried can be misleading since the truck can travel longer distances and carry more load, for example. To better evaluate the efficiency, a parameter called here as CE (collection efficiency) was calculated by a simple division between the amount of load (m^3) and traveled distance (km).

In Fig. 4a, CE obtained for regular selection of dumpsters using different upper limits is exhibited. This result demonstrates that 90% is the collection level that returns the highest efficiency. However, traditional collection simulation is still 11.94% more efficient than the highest efficiency obtained using only levels to select dumpsters suitable for collection.

Using dynamic selection of dumpsters for the collection returned the results of CE exhibited in Fig. 4b. Despite already confirming with previous results that 90% as maximum level returns the highest efficiency, simulations for dynamic selection considering 70% and 80% were performed here as proof of concept. Results obtained for CE demonstrated that the strategy can increase collection efficiency even for C_{max} considered less efficient. For instance, efficiencies in dynamic approaches were higher than TC's efficiency, and best result (90% dynamic) was 50.45% higher than TC result.

Daily distances traveled and load carried in best result (90% dynamic) are shown along with the worst results (70% regular) in Fig 5. Graph with daily distances traveled shows that collection days in 70% regular and 90% dynamic have about the same distance traveled (see days 4 and 6, for example). The real improvement can be observed by comparing the amount of collection days and

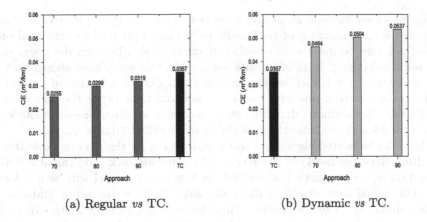

(a) Regular *vs* TC.   (b) Dynamic *vs* TC.

Fig. 4. CE obtained for regular (a) and dynamic (b) selection of dumpsters for collection route *vs TC*.

the daily load carried per collection day. In this regard, 90% dynamic has 58% fewer trips and 112.37% more average load collected per collection day compared to 70% regular. For instance, the average load collected per collection day in 90% dynamic approach is also 43.49% higher with 47% fewer trips compared to traditional collection.

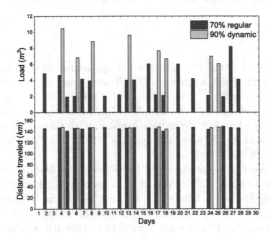

Fig. 5. Total distance traveled and load carried for worst (70% regular) and best (90% dynamic) approach.

5 Conclusions and Future Work

Maximum level of waste in dumpsters to increase collection efficiency was determined to be 90% according to results. In initial tests, the simple selection of dumpsters for collection according to level was not enough to overcome results observed with simulation of traditional collection. This scenario changed once a method considering the forecasting of waste level after two days was used (dynamic selection). In brief, results obtained in this work have shown that (i) the best level for waste collection and (ii) strategies for selection of dumpsters to be collected are a powerful and necessary tool to improve the collection of solid waste. Furthermore, dynamic selection considers 2-days forecast waste level, which significantly reduces the probability of overfilled dumpsters.

More sophisticated algorithms will be considered for the dynamic selection of the dumpsters that need collection (*e.g.* GA) in future work. Additionally, if data regarding waste production is available, a forecasting model will be developed based on neural networks to predict daily waste levels in dumpsters and achieve higher similarity to real scenarios. After developing a suitable methodology for optimizing routes, the last step will be the WSN assembly to collect real-time information regarding waste level inside dumpsters for better decision-making.

Acknowledgements. Adriano Silva was supported by FCT-MIT Portugal PhD grant SFRH/BD/151346/2021, and Thadeu Brito was supported by FCT PhD grant SFRH/BD/08598/2020. This work was financially supported by UIDB/057 57/2020 (CeDRI), UIDB/00690/2020 (CIMO), LA/P/0045/2020 (ALiCE), UID B/500 20/2020, UI- DP/50020/2020 (LSRE-LCM) and funded by national funds through FCT/MCTES (PIDDAC). Jose L. Diaz de Tuesta acknowledges the financial support through the program of Attraction de Talento of Atraccion al Talento of the Comunidad de Madrid (Spain) for the individual research grant 2020-T2/AMB-19836.

References

1. Anagnostopoulos, T., et al.: Challenges and opportunities of waste management in IoT-enabled smart cities: a survey. IEEE Trans. Sustain. Comput. **2**(3), 275–289 (2017). https://doi.org/10.1109/TSUSC.2017.2691049
2. Tirkolaee, E.B., Abbasian, P., Soltani, M., Ghaffarian, S.A.: Developing an applied algorithm for multi-trip vehicle routing problem with time windows in urban waste collection: a case study. Waste Manage. Res. **37**(1), 4–13 (2019). https://doi.org/10.1177/0734242X18807001
3. Barbucha, D.: An agent-based guided local search for the capacited vehicle routing problem. In: O'Shea, J., Nguyen, N.T., Crockett, K., Howlett, R.J., Jain, L.C. (eds.) KES-AMSTA 2011. LNCS (LNAI), vol. 6682, pp. 476–485. Springer, Heidelberg (2011). https://doi.org/10.1007/978-3-642-22000-5_49
4. Elshaer, R., Awad, H.: A taxonomic review of metaheuristic algorithms for solving the vehicle routing problem and its variants. Comput. Indus. Eng. **140**, 106242 (2020). https://doi.org/10.1016/j.cie.2019.106242
5. Erdinç, O., Yetilmezsoy, K., Erenoğlu, A.K., Erdinç, O.: Route optimization of an electric garbage truck fleet for sustainable environmental and energy management. J. Clean. Prod. **234**, 1275–1286 (2019). https://doi.org/10.1016/j.jclepro.2019.06.295
6. Esmaeilian, B., Wang, B., Lewis, K., Duarte, F., Ratti, C., Behdad, S.: The future of waste management in smart and sustainable cities: a review and concept paper. Waste Manage. **81**, 177–195 (2018). https://doi.org/10.1016/j.wasman.2018.09.047

7. Ferreira, J.A.O.: Optimização do Processo de Recolha de Resíduos: Desenvolvimento de Ferramentas de Investigação Operacional para o Problema de Orientação de Equipas com Multi-Restrições. Ph. D. thesis, Universidade do Minho (Portugal) (2015)

8. Ferrer, J., Alba, E.: BIN-CT: Urban waste collection based on predicting the container fill level. Biosystems **186**, 103962 (2019). https://doi.org/10.1016/j.biosystems.2019.04.006

9. Grakova, E., Slaninová, K., Martinovič, J., Křenek, J., Hanzelka, J., Svatoň, V.: Waste collection vehicle routing problem on HPC infrastructure. In: Saeed, K., Homenda, W. (eds.) CISIM 2018. LNCS, vol. 11127, pp. 266–278. Springer, Cham (2018). https://doi.org/10.1007/978-3-319-99954-8_23

10. Gruler, A., Quintero-Araújo, C.L., Calvet, L., Juan, A.A.: Waste collection under uncertainty: a simheuristic based on variable neighbourhood search. Europ. J. Indus. Eng. **11**(2), 228–255 (2017)

11. Hatamleh, R., Jamhawi, M., Al-Kofahi, S., Hijazi, H.: The use of a GIS system as a decision support tool for municipal solid waste management planning: the case study of al Nuzha district, Irbid, Jordan. Procedia Manufact. **44**, 189–196 (2020). https://doi.org/10.1016/j.promfg.2020.02.221

12. Karakatič, S.: Optimizing nonlinear charging times of electric vehicle routing with genetic algorithm. Expert Syst. Appl. **164**, 114039 (2021). https://doi.org/10.1016/j.eswa.2020.114039

13. Khachay, M., Ogorodnikov, Y.: PTAS For the Euclidean capacitated vehicle routing problem with time windows. In: Matsatsinis, N.F., Marinakis, Y., Pardalos, P. (eds.) LION 2019. LNCS, vol. 11968, pp. 224–230. Springer, Cham (2020). https://doi.org/10.1007/978-3-030-38629-0_18

14. Le, T.D.C., Nguyen, D.D., Oláh, J., Pakurár, M.: Optimal vehicle route schedules in picking up and delivering cargo containers considering time windows in logistics distribution networks: a case study. Prod. Eng. Archives **26**(4), 174–184 (2020). https://doi.org/10.30657/pea.2020.26.31

15. Lu, X., Pu, X., Han, X.: Sustainable smart waste classification and collection system: a bi-objective modeling and optimization approach. J. Clean. Prod. **276**, 124183 (2020). https://doi.org/10.1016/j.jclepro.2020.124183

16. Ma, Y., Zhang, W., Feng, C., Lev, B., Li, Z.: A bi-level multi-objective location-routing model for municipal waste management with obnoxious effects. Waste Manage. **135**, 109–121 (2021). https://doi.org/10.1016/j.wasman.2021.08.034

17. Mahmood, I., Idwan, S., Matar, I., Zubairi, J.A.: Experiments in routing vehicles for municipal services. In: 2018 International Conference on High Performance Computing & Simulation (HPCS), pp. 993–999. IEEE (2018). https://doi.org/10.1109/HPCS.2018.00156

18. Mofid-Nakhaee, E., Barzinpour, F.: A multi-compartment capacitated arc routing problem with intermediate facilities for solid waste collection using hybrid adaptive large neighborhood search and whale algorithm. Waste Manage. Res. **37**(1), 38–47 (2019). https://doi.org/10.1177/0734242X18801186

19. Nirde, K., Mulay, P.S., Chaskar, U.M.: IoT based solid waste management system for smart city. In: 2017 International Conference on Intelligent Computing and Control Systems (ICICCS), pp. 666–669. IEEE (2017). https://doi.org/10.1109/ICCONS.2017.8250546

20. do Nordeste, R.: Relatório de Sustentabilidade 2019. https://www.residuosdonordeste.pt/documentos/ (2019)

21. O'Dwyer, E., Pan, I., Acha, S., Shah, N.: Smart energy systems for sustainable smart cities: current developments, trends and future directions. Appl. Energy **237**, 581–597 (2019). https://doi.org/10.1016/j.apenergy.2019.01.024

22. Pardalos, P.M.: Artificial intelligence, machine learning, and optimization tools for smart cities: designing for sustainability. Springer Nature (2022). https://doi.org/10.1007/978-3-030-84459-2

23. Paul, A., Kumar, R.S., Rout, C., Goswami, A.: Designing a multi-depot multi-period vehicle routing problem with time window: hybridization of tabu search and variable neighbourhood search algorithm. Sadhana **46**(3), 1–11 (2021). https://doi.org/10.1007/s12046-021-01693-2

24. Ramalho, M.S., Rossetti, R.J., Cacho, N., Souza, A.: SmartGC: a software architecture for garbage collection in smart cities. Int. J. Bio-Inspired Comput. **16**(2), 79–93 (2020)

25. Rassia, S., Pardalos, P.: Smart city networks. Springer (2017).https://doi.org/10.1007/978-3-319-61313-0

26. Rassia, S.T., Pardalos, P.M.: Sustainable environmental design in architecture: impacts on health, vol. 56. Springer Science & Business Media (2012). https://doi.org/10.1007/978-1-4419-0745-5

27. Rassia, S.T., Pardalos, P.M. (eds.): Future city architecture for optimal living. SOIA, vol. 102. Springer, Cham (2015). https://doi.org/10.1007/978-3-319-15030-7

28. Shah, P.J., Anagnostopoulos, T., Zaslavsky, A., Behdad, S.: A stochastic optimization framework for planning of waste collection and value recovery operations in smart and sustainable cities. Waste Manage. **78**, 104–114 (2018). https://doi.org/10.1016/j.wasman.2018.05.019

29. Vu, H.L., Bolingbroke, D., Ng, K.T.W., Fallah, B.: Assessment of waste characteristics and their impact on GIS vehicle collection route optimization using ANN waste forecasts. Waste Manage. **88**, 118–130 (2019). https://doi.org/10.1016/j.wasman.2019.03.037

30. Yousefloo, A., Babazadeh, R.: Designing an integrated municipal solid waste management network: a case study. J. Clean. Prod. **244**, 118824 (2020). https://doi.org/10.1016/j.jclepro.2019.118824

Single MCMC Chain Parallelisation on Decision Trees

Efthyvoulos Drousiotis[1](✉) and Paul G. Spirakis[2,3]

[1] Department of Electrical Engineering and Electronics, University of Liverpool,
Liverpool L69 3BX, UK
E.Drousiotis@liverpool.ac.uk
[2] Department of Computer Science, University of Liverpool, Liverpool L69 3BX, UK
spirakis@liverpool.ac.uk
[3] Department of Computer Engineering and Informatics, University of Patras,
26504 Patras, Greece

Abstract. Decision trees are highly famous in machine learning and usually acquire state-of-the-art performance. Despite that, well-known variants like CART, ID3, random forest, and boosted trees miss a probabilistic version that encodes prior assumptions about tree structures and shares statistical strength between node parameters. Existing work on Bayesian decision trees depend on Markov Chain Monte Carlo (MCMC), which can be computationally slow, especially on high dimensional data and expensive proposals. In this study, we propose a method to parallelise a single MCMC decision tree chain on an average laptop or personal computer that enables us to reduce its run-time through multi-core processing while the results are statistically identical to conventional sequential implementation. We also calculate the theoretical and practical reduction in run time, which can be obtained utilising our method on multi-processor architectures. Experiments showed that we could achieve 18 times faster running time provided that the serial and the parallel implementation are statistically identical.

Keywords: Parallel algorithms · Machine learning · MCMC decision tree

1 Introduction

In Bayesian statistics, it is a common problem to collect and compute random samples from a probability distribution. Markov Chain Monte Carlo (MCMC) is an intensive technique commonly used to address this problem when direct sampling is often arduous or impossible. MCMC using Bayesian inference is often used to solve problems in biology [13], forensics [12], education [5], and chemistry [6], among other areas making it one of the most widely used algorithms when a collection of samples from a probability distribution is needed. Monte Carlo applications are generally considered embarrassingly parallel since each chain can run independently on two or more independent machines or cores. Despite that, the main problem is that each chain is not embarrassingly parallel, and

D. E. Simos et al. (Eds.): LION 2022, LNCS 13621, pp. 191–204, 2022.
https://doi.org/10.1007/978-3-031-24866-5_15

when the feature space and the proposal are computationally expensive, we can not do much to improve the running time and get results faster. When we have to handle huge state-spaces and complex compound states, it takes significant time for an MCMC simulation to converge on an adequate model not only in terms of the number of iterations required but also the complexity of the calculations occurring in each iteration(such as searching for the best features and tree shape of a decision tree). For example, running an MCMC on a single chain Decision tree for a dataset of 400000 datapoints and 15 features took upwards of 6 h to converge when run on a 2.3 − 5.10 GHz Intel Core i7-10875H. In [3], an approach aiming to parallelise a single chain is presented, and the improvement achieved is at its best 2.2 times faster. The functionality of this kind of solution is therefore limited as in real-time, and life applications run time is critical. The work presented in this paper aims to find methods to significantly reduce the MCMC Decision tree's runtime by emphasising on the implementation of MCMC rather than the statistical algorithm itself. We aim to reduce significantly and up to an order of magnitude the run time of the MCMC Decision Tree on a single laptop or personal computer which is going to make the algorithm widely applicable and suitable for non tecinacl users. The remainder of this paper is organised as follows. Section 2 explains the MCMC in General and the Most Recent Work. Section 3 presents the MCMC in Decision trees. Our method is outlined in Sect. 4, with the possible theoretical improvements. We introduce the case study in which we applied our method and reviewed results in Sect. 5. Section 6 concludes the paper.

2   Markov Chain Monte Carlo in General and Most Recent Work

One of the most widely used algorithms is the Metropolis [9] and its general-isation (see Algorithm 1), the Metropolis-Hastings sampler (MH) [8]. Given a partition of the state vector into components, i.e., $x = (x_1, ..., x_k)$, and that we wish to update the $i_t h$ component, the Metropolis-Hastings update proceeds as follows. We first have a density for producing candidate observation x', such that $x'_i = x_i$, which is denoted by $q(x, x')$. Given the chains ergodic condition, the definition of q is arbitrary, and it has a stationary distribution π which is selected so that the observations may be generated relatively easily. After the new state generation $x' = (x_1, ..., x_{i-1}, x_i, x_{i+1}, ..., x_k)$ from density $q(x, x')$, the new state is accepted or rejected using the Rejections Sampling principle with acceptance probability $\alpha(x, x')$ given by Eq. 1. If the proposed state is rejected, the chain remains in the current state.

It is worth mentioning that acceptance probability in this form is not unique, considering there are many acceptance functions that supplies a chain with the required properties. Nevertheless, Peskun(1973) [10] proved that MH is the optimal one where the proper states are rejected least often, which maximises the statistical efficiency meaning that more samples are collected with fewer iterations.

$$a(\chi, \chi') = \min(1, \frac{\pi(x)}{\pi(x')} \frac{q(\chi|\chi')}{q(\chi'|\chi)}) \tag{1}$$

On a Markov process, the next step depends on the current state, which makes it hard for a single Markov chain to be processed contemporaneously by several processing elements. Byrd [2]. Proposed a method to parallelise a single Markov chain(Multithreading on SMP Architectures), where we consider backup move "B" in a separate thread of execution as it is not possible to determine whether move "A" will be accepted. If "A" is accepted, the backup move B - whether accepted or rejected - must be discarded as it was based upon a now supplanted chain state. If "A" is rejected, control will pass to "B", saving much of the real-time spent considering "A" had "A" and "B" been evaluated sequentially. Of course, we may have as many concurrent threads as desired.

At this point, it is worth mentioning that the single chain parallelisation can become quickly problematic as the efficiency of the parallelisation is not guaranteed, especially for computationally cheap proposal distributions. Also, we need to consider that nowadays, computers make serial computations much faster than in 2008, when the single parallelisable chain was proposed.

Another way of making faster MCMC applications is to reduce the convergence rate by requiring fewer iterations. Metropolis-Coupled MCMC($(MC)^3$) utilised multiple MCMC [1] chains to run at the same time, while one chain is treated as the "cold" where its parameters are set to normal while the other chains are treated as "hot", which are expected to accept the proposed moves. The space will be explored faster through the "hot" chains than the "cold" as they are more possible to make disadvantageous transitions and not to remain at near-optimal solutions. The speedup increased when more chains and cores were added.

Our work is focused on achieving a faster execution time of the MCMC algorithm on Decision trees through multiprocessor architectures. We aim to reduce the number of iterations while the number of samples collected is not affected. Multi-threading on SMP Architectures and $(MC)^3$ differs from our work as the former targets rejected moves as a place for optimisation, and the latter requires communication between the chains. Moreover, the aims are different as $(MC)^3$ expands the combination of the chain, enhancing the possibilities of discovering different solutions and assisting avoid the simulation getting stuck in local optima.

2.1 Probabilistic Trees Packages and Level of Parallelism

Most of the existing probabilistic tree packages are only supported by the R programming language.

BART [4] software included in the CRAN package[1] supports multi threading based on OpenMP, where there are numerous exceptions for operating systems, so it is difficult to generalise. Generally, Microsoft Windows lacks OpenMP detection since the GNU autotools do not natively exist on this platform and

---

[1] https://cran.r-project.org/web/packages/BART/index.html.

Apple macOS since the standard Xcode toolkit is also not provided. The parallel package provides multi-threading via forking, only available in Unix. BART under CRAN uses parallelisation for the predict function and running concurrent chains.

BartMachine[2], which is written in Java and its interface is provided by rJava package, which requires Java Development Kit(JDK), provides multi-threading features similar to BART. BartMachine is recommended only for those users who have a firm grounding in the java language and its tools to upgrade the package and get the best performance out of it. Similar to BART, its parallelisation is based on running concurrent chains.

The rest of the available packages, BayesTree[3],dbarts[4],Bartpy[5],XBART[6] and imptree[7] does not support any kind of parallelisation.

Concurrent chains can not solve the problem of long hours of execution time. For example, if a single chain needs 50 h to execute, 5 chains will still need 50 h if run concurrently. In contrast, in our case, a chain that serially needs 50 h now takes approximately 2 h for each chain. Moreover, we can run concurrent chains where each chain is parallelised. If our implementation is compared to a package like BartMachine and BART, the runtime improvement we achieved is around 18 times faster, and if we compare it with a package that does not offer any parallelisation like most of the existing ones, the run time improvement for 5 chains is around 85 times faster.

3   Markov Chain Monte Carlo in Decision Tree

A decision tree typically starts with a root node, which branches into possible outcomes. Each of those outcomes leads to additional decision nodes, which branch off into other possibilities ending up in leaf nodes. This gives it a treelike shape.

Our model describes the conditional distribution of y given x, where x is a vector of predictors $[x = (x_1, x_2, ..., x_p)]$. The main components of the $tree(T)$ includes the depth of the $tree(d(T))$, θ which is the set of features($k(T)$) and the set of thresholds($c(T)$) for each node, and the possibilities $p(Y|T, \theta, x)$ for each leaf node($L(T)$). If x lies in the region corresponding to the $i_t h$ terminal node, then $y|x$ has distribution $f(y|\theta_i)$, where f represents a parametric family indexed by θ_i. The model is called a probabilistic classification tree, according to the quantitative response y.

[2] https://cran.r-project.org/web/packages/bartMachine/index.html.

[3] https://cran.r-project.org/web/packages/BayesTree/BayesTree.pdf.

[4] https://cran.r-project.org/web/packages/dbarts/index.html.

[5] https://pypi.org/project/bartpy/.

[6] https://jingyuhe.com/xbart.html.

[7] https://cran.r-project.org/web/packages/imptree/index.html.

As Decision Trees are identified by (θ, T), a Bayesian analysis of the problem proceeds by specifying a prior probability distribution $p(\theta, T)$. Because θ indexes the parametric model for each T, it will usually be convenient to use the relationship

$$p(Y_1 :_N, T, \theta | x_1 :_N) = p(Y | T, \theta, x) p(\theta | T) p(T) \tag{2}$$

In our case it is possible to analytically obtain Eq. 2 and calculate the posterior of T as follows:

$$p(Y | T, \theta, x) = \prod_{i=1}^{N} p(Y_i | x_i, T, \theta) \tag{3}$$

$$p(\theta | T) = \prod_{j \in (T)} p(\theta_j | T) = \prod_{j \in (T)} p(k_j | T) p(c_j | k_j, T) \tag{4}$$

$$p(T) = \frac{a}{(1+d)^\beta} \tag{5}$$

Equation 3 describes the product of the probabilities of every data point(Y_i) classified correctly given the datapoints features(x_i), the tree structure(T), and the features/thresholds(θ) on each node on the tree. Equation 4 describes the product of possibilities of picking the specific feature(k) and threshold(c) on every node given the tree structure(T). Equation 5 is used as the prior for tree T_i. This formula is recommended by [4] and three aspects specify it: the probability that a node at depth $d(= 0.1.2....)$ is nonterminal, the parameter $a \in 0, 1$ which controls how likely a node would split, with larger α values increasing the probability of split, and the parameter $\beta > 0$ which controls the number of terminal nodes, with larger values of β reducing the number of terminal nodes. This feature is crucial as this is the penalizing feature of our probabilistic tree which prevents it from overfitting and allowing convergence to the target function $f(X)$ [11], and it puts higher probability on "bushy" trees, those whose terminal nodes do not vary too much in depth.

An exhaustive evaluation of Eq. 2 over all trees T will not be feasible, except in trivially small problems, because of the sheer number of possible trees, which makes it nearly impossible to determine precisely which trees have the largest posterior probability.

Despite these limitations, Metropolis-Hastings algorithms can still be used to explore the posterior. Such algorithms simulate a Markov chain sequence of trees such as:

$$T_0, T_1, T_2, ..., T_n \tag{6}$$

which are converging in distribution to the posterior $p(Y | T, \theta, x) p(\theta | T) p(T)$ in Eq. 2.

Because such a simulated sequence will tend to gravitate toward regions of higher posterior probability, the simulation can be used to search for high-posterior probability trees stochastically. We next describe the details of such algorithms and their implementation.

3.1 Specification of the Metropolis-Hastings Search Algorithm on Decision Trees

The Metropolis-Hastings(MH) algorithm for simulating the Markov chain in Decision trees (see Eq. 7) is defined as follows. Starting with an initial tree T_0, iteratively simulate the transitions from T_i to $T_i + 1$ by these two steps:

1. Generate a candidate value T' with probability distribution $q(T_i, T')$.
2. Set $T_{i+1} = T'$ with probability

$$a(T_i, T') = \min(1, \frac{\pi(Y_1 :_N, T', \theta'|x_1 :_N)}{\pi(Y_1 :_N, T, \theta|x_1 :_N)} \frac{q(T, \theta|T', \theta')}{q(T', \theta'|T, \theta)}) \qquad (7)$$

Otherwise set $T_{i+1} = T_i$.

To implement the algorithm, we need to specify the transition kernel q. We consider kernels $q(T, T')$, which generate T' from T by randomly choosing among four steps:

- Grow(G) : add a new $D(T)$ and choose uniformly a $k(T)$ and a $c(T)$
- Prune(P) : choose uniformly a $D(T)$ to become a leaf
- Change(C) = choose uniformly a $D(T)$ and change randomly a $k(T)$ and a $c(T)$
- Swap(S) = choose uniformly two $D(T)$ and swap their $k(T)$ and $c(T)$

The rules are chosen by picking a number uniformly between 0 and 1 and each action have its own interval. For example, $p(G) = 0.3, p(P) = 0.3, p(C) = 0.2, p(S) = 0.2, [0, 0.3, 0.6, 0.8, 1]$

The probabilities (see Eq. 8) represent the sum of the probabilities of every accepted forward move. P(G), p(P), p(C), p(S) are set by the user who chooses how often each move wants to be proposed.

$$q(T', \theta'|T, \theta) = q(T'|T)q(\theta'|T') = \sum_a q(a)q(T'|T, a)q(\theta'|T', \theta, a) \qquad (8)$$

where :

$$q(G)q(T'|T, G)q(\theta'|T', \theta, G) = p(G) \times \frac{1}{c} \times \frac{1}{k} \times \frac{1}{|L(T)|} \qquad (9)$$

$$q(P)q(T'|T, P)q(\theta'|T', \theta, P) = p(P) \times \frac{1}{|D(T)| - 1} \qquad (10)$$

$$q(C)q(T'|T, C)q(\theta'|T', \theta, c) = p(C) \times \frac{1}{|D(T)|} \times \frac{1}{c} \times \frac{1}{k} \qquad (11)$$

$$q(S)q(T'|T, S)q(\theta'|T', \theta, S) = p(S) \times \frac{1}{(|D(T)|(|D(T)| - 1))/2} \qquad (12)$$

Equation 9 can be described as the possibility of proposing the grow move including the probability of choosing the specific feature(k), threshold(c) and leaf node($|L(T)|$) to grow. P(G) is multiplied by the number of features(k), the unique number of datapoints(c) and the number of leaf nodes($|L(T)|$). For example, given a dataset with 100 unique datapoints(c), 5 features(k), a tree structure(T) with 7 leaf nodes($|L(T)|$) and a $p(G) = 0.3$ Eq. 9 will be $0.3 \times \frac{1}{100} \times \frac{1}{5} \times \frac{1}{7}$.

Equation 10 is the possibility of proposing the prune move, where p(P) is multiplied by the number of decision nodes subtracting one($(|D(T)| - 1)$ we are not allowed to prune the root node). In practise, given a $p(P) = 0.3$ and a tree structure(T) with 7 decision nodes($|D(T)|$) Eq. 10 will be $0.3 \times \frac{1}{10-1}$

Equation 11 is the possibility of proposing the change move where p(C) is multiplied by the number of decision nodes($|D(T)|$), the number of features(k), and the number of unique datapoints(c). For example, given a dataset with with 100 unique datapoints(c), 5 features(k), a tree structure(T) with 12 decision nodes($|D(T)|$) and a $p(G) = 0.2$ Eq. 9 will be $0.2 \times \frac{1}{100} \times \frac{1}{5} \times \frac{1}{12}$.

Equation 12 is the possibility of proposing the swap move where p(S) is multiplied by the number of paired decision nodes($|D(T)|$).In practise, given a tree structure(T) with 12 decision nodes($|D(T)|$) and a $p(S) = 0.2$ Eq. 12 will be $0.2 \times \frac{1}{((12)(12-1))/2}$

Theorem 1. *Transition kernel(see Eq. 13) yields a reversible Markov chain, as every step from T to T' has a counterpart that can move from T' to T.*

$$q(T', \theta'|T, \theta) \tag{13}$$

Proof. Assume a Markov chain, starting from its unique invariant distribution π. Now, take into consideration that for every sample $T_0, T_1, ..., T_n$ have the same joint probability mass function(p.m.f) as their time reversal $T_n, T_{n-1}, ..., T_0$, so as we can call the Markov chain reversible, as well as its invariant distribution π is reversible. This can be explained as a simulation of a reversible chain that looks the same if it runs backward.

The first thing we have to look for is if the Markov chain starts at π, and it can be checked by Eq. 14

$$
\begin{aligned}
&P(T_k = i|T_{k+1} = j, T_{k+2} = i_{k+2}, ..., T_n = i_n) \\
&= \frac{P(T_k = i, T_{k+1} = j, T_{k+2} = i_{k+2}, ..., T_n = i_n)}{P(T_{k+1} = j, T_{k+2} = i_{k+2}, ..., T_n = i_n)} \\
&= \frac{\pi P_{ij} P_{ji_{k+2}} .... P_{i_{n-1}i_n}}{\pi P_{ji_{k+2}} .... P_{i_{n-1}i_n}} \\
&= \frac{\pi P_{ij}}{\pi_j}
\end{aligned} \tag{14}
$$

Equation 14 is only dependent on i and j where this expression for reversibility must be the same as the forward transition probability $P(X = T_{k+1} = i|X = T_k = j) = P_{ji}$. If, both original and the reverse Markov chains have the same transition probabilities, then their p.f.m must be the same as well.

An example for our probabilistic tree is the following:

Assume a tree structure(T) with 5 leaf nodes($|L(T)|$) and 11 decision nodes($|D(T)|$) sampling from a given dataset with 4 features(c) and 100 unique datapoints(k) for each feature.

If for example the forward proposal($q(T', \theta'|T, \theta)$) = ("*change*") with p(C) = 0.2, we end up with the following equation: $0.2 \times \frac{1}{11} \times \frac{1}{4} \times \frac{1}{100}$. At the same time the reverse proposal(going from the current position to the previous) $(q(T, \theta|T', \theta'))$ equation looks exactly the same as the forward proposal. Given the above practical example we have strengthen our proof which shows that $(q(T', \theta'|T, \theta))$ = $(q(T, \theta|T', \theta'))$, which shows in practise the reverse transition kernel nature of our model.

Algorithm 1. The General Metropolis-Hashting Algorithm

Initialize X_0
for $i = 1$ to N do
    sample χ' from $q(\chi'|\chi_{t-1})$
    Calculate $\alpha(\chi_i, \chi_{t-1}) = \min(1, \frac{\pi(x)}{\pi(x')} \frac{q(\chi|\chi')}{q(\chi'|\chi)})$
    Draw u from $u[0, 1]$
    if $u < \alpha(\chi_i|\chi_{t-1})$ then
        $\chi_i = \chi'$
    else
        $\chi_i = \chi_{t-1}$
    end if
end for

---

4   Parallelising a Single Decision Tree MCMC Chain

Given a Decision's Tree MCMC chain with N iterations, we propose a method that utilises C number of cores aiming to enhance the running time of a single chain by at least an order of magnitude. As stated in Section 1 and Section 2, at each iteration, a new sample χ' is drawn from the proposal distribution. Our method requires sampling from C number of cores, $S(C = S)$ number of samples in parallel. We then accept the sample with the greatest $a(T_i, T')$ and repeat the same method until the Markov Chain converges to a stationary distribution. In our method, we check the Markov chain convergence when the F1-score fluctuates less than $\pm 3\%$ for at least 100 iterations. Once the Chain has converged, we proceed to the second phase of our method. We now keep producing samples using C cores, but we can now collect more than one sample which satisfies $a(T_i, T') >= u$. (u is a random uniform number$[0, 1]$), otherwise we collect T_i From this point, we will propose new samples from the sample with the greatest $a(T_i, T')$ until we are happy with the number of samples collected. Using this method, we can collect the same number of samples and explore the feature space as effectively as the serial implementation, but 18 times faster using an average laptop or personal computer.

Our algorithm reduces the number of iterations and explores the feature space faster as we use more cores. This provides us with a significant run time improvement up to 18 times faster when the feature space is big and the proposal is expensive. The following sections will evaluate the running time improvement and the quality of the samples produced.

Algorithm 2. Single Chain parallelisation on MCMC Decision Trees

---
Initialize T_0
for $i = 1$ to N **do**
    sample C number of T' from $Q(T^{j'}|T^j{}_{t-1})$
    Calculate in parallel $\alpha(T^j{}_i|T^j{}_{t-1}) = \min(1, \frac{p(Y_{1:N},T',\theta'|x_{1:N})}{p(Y_{1:N},T,\theta|x_{1:})} \frac{q(T,\theta|T',\theta')}{q(T',\theta'|T,\theta)})$ for each sample
    Store every sampled posterior $\alpha(T^j{}_t|T^j{}_{t-1})$ value
    Until converge $T' = \max \alpha(T^j{}_i|T^j{}_{t-1})$
    if Markov Chain converged **then**
        Draw u from $uniform[0,1]$
        for $j = 1$ to j **do**   ▷ For loop run in parallel
            if $\alpha(T^j{}_i|T^j{}_{t-1}) > u$ **then**
                Collect Sample T'
                $T' = \max \alpha(T^j{}_i|T^j{}_{t-1})$
            else
                Collect Sample T
            end if
        end for
    end if
end for

Theoretical Gains. Using C cores simultaneously, the programme cycle consists of repeated "steps," each performing the equivalent of between 1 and n iterations. We need to calculate the number of iterations based on the acceptance rate to produce the same number of samples(S) when we increase C. The moves are considered in parallel, where they are accepted or rejected. Given that the average probability of a single arbitrary move being rejected is p_r, the probability of the i_{th} in every single concurrent core is pr. This step continues for i iterations where Eqs. 15, 16, and 17 show the iterations needed, run time, and speedup improvement, respectively, given a time(t) in minutes for each iteration. Theoretical speedup (see Fig. 1) were plotted for varying cores.

$$i = \frac{S/C}{pr} \tag{15}$$

$$Runtime = \frac{i}{t} \tag{16}$$

$$Speedup = \frac{Runtime}{C} \tag{17}$$

For example, given a single MCMC Decision Tree chain running for 10000 iterations and an acceptance rate of 70%, after 3000 burn-in iterations, we end up with 4900 samples.

For the parallel MCMC chain, given the same settings as the serial one, with a 30% burn-in period and 25 cores, we will collect the same number of samples with 500 iterations. This provides us with a 25 times faster execution time. Algorithm 2 indicates the part implemented on a parallel environment, and Fig. 3 the maximum theoretical benefits from utilising our method. Considering the communication overhead, the parts of the algorithm that are not parallelised, and the fact that the cores does not receive constant utilisation, in practise the speedups of this order are not achievable. Therefore, we will test it in practise and find out how it performs on real-life scenarios respect to the accuracy and the runtime improvement

5 Results

5.1 Quality of the Samples Between Serial and Parallel Implementation

We have used the Wine dataset from scikit learn datasets[8] repository as well as Pima Indians Diabetes and Dermatology from UCI machine learning repository[9], which are publicly available, to examine the quality of the samples on several testing hypotheses, including the different number of cores per iteration given the average F1-Score. Precision and Recall were also calculated for more depth and detailed insights about the performance and quality of the samples. The results, including the F1-Score, Precision, Recall, and Accuracy(see Tables 2, 3 and 4) produced through 25-Fold Cross-Validation, ensure that every observation from the original dataset has the possibility of appearing in the training and test sets and also reduce any statistical error. Before any performance comparison, we need to examine whether the samples produced for each test case(25 cores and 40 cores) have any statistical difference from the serial implementation. We next examine if extracted samples by utilising 25 and 40 cores are representative of the family of the data they come. We use as ground truth data the F1-scores on each sample collected on every fold of the serial implementation, and we compare them with the corresponding collected samples from the other two test cases. In order to check any statistical difference between the samples, we performed the two-sample t-test for unpaired data [7], which is defined as follows:

$$T = \frac{Y_1 - Y_2}{\sqrt{\frac{s_1^2}{N_1} + \frac{s_2^2}{N_2}}} \tag{18}$$

---

[8] https://scikit-learn.org/stable/datasets/toydataset.html.
[9] https://archive.ics.uci.edu/ml/index.php.

$$|T| > t_{1-a/2,\nu} where : \nu = \frac{\left(\frac{s_1^2}{n_1} + \frac{s_2^2}{n_2}\right)^2}{\frac{(s_1^2/n_1)^2}{n_1 - 1} + \frac{(s_2^2/n_2)^2}{n_2 - 1}} \quad (19)$$

Formula 18 is used to calculate the t-Test statistic equation where N_1 and N_2 are the sample sizes, Y_1 and Y_2 are the sample means, and s_1^2 and s_2^2 are the sample variances. The null hypothesis is rejected when Eq. 19 holds, which is the critical value of the t distribution with ν degrees of freedom.

For our first dataset(Wine), we first examine the serial implementation with the parallel using 25 cores. The absolute value of the t-Test, 0.62, is less than the critical value of 1.964, so we prove the null hypothesis and conclude that the samples drawn by using 25 cores have not any statistical difference at the 0.05 significance level. We then compare the serial implementation with the parallel using 40 cores. In this case, the absolute value of the t-Test, 0.63, is less than the critical value of 1.964, so we prove the null hypothesis and conclude that the samples drawn by using 40 cores have not any statistical difference at the 0.05 significance level. We also examine the parallel implementations between them(using 25 and 40 cores accordingly). In this case, the absolute value of t-Test, 0.64, is less than the critical value of 1.964, so we reject the null hypothesis and conclude that the samples drawn using 40 cores(in comparison with the samples drawn with 25 cores) have not a statistical difference at the 0.05 significance level. The results for the rest of the datasets are presented explicitly in Table 1.

Table 1. Datasets critical values

Datasets	1 vs 25 cores	1 vs 40 cores	25 vs 40 cores	Critical T value
Pima Indians Diabetes	0.51	0.63	0.63	1.97
Dermatology	0.73	0.77	0.77	1.97
Wine	0.62	0.64	0.65	1.97

T-test proves that if we use up to 40 cores for sampling(rarely laptops and personal computers have more than 40cores), the quality of the samples are the same, ending up with statistically same samples as the serial implementation. Tables 1, 2, and 3 shows that when we sample in parallel using up to 40 cores, the accuracy and the F1-score remain on the same levels as the serial implementation. Tables 1, 2, 3, 4 indicate that a single chain on MCMC Decision Trees can not be an embarrassingly parallel algorithm as we can only improve the running time of a single chain by utilising a specific number of cores. The running time improvement we achieved($\times 18$ faster) is the maximum run-time enhancement we can achieve on an MCMC decision tree to maintain the high metrics produced by the serial implementation. If we try to extract up to 40 samples per iteration, it is highly probable to get samples that does not affect the final results negatively. According to our results, the maximum number of cores can that be used is 40. Furthermore, Precision, Recall, and F1-score metrics indicate that no overfit is observed even when more than 3 labels exist, proving the samples' quality even in multi class classification problems.

Table 2. Results for wine dataset

Labels	Precision			Recall			F1-score		
	1 core	20cores	40cores	1 core	20cores	40cores	1 core	20cores	40cores
0	0.85	0.88	0.88	1.00	1.00	1.00	0.92	0.94	0.94
1	1.00	1.00	1.00	0.78	0.78	0.78	0.88	0.88	0.88
2	1.00	0.93	0.93	1.00	1.00	1.00	1.00	0.96	0.96
Accuracy							**0.93**	**0.93**	**0.93**

Table 3. Results for Pima Indian Diabetes Datasets

Labels	Precision			Recall			F1-score		
	1 core	20cores	40cores	1 core	20cores	40cores	1 core	20cores	40cores
0	0.83	0.84	0.84	0.86	0.83	0.83	0.84	0.84	0.84
1	0.65	0.63	0.63	0.61	0.65	0.86	0.63	0.64	0.64
Accuracy							**0.78**	**0.78**	**0.78**

Table 4. Results for dermatology datasets

Labels	Precision			Recall			F1-score		
	1 core	20cores	40cores	1 core	20cores	40cores	1 core	20cores	40cores
0	0.93	0.93	0.93	1.00	1.00	1.00	0.97	0.97	0.97
1	0.94	0.94	0.93	1.00	1.00	1.00	0.97	0.97	0.96
2	1.00	1.00	1.00	0.96	0.96	0.96	0.98	0.98	1.00
3	0.94	0.94	0.92	0.94	0.94	0.95	0.94	0.94	0.95
4	1.00	1.00	1.00	1.00	1.00	1.00	1.00	1.00	1.00
5	1.00	1.00	1.00	0.67	0.67	0.67	0.80	0.80	0.80
Accuracy							**0.96**	**0.96**	**0.96**

5.2 Practical Gains

Figure 1 presents the theoretical and practical speedup achieved given the number of cores used demonstrating a remarkable runtime improvement, especially when the feature space is ample and the proposal is expensive. Figure 1 demonstrates that in practise, theoretical speedups of this order can not be achieved for various reasons, including communications overhead, and as well as the cores do not receive constant utilisation. The practical improvement achieved used the novel method we proposed, speeds up the process up to 18 times depending on the number of cores the user may choose to utilise. Moreover, Fig. 1 demonstrates that even if we use more than 25 cores, the speedup achieved is the same, because of the architecture of the cores. When we run on a local machine(laptop or personal computer) a medium size dataset(500,000 entries), the memory is not enough to run every single parallel tree on a separate core. Given that, we have to wait for a core to finish the task, in order to allocate its memory to another core. Given that, we end up that it is not always beneficial to use more cores, as faster execution time and speedup is not guaranteed. Scaling up to 25 cores is the ideal, having in mind that any number above that, might not benefit the run time. To the best of our knowledge it is the first time where a single chain in general, specifically on decision trees, is parallelised with our proposed method.

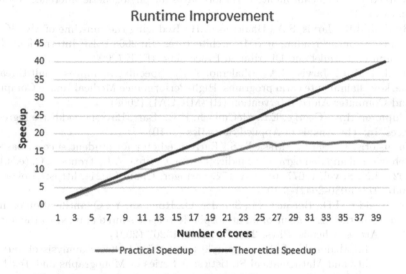

Fig. 1. Speedup achieved by utilising different number of cores

6   Conclusion

Our novel proposed method for parallelising a single MCMC Decision tree chain takes advantage of multicore machines without altering any properties of the Markov Chain. Moreover, our method can be easily and safely used in conjunction with other parallelisation strategies, i.e., where each parallel chain can be processed on a separate machine, each being sped up using our method.

Furthermore, our approach can be applied to any MCMC Decision tree algorithm which needs to process hundreds of thousands of data given an expensive proposal where an execution time of 18 times faster can be easily achieved. As multicore technology improves, CPUs with multiple processing cores will provide speed-ups closer to the theoretical limit calculated. By taking advantage of the improvements in modern processor designs our method can help make the use of MCMC Decision tree-based solutions more productive and increasingly applicable to a broader range of applications. Future work includes on expanding our method on a High Performance Computer(HPC), servers, and cloud which are build for this kind of tasks to compare and demonstrate possible runtime improvements, and discover the merits of such technologies. Moreover, we plan to implement more MCMC single chain parallelisation techniques, including data partitioning, and conduct experiments with various size and shapes datasets, to find the most effective technique, given the type and shape of the dataset.

References

1. Altekar, G., Dwarkadas, S., Huelsenbeck, J.P., Ronquist, F.: Parallel metropolis coupled markov chain monte carlo for Bayesian phylogenetic inference. Bioinformatics (2004)
2. Byrd, J.M.R., Jarvis, S.A., Bhalerao, A.H.: Reducing the run-time of MCMC programs by multithreading on SMP architectures. In: 2008 IEEE International Symposium on Parallel and Distributed Processing. IEEE (2008)
3. Byrd, J.M.R., Jarvis, S.A., Bhalerao, A.H.: Speculative moves: multithreading markov chain monte carlo programs. High-Performance Medical Image Computing and Computer Aided Intervention (HP-MICCAI) (2008)
4. Chipman, H.A., George, E.I., McCulloch, R.E.: Bart: Bayesian additive regression trees. In: The Annals of Applied Statistics (2010)
5. Drousiotis, E., Shi, L., Maskell, S.: Early predictor for student success based on behavioural and demographic indicators. In: Cristea, A.I., Troussas, C. (eds.) ITS 2021. LNCS, vol. 12677, pp. 161–172. Springer, Cham (2021). https://doi.org/10.1007/978-3-030-80421-3_19
6. Le Brazidec, J.D., Bocquet, M., Saunier, O., Roustan, Y.: Quantification of uncertainties in the assessment of an atmospheric release source applied to the autumn 2017. Atmos. Chemis. Phys. 21(17), 13247–13267 (2021)
7. Fisher, L., Mcdonald, J.: 3-two-sample t-test. Fixed effects analysis of variance. Probability and Mathematical Statistics: A Series of Monographs and Textbooks (1978)
8. Hastings, W.K.: Monte carlo sampling methods using Markov chains and their applications (1970)
9. Metropolis, N., Rosenbluth, A.W., Rosenbluth, M.N., Teller, A.H., Teller, E.: Equation of state calculations by fast computing machines. J. Chem. Phys. 21(6), 1087–1092 (1953)
10. Peskun, P.H.: Optimum monte-carlo sampling using Markov chains. Biometrika 60(3), 607–612 (1973)
11. Ročková, V., Saha, E.: On theory for Bart. In: The 22nd International Conference on Artificial Intelligence and Statistics. PMLR (2019)
12. Taylor, D., Bright, J.-A., Buckleton, J.: Interpreting forensic DNA profiling evidence without specifying the number of contributors. Forensic Sci. Int. Genet. 13, 269–280 (2014)
13. Valderrama-Bahamóndez, G.I., Fröhlich, H.: MCMC techniques for parameter estimation of ODE based models in systems biology. Front. Appl. Math. Stat. 5, 55 (2019)

An Extension of NSGA-II for Scaling up Multi-objective Spatial Zoning Optimization

Mohadese Basirati[1,2(✉)], Romain Billot[1], and Patrick Meyer[1]

[1] IMT Atlantique, Lab-STICC, UMR CNRS 6285, F-29238 Brest, France
{mohadese.basirati,romain.billot,patrick.meyer}@imt-atlantique.fr
[2] Mines Saint-Etienne, Univ Clermont Auvergne, INP Clermont Auvergne, CNRS,
UMR 6158 LIMOS, Saint-Etienne, France

Abstract. Among decision problems in spatial management planning, marine spatial planning (MSP) has lately gained popularity. One of the difficulties in MSP is to determine the best place for a new activity while taking into account the locations of current activities. This paper presents the results of the extension of one multi-objective evolutionary-based algorithm (MOEA), non-dominated sorting genetic algorithm-II (NSGA-II) solved the multi-objective spatial zoning optimization problem. The proposed algorithm aims to maximize the interest of the area of the zone dedicated to the new activity while maximizing its spatial compactness. The extended NSGA-II, unlike the traditional one, makes use of a different stop condition, four crossover operators, three mutation operators, and repairing operators. This algorithm is developed for the raster data and it computes solutions for the multi-objective spatial zoning optimization model at a large scale. The proposed NSGA-II has revealed a good performance in comparison with the exact method tested on a small scale. To improve the performance of the algorithm, its parameters are calibrated and tuned using the Multi-Response Surface Methodology (MRSM) method. Analysis of variance (ANOVA) was used to determine the effective and non-effective factors and correctness of the regression models. Finally, conclusions are made and future research works are recommended.

Keywords: Multi-objective spatial zoning optimization · Evolutionary algorithms · NSGA-II · Multi-response surface methodology · Marine spatial planning · Raster

1 Introduction

A wide range of decision problems in the spatial planning strategy, such as land-use planning [6], and marine spatial planning (MSP) [2] make use of spatial data. Formally, MSP decision problems are optimization problems in which we must identify the optimal position and form of a spatial region for a new activity, given certain restrictions. Due to its complexity, the problem is formulated as a multi-objective optimization problem (MOOP), resulting into a group of solutions known as Pareto optimal solutions, rather than a single solution [1,3]. In a

© Springer Nature Switzerland AG 2022
D. E. Simos et al. (Eds.): LION 2022, LNCS 13621, pp. 205–219, 2022.
https://doi.org/10.1007/978-3-031-24866-5_16

recent paper, an exact mathematical zoning model for MSP is proposed as a Multi-Objective Integer Linear Program [2]. Nevertheless, most of the linear models could not be scaled up due to computational hardness [9]. Likewise, many industrial and scientific optimization problems are intractable in terms of computing optimal solutions. In reality, "good" results provided using heuristic or meta-heuristic algorithms are frequently sufficient [13].

The contributions of this paper can be summarized as follows:

1. To address the computational complexity issue of the exact method for the large-scale raster-based zoning problem in MSP, we proposed a new version of a population-based MOEA (NSGA-II) which is a Pareto approach. Innovations are proposed for initialization process, stopping condition, chromosome encoding, crossover, mutation, check and repair operators, constraint handling strategies, and algorithm structure on raster data. The proposed NSGA-II is used to simultaneously optimize the interestingness and compactness objectives of the new activity's zone.
2. Multi-Response Surface Methodology for parameters tuning: we set up a design experiment (DOE) as Box-Behnken design(BBD) for the algorithm which implements a multi-response regression model for three different map sizes of the problem in order to determine the optimal value of the algorithm parameters. Moreover, the effectiveness of all models are validated by Analysis of Variance (ANOVA).

The article is structured as follows. In Sect. 2 we recall the problem setting recently published. In Sect. 3, we present the NSGA-II algorithm for the problem at hand to find the optimal solutions. In Sect. 4, we describe an experimental setting to solve the proposed algorithm, while in Sect. 5, we propose the computational results for artificially generated synthetic instances. Conclusions are drawn in Sect. 6.

2   Related Work

As stated in the recent paper [2], one of the sub-topics of MSP is the zoning problem: the objective is to identify the ideal zone of the new activity while a certain number of human activities already exist and are considered unchangeable in a specified maritime region.

Figure 1 recalls the problem definition introduced in [2] in a fictive maritime area along with all its various elements.

As shown in Fig. 1 these existing activities are shipping lanes, ports, and restricted areas. The goal of this problem is to find the best place for a new activity that optimizes both its interest and compactness. In the meanwhile, it must observe the minimum and maximum distances from existing activities without overlapping with them. For example, zone A is quite compact and interesting, while meeting all minimal distance constraints. However, area B disobeys a maximal distance constraint to the closest port $(d_{p'}^{\leq})$ and a minimal distance to the restricted windmill farm $(d_{r'}^{\geq})$. While it is a completely compact zone but less

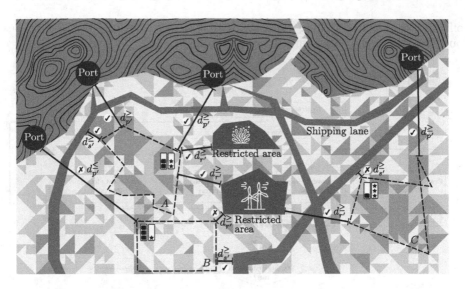

Fig. 1. Problem definition [2]

interesting. Likewise, the area C violates the minimal distance constraint to the shipping lane ($d_{s'}^{\geq}$), whereas it is less compact but utterly interesting. Finally, as a result of these mentioned restrictions, among all three proposed zones, only the area A is acceptable with respect to simultaneously objectives and constraints. We refer the reader to [2] for a comprehensive formulation and exact resolution of the problem, while the rest of this paper is dedicated to the development of a meta-heuristic approach able to cope with large-scale problems.

The starting point in our problem is spatial data in which the information describes different elements along with their locations on or near the earth's surface. In this case, we chose to use raster data presented as a regular grid of cells or pixels. All related formulations are compiled with this data structure.

3   Problem Resolution

3.1   Non-dominated Sorting Genetic Algorithm-II (NSGA-II)

High computational cost and hardness limitations are the main issues in the exact solvers for generating the exact solution sets. To cope with this issue, implementing MOEAs such as single solution-based or population-based multi-objective meta-heuristic algorithms can be applied. NSGA-II as a population-based algorithms, is selected and implemented for this problem. However, to enhance diversification and intensification of the proposed solutions, we selected the population-based algorithm, NSGA-II [4]. Two key challenges in MOOPs are (1) computing complexity and (2) non-elitism approach. To deal with the computational complexity, NSGA-II employs a fast non-dominated sorting strategy. To address constrained multi-objective optimization problems, NSGA-II employs an effective constraint-handling approach. The flowchart shown in Fig. 2, proposes our NSGA-II architecture for solving the problem.

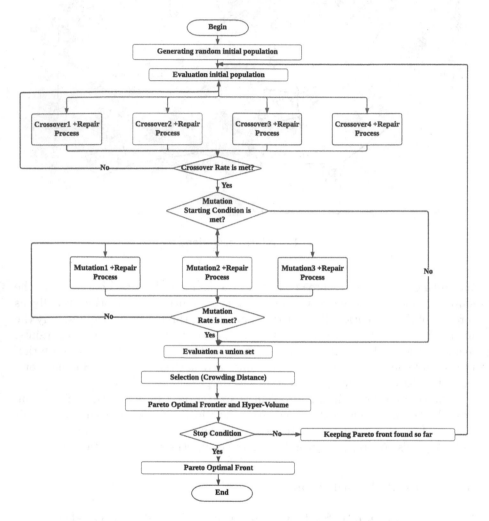

Fig. 2. The flowchart of innovative structure of the proposed NSGA-II for SZOP

During the search process, the proposed NSGA-II uses a different initialization process, stop criterion, four crossover operators, and three mutation operators as opposed to the traditional NSGA-II [5]. In addition, offspring chromosomes, which are generated by the four crossovers, and three mutations, can have a chance to compete with parent chromosomes for survival from one generation to the next. Moreover, the proposed NSGA-II has two check and repair mechanisms to automatically prevent its search process to get stuck in the local optima or infeasible solutions. In other words, the proposed NSGA-II is capable of preventing repetitive solutions by generating solutions with different structures and not throwing away the non-feasible solutions: it is able to repair those which need a minor modification to make them feasible/acceptable. It should be noted that the two notations p_m, and N_{pop} in Fig. 2 represent (1) the probability of mutation, and (2) the population size, respectively. In addition, the details of the proposed NSGA-II components will be presented in the subsequent sections.

In this flowchart, after generating the random initial population including the feasible different zones with respect to the solution and population size, the current zones set are evaluated to make a first memory of the objective functions. By doing so, the main loop of this algorithm begins. By considering the stop criterion, at the first step, one out of four crossovers is selected randomly to generate non-iterative and acceptable offspring and this process repeats until reaching the crossover rate. Afterwards, if the mutation probability is met, the same process in a similar way as crossover is carried out for three mutation operators to make offspring. Next, all populations and achieved offspring combined to make a union set are evaluated. In this step, because the size of a union set should be the same as that of the initial population, all zones may not be in a union set. As a result, crowding sorting is used to complete it by adding an incomplete front in the crowding distance technique, in which the required population is created by the top of the front elements without sacrificing good solutions (elitism). Moreover, to calculate the stop criterion, which is the number of non-improved Hypervolume values, it needs to achieve the Pareto front in each iteration. Therefore, in case of not meeting the stop criterion, the final selected solution set is stored and gives rise to the next generation. Otherwise, the final Pareto front is reported.

3.2   Solution Encoding Schema

The structure of the solution representation is one of the first steps in the successful implementation of meta-heuristic algorithms. For our problem, the constraints are as follows:

- Each solution should have a fixed number of cells (the required solution size).
- Each solution should be without any hole.
- Each solution should be in the feasible area without crossing other existing activities.
- The shape and structure of each solution should be compact without any interruption.

With respect to the mentioned elements, the representation form is a multi-dimensional matrix, $N_{pop} \times [1 \times solution\_size]$ including N_{pop} arrays(chromosomes) $[1 \times solution\_size]$. Each chromosome includes the cell coordinates (x,y) of the proposed new activity zone in the related problem.

3.3   The Initialization Operators

Generating a well-initialized population could reduce the convergence rate to reach the optimal Pareto Front in less iterations. The used algorithm for generating the random population is like circle filling on a grid by bounding box in which instead of checking distances on the entire possible grids, we are saving a lot of time by checking a much smaller area without looking at the rest of the grid. This algorithm is able to generate compact zones with enough diversification which meet all constraints mentioned in the Sect. 3.2. The steps of this algorithm could be summarized as follows:

1. Defining the square bounding box
2. Gathering all cells inside of this box
3. Selecting those cells which meet all following conditions
   - Being inside of the circle starting with radius 1 (if it is less or equal than the radius then mark it)
   - Being feasible
   - Not being repetitive
4. Repeating Step 3 by reaching the upper-bound solution size and increasing the radius in each iteration up to the max predefined radius (8)

3.4   Crossover Operators

Single-Point Vertical Cutting Crossover (Crossover-1). To thoroughly explore the search space of the problem, the proposed NSGA-II employs four crossovers, namely, crossover 1, crossover 2, crossover 3, and crossover 4, applied to three different parts of the chromosome. For each crossover, we need to select two parents as the inputs for this operator. Therefore binary tournament selection [5] is used to select two selective parents. As shown in Fig. 3, two parents which are representing two zones along with their encoded chromosomes are in purple and yellow. Next, as can be seen in Fig. 3, one cut-point cell is randomly selected by the length of each chromosome. Next, the middle cell of these two cut-point cells is made, which is called "**C**" in red. By doing so, each parent is divided into three parts; before cut-point, cut-point, and after the cut-point, respectively. Initially, both parents are sorted based on x-coordinate, hence this division is done vertically, that is why this crossover is called "**single-point vertical cutting crossover**". Having found the middle cell, the two other parts are vertically swapped and transformed into the new center point, that is, "**left-hand**" side of "**Parent-1**" and "**right-hand**" side of "**Parent-2**" are shifted to the middle cell that forms the "**Offspring-1**". On the other side, the "**left-hand**" side of "**Parent-2**" and the "**right-hand**" side of "**Parent-1**" are replaced with the same middle cell, producing another offspring, "**Offspring-2**".

Single-Point Horizontal Cutting Crossover (Crossover-2). In this crossover, unlike crossover-1, the cutting direction is changed from vertical to horizontal. In other words, two parent chromosomes are sorted based on y-coordinate.

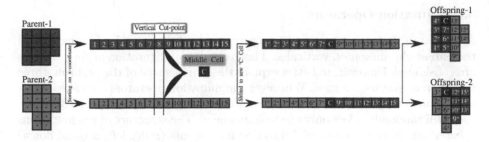

Fig. 3. Single-point vertical cutting crossover (Crossover-1)

Semi-Proportional Vertical Single-Point Cutting Crossover (Crossover-3). By implementing the first two crossovers, the main focus is making well-diversified offspring from the structure point of view in the parent neighbourhoods. However, in the next two crossovers, the aim is to generate well-diversified located offspring in the adjacency of parents. Therefore, the cutting type in crossovers 3 and 4 is the same as crossover-1 (vertical single-point cutting). Nevertheless, after randomly selecting the cutting cells in the parents, the new rule is applied to find the cell "**C**" in the offspring. The approach to find the cell "**C**" is based on the first objective function, namely, the interestingness value of parents.

In Crossover-3, an indicator called "**proportion** is used, which calculates the difference between the first objective function values of two parent chromosomes. Three different scenarios are considered for the proportion value as follows:

1. **Zero proportion**: when the first objective function value of both parents is equal, the middle cell in the distance between the selected cutting cells of parents $(1/2 \times A)$ is chosen as a new "C" cell.
2. **Negative proportion**: when the first objective function value of "**Parent-1**" is less than "**Parent-2**", a new "C" cell is pulled toward the "**Parent-2**" located in the distance $(2/3 \times A)$ from the "**Parent-1**" making one of the offspring. However, the other offspring is generated as before in the middle distance.
3. **Positive proportion**: The movement direction of the "C" cell here is exactly in contrary with the negative proportion.

Full-Proportional Vertical Single-Point Cutting Crossover (Crossover-4). The only difference between crossover 3 and 4 is that in case of positive and negative proportions, both offspring intend to get closer toward the parent with higher objective function. Therefore, we could name it as full-proportional vertical single-point cutting crossover. All 4 crossovers are implemented iteratively through a loop. In each iteration, the check and repair operators check the feasibility of the offspring (explained in Sect. 3.6). If each offspring is validated, it will be added to the list of offspring. This insertion will continue until the crossover rate is reached.

3.5   Mutation Operators

As shown in the crossover operators, to exploit well-diversified solutions around the parent chromosomes, vertical and horizontal reconfiguration of the solutions are considered. However, to better explore the search space of the problem, three mutation operators are used. Whenever the mutation operators are called, after a predefined number of iterations during the main loop of the proposed MOEAs, they will randomly select only one to implement. The structure of the mutations is based on the combination of 4-direction movements (right, left, up, and down) and rotational symmetry of the solution through the solution space (90°, 180°, 270°). Like the crossover operators in Sect. 3.4, the inputs of each mutation are two parent chromosomes (binary tournament selection).

In mutation 1, each chromosome is moved to the other part of the search space, where the moving step between the list of 4-directions is randomly selected for each parent. Meanwhile, on the one hand, in mutation 2, in addition to 4-direction movement, each gen (x,y) coordinate in each parent chromosome should be rotated counterclockwise by a given angle around a given origin. On the other hand, in mutation 3, each parent chromosome is only influenced by the rotation process. After checking the feasibility and repairing the output of each mutation, there could be at maximum two offspring chromosomes.

3.6   Check and Repair Operators

Some issues could happen in the making process of the solutions. One issue is generating the new solutions in the feasible solution space neither outside of that nor crossing with the existing activities. The other issue is reproducing the solutions well compact (*i.e.* without any hole). To solve these issues, two different check and repair operators are used, called **"check-and-repair"** and **"compacity-improver"**. For the first issue, three different scenarios and one scenario for the second issue could happen as follows:

– Scenarios related to the first issue:
  1. The chromosomes are outside of the solution space (map)
  2. The chromosomes are inside of the solution space but have overlap with the existing activities.
– Scenarios related to the second issue:
  1. The chromosomes include holes or interruptions.

For each mentioned scenario some solutions are suggested hereafter;

– The proposed approach for the first issue using **"check-and-repair"** operator:
  1. Generating and replacing a totally new chromosome by a random population generator.

2. Counting the number of overlapped cells; if less than 5, the searching process begins to find alternatives in their adjacent cells while keeping or improving the compacity; otherwise, it is totally deleted. All feasible and non-iterative 4-direction ((-1, 0), (0, 1), (0, -1), (1, 0)) neighbours of the overlapped cell are gathered.

- The proposed approach for the second issue using "**compacity-improver**" operator:
  1. Bounding the "0-1" solution matrix by value "2" .
  2. Exploring the rows and columns without any "1" or any single "0" fully surrounded by "1".
  3. Removing the found zero rows/columns and exchanging the single surrounded "0" element with one of the feasible "1" in the outer layer of the matrix.

These two repairing operators are implemented in the body of all crossovers and mutations, whenever they are called in the main loop of the NSGA-II.

3.7 Evaluation and Selection Operators

Fitness value of each chromosome (quality of a solution) is calculated by two different objective functions. The first objective function is calculated by summing the interest value of zone's cells, and the compacity value is calculated by the representative method for the measurement of compactness called the "**Normalized Discrete Compactness (NDC)** measure, suggested by [15].

Through the selection process, each time, the population is sorted into different non-dominance levels. Each solution is assigned a fitness equal to its non-dominance level ("1" will be assigned to the first non-dominated front). Accordingly, even though the objectives are maximization, by multiplying into -1, the assumption of the problem is changed to the minimization of the fitness functions.

The process of non-dominated sorting and filling the population steps according to the crowding distance can be carried out simultaneously in this procedure until the population size requirement is met. As a result, a non-dominated front finding operator was applied each time to see if the obtained solution could be included into the Pareto set. If this is not possible, there is no need to sort any further. However, if the number of identified solutions exceeds the population size, the extra number of solutions will be deleted using the crowding-distance measure from the last front that could not be fully accommodated.

3.8 Stop Condition

In the MOEAs different stopping conditions may be used, 1) fixed number of iterations, 2) convergence to a solution with a given quality [12]. We devised a new stop criterion that may partially address the weaknesses of redundant generations and reduce the ratio of solution quality to computing time. When the algorithm has executed a certain number of iterations without improvement, this condition is used to end the operation. It is based on the HV value, the diversity and convergence control measure of MOEAs Pareto set, during a given number of iterations.

4   Response Surface Methodology for Parameters Tuning

This section calibrates the NSGA-II using a parameter tuning procedure. To do so, the response surface technique (RSM) is used to optimize the parameters of MOEAs, which have a significant impact on the quality of the solution obtained. However, most past RSM-based applications have only dealt with single-response problems, multi-response situations receiving less attention.

4.1   Multi-Response RSM (MRRSM) Optimization

Multi-Response RSM (MRRSM) Optimization problems can be divided into three groups; Desirability viewpoints, Priority based methods, and Loss function [8].

Although the MRRSM's responses may have curvature over the search ranges of the significant factors, BBD with one central point is chosen to run the experiments [10]. There are $k = 4$ factors in three levels, i.e. low, medium and high, signed by -1 , 0 , and $+1$, respectively. Using the initial data in Table 1, the coded NSGA-II is executed based on the BBD for four factors in three levels with one center point shown in Table 2. This design yields an experiment with 25 runs for three different map sizes.

Table 1. Data generation parameters

Parameter name	Possible values	Description
n_{row}	55, 300, 1000	Number of rows of the raster grid
n_{col}	55, 300, 1000	Number of columns of the raster grid
n_p	6, 8, 10	Number of ports
n_s	6, 7, 8	Number of shipping lanes
n_a	3, 4, 5	Number of protected area
n_w	2, 3, 4	Number of windmill farms
$solution\_size$	15	Size of a solution

To make a balance between robustness and optimization for multiple response problems, we make a hybrid method by using the loss function of Taguchi method to compact and compute multi-responses.

There are two kinds of factors in the loss function of Taguchi method; noise factors N, and controllable factor S. Since MOEAs have multiple runs to obtain better solutions, signal-to-noise ratio (S/N) is used in this research to analyze the results.

In this article, three measures are utilized to evaluate NSGA-II in comparison with the exact method: Hypervolume (HV) [11], number of Pareto solutions

Table 2. Search range of algorithm parameters

Algorithm	Actual values	Coded values	Low(-1)	Medium(0)	High(+1)
NSGA-II	Population_size (N_{pop})	x1	100	150	200
	Crossover_rate (R_c)	x2	0,4	0,6	0,8
	Mutation_rate (R_m)	x3	0,1	0,4	0,7
	Mutation_probability (P_m)	x4	0,25	0,5	0,75

(NPS), and best solution (Best Sol). To determine best solution, for each set of solution, both objective functions values, weighted 0.5 each, are added together. After that, the optimal solution, the maximum, is chosen as Best Sol (similar to how the simple additive weighting algorithm (SAWA) in multi-criteria decision making (MCDM) handles [16]). As the problem is maximization, the the greater HV, NPS, and Best Sol values, the more efficient.

According to the goal of our experiment, among four different formulations of the Taguchi method, we selected the first type, the larger is better [7], whose aim is to find the maximum S/N defined in Eq. 1:

$$\frac{S}{N} = -10 \log(\frac{1}{n} \sum_{i=1}^{n} \frac{1}{sum_i^2}) \tag{1}$$

where sum_i is the response in the Taguchi method, and n is the number of replications($n = 3$). Now, S/N is our MRRSM response. To establish the significance of the individual process parameters and their interactions, a regression equation can be formed. It estimates the correlation between the response and the input process parameters.

Finally ,the obtained metrics are combined into a single number to calculate the S/N (using SAWA in the MCDM approach with equal weight [16]).

The developed algorithms and the experimental DOE tests are coded, respectively, in Python, version 3.8 and R version 4.1.2 to estimate the response functions. All experimental tests are implemented on the Openstack virtual system with 20 VCPU, Disk 10 GB with 30 GB RAM running Linux/Ubuntu 20.04.1 LTS.

In this paper, we used the combination of stepwise regression and cross-validation which returns the best performing model as a feature selection technique for all regression models. All final models are solved within the coded parameters, and the optimum combinations of the parameters (a stationary point in the original units) are achieved by the algorithm. Statistical significance was checked by F-value and p-value (significant probability value less than 0.05). Based on the ANOVA results, the quality of fit of the polynomial model was evaluated using the determination coefficients R-squared, adjusted R-squared, p-value, and the acceptable stationary point in original units. Finally, the optimal combination, which is the stationary point based on the generated response surface model, is obtained.

4.2   Final Tuned Parameters

Having found the best fitted MRSM regression model for three different map sizes, the optimal values of the tuned parameters are given in Table 3. Note that depending on problem size in terms of map size, we fit one individual MRSM model leading to different tuned parameters.

Table 3. Tuned parameters for NSGA-II

Solving methodologies	Parameters	Size		
		Small	Medium	Large
NSGA-II	N_{pop}	157	179	150
	R_c	0,6	0,67	0,69
	R_m	0,44	0,43	0,46
	P_m	0,5	0,5	0,5

Moreover, to determine the best value for the stop criteria, for each size of the problem, we run 8 different problems with the fixed number of iterations 3000, to see the trend of HV value for estimation. Next, the average non-improvement number of iterations 600 for each size is considered as a stop condition.

5   Results

5.1   Performance Measures

To assess the performance of NSGA-II rather than the exact method, quality indicators are provided. They are searching for two complimentary performance goals: 1) convergence to the ideal Pareto front and 2) diversity of options along the front [14].

Therefore, in addition to the metrics defined in Sect. 4, iteration number and three other performance metrics are applied. Among the explained metrics, *NPS*, *BestSol*, and *HV* are considered as the diversity-based, convergence-based, and hybrid category of the quality indicators which combines convergence and diversity measures, respectively.

Corresponding to the following three performance metrics, the higher, the better of the solution quality we have.

Mean Ideal Distance (MID):

This measure presents the closeness between Pareto solution and the ideal point (0, 0) which is a convergence-based indicator shown as in Eq. 2:

$$MID = \frac{\sum_{i=1}^{n} c_i}{n} \tag{2}$$

where n is the number of non-dominated set and $c_i = \sqrt{f_{1i}^2 + f_{2i}^2}$, and f_{1i}, f_{2i} are the value of ith non-dominated solution for first and second objective function respectively.

Spread of non-dominance solution (SNS):
The spread of a non-dominance solution is a diversity-based indicator that assesses the uniformity of the generated solution distribution in terms of dispersion and extension. The formulation of this indicator is illustrated in Eq. 3.

$$SNS = \sqrt{\frac{\sum_{i=1}^{n}(MID - c_i)^2}{n-1}} \qquad (3)$$

The rate of achievement to two objectives simultaneously (RAS):
The balance in reaching to objective functions is another convergence-based quality metrics. In the following Eq. 4 $F_i = min(f_{1i}, f_{2i})$.

$$RAS = \frac{\sum_{i=1}^{n} |\frac{f_{1i}-F_i}{F_i}| + |\frac{f_{2i}-F_i}{F_i}|}{n} \qquad (4)$$

5.2 Comparison Analysis

To calculate all these metrics, first, 8 different randomly selected datasets as different sets of inputs are tested. Then, to evaluate a much more robust comparison, each instance is implemented 30 times and each digit is the median of 30 runs of each problem with its respective method.

To prove the validity of the algorithms, we need to show the gap between the optimal and NSGA-II solutions for a small map size. As shown in Table 4 according to each measure, NSGA-II achieved promising values and pretty well close to exact solutions in small size.

Table 4. The comparison between NSGA-II and the optimal solution

Problems	Map Size	HV		Best Sol		MID		SNS		RAS	
		Exact	NSGA-II	Exact	NSGA-II	Exact	NSGA-II	Exact	NSGA-II	Exact	NSGA-II
1	Small	5,866	5,5173	0,563	0,5315	78,859	76,0040	3,038	2,215	0,995	0,9903
2		6,310	5,7088	0,557	0,5210	85,126	79,5036	4,936	1,730	0,996	0,9909
3		6,313	5,6567	0,558	0,5205	84,002	79,0040	6,725	1,368	0,995	0,9904
4		6,310	5,7088	0,557	0,5214	85,112	79,1463	5,039	1,891	0,996	0,9904
5		6,316	5,7119	0,557	0,5220	85,126	79,3371	4,814	2,060	0,996	0,9901
6		6,313	5,7149	0,557	0,5205	85,751	79,3374	5,158	1,869	0,996	0,9904
7		6,310	5,6299	0,558	0,5205	85,223	79,0040	5,606	2,060	0,996	0,9901
8		6,310	5,7149	0,557	0,5181	85,112	79,3368	5,314	2,079	0,996	0,9901
Average		**6,2556**	**5,6704**	**0,5580**	**0,5219**	**84,2891**	**78,8342**	**5,0789**	**1,9090**	**0,9958**	**0,9903**

Finally, in addition to the small size, the results of NSGA-II for medium and large problem sizes with respect to the performance metrics are shown in Table 5.

Table 5. The Result of NSGA-II

Problems	Map size	HV	Best Sol	MID	SNS	RAS
1	Medium	5,7088	0,5163	78,4038	0,8989	0,9901
2		5,7088	0,5196	77,8921	1,5506	0,9901
3		5,7088	0,5172	78,3789	0,8721	0,9901
4		5,7057	0,5154	78,4323	0,8692	0,9901
5		5,7057	0,5154	78,4038	0,9147	0,9901
6		5,8636	0,5356	78,4328	2,5409	0,9898
7		5,7057	0,5154	78,3675	0,8647	0,9899
8		5,6451	0,5163	78,3789	0,8421	0,9899
Average		**5,7190**	**0,5189**	**78,3363**	**1,1691**	**0,9900**
1	Large	5,8958	0,5129	80,6714	1,2377	0,9901
2		5,9640	0,5145	80,6714	1,4120	0,9897
3		5,8881	0,5168	80,2547	1,1655	0,9895
4		5,8881	0,5155	80,6714	1,2235	0,9894
5		5,8138	0,5054	80,6463	0,4633	0,9888
6		5,8850	0,5129	80,3623	1,0260	0,9895
7		5,8942	0,5103	81,0039	1,0945	0,9900
8		5,8881	0,5103	80,5050	0,8421	0,9897
Average		**5,8896**	**0,5123**	**80,5983**	**1,0581**	**0,9896**

6   Conclusion

In this paper, a novel population-based meta-heuristics algorithm is proposed to solve one of the challenges mentioned in [2]. The extended NSGA-II thoroughly addresses the problem size restriction in an enough fast way while providing solutions that are close to optimality. On a small scale, our study represents different performance measures to prove the validation of the proposed NSGA-II in comparison with the exact solutions. Moreover, having tuned the proposed algorithm by MRSM for three different problem sizes, the results for the large-scale are represented. Another extension of this study, however, would be to determine the ideal site for numerous new activities at the same time, which would be one of the viewpoints of this effort. In the event of disagreement, we intend to provide certain negotiation-based algorithms to find a solution that is acceptable to all parties.

References

1. Basirati, M., Akbari Jokar, M.R., Hassannayebi, E.: Bi-objective optimization approaches to many-to-many hub location routing with distance balancing and hard time window. Neural Comput. Appl. **32**(17), 13267–13288 (2020)

2. Basirati, M., Billot, R., Meyer, P., Bocher, E.: Exact zoning optimization model for marine spatial planning (MSP). Front. Marine Sci. **8**, 726187 (2021)
3. Censor, Y.: Pareto optimality in multiobjective problems. Appl. Math. Optim. **4**(1), 41–59 (1977)
4. Deb, K.: Multi-objective optimisation using evolutionary algorithms: an introduction. In: Wang, L., Ng, A., Deb, K. (eds.) Multi-objective Evolutionary Optimisation for Product Design and Manufacturing. Springer, London (2011). https://doi.org/10.1007/978-0-85729-652-8_1
5. Deb, K., Agrawal, S., Pratap, A., Meyarivan, T.: A fast elitist non-dominated sorting genetic algorithm for multi-objective optimization: NSGA-II. In: Schoenauer, M., et al. (eds.) PPSN 2000. LNCS, vol. 1917, pp. 849–858. Springer, Heidelberg (2000). https://doi.org/10.1007/3-540-45356-3_83
6. Gwaleba, M.J., Chigbu, U.E.: Participation in property formation: Insights from land-use planning in an informal urban settlement in tanzania. Land Use Policy **92**, 104482 (2020)
7. Heckert, N.A., et al.: Handbook 151: Nist/sematech e-handbook of statistical methods. In: e-Handbook of Statistical Methods, pp. 2 (2002)
8. Hejazi, T.H., Bashiri, M., Dı, J.A., Noghondarian, K., et al.: Optimization of probabilistic multiple response surfaces. Appl. Math. Model. **36**(3), 1275–1285 (2012)
9. Lokman, B., Köksalan, M., Korhonen, P.J., Wallenius, J.: An interactive approximation algorithm for multi-objective integer programs. Comput. Oper. Res. **96**, 80–90 (2018)
10. Myers, R.H., Montgomery, D.C., Vining, G.G., Borror, C.M., Kowalski, S.M.: Response surface methodology: a retrospective and literature survey. J. Qual. Technol. **36**(1), 53–77 (2004)
11. Paquete, L., Schulze, B., Stiglmayr, M., Lourenço, A.C.: Computing representations using hypervolume scalarizations. Comput. Oper. Res. **137**, 105349 (2022)
12. Sidi, M.O., Kadrani, A., Quilot-Turion, B., Lescourret, F., Génard, M.: Compromising NSGA-II performances and stopping criteria: case of virtual peach design. In: International Conference on Metamaterials, Photonic Crystals and Plasmonics, p. 2 (2012)
13. Stewart, T.J., Janssen, R., van Herwijnen, M.: A genetic algorithm approach to multiobjective land use planning. Comput. Oper. Res. **31**(14), 2293–2313 (2004)
14. Talbi, E.G.: Metaheuristics: from design to implementation, vol. 74. John Wiley & Sons (2009)
15. Wenwen, L., Goodchild, F., Church, R.: An efficient measure of compactness for 2d shapes and its application in regionalization problems. Int. J. Geograph. Info Sci. **27**(6), 1227–1250 (2013)
16. Zanakis, S.H., Solomon, A., Wishart, N., Dublish, S.: Multi-attribute decision making: A simulation comparison of select methods. Eur. J. Oper. Res. **107**(3), 507–529 (1998)

Competitive Supply Allocation in a Distribution Network Under Overproduction

Alexander Krylatov[1,2]($\boxtimes$) (iD), Yulia Lonyagina[1], and Anastasiya Raevskaya[1] (iD)

[1] Saint Petersburg State University, Saint Petersburg, Russia
a.krylatov@spbu.ru, aykrylatov@yandex.ru
[2] Institute of Transport Problems RAS, Saint Petersburg, Russia

Abstract. The paper aims to deal with the reallocating supply problem that result from the order promising process under overproduction. To this end, we develop a competitive distribution model to facilitate decision-making for order managers and to provide an intelligent support tool. The basis of the distribution model structure is a non-linear constrained optimization program that intends to minimize the costs of competing suppliers in case of an overproduction strategy. We obtain explicit conditions for orders relocation under affine delivery costs. An explicit form of conditions on the current delivery pattern will allow one to develop intelligent tools for decision-making support in the field of order management.

Keywords: Nonlinear optimization · Resource allocation problem · Distribution network

1 Introduction

The order penetration point defines the stage in the manufacturing value chain where a particular product is linked to a specific customer order through different product delivery strategies, such as make-to-stock, assemble-to-order, make-to-order and engineer-to-order [13]. In this paper, we study the case of an overproduction strategy for supplier to avoid shortage. During the order promising process, distributors normally make commitments with customers about the quantities and dates of orders. However, unexpected events may happen that could lead to a shortage supply. Researchers pointed out several causes of these unexpected events: (i) arrival of more priority customer orders that require already reserved products; (ii) delays in raw materials or components; (iii) machine breakdowns; (iv) workers absenteeism, among others [4]. These events might lead to the possibility of making partial or delayed deliveries, i.e., the shortage situation. In particular, cyclical industries face alternating periods of undersupply, when buyers

The work was supported by a grant from the Russian Science Foundation (No. 19-71-10012 Multi-agent systems development for automatic remote control of traffic flows in congested urban road networks).

know that a shortage is imminent and rationing will occur. Thus, suppliers can follow overproduction strategy to avoid partial, delayed and cancelled deliveries.

In practice, when supply delivery time increases, customers make multiple orders with the same supplier or with different suppliers. Such multiple orders may overload the capacity of a distribution network and increase lead-time. In the literature, there are studies of shortage gaming as a leading contributor to the bullwhip phenomenon [15]. Researchers considered shortage decision policies, investigated an integrated production and maintenance planning model with time windows and shortage costs [2,11]. Machine learning techniques for reducing underproduction costs and overproduction costs were developed [6]. In this paper, we develop the distribution model that intends to minimize the costs of competing suppliers in case of an overproduction strategy. We show that the side effect of this strategy is the relocation of order deliveries in a distribution network. The basis of the model structure is a non-linear constrained optimization program.

Samuelson constructed the net social pay-off function and offered the first mathematical formulation of equilibrium commodity flow assignment problem in a network of finished goods in a form of constrained optimization program [16]. The supply-demand allocation pattern, which satisfies this program, is called equilibrium. Researchers generalized this model for the network of multicommodity goods and, nowadays, this model is called spatial equilibrium model [18]. Worth mentioning that this model takes into account relationships between supply, demand and logistics costs. The study of general optimality conditions for this program is given in [5].

Today the problem of supply allocation in distribution networks is highly urgent. In particular, researchers discuss on implementation of this model when investigate actual transportation networks [19]. Some of them concentrate on spatial models under imperfect competition, others study integration of distribution networks under perfect competition [3,9,20]. On the one hand, the nonidentity of equilibrium models for distribution networks and integrative models with non-zero commodity flow is pointed out [1,12]. On the other hand, equilibrium spatial models demonstrate explainability and methodological potential for the analysis of commodity flows and pricing in logistic networks [7,17].

Recently, the conditions on active flows in a network of homogeneous goods were obtained explicitly under linear mappings of elastic demand and supply [8]. However, when manufacturing faces such uncertainties as overproduction (shortage), supply (demand) can no longer consider elastic. In this paper, we study the reallocating supply problem that result from the order promising process under overproduction. We develop a competitive distribution model to facilitate decision-making for order managers and to provide an intelligent support tool. Section 2 contains the basis of the distribution model in a form of the non-linear constrained optimization program that intends to minimize the costs of competing actors suppliers and customers. In Sect. 3, we obtained explicit conditions for orders relocation under affine delivery costs. An explicit form of conditions on the current delivery pattern will allow one to develop intelligent tools for decision-making support in the field of order management. Section 4 discusses on strategies of suppliers under overproduction. Conclusions are given in the last section of the paper.

2   Equilibrium Flow Allocation in a Single-Commodity Network

Consider the set of suppliers M and the set of customers N, which are associated with commodity production, distribution, and consumption. We denote by s_i the supply of $i \in M$, and by λ_i – the price of a unit of the ith supply, $\lambda = (\lambda_1, \ldots, \lambda_m)^T$. By d_j we denote the demand of $j \in N$, and by μ_j – the price of a unit of the jth demand, $\mu = (\mu_1, \ldots, \mu_n)^T$. Finally, let $x_{ij} \geq 0$ be the commodity volume between a pair (i, j), while $c_{ij}(x_{ij})$ is the delivery cost of a unit of x_{ij}. Let us also introduce the indicator of delivery status:

$$\delta_{ij} = \begin{cases} 1 \text{ for } x_{ij} > 0, \\ 0 \text{ for } x_{ij} = 0, \end{cases} \quad \forall (i, j) \in M \times N.$$

Definition. The allocation pattern x is called *equilibrium* if

$$\begin{aligned} \lambda_i + c_{ij}(x_{ij}) = \mu_j \text{ for } & x_{ij} > 0, \\ \lambda_i + c_{ij}(x_{ij}) \geq \mu_j \text{ for } & x_{ij} = 0, \end{aligned} \quad \forall (i, j) \in M \times N.$$

Thus, if the sum of the supplier's price and the delivery costs for a customer exceeds his/her demand price, then the supplier will face with the cancelled delivery.

An equilibrium allocation pattern can be obtain as a solution of the following optimization problem [10, 14]:

$$\min_x \sum_{i \in M} \sum_{j \in N} \int_0^{x_{ij}} c_{ij}(u) du$$

subject to

$$\sum_{j \in N} x_{ij} = s_i \quad \forall i \in M,$$

$$\sum_{i \in M} x_{ij} = d_j \quad \forall j \in N,$$

$$x_{ij} \geq 0 \quad \forall i, j \in M \times N,$$

under

$$\sum_{i \in M} s_i = \sum_{j \in N} d_j.$$

In this paper, we develop a competitive distribution model based on the above non-linear constrained optimization program. We show that this model facilitates decision-making for order managers and provides an intelligent support tool. To this end, we obtain explicit conditions for orders relocation under affine delivery costs. Obtained supply relocation policy in a distribution network under overproduction is appeared to allow order manager relocate supply among customers in order to avoid cancelled deliveries. This relocation guarantees minimum costs for all customers caused by the unexpected shortage.

3   Competitive Supply Allocation in a Distribution Network Under Overproduction

Let us study a competitive supply allocation in a distribution network modelled by a single-commodity network with m suppliers and one customer (i.e., $|M| = m$ and $|N| = 1$). We assume that available supply is more than the overall demand:

$$d < \sum_{i \in M} s_i.$$

In other words, a distributor faces competitive supply relocation in a distribution network under overproduction. Thus, we introduce $\Delta > 0$ as an overproduction value:

$$\sum_{i \in M} s_i - d = \Delta, \tag{1}$$

while $\epsilon_i \geq 0$ as the difference between i-th demand and its actual delivery volume, $i, \ i \in M, \ \epsilon = (\epsilon_1, \epsilon_2, \ldots, \epsilon_m)$:

$$\sum_{i \in M} (s_i - x_i) = \sum_{i \in M} \epsilon_i = \Delta. \tag{2}$$

In terms of a single-commodity network, the allocation pattern x^*, which satisfies the following optimization problem:

$$x^* = \arg \min_x \sum_{i \in M} \int_0^{x_i} c_i(u)du, \tag{3}$$

subject to

$$\sum_{i \subset M} x_i = d,$$
$$x_i = s_i - \epsilon_i, \quad \forall i \in M,$$
$$x_i \geq 0, \quad \forall i \in M, \tag{4}$$
$$\epsilon_i \geq 0, \quad \forall i \in M,$$
$$\sum_{i \in M} \epsilon_i = \Delta.$$

is the equilibrium deliveries allocation under overproduction.

Within the present paper, we examine equilibrium allocation in a case of affine delivery functions. In other words, we assume that

$$c_i(z) = c_i^0 + k_i z, \quad c_i^0 \geq 0, \quad k_i > 0, \quad \forall i \in M, \tag{5}$$

i.e., delivery costs increase when the volume of the order increases.

Lemma 1. *Equilibrium deliveries allocation of overproduction in problem (3)–(4) under affine delivery costs (5) is obtained be the following pattern:*

$$x_i = \begin{cases} \frac{\mu - \lambda_i - c_i^0}{k_i}, & if \ \mu - \lambda_i > c_i^0, \\ 0, & if \ \mu - \lambda_i \leq c_i^0, \end{cases} \quad \forall i \in M, \tag{6}$$

where λ and μ satisfy

$$\begin{cases} \sum\limits_{i \in M} \frac{\mu - \lambda_i - c_i^0}{k_i} \delta_i = d, \\ \frac{\mu - \lambda_i - c_i^0}{k_i} \delta_i = s_i - \epsilon_i, \quad \forall i \in M \\ \sum\limits_{i \in M} \epsilon_i = \Delta \\ \lambda_i = \eta, \quad if \ \epsilon_i > 0 \\ \lambda_i = \eta + \beta_i, \quad if \ \epsilon_i = 0 \end{cases} \tag{7}$$

Proof. Since goal function (3) is convex as well as the restriction set (4), then Karush–Kuhn–Tucker (KKT) conditions are necessary and sufficient. Let us study the Lagrangian of problem (3)–(4):

$$L = \sum_{i \in M} \int_0^{x_i} (c_i^0 + k_i u) du + \mu \left(d - \sum_{i \in M} x_i \right) +$$

$$+ \sum_{i \in M} \lambda_i (x_i - s_i + \epsilon_i) + \sum_{i \in M} (-\alpha_i x_i) +$$

$$+ \sum_{i \in M} (-\beta_i \epsilon_i) + \eta \left(\Delta - \sum_{i \in M} \epsilon_i \right).$$

We differentiate this Lagrangian with respect to x_i and ϵ_i, $i \in M$, and equate the results to zero:

$$\frac{\partial L}{\partial x_i} = c_i^0 + k_i x_i - \mu + \lambda_i - \alpha_i = 0 \quad \forall i \in M, \tag{8}$$

$$\frac{\partial L}{\partial \epsilon_i} = \lambda_i - \beta_i - \eta = 0 \quad \forall i \in M. \tag{9}$$

According to complementary slackness,

$$- \alpha_i x_i = 0, \quad \forall i \in M, \tag{10}$$

$$- \beta_i \epsilon_i = 0, \quad \forall i \in M. \tag{11}$$

Using (10), due to (8) we obtain:

$$x_i = \begin{cases} \frac{\mu - \lambda_i - c_i^0}{k_i}, & if \ \mu - \lambda_i > c_i^0, \\ 0, & if \ \mu - \lambda_i \leq c_i^0, \end{cases} \quad \forall i \in M,$$

that leads to (6). Moreover, due to (9) and (11), we obtain:

$$\lambda_i = \begin{cases} \eta, & \text{if } \epsilon_i > 0, \\ \eta + \beta_i, & \text{if } \epsilon_i = 0, \end{cases} \quad \forall i \in M.$$

However, since $\sum_{i \in M} x_i = d$ and $x_i = s_i - \epsilon_i$, $i \in M$, then:

$$\sum_{i \in M} x_i \delta_i = d, \tag{12}$$

$$x_i \delta_i = s_i - \epsilon_i \quad \forall i \in M. \tag{13}$$

Therefore, taking into account $\sum_{i \in M} \epsilon_i = \Delta$, when one substitutes the expression of x_i, $i \in M$, into (12)–(13), we obtain (7). $\qquad\square$

Without loss of generality, we order suppliers as follows:

$$c_1^0 + k_1 s_1 \geq c_2^0 + k_2 s_2 \geq \cdots \geq c_m^0 + k_m s_m. \tag{14}$$

Theorem 1. *If there is $\bar{m}$ such that*

$$\begin{cases} \sum\limits_{i=1}^{\bar{m}} \dfrac{(c_i^0 + k_i s_i) - (c_\tau^0 + k_\tau s_\tau)}{k_i} < \Delta, & \forall \tau = 1, \ldots, \bar{m}, \\ \sum\limits_{i=1}^{\bar{m}} \dfrac{(c_i^0 + k_i s_i) - (c_\tau^0 + k_\tau s_\tau)}{k_i} \geq \Delta, & \forall \tau = \bar{m}+1, \ldots, m, \end{cases}$$

then

$$x_i = \begin{cases} 0, & \text{if } c_i^0 \geq \dfrac{\sum\limits_{l=1}^{\bar{m}} \frac{c_l^0 + s_l k_l}{k_l} - \Delta}{\sum\limits_{l=1}^{\bar{m}} \frac{1}{k_l}}, \\[4ex] \dfrac{\sum\limits_{l=1}^{\bar{m}} s_l - \Delta + \sum\limits_{l=1}^{\bar{m}} \frac{c_l^0 - c_i^0}{k_l}}{k_i \sum\limits_{l=1}^{\bar{m}} \frac{1}{k_l}}, & \text{if } c_i^0 < \dfrac{\sum\limits_{l=1}^{\bar{m}} \frac{c_l^0 + s_l k_l}{k_l} - \Delta}{\sum\limits_{l=1}^{\bar{m}} \frac{1}{k_l}}, \end{cases} \quad \forall i = \overline{1, \bar{m}},$$

and $x_i = s_i$ for all $i = \overline{\bar{m}+1, m}$.

Proof. **I.** Let us introduce $\bar{M} \subseteq M$ such that $\epsilon_i > 0$ for all $i \in \bar{M}$. We summarize equalities $\frac{\mu - \lambda_i - c_i^0}{k_i} = s_i - \epsilon_i$, $i \in M$, for all $i \in \bar{M}$:

$$\sum_{i \in \bar{M}} \frac{\mu - \lambda_i - c_i^0}{k_i} = \sum_{i \in \bar{M}} s_i - \Delta,$$

due to $\lambda_i = \eta$, for all $i \in \bar{M}$, we obtain:

$$\sum_{i \in \bar{M}} \frac{\mu - \eta - c_i^0}{k_i} = \sum_{i \in \bar{M}} s_i - \Delta$$

or

$$\eta = \frac{\sum\limits_{i \in \bar{M}} \frac{\mu - c_i^0}{k_i} - \sum\limits_{i \in \bar{M}} s_i + \Delta}{\sum\limits_{i \in \bar{M}} \frac{1}{k_i}}. \tag{15}$$

II. Since

$$\frac{\mu - \lambda_i - c_i^0}{k_i} = s_i, \quad \forall i \in M \backslash \bar{M},$$

or, in a matrix form,

$$\begin{pmatrix} -\frac{1}{k_{i_1}} & 0 & \cdots & 0 \\ 0 & -\frac{1}{k_{i_2}} & \cdots & 0 \\ \vdots & \vdots & \ddots & \vdots \\ 0 & 0 & \cdots & -\frac{1}{k_{i_{\bar{m}}}} \end{pmatrix} \begin{pmatrix} \lambda_{i_1} \\ \lambda_{i_2} \\ \vdots \\ \lambda_{i_{\bar{m}}} \end{pmatrix} = \begin{pmatrix} s_{i_1} - \frac{\mu - c_{i_1}^0}{k_{i_{1_0}}} \\ s_{i_2} - \frac{\mu - c_{i_2}^0}{k_{i_2}} \\ \vdots \\ s_{i_{\bar{m}}} - \frac{\mu - c_{i_{\bar{n}}}^0}{k_{i_{\bar{m}}}} \end{pmatrix},$$

then

$$\lambda_i = \mu - c_i^0 - s_i k_i, \quad \forall i \in M \backslash \bar{M}. \tag{16}$$

According to

$$\begin{aligned} \lambda_i = \eta, & \quad \text{if } \epsilon_i > 0, \\ \lambda_i = \eta + \beta_i, & \quad \text{if } \epsilon_i = 0, \end{aligned} \quad \forall i \in M,$$

that is

$$\begin{aligned} \lambda_i = \eta, & \quad \forall i \in \bar{M}, \\ \lambda_i = \eta + \beta_i, & \quad \forall i \in M \backslash \bar{M}, \end{aligned}$$

one can see $\lambda_i \geq \eta$, for all $i \in M \backslash \bar{M}$. Thus, due to (15) and (16), we obtain:

$$\mu - c_\tau^0 - s_\tau k_\tau \geq \frac{\sum\limits_{i \in \bar{M}} \frac{\mu - c_i^0}{k_i} - \sum\limits_{i \in \bar{M}} s_i + \Delta}{\sum\limits_{i \in \bar{M}} \frac{1}{k_i}}, \quad \forall \tau \in M \backslash \bar{M},$$

while, since $\sum\limits_{i \in \bar{M}} \frac{1}{k_i} > 0$, then

$$\left(\mu - c_\tau^0 - s_\tau k_\tau\right) \sum\limits_{i \in \bar{M}} \frac{1}{k_i} \geq \sum\limits_{i \in \bar{M}} \frac{\mu - c_i^0}{k_i} - \sum\limits_{i \in \bar{M}} s_i + \Delta, \quad \forall \tau \in M \backslash \bar{M},$$

or

$$\sum\limits_{i \in \bar{M}} \frac{\mu - c_\tau^0 - s_\tau k_\tau}{k_i} \geq \sum\limits_{i \in \bar{M}} \frac{\mu - c_i^0 - s_i k_i}{k_i} + \Delta, \quad \forall \tau \in M \backslash \bar{M},$$

Eventually, we obtain:

$$\sum\limits_{i \in \bar{M}} \frac{\left(c_i^0 + k_i s_i\right) - \left(c_\tau^0 + k_\tau s_\tau\right)}{k_i} \geq \Delta, \quad \forall \tau \in M \backslash \bar{M}. \tag{17}$$

III. Since $x_\tau < s_\tau$ for all $\tau \in \bar{M}$, then for $\tau \in \bar{M}$ either $s_\tau > x_\tau = 0$ or $s_\tau > x_\tau > 0$. Thus, according to (6), we obtain

$$
\begin{aligned}
\lambda_\tau &= \mu - c_\tau^0 - k_\tau x_\tau, \quad \text{if } x_\tau > 0, \\
\lambda_\tau &\geq \mu - c_\tau^0, \quad\quad\quad \text{if } x_\tau = 0,
\end{aligned} \quad \forall \tau \in \bar{M}.
$$

Taking into account $k_\tau s_\tau > 0$, for all $\tau \in N$, we can re-write this system as follows:

$$
\begin{aligned}
\lambda_\tau &= \mu - c_\tau^0 - k_\tau x_\tau, \quad \text{if } x_\tau > 0, \\
\lambda_\tau &> \mu - c_\tau^0 - k_\tau s_\tau, \quad \text{if } x_\tau = 0,
\end{aligned} \quad \forall \tau \in \bar{M}.
$$

Since $\lambda_\tau = \eta$ and $x_\tau < s_\tau$, for all $\tau \in \bar{M}$, then

$$
\eta > \mu - c_\tau^0 - k_\tau s_\tau, \quad \forall \tau \in \bar{M}.
$$

Due to (15), we obtain:

$$
\mu - c_\tau^0 - k_\tau s_\tau < \frac{\sum\limits_{i \in \bar{M}} \frac{\mu - c_i^0}{k_i} - \sum\limits_{i \in \bar{M}} s_i + \Delta}{\sum\limits_{i \in \bar{M}} \frac{1}{k_i}} \quad \forall \tau \in \bar{M},
$$

i.e.,

$$
\sum_{i \in \bar{M}} \frac{(c_i^0 + k_i s_i) - (c_\tau^0 + k_\tau s_\tau)}{k_i} < \Delta \quad \forall \tau \in \bar{M}. \tag{18}
$$

IV. If customers are ordered according to (14), then

$$
\left(c_i^0 + k_i s_i\right) - \left(c_\tau^0 + k_\tau s_\tau\right) = t_i(s_i) - t_\tau(s_\tau) \begin{cases} \geq 0, & \text{if } i < \tau, \\ \leq 0, & \text{if } i > \tau. \end{cases}
$$

If $\tau = 1$, then

$$
\frac{c_i(s_i) - c_1(s_1)}{k_i} \leq 0, \quad \forall i \in M.
$$

Since $\Delta > 0$, then $\tau = 1 \notin M \backslash \bar{M}$, i.e., $\tau = 1 \in \bar{M}$. If $\tau = 2$, then

$$
\begin{cases} c_1(s_1) - c_2(s_2) \geq 0, \\ c_i(s_1) - c_2(s_2) \leq 0, \quad \forall j = 2, \ldots, m. \end{cases} \tag{19}
$$

Hence, either

$$
\frac{c_1(s_1) - c_2(s_2)}{k_1} \geq \Delta
$$

or

$$
\frac{c_1(s_1) - c_2(s_2)}{k_1} < \Delta.
$$

If $\frac{c_1(s_1) - c_2(s_2)}{k_1} \geq \Delta$, then, due to (14),

$$
\Delta \leq \frac{c_1(s_1) - c_2(s_2)}{k_1} \leq \frac{c_1(s_1) - c_i(s_i)}{k_1}, \quad \forall j = 2, \ldots, m.
$$

Thus, we obtain:

$$\bar{M} = \{1\}, \quad M\backslash\bar{M} = \{2,\ldots,n\}. \tag{20}$$

However, if $\frac{c_1(s_1)-c_2(s_2)}{k_1} < \Delta$, then, due to (19), $\tau = 2 \notin M\backslash\bar{M}$, i.e., $\tau = 2 \in \bar{M}$. Such a chain of reasoning leads to the existence of required $\bar{m}$, $1 \leq \bar{m} \leq n$.

V. Due to $x_i = s_i - \epsilon_i$ and $\epsilon_i = 0$ for all $i = \overline{\bar{m}+1,m}$, then

$$x_i = s_i, \quad i = \overline{\bar{m}+1,m}.$$

Moreover, due to (6) under $\lambda_i = \eta$ for all $i = \overline{1,\bar{m}}$, we have

$$x_i = \begin{cases} 0, & \text{if } c_i^0 \geq \dfrac{\sum\limits_{l=1}^{\bar{m}} \frac{c_i^0+s_l k_l}{k_l} - \Delta}{\sum\limits_{l=1}^{\bar{m}} \frac{1}{k_l}} \\[2em] \dfrac{\mu-\eta-c_i^0}{k_i}, & \text{if } c_i^0 < \dfrac{\sum\limits_{l=1}^{\bar{m}} \frac{c_i^0+s_l k_l}{k_l} - \Delta}{\sum\limits_{l=1}^{\bar{m}} \frac{1}{k_l}} \end{cases} \quad i = \overline{1,\bar{m}},$$

which is the required expression, due to (15).   □

Theorem 1 gives the supply relocation policy in a distribution network under overproduction. In other words, the supply can be relocated among customers in such a way to avoid cancelled orders. This relocation guarantees minimum costs for all customers caused by the unexpected shortage.

4   Strategies of Suppliers Under Overproduction

Let us consider the order management policy of suppliers in case of an overproduction strategy (Fig. 1).

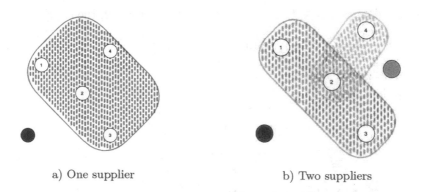

a) One supplier   b) Two suppliers

Fig. 1. Order management: strategies of suppliers under overproduction

Assume that the green supplier has four customers (Fig. 1a). We tend to study risks that can arise when the yellow supplier appears to offer its overproduction (Fig. 1b). One can see that customers 1, 2, and 3 are located quite close to the green supplier, while customers 2, 3, and 4 are located quite close to the yellow one. According to Theorem 1, if

$$c_\tau^0 \geq \frac{\sum_{l=1}^{\bar{m}} \frac{c_l^0 + s_l k_l}{k_l} - \Delta}{\sum_{l=1}^{\bar{m}} \frac{1}{k_l}}, \tag{21}$$

then the customer τ can cancel order and change the supplier for the closer one. Moreover, according to Theorem 1, $x_i = s_i$ for $i = \overline{\bar{m}+1, m}$, while

$$c_1^0 + k_1 s_1 \geq c_2^0 + k_2 s_2 \geq \cdots \geq c_m^0 + k_m s_m, \tag{22}$$

and

$$\begin{cases} \sum_{i=1}^{\bar{m}} \frac{(c_i^0 + k_i s_i) - (c_\tau^0 + k_\tau s_\tau)}{k_i} < \Delta, & \forall \tau = 1, \ldots, \bar{m}, \\ \sum_{i=1}^{\bar{m}} \frac{(c_i^0 + k_i s_i) - (c_\tau^0 + k_\tau s_\tau)}{k_i} \geq \Delta, & \forall \tau = \bar{m}+1, \ldots, m. \end{cases} \tag{23}$$

In other words, a customer choose the closest suppliers with small orders rather than large order from distant supplier. Moreover, if

$$c_i^0 < \frac{\sum_{l=1}^{\bar{m}} \frac{c_l^0 + s_l k_l}{k_l} - \Delta}{\sum_{l=1}^{\bar{m}} \frac{1}{k_l}}, \tag{24}$$

then

$$x_i = \frac{\sum_{l=1}^{\bar{m}} s_l - \Delta + \sum_{l=1}^{\bar{m}} \frac{c_l^0 - c_i^0}{k_l}}{k_i \sum_{l=1}^{\bar{m}} \frac{1}{k_l}}, \tag{25}$$

for all $i = \overline{1, \bar{m}}$ from (23), where (25) is the value of a partially confirmed order. Hence, a supplier has the following set of risks (Table 1).

Table 1. Scenarios for decision-making support.

Evaluation	Scenario	Risk
Inequality (21) holds	Delivery cost exceeds the equilibrium Value for the given distribution network	Cancelled Order
Inequality (24) holds	Delivery cost is less than the equilibrium Value for the given distribution network	Partial order Confirmation

Therefore, if suppliers follow an overproduction strategy in order to avoid the shortage, they can face several risks raised as side effects of this strategy. The first risk is the cancelled order. In other words, if inequality (21) holds, the customer can cancel his/her order and choose another supplier. The second risk is partial order confirmation. Indeed, if inequality (24) holds, the customer can confirm the part of the order and choose another supplier for the rest.

5   Conclusion

The paper aimed to deal with the reallocating supply problem that result from the order promising process under overproduction. To this end, we developed a competitive distribution model to facilitate decision-making for order managers and to provide an intelligent support tool. The basis of the distribution model structure was a non-linear constrained optimization program that intends to minimize the costs of competing suppliers in case of an overproduction strategy. We obtained explicit conditions for orders relocation under affine delivery costs. An explicit form of conditions on the current delivery pattern will allow one to develop intelligent tools for decision-making support in the field of order management.

References

1. Barrett, C., Li, J.: Distinguishing between equilibrium and integration in spatial price analysis. Am. J. Agr. Econ. **84**(2), 292–307 (2002)
2. Barron, Y., Hermel, D.: Shortage decision policies for a fluid production model with map arrivals. Int. J. Prod. Res. **55**(14), 3946–3969 (2017)
3. Bramoulle, Y., Kranton, R.: Public goods in networks. J. Econ. Theory **135**, 478–494 (2007)
4. Esteso, A., Mula, J., Campuzano-Bolarín, F., Diaz, M., Ortiz, A.: Simulation to reallocate supply to committed orders under shortage. Int. J. Prod. Res. **57**(5), 1552–1570 (2019)
5. Florian, M., Los, M.: A new look at static spatial price equilibrium models. Reg. Sci. Urban Econ. **12**(4), 579–597 (1982)
6. Ji, B., Ameri, F., Cho, H.: A non-conformance rate prediction method supported by machine learning and ontology in reducing underproduction cost and overproduction cost. Int. J. Prod. Res. **59**(16), 5011–5031 (2021)
7. Kiselev, A., Yurchenko, N.: Game equilibria and transition dynamics in a dyad with heterogeneous agents. Autom. Remote. Control. **82**(3), 549–564 (2021)
8. Krylatov, A., Lonyagina, Y.: Equilibrium flow assignment in a network of homogeneous goods. Autom. Remote. Control. **83**(5), 805–827 (2022)
9. McNew, K.: Spatial market integration: definition, theory, and evidence. Agricult. Resour Econ. Rev. **25**(1), 1–11 (1996)
10. Nagurney, A.: Network economics: a variational inequality approach. Kluwer Academic Publishers, The Netherlands (1993)
11. Najid, N., Alaoui-Selsouli, M., Mohafid, A.: An integrated production and maintenance planning model with time windows and shortage cost. Int. J. Prod. Res. **49**(8), 2265–2283 (2011)

12. Novikov, D.A.: Games and networks. Autom. Remote. Control. **75**(6), 1145–1154 (2014). https://doi.org/10.1134/S0005117914060149
13. Olhager, J.: Strategic positioning of the order penetration point. Int. J. Prod. Econ. **85**(3), 319–329 (2003)
14. Patriksson, M.: The Traffic Assignment Problem: Models and Methods. Dover Publications, New York (1994)
15. Samuel, C., Mahanty, B.: Shortage gaming and supply chain performance. Int. J. Manuf. Technol. Manage. **5**(5/6), 536–548 (2003)
16. Samuelson, P.: Spatial price equilibrium and linear programming. Am. Econ. Rev. **42**(3), 283–303 (1952)
17. Stephens, E., Mabaya, E., von Cramon-Taubadel, S., Barrett, C.: Spatial price adjustment with and without trade. Oxford Bull. Econ. Stat. **74**(3), 453–469 (2012)
18. Takayama, T., Judge, G.: Equilibrium among spatially separated markets: a reformulation. Econometrica **32**(4), 510–524 (1964)
19. Vasin, A., Grigoryeva, O., Tsyganov, N.: A model for optimization of transport infrastructure for some homogeneous goods markets. J. Global Optim. **76**(3), 499–518 (2020)
20. Vasin, A.A., Daylova, E.A.: Two-node market under imperfect competition. Autom. Remote. Control. **78**(9), 1709–1729 (2017). https://doi.org/10.1134/S0005117917090144

Safe-Exploration of Control Policies from Safe-Experience via Gaussian Processes

Antonio Candelieri$^{(\boxtimes)}$ ⑩, Andrea Ponti⑩, and Francesco Archetti⑩

University of Milano-Bicocca, Milan, Italy
{antonio.candelieri,francesco.archetti}@unimib.it,
a.ponti5@campus.unimib.it

Abstract. Control of many real-life systems strongly relies on the knowledge of a domain expert, who usually adopts a *safe* control policy to deal with uncertainty. The term *safe* means that the policy is aimed at avoiding system's disruptions or relevant deviations from the desired behaviour, usually at the cost of sub-optimal performances. This paper proposes a statistically-sound approach which exploits the collected experience to *safe-explore* new policies by assuming a reasonable risk in terms of safety while improving performances. Gaussian Process regression is the core of the approach, providing a probabilistic approximation of both system's dynamics and performances, depending on historical data related to the application of the safe policy. Being a probabilistic model, Gaussian Process provides both an estimate of the level of safety and, more important, the associated predictive uncertainty, which is crucial for implementing the safe-exploration of new efficient policies. The approach allows to avoid the typically expensive implementation of a *digital twin* of the system, required in the case of simulation-optimization approaches, as well as the formulation as a stochastic programming problem. Results on two case studies, inspired by real-life systems, are presented, showing an improvement in terms of performances with respect the initial safe policy, with reasonable safety of the systems.

Keywords: Test · Gaussian processes · Safe-exploration · Optimal control

1 Introduction

1.1 Motivation

Many dynamic real-life systems are usually controlled depending on a policy usually aimed at ensuring *safety* of the system itself, at the cost of settling for sub-optimal performances. Depending on the specific system, *safety* is associated to different aspects. Some examples of safety are: ensuring a satisfactory level/quality of service for a utility (e.g., energy, water, oil/gas distribution networks), avoiding disruptions in controlling a robot/plant (e.g., manufacturing system), or avoiding discomfort to people (e.g., house cooling/heating/lighting

© Springer Nature Switzerland AG 2022
D. E. Simos et al. (Eds.): LION 2022, LNCS 13621, pp. 232–247, 2022.
https://doi.org/10.1007/978-3-031-24866-5_18

system). Moreover, most of these systems are inevitably affected by uncertainty, making difficult to optimize control's performances under safety constraints.

Typical solutions include the mathematical formalization – most of the time as a rough approximation – of all the equations regulating the system, along with a possible model of the uncertain components, for instance depending on historical observations. Then, the resulting mathematical problem can be addressed through mathematical programming, typically *stochastic optimization* approaches [1–3].

Another possibility is to implement a *digital twin* of the system, usually offering a more accurate model of the system compared to mathematical programming. Then, *simulation-optimization* is used to search for optimal control policies by simulating a large number of *scenarios* according to the estimated distribution of the uncertain components [4–6]. An efficient search mechanism is crucial in the case that the simulation of a single scenario is expensive in terms of computational resources and/or time [7,8]. A relevant practical issue is that the realization of an accurate digital twin could a really expensive task depending on the complexity of the system to control.

However, mathematical programming and simulation-optimization do not make any efficient usage of historical data collected by applying a safe control policy. Typically, historical observations are just used to estimate the unknown distribution of the uncertainty components, enabling stochastic and/or robust optimization approaches. More recently, research has focused on exploiting the *safe experience*'s data to approximate the system's behaviour in response to the control actions. Although safe experience's data provides just a partial knowledge about the system to control, Machine Learning (ML) methods allows to generalize over unseen data – i.e., new control actions – while dealing with two different sources of uncertainty: the one related to ML algorithm's prediction error and the uncertain components affecting the system.

This section concludes with an overview on related works in different application domains and the specific contribution of the paper. Then, Sect. 2 provides the general definition of the considered setting, that is the optimal safe control of a dynamic system; Sect. 3 details the proposed approach; Sect. 4 presents the two case studies considered in the study; Sect. 5 summarizes relevant results; finally, Sect. 6 provides conclusions, limitations, and perspectives.

1.2   Related Works

The interlink between *safety* and *optimality* has been studied in different research communities, relatively to optimal learning and optimal control. For instance, in [9] a Safe active learning algorithm was proposed to learn a model of a system's performance metric under the constraint that the system always stays in a safe operation. In [10] an approach mixing safe active learning and Bayesian optimization is proposed to address the optimal calibration of a PID (proportional-integrative-derivative) controller for a high pressure fuel supply system of an engine. The combination between safe active learning and Bayesian optimization has been successively and extensively investigated, such as in [11–15], proving that sample efficiency of Bayesian optimization [16,17], along with safe active learning, allows to suitably address the safe optimal tuning of a controller's hyperparameters, typically a PID controller. Most of these approaches use Gaussian Process regression [18,19] to obtain a probabilistic model of the objective function to be optimized under safety constraints. Another important application domain is robotics, where optimizing a control policy is more sophisticated than tuning few controller's hyperparameters. Many research works use GP regression to implement data-efficient approaches, also in the reinforcement learning setting [20–22]. Finally, [23] is Lipschitz safe optimization method to adopt in the case that an estimation of the uncertainty is available.

1.3 Contribution

The main contributions of this paper can be summarized as follows:

- using experience's data from a safe policy to model the system's dynamic, including performance metric and safety, without requiring the – usually expensive – realization of a digital twin of the system
- adopting Gaussian Process regression to obtain a probabilistic model of the system's dynamic while dealing with different sources of uncertainty, that are prediction error, partial knowledge (i.e., data are only collected according to a safe policy), and uncertainty affecting system's response.
- testing the approach for safe-exploring new control policies, starting from safe experience related to a previously adopted safe – and sub-optimal – policy, for two test cases inspired by real-life systems. Improvement in terms of performances and incurred risk in term of safety are analysed.

2 Reference Problem

2.1 Control of a Dynamic System

The general setting considered in the paper refers to the control of a dynamic system S under uncertainty. A graphical representation is provided in Fig. 1, where π is the control policy applied by the controller, $a(t)$ is the action performed accordingly to π, that is $a(t) = \pi(t, s(t))$, $s(t)$ is the system's output, $\xi(t)$ is the uncertain component, and λ is a set of system and/or control specific parameters (e.g., a target reference for $s(t)$ in some applications).

It is important to remark that the response of the system has some delay δt due to the need to process the input, and it also depends on both the system's current value (aka *state*), the control action $a(t)$, and the *realization* $\xi(t)$ of uncertain component. Formally:

$$s(t + \delta t) = f\big(s(t), a(t), \xi(t)\big).$$

Furthermore, $\xi(t)$ is observable only a-posteriori – specifically after the application of $a(t)$ – and its distribution, $\xi(t) \sim P(\theta_t)$, is unknown. On the contrary,

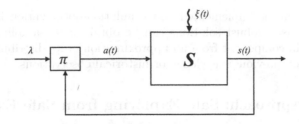

Fig. 1. Control of a dynamic system under uncertainty: a schematic representation.

we assume that it is possible to observe an *instantaneous regret* in response to the control action $a(t)$ – namely, $r(t, s(t), a(t), \xi(t))$ – along with the compliance to *operational* and *safety* constraints, overall denoted by

$$g_i\big(t, x(t), s(t), \xi(t)\big) > 0 \tag{1}$$

with $i = 1, ..., n_g$ and n_g the number of constraints. To better understand the difference between *operational* and *safety* constraints, consider the following example: the min-max range for $a(t)$ is an operational constraint because it is associated to the physical capabilities of the controller, while a safety constraint is any constraint associated to the disruption of the system or a critical deviation of its behaviour from the desired one (e.g., $s(t)$ outside a certain range).

It is important to remark that we are not considering *chance-constrained optimization*. Indeed, chance constrained optimization ensures that the probability of meeting a certain constraint is above a given threshold p_i – that is $P\big(g_i(t, x(t), s(t), \xi(t)) > 0\big) > p_i$. Instead, we want that all the constraints (1) are always met: in chance constrained terms this means that $p_i = 1 \; \forall i = 1, ..., n_g$.

2.2   Problem Formulation and Usual Solving Methods

As far as optimal control is concerned, the aim is to search for an optimal policy π^* minimizing the *cumulative regret*, that is the regret accumulated over a given time interval $[0, T]$ in response to the control policy, while satisfying all the operational and safety constraints. Formally:

$$\pi^* = \arg\min_{\pi \in \Pi} \int_0^T r\big(t, s(t), a(t), \xi(t)\big)\, dt \tag{2}$$

$$s.t. \quad g_i\big(t, h(t), a(t), \xi(t)\big) \qquad > 0 \quad \forall i = 1, ..., n_g; \forall t = 0, ..., T \tag{3}$$

where it is important to remark that $a(t) = \pi(t, s(t))$, and with Π denoting the search space of all the possible control policies.

Solving the problem (2–3) requires to explicit both the regret function and all the constraints, while dealing with the possible realizations of the uncertain component. When a mathematical formalization is possible – even if roughly approximated – stochastic approximation methods can be used, otherwise a digital twin of the system can be implemented and simulation-optimization is adopted. In both the two cases, robust solutions can be obtained by sampling realizations of the uncertain component from an approximation of its distribution, which is usually based on previous knowledge or historical observations.

3   Novel Approach: Safe-Exploring from Safe Experience

Many real-life systems are controlled through policies that are typically *safe* but largely sub-optimal. The (safe) experience collected by these *safe-by-design* policies is anyway precious. It is important to remark that the term *safe-by-design* is here used to denote a very simple policy aimed at generating actions satisfying all the constraints (2) by ignoring the resulting cumulative regret (3). The proposed approach starts from the safe experience to explore new and more efficient policies – in terms of regret – while facing some risk in terms of safety. Specifically, we use Gaussian Processes (GPs) as probabilistic models to enable safe exploration based on the estimation of the trade-off between risk and benefit.

First, the safe experience collected according to a safe-by-design policy, π_0, can be represented by a set of tuples, each one denoted as follows:

$$< t, s(t), a(t), s(t + \Delta t), r(t) >$$

when a real-life system is considered, the continuous-time assumption can be relaxed according to the intrinsically discrete nature of data acquisition process (e.g., sensors' sampling rate). Moreover, the system's response to the control action is not immediate. Thus, we can consider a constant delay Δt between the control action $a(t)$ and the observation of the system's response $s(t + \Delta t)$.

Information collected into the tuples can be reorganized into two sets of data, namely the **transition dynamic dataset** $\mathcal{D}_{\pi_0}$ and the **instantaneous regret dataset** $\mathcal{R}_{\pi_0}$:

$$\mathcal{D}_{\pi_0} = \left\{ \left(\mathbf{x}^{(i)}, s'^{(i)} \right) \right\}_{i=1:N_0} \quad ; \quad \mathcal{R}_{\pi_0} = \left\{ \left(\mathbf{x}^{(i)}, r^{(i)} \right) \right\}_{i=1:N_0} \quad (4)$$

where N_0 is the number of tuples collected over time in response to the application of π_0 (i.e., N_0 is the experience size), $\mathbf{x}^{(i)} = \left(t^{(i)}, s(t)^{(i)}, a(t)^{(i)} \right)$, $s'^{(i)} = s(t + \Delta t)^{(i)}$, and $r^{(i)}$ is the observed regret associated to the ith tuple.

A GP regression model is fitted on $\mathcal{D}_{\pi_0}$ to approximate the transition dynamic of the system $\boldsymbol{S}$. As a probabilistic regression model, this GP provides both a prediction of the system's response, $s' = \mu_{\mathcal{D}}(\mathbf{x})$, and its associated uncertainty $\sigma_{\mathcal{D}}(\mathbf{x})$, for every input $\mathbf{x}$ – where $\mathbf{x}$ is interpreted as "performing $a(t)$, at time t, given $s(t)$". Using expanded notation allows for clearly visualizing time-dependency: $s' = s(t + \Delta t) \approx \mu_{\mathcal{D}}(t, s(t), a(t))$, with associated uncertainty $\sigma_{\mathcal{D}}(t, s(t), a(t))$.

Analogously, a GP regression model is fitted on $\mathcal{R}_{\pi_0}$ to estimate the instantaneous regret expected for a given $\mathbf{x}$. The predicted regret is given by the GP's posterior mean $\mu_{\mathcal{R}}(\mathbf{x})$ with associated uncertainty $\sigma_{\mathcal{R}}(\mathbf{x})$.

The new improved policy, π^+, consists in solving, at each time t, and for the given $s(t)$, the following optimization problem:

$$a^+(t) = \arg\min_{a \in \mathcal{A}} \left\{ \mu_{\mathcal{R}}(\mathbf{x}|t, s(t)) - \beta_{\mathcal{R}}\, \sigma_{\mathcal{R}}(\mathbf{x}|t, s(t)) \right\} \tag{5}$$

$$s.t. \quad \gamma_j\left(\mu_{\mathcal{D}}(\mathbf{x}|t, s(t)), \sigma_{\mathcal{D}}(\mathbf{x}|t, s(t)) \right) > 0 \quad \forall j = 1, ..., n_\gamma; \forall t = 0, ..., T \tag{6}$$

where the notation $(\mathbf{x}|t, s(t))$ denotes that t and $s(t)$ are given and, consequently, the optimization is only performed on $a \in \mathcal{A}$, with $\mathcal{A}$ the set of all the possible control actions. This is important because it means that while GP is defined on a d-dimensional space, the optimization is performed on a q-dimensional space, with $q < d$. Specifically, $q = 1$ in the case that $a(t)$ is a scalar.

Here, our important assumption is that the problem can be solved step-wise, and that an (estimated) safe action at a step $\bar{t}$ is sufficient to guarantee the existence of safe actions at successive steps $t > \bar{t}$. If valid, this avoids expensive look-ahead and/or rollout procedures, leading to short computational time for solving (5-6).

The problem (5-6) is therefore a re-formulation of the original problem (2-3), where **partial knowledge** about the systems' transition dynamic and instantaneous regret – collected according to the application of the safe policy π_0 – is included through the two GP regression models.

To make a parallel, the two GPs are the counterparts of the equations of a mathematical (approximated) formalization as well as of a (partial) digital twin of the system.

Precisely, we propose to adopt an *optimistic-in-face-of-uncertainty* approach to optimize the objective function (5). Indeed, we add an *uncertainty bonus*, $\sigma_{\mathcal{R}}(\mathbf{x})$, to our estimate of the instantaneous regret, $\mu_{\mathcal{R}}(\mathbf{x})$, suitably weighted by $\beta_{\mathcal{R}}$. If $\beta_{\mathcal{R}} = 0$ then we assume that the GP's prediction is the best approximation of the instantaneous regret, otherwise we give a chance to exploration. This is also known as GP Confidence Bound (GP-CB) an it is a widely and successfully applied *acquisition function* in Bayesian optimization [16,17], with different suitably scheduling of $\beta_{\mathcal{R}}$ proposed to guarantee convergence to global optimum in the global optimization setting [24–27].

It is important to remark that also constraints – all or some of them – must be re-formulated according to the partial transition dynamic modelled by the associated GP, possibly leading, in the more general case, to a different number of constraints. Both prediction, $\mu_{\mathcal{D}}(\mathbf{x})$, and uncertainty, $\sigma_{\mathcal{D}}(\mathbf{x})$, must be considered to implement safety evaluation mechanisms with different risk attitudes. Contrary to $\beta_{\mathcal{R}}$, the multiplier $\beta_{\mathcal{D}}$ is here used to weight the uncertainty associated to the constraints satisfiability, especially safety constraints. Thus, it is not related to optimality, just like in some approaches proposed for a different optimization setting – namely multiple information source optimization [28,29] – where it is used to tune a sort of *reliability* measure.

The formulation of the two case studies considered in this paper will make easier to understand how to reformulate objective function and constraints in terms of GPs and according to the specific target problem.

4   Case Studies

4.1   Case Study 1: Optimal Control of a Water Tank

System Description. The first case study refers to the control of a water tank in a water distribution network. The system is sketched in Fig. 2 and it is characterized by the following mass conservation equation:

$$s(t + \Delta t) = s(t) + a[t : t + \Delta t] - \xi[t : t + \Delta t] \tag{7}$$

where $[t : t + \Delta t]$ denotes the time interval from t to $t + \Delta t$. Thus, $s(t + \Delta t)$ is the amount of water into the tank at the end of the time interval, $s(t)$ is the amount of water into the tank at the beginning of the time interval (i.e., when the control action is decided), $a[t : t + \Delta t]$ is the amount of water pumped into the tank within that time interval (i.e., the control action), and $\xi[t : t + \Delta t]$ is the amount of water supplied to match the demand in that time interval.

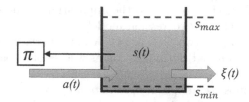

Fig. 2. Control of a water tank: a schematic representation. Notation follows that used in the general schema in Fig. 1.

We consider the case that the control is performed on a hourly basis, that is $\Delta t = 1$ [hour], over an entire day, that is $T = 24$ [hours]. Furthermore, to simplify notation, we rewrite (7) as:

$$s_t = s_{t-1} + a_{t-1} - \xi_{t-1} \qquad \forall t = 0, ..., T - 1 \tag{8}$$

where a_{t-1} and ξ_{t-1} are intended as $a[t - 1 : t]$ and $\xi[t - 1 : t]$, respectively.

Optimal Control Problem. The specific problem can be formalized as:

$$\min_{a_t \in \Re} \sum_{t=0}^{T-1} c(t, a_t) \tag{9}$$

$$\text{s.t.} \quad s_t = s_{t-1} + a_{t-1} - \xi_{t-1} \qquad \forall t = 0, ..., T-1 \tag{10}$$

$$s_{min} \le s_t \le s_{max} \qquad \forall t = 0, ..., T \tag{11}$$

$$a_{min} \le a_t \le a_{max} \qquad \forall t = 0, ..., T-1 \tag{12}$$

where the instantaneous regret $r(t, s_t, a_t, \xi_t)$ is simply given by $c(t, a_t)$, that is the cost for pumping an amount a_t of water within a hour; it explicitly depends on t because energy price usually changes over the day (i.e., Time-of-Use tariff). Then, (10) is the mass-balance equation (i.e., transition dynamics of the system), (11) is a *safety constraint*, and (12) is an operational constraint (i.e., min-max amount of water which can be supplied in a hour). It is important to remark again that ξ_t is the uncertain component, only observable a-posteriori and whose distribution is unknown. Furthermore, we assume that the system's dynamic is also unknown: we have just reported all the equations for illustrative purposes.

Experimental Setup. The case study was instantiated by setting $s_{min} = 5\text{m}^3$, $s_{max} = 50\text{m}^3$, $a_{min} = 0\text{m}^3$, $a_{max} = 10\text{m}^3$. The safe-by-design policy π_0 consists in $a_t = \min\{a_{min}, s_{max} - s_{t-1}\}$ and it has been applied on 365 days with water demand sampled from a typical real-life daily water demand pattern. The 365 samples are shown in Fig. 3: average and standard deviation of the base pattern can be easily identified. Water demand is the uncertain component and its distribution is therefore considered as unknown.

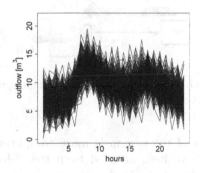

Fig. 3. Daily water demand patterns related to 365 days.

With respect to $c(t, a_t)$, we assumed a non-linear relation between the amount of water supplied in a hour and the consumed energy, that is $E_t = a_t^3$, leading to:

$$c(t, a_t) = \frac{price_t \cdot E_t}{\eta} = \frac{price_t \cdot a_t^3}{\eta} \tag{13}$$

with $\eta = 0.85$ the pump's efficiency and

$$price_t = \begin{cases} 1€/kWh & \text{if } t \in [1;6] \cup [23;24] \\ 2€/kWh & \text{if } t \in [9;18] \cup [21;23] \\ 3€/kWh & \text{if } t \in [7;8] \cup [19;20] \end{cases} \qquad (14)$$

Instead of solving (9-12) via stochastic programming or simulation-optimization, the proposed approach starts from the available safe – and sub-optimal – control policy π_0, such that $a_t = \min\{a_{min}, s_{max} - s_{t-1}\}$. The two associated datasets $\mathcal{D}_{\pi_0}$ and $\mathcal{R}_{\pi_0}$ are used to fit the two GPs and to obtain the associated $\mu_{\mathcal{D}}(\mathbf{x})$, $\sigma_{\mathcal{D}}(\mathbf{x})$, $\mu_{\mathcal{R}}(\mathbf{x})$, and $\sigma_{\mathcal{R}}(\mathbf{x})$. The the optimization problem is formulated as:

$$\min_{a_t \in [a_{min}, a_{max}]} \left\{ \mu_{\mathcal{R}}(\mathbf{x}|t, s_t) - \beta_{\mathcal{R}} \, \sigma_{\mathcal{R}}(\mathbf{x}|t, s_t) \right\} \qquad (15)$$

$$s.t. \quad \mu_{\mathcal{D}}(\mathbf{x}|t, s_t) + \beta_{\mathcal{D}} \, \sigma_{\mathcal{D}}(\mathbf{x}|t, s_t) \leq s_{max} \qquad (16)$$

$$\mu_{\mathcal{D}}(\mathbf{x}|t, s_t) - \beta_{\mathcal{D}} \, \sigma_{\mathcal{D}}(\mathbf{x}|t, s_t) \geq s_{min} \qquad (17)$$

In the case that a feasible solution of the problem does not exist, then a_t is chosen according to the safe-by-design policy π_0.

4.2   Case Study 2: Optimal Control of a House Heating System

System Description. The second case study is the control of a house heating system. A graphical representation of the case study is sketched in Fig. 4.

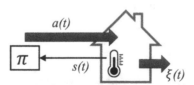

Fig. 4. Control of a house heating system: a schematic representation. Notation follows that used in the general schema in Fig. 1.

Transition dynamics of the system is more complicated than the previous case study: equations have been adapted from the "Model A House Heating System" example available at the Mathworks website[1].

The *changing in-house temperature* equation is:

$$\frac{ds}{dt} = \frac{1}{m\kappa} \left(\frac{dQ_G}{dt} - \frac{dQ_L}{dt} \right) \qquad (18)$$

---

[1] https://www.mathworks.com/help/simulink/ug/model-a-house-heating-system.html.

where m is the mass of air in the house, κ is the heat capacity and $\frac{dQ_G}{dt}$ and $\frac{dQ_L}{dt}$ are the *rate of heat gain* equation and the *rate of heat loss* equation, respectively. More specifically, they are computed as follows:

$$\frac{dQ_G}{dt} = M\kappa\left(a(t) - s(t)\right); \qquad \frac{dQ_L}{dt} = \frac{\left(s(t) - \xi(t)\right)}{R}$$

with M the mass of air of the heater, R the thermal resistance, $s(t)$ the current in-house temperature, and $\xi(t)$ the outside temperature. Finally, the system's dynamics can be synthesized by the following equation:

$$s(t + \Delta t) = s(t) + \frac{\Delta t}{m\kappa}\left(M\kappa\left(a(t) - s(t)\right) + \frac{\left(s(t) - \xi(t)\right)}{R}\right) \qquad (19)$$

Optimal Control Problem. The objective function of this problem is

$$\sum_{t \in [0;T]} c\left(t, a(t)\right) \qquad (20)$$

with a time step equal to $\Delta t = 1$ [minute]. The constraints are given by:

- the system's transition dynamic equation (19);
- the operational constraint $a_{min} \leq a(t) \leq a_{max}$ $\forall t \in \{0, \Delta t, ..., T - \Delta t\}$
- the safety constraints are related to keep an in-house target temperature $\bar{s} = 18°C$ by
  - reaching it within 30 min;
  - not exceeding a sovraelongation of $1°C$ (i.e., $s(t) <= 19°C$ for $t <= 30$ [minutes]);
  - and keeping $s(t) \in [\bar{s} - 0.5°C; \bar{s} + 0.5°C]$ $\forall t = 0, ..., T$
- the Time-of-Use tariff:

$$price_t = \begin{cases} 1€/°C & \text{if } t \in [0; 7] \cup [21; 24] \\ 10€/°C & \text{if } t \in (7; 10] \cup (17; 21] \\ 5€/°C & \text{if } t \in (10; 17] \end{cases} \qquad (21)$$

Experimental Setup. The values of all the parameters characterizing the in-house heating system are taken from the Mathworks' example and reported here for the sake of completeness: $m = 1470$ [kg], $\kappa = 1005.4$, [Joule/°C kg], $M = 3600$ [kg hour], $R = 4.329 \cdot 10^{-7}$[°C hour/Joule].

For the outside temperature, $\xi(t)$, we have considered the freely available Yosemite temperature data[2]: 60 days of temperature measures, with a 1 sample every minute. All the values were converted from Fahrenheit to Celsius degrees.

---

[2] https://github.com/facebook/prophet/blob/main/examples/
example_yosemite_temps.csv.

In this case, the safe-by-design policy π_0 consists of a PID (proportional-integrative-derivative) controller, such that

$$a(t) = \min \left\{ K_p e(t) + K_i \sum_{i \in [0;t]} e(t) + K_d \frac{e(t) - e(t-1)}{\Delta t}, \quad a_{max} \right\} \tag{22}$$

where $e(t) = \bar{s} - s(t)$, $a_{max} = 70°$ C and the PID's parameters have been manually set to $K_p = 43$, $K_i = 0.17$ and $K_d = 0$ to meet all the constraints.

Figure 5 shows: (left) the outside temperature data, (middle) the associated control actions performed by the PID controller, and (right) the resulting in-house temperature, separately for the 60 days considered

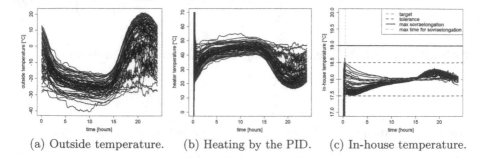

(a) Outside temperature.   (b) Heating by the PID.   (c) In-house temperature.

Fig. 5. Safe control policy of the in-house heating system.

It is important to remark that, contrary to most of the existing approaches which combines safe active learning and Bayesian optimization to efficiently tune the PID's parameters [10,13–15], the proposed approach aims at searching for a new policy which could be more sophisticate than (22).

5 Experiments and Results

5.1 Results on Case Study 1

Figure 6 shows the approximation (i.e., the GP model) of the water tank's dynamics at $t = 1$, with $\mu_{\mathcal{D}}(\mathbf{x})$ and $\sigma_{\mathcal{D}}(\mathbf{x})$ respectively on the left and right hand side. the 365 blue dots represent the safe experience, that is the actions performed according to both π_0 and the realizations of $\xi(t)$ for the 365 days considered.

More in detail, the picture shows that many days started with an amount of water into the tank $s(t-1) \leq 40m^3$, leading to pump the maximum possible amount of water (i.e., $a(t) = 10m^3$). For all the other days, starting with $s(t-1) > 40$, the actions implied by π_0 are given by $s_{max} - s(t-1)$.

Figure 7 shows how actions change over time due to a higher withdrawal of water from the tank over the day. As an example, $t = 11$ is reported.

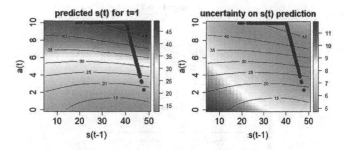

Fig. 6. Time $t = 1$: predicted amount of water $s(t)$ depending on $s(t - 1)$ and in response to $a(t)$ (left), and uncertainty about the prediction (right).

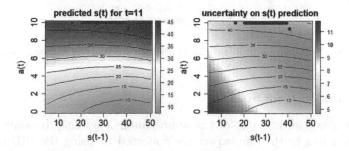

Fig. 7. Time $t = 11$: predicted amount of water $s(t)$ depending on $s(t - 1)$ and in response to $a(t)$ (left), and uncertainty about the prediction (right).

Figure 8 provides an example of the optimization problem (15–17) solved for a given day and at a certain hour (i.e., $t = 17$). The top shows the approximation of the instantaneous regret – specifically Eq. (15) with $\beta_{\mathcal{R}} = 1$ (dashed blue line) – to be minimized under the (approximated) safety constraints depicted in the bottom – specifically Eqs. (16–17) with $\beta_{\mathcal{D}} = 3$ (i.e., 99.7% of the possible realizations of $s(t)$, under the Gaussian assumption). The horizontal dashed line represents the optimal and safe action identified.

The approach was tested over the first 30 days of the 365 to compute the improvement in terms of performances (i.e., cost reduction) with respect to the risk for safety. **Results show a reduction of the daily cost equal to 3.7%, on average, but with a safety violation for one out of the 30 days.**

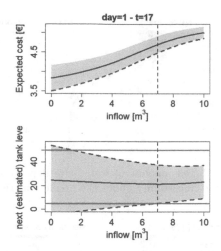

Fig. 8. Optimization problem (15-17) solved for a given day and time t.

5.2   Results on Case Study 2

Figure 9 presents an example of the approximated systems dynamic (i.e., GP model) according to the safe experience collected by using the PID controller. As an example we have selected one of the first minutes to highlight that when the in-house temperature $s(t)$ is very low then the control action is the maximum possible heating, that is $a(t) = 70\,°C$, otherwise a lower heating is selected. Blue points are used to depict safe experience data.

At every time step, more precisely every minute, an optimization problem is solved aimed at minimizing the most optimistic estimation of the instantaneous regret, that is $\mu_\mathcal{R}(\mathbf{x}|t, s(t)) + \beta_\mathcal{R}\sigma_\mathcal{R}(\mathbf{x}|t, s(t))$ (with $\beta_\mathcal{R} = 1$), under the constraints listed above and whose fulfilment is evaluated with respect to the prediction $s(t + \Delta t) \simeq \mu_\mathcal{D}(\mathbf{x}|t, s(t))$, uncertainty $\sigma_\mathcal{D}(\mathbf{x}|t, s(t))$, and $\beta_\mathcal{D} = 3$.

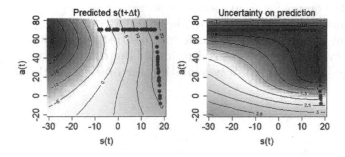

Fig. 9. House heating system: an example of the approximated system dynamic for a given t (on the left) and associated predictive uncertainty (on the right).

Although this second case study is more complicated than the first one, especially in terms of constraints, the proposed approach has provided analogous results. Specifically, a**cost reduction of around 4% is achieved with few violations of the constraints, over 60 days**.

5.3 Software and Data

Software and data will be published on github if the paper is accepted.

6 Conclusions, Limitations, and Perspectives

Empirical results on the two case studies prove that the proposed approach can explore new and more efficient policies starting from safe experience data, without requiring stochastic programming or an expensive a digital twin. However, results also show there is the risk to incur into safety violations. This is due to two different aspects: the first is that β_D should be chosen to set a suitable *margin* on wrong predictions of the next state; the second is related to the partial knowledge of the system dynamic and to the fact that the new policy could lead to states significantly different from any other in the safe experience data, without any chance to avoid *unsafe* actions. Ongoing works aims at (i) defining a mechanism to identify a suitable β_D, (ii) evaluating safety not only step-wise but also looking ahead – considering the additional computational cost – and (iii) extending the safe experience by including the new one collected in response to the new policy. The last point is really important, especially in terms of an analysis of the convergence towards the optimum and safest policy for the specific target problem, under the realizations of the uncertain component $\xi(t)$.

References

1. Lu, Q., et al.: Stochastic programming for floodwater utilization of a complex multi-reservoir system considering risk constraints. J. Hydrol. **599**, 126388 (2021)
2. Han, D., Lee, J.H.: Two-stage stochastic programming formulation for optimal design and operation of multi-microgrid system using data-based modeling of renewable energy sources. Appl. Energy **291**, 116830 (2021)
3. Lima, R.M., Conejo, A.J., Giraldi, L., LeMaitre, O., Hoteit, I., Knio, O.M.: Risk-averse stochastic programming vs. adaptive robust optimization: a virtual power plant application. INFORMS J. Comput. **34**, 1795–1818 (2022)
4. Rachih, H., Mhada, F., Chiheb, R.: Simulation optimization of an inventory control model for a reverse logistics system. Dec. Sci. Lett. **11**(1), 43–54 (2022)
5. Chakraei, I., Safavi, H.R., Dandy, G.C., Golmohammadi, M.H.: Integrated simulation-optimization framework for water allocation based on sustainability of surface water and groundwater resources. J. Water Resour. Plan. Manag. **147**(3), 05021001 (2021)
6. Tordecilla, R.D., Juan, A.A., Montoya-Torres, J.R., Quintero-Araujo, C.L., Panadero, J.: Simulation-optimization methods for designing and assessing resilient supply chain networks under uncertainty scenarios: A review. Simul. Model. Pract. Theory **106**, 102166 (2021)

7. Candelieri, A., Galuzzi, B., Giordani, I., Archetti, F.: Learning optimal control of water distribution networks through sequential model-based optimization. In: International Conference on Learning and Intelligent Optimization, pp. 303–315 (2020)
8. Candelieri, A., Ponti, A., Archetti, F.: Data efficient learning of implicit control strategies in water distribution networks. In: 2021 IEEE 17th International Conference on Automation Science and Engineering (CASE), pp. 1812–1816 (2021)
9. Schreiter, J., Nguyen-Tuong, D., Eberts, M., Bischoff, B., Markert, H., Toussaint, M.: Safe exploration for active learning with gaussian processes. In: Bifet, A., et al. (eds.) ECML PKDD 2015. LNCS (LNAI), vol. 9286, pp. 133–149. Springer, Cham (2015). https://doi.org/10.1007/978-3-319-23461-8_9
10. Schillinger, M., Hartmann, B., Skalecki, P., Meister, M., Nguyen-Tuong, D., Nelles, O.: Safe active learning and safe Bayesian optimization for tuning a PI-controller. IFAC-PapersOnLine **50**(1), 5967–5972 (2017)
11. Sui, Y., Zhuang, V., Burdick, J., Yue, Y.: Stagewise safe bayesian optimization with gaussian processes. In: International Conference on Machine Learning, pp. 4781–4789. PMLR (2018)
12. Kirschner, J., Mutny, M., Hiller, N., Ischebeck, R., Krause, A.: Adaptive and safe Bayesian optimization in high dimensions via one-dimensional subspaces. In: International Conference on Machine Learning, pp. 3429–3438. PMLR (2019)
13. Fiducioso, M., Curi, S., Schumacher, B., Gwerder, M., Krause, A.: Safe contextual Bayesian optimization for sustainable room temperature PID control tuning. arXiv preprint arXiv:1906.12086 (2019)
14. Berkenkamp, F., Krause, A., Schoellig, A. P.: Bayesian optimization with safety constraints: safe and automatic parameter tuning in robotics. Mach. Learn. 1–35 (2021)
15. König, C., Turchetta, M., Lygeros, J., Rupenyan, A., Krause, A.: Safe and efficient model-free adaptive control via Bayesian optimization. In: 2021 IEEE International Conference on Robotics and Automation (ICRA), pp. 9782–9788. IEEE (2021)
16. Frazier, P.I.: Bayesian optimization. In: Recent Advances in Optimization and Modeling of Contemporary Problems, pp. 255–278. Informs (2018)
17. Archetti, F., Candelieri, A.: Bayesian Optimization and Data Science. Springer, Cham (2019). https://doi.org/10.1007/978-3-030-24494-1
18. Williams, C.K., Rasmussen, C.E.: Gaussian Processes for Machine Learning. MIT Press, Cambridge (2006)
19. Gramacy, R.B.: Surrogates: Gaussian Process Modeling, Design, and Optimization for the Applied Sciences. Chapman and Hall/CRC, Boca Raton (2020)
20. Deisenroth, M.P., Fox, D., Rasmussen, C.E.: Gaussian processes for data-efficient learning in robotics and control. IEEE Trans. Pattern Anal. Mach. Intell. **37**(2), 408–423 (2013)
21. Bischoff, B., et al.: Policy search for learning robot control using sparse data. In: 2014 IEEE International Conference on Robotics and Automation (ICRA), pp. 3882–3887. IEEE (2014)
22. Kamthe, S., Deisenroth, M.: Data-efficient reinforcement learning with probabilistic model predictive control. In: International Conference on Artificial Intelligence and Statistics, pp. 1701–1710. PMLR (2018)
23. Sergeyev, Y.D., Candelieri, A., Kvasov, D.E., Perego, R.: Safe global optimization of expensive noisy black-box functions in the δ-Lipschitz framework. Soft. Comput. **24**(23), 17715–17735 (2020). https://doi.org/10.1007/s00500-020-05030-3

24. Srinivas, N., Krause, A., Kakade, S.M., Seeger, M.W.: Information-theoretic regret bounds for gaussian process optimization in the bandit setting. IEEE Trans. Inf. Theory **58**(5), 3250–3265 (2012)

25. De Ath, G., Everson, R. M., Fieldsend, J. E., Rahat, A. A.: ε-shotgun: ε-greedy batch Bayesian optimisation. In: Proceedings of the 2020 Genetic and Evolutionary Computation Conference, pp. 787–795 (2020)

26. De Ath, G., Everson, R.M., Rahat, A.A., Fieldsend, J.E.: Greed is good: exploration and exploitation trade-offs in Bayesian optimisation. ACM Trans. Evol. Learn. Optim. **1**(1), 1–22 (2021)

27. Berk, J., Gupta, S., Rana, S., Venkatesh, S.: Randomised Gaussian process upper confidence bound for Bayesian optimisation. In: Proceedings of the 29th International Conference on Artificial Intelligence, pp. 2284–2290 (2021)

28. Candelieri, A., Archetti, F.: Sparsifying to optimize over multiple information sources: an augmented Gaussian process based algorithm. Struct. Multidiscip. Optim. **64**(1), 239–255 (2021). https://doi.org/10.1007/s00158-021-02882-7

29. Candelieri, A., Perego, R., Archetti, F.: Green machine learning via augmented Gaussian processes and multi-information source optimization. Soft. Comput. **25**(19), 12591–12603 (2021). https://doi.org/10.1007/s00500-021-05684-7

Bayesian Optimization in Wasserstein Spaces

Antonio Candelieri[1]($\boxtimes$), Andrea Ponti[1,3,4], and Francesco Archetti[2,4]

[1] Department of Economics, Management and Statistics,
University of Milano-Bicocca, Milan, Italy
`antonio.candelieri@unimib.it`
[2] Department of Computer Science, Systems and Communication,
University of Milano-Bicocca, Milan, Italy
[3] Oaks s.r.l., Milan, Italy
[4] Consorzio Milano Ricerche, Milan , Italy

Abstract. Bayesian Optimization (BO) is a sample efficient approach for approximating the global optimum of black-box and computationally expensive optimization problems which has proved its effectiveness in a wide range of engineering and machine learning problems. A limiting factor in its applications is the difficulty of scaling over 15–20 dimensions. It has been remarked that global optimization problems often have a lower intrinsic dimensionality which can be exploited to construct a feature mapping the original problem into low dimension manifold. In this paper we take a novel approach mapping the original problem into a space of discrete probability distributions endowed with a Wasserstein metric. In this new approach both the Gaussian process model and the acquisition function work in a 1-dimensional Wasserstein (WST) space. The results in the WST space are then mapped back to the original space using a neural network. Computational results show that, at least for high dimension additive test functions, the exploration in the Wasserstein space is significantly more effective.

Keywords: Bayesian optimization · Wasserstein distance · Gaussian processes

1 Introduction

Bayesian Optimization (BO) is a sample efficient approach for approximating the global optimum of black-box and computationally expensive optimization problems which has proved its effectiveness in a wide range of engineering design and machine learning problems. A limiting factor in its applications is the difficulty of scaling over 15–20 dimensions. The issue of BO in high dimensional problems has been addressed translating it into low-dimensional problems defined on subsets of variables. (Kandasamy et al. 2015; Moriconi et al. 2020a) or exploiting a lower intrinsic dimensionality. To tackle the issue of high dimensionality we take a novel approach embedding the original problem into a space of discrete probability distributions endowed with a Wasserstein metric. We consider the optimization of a black-box, expensive, multi-extremal function $f(x)$:

$$f(x) : x \in X \subset \mathbb{R}^d \to R \qquad (1)$$

© Springer Nature Switzerland AG 2022
D. E. Simos et al. (Eds.): LION 2022, LNCS 13621, pp. 248–262, 2022.
https://doi.org/10.1007/978-3-031-24866-5_19

where $\mathcal{X}$ is the search space and neither gradient nor convexity information are available. Moreover, we assume that:

- $f(x)$ is some unknown composition of functions, that is:

$$f(\mathbf{x}) = C(h_1(\mathbf{x}), \dots, h_M(\mathbf{x})) \tag{2}$$

- Also the mapping between $x \in \mathcal{X} \subset \mathfrak{R}^d$ and $h(x) \in \mathfrak{R}^M$ is unknown. Only the value of each component function, $h_m(\mathbf{x})$ with $m = 1, \dots, M$, is observable.

Another approach for acquisition function in high dimensional spaces has been proposed in (Candelieri et al. 2020). Some classes of problems belong to this setting, as first observed in (Kandasamy et al. 2015) where the additive structure in high dimensional problems has been leveraged into the algorithm Add-GP-UCB. Also the scalarization approach to multi-objective optimization (Zhang and Golovin, 2020) is a specific case of the model in Eq. 2. Another case, considered in (Astudillo and Frazier 2021) assumes that $h_1, \dots, h_m$ can be arranged in a directed acyclic network and also fits in Eq. 2. In the above papers the composition function $C(\cdot)$ is assumed known.

Instead, in the approach proposed in this paper, no assumptions about $C(\cdot)$ are postulated. This argument is related to the approach proposed in (Astudillo and Frazier 2022) which leverages internal information of composite objective functions where one can observe and selectively evaluate individual constituents. The knowledge of the vector $h(x)$ is extremely important and valuable because its elements are informative of the value of $f(x)$, significantly more than x. We assume that $h(x)$ can be observed.

Let denote with $\psi : \mathfrak{R}^M \to \mathcal{W}$, a function transforming the values of the $h(x)$'s into a histogram, $x \Rightarrow h(x) = (h_1(x), \dots, h_M(x)) \Rightarrow \psi(h(x)) = H(x)$ which is the histogram associated to x. A general constructive procedure for the computation of $\psi(\cdot)$ is as follows.

- Given the function $f(x) = C(h_1(x), \dots, h_M(x))$ for $i = 1, \dots, M$ we assume without, loss of generality, that a and b can be a lower bound a an upper bound respectively given by $\min_i \min_x h_i(x)$ and $\max_i \max_x h_i(x)$ that is split into K equi-intervals $\Delta_k = t_{k+1} - t_k$: $t_0 = a$ and $t_K = b$. The histogram $H(x)$ has bins Δ_k and weights w_k. The weight w_k associated to bin Δ_k is given by the cardinality of the set

$$w_k = |\{i : h_i(x) \in \Delta_k\}| \tag{3}$$

- In the specific case $f(x) = \sum_{i=1}^{M} \lambda_i h_i(x)$ the same procedure associates for each λ vector a histogram to x.

Other situations which fit into the above model and yield naturally a distributional representation of a candidate solution is when the data generating process is a simulation model. This is the case in optimizing the average performance of a system where $f(x, w)$ is the value of $f(x)u$ nder the environmental condition w and $p(w)$ represents the "relevance" of condition w (probability of its occurrence or the fraction of time that condition w occurs). A closely related instance is when optimizing the expected performance of systems modelled by a discrete event simulation $f(x, w)$ where w is a random variable. In this case $f(x) = \mathbb{E}(f(x, w))$.This is the case of hyperparameter optimization in a machine learning algorithm based on k-fold cross validation. In this application $f(x, w)$ is the any loss function (predictive accuracy, fairness, explainability) on fold w using hyperparameter configuration x, $N = k$, and our goal is to minimize the loss function $\sum_w f(x, w)$. Also in this case it might be expedient rather than fit a model to the data to consider directly the empirical distribution.

In these cases, the procedure is the same as in Eq. 3: a candidate solution x generates a set (later called sample) of values $f(x, w_j)$ with $j = 1, \ldots N$ where N can be the number of environmental conditions or scenarios rather than the number of observations of the stochastic elements in the simulation This sample $f(x, w_j)j = 1, \ldots N$ is "bucketed" into an histogram with support in the range (f^l, f^u), where f^l and f^u are a lower and upper bound of f, subdivided into K equal subintervals (bins).

The key idea is to represent a point x, upon the evaluation of $f(x)$, as a histogram which encodes the information gathered about the objective function. The space of histograms is structured as a metric space that is a non-empty set with a metric defined on the set. The metric is a function which defines a distance between two elements of the set (points in the space) with the properties of positivity, symmetry, and triangle inequality. Since the points in this metric space are histograms, the distance between them is a distance between probability distributions. There is a plethora of distances between distributions (Sriperumbudur et al. 2010). The distributional distances commonly used in machine learning are the entropy-based ones like Kullback-Leibler (KL) and Jensen-Shannon (JS) whose application to gauge network dissimilarity between networks has been proposed in (Schieber et al. 2017). Some of them do not satisfy the requirements of non-negativity, symmetry and triangle inequality and are therefore not true metrics. In this paper we focus on the Wasserstein (WST) distance and embed the optimization problem in the metric space whose points are histograms equipped with the Wasserstein distance, which we call Wasserstein space. Wasserstein distance, also known as the optimal transport distance, is a mathematically principled method to align probability distributions. Originated by a paper of Monge (Monge 1781) it received its linear programming formulation in Kantorovich (Kantorovich 1942). A complete mathematical formulation is in (Villani 2009) while (Peyre and Cuturi 2019) offer a complete review of the recent theoretical and computational advances. It has many versions, both continuous and discrete, and its versatility has warranted attention also beyond its mathematical foundations Notably it has found applications in machine learning from shape analysis (Gangbo and McCann 2000) to image interpolation, domain adaptation (Redko et al. 2019), parameter estimation in simulation models (Öcal et al. 2019), structured data on graphs (Vayer et al. 2018) active learning (Frogner et al. 2019) and adversarial networks (Ariovsky et al. 2017).

WST distance, also called optimal transport distance, lifts a ground metric between bins to a metric between histograms and this makes it the method of choice to compare histograms. The optimal transport distance, has many important properties. It is a "weak" distance meaning that WST spaces are very large, and it allows to compare singular distributions (e.g., discrete ones) whose supports do not overlap and quantify the spatial shift between supports. Their representational capability has been shown by embedding a variety of complex objects like images, networks and words. An explanation of the interest in the Wasserstein distance is that Euclidean embeddings of data are flawed as they account for the correspondence of each feature independently of the other features. This limits the ability of Euclidean embedding to capture complex relationships between inputs due to limiting assumptions about neighbourhood sizes and connectivity. At the contrary the Wasserstein distance is cross-feature and provides geometrically meaningful distances to compare discrete distributions. Bayesian Optimization algorithms have so far largely focused on problems where inputs are represented as numerical and categorical variables in Euclidean spaces.

In this paper we extend the distributional approach to Bayesian optimization encoding the geometry of the data generated in the sequential optimization process and performing the search in the Wasserstein space. In this paper we focus on the relation between WST distance and Bayesian optimization inducing, through the function $\psi(h(x))$, a non-Euclidean structure in the search space and consequently in the representation of the data sequentially acquired during the optimization process. The key advantage of Bayesian Optimization over other black-box learning and optimization algorithms is sample efficiency which might be not enough in high dimensional spaces. This observation begets the key research question we address in this paper: can the sample efficiency of Bayesian optimization for high dimensional problems be improved embedding the optimization in the Wasserstein space?

A GP is completely defined by its mean $\mu(x)$ and covariance function, where the latter is typically a kernel function $k(x, x')$: $\mathcal{GP}(\mu(x), k(x, x'))$.

Here we consider, just for the sake of simplicity, the Squared Exponential kernel: $k(x, x') = e^{-\frac{\|x-x'\|^2}{2\ell^2}}$ which translates in the Wasserstein space into $k(w, w') = e^{-\frac{\|w-w'\|^2}{2\ell^2}}$ where w and w' are the histograms associated to $(x, f(x))$ and $\left(x', f\left(x'\right)\right)$.

Learning the value of the length scale ℓ in the WST space is not a problem as we can use MLE. The main difference in training the GP model in $\mathcal{W}$ is a different definition of the function evaluation dataset:

$$D_{1:n} = (X_{1:n}, W_{1:n}, y_{1:n}) \tag{4}$$

where:

- $X_{1:n} = \left\{x^{(i)}\right\}_{i=1:n}$ are the locations queried.
- $W_{1:n} = \left\{w^{(i)}\right\}_{i=1:n}$ are the associated histograms (in $\mathcal{W}$).
- $y_{1:n} = \left\{y^{(i)}\right\}_{i=1:n}$ are the observed function values.

The WST Bayesian Optimization framework is based on two building blocks.

The first performs the search in the WST space. It this paper we use the squared exponential in the Wasserstein space $K_{SE}(w, w') = e^{-\frac{\left\|w-w'\right\|^2}{2\sigma^2}}$ and the L/UCB acquisition also in the Wasserstein space. The second block learns a neural network which maps the results of the first back to the original search space $\mathcal{X}$. The WST algorithm, called RBF kernel - W, has been tested vs. basic BO, called RBF kernel - $\mathcal{X}$: the computational results show that the WST algorithm outperforms basic BO already in 10 dimensions and that the competitive edge increases substantially as the dimension of the search space increases.

1.1 Related Works

The use of the WST distance in optimization problems is still a sparsely explored field. Important results from the mathematical programming community, (Mohajerin Esfahani and Kuhn 2018) and (Nguyen et a. Nguyen et al., 2021), use the Wasserstein metric for distributionally robust optimization (DRO). (Kuhn et al. 2019) focus Wasserstein based DRO onto machine learning and carry out parameter estimation by minimizing empirical risk of any data measure within given value of the WST distance of the input data. The issue of DRO is analysed also in (Liu 2022) who propose a Wasserstein barycentric ambiguity set. (Pflug and Pichler 2012) use the WST distance for multistage stochastic optimization models. (Cohen et al. 2020) introduce a practical algorithm that treats the barycenter as a parametric model and can be applied to high dimensions. (Moriconi et al. 2020b) use quantile Gaussian processes. Closer to the focus of our paper is (Kandasamy et al. 2018) which use a kernel induced by Wasserstein distance in a Bayesian optimization framework to search the space of neural network architectures. It must be remarked that in the case of multivariate distributions the construction of positive definite kernels on sets of probability measures is not straightforward. The traditional approach for learning from distribution is to consider reproducing kernel Hilbert spaces (RKHS). The kernels associated to probability distributions, in particular the Hilbertian kernel on probability measures have been first proposed in (Hein et al. 2005).

A general solution to the problem in the setting of Hilbert spaces has been provided in (Peyré and Cuturi 2019). Specific positive definite kernels are designed in order to map distributions into a reproducing kernel Hilbert space and extend kernel methods to probability measures. Results are strongly dependent on the dimension of the histograms.

(De Plaen et al. 2020) gives an approximation result which shows that as the parameter sigma in goes to 0, the smallest eigenvalue of the kernel matrix tends to 1 therefore yielding a positive definite kernel. A distributional distance-based learning has been shown to be very effective also over discrete structures (Ponti et al. 2021a; Ponti et al. 2021b).

The major problem in extending the kernel to WST spaces is that for $K_w(w, w') = e^{-\frac{w_2^2(w,w')}{2\sigma^2}}$ to be a kernel, any resulting squared exponential kernel matrix built with the Wasserstein distance must be positive definite. This in general cannot be guaranteed because Wasserstein spaces are not Hilbertian and therefore a Wasserstein based kernel $K_W(w, w')$ is not guaranteed, but in some specific conditions, to be positive definite.

One such condition is when the histograms is univariate which is met by our problem and in general by single objective problems. Single objective optimizations generates 1-d histograms so that we are in the condition in which the resulting kernel $K_W(w, w')$ is definite positive.

1.2 Our Contributions

- A first contribution of this paper is to show that a distributional representation of points in the search space of an optimization problem as histograms, which encode the information about the objective function, can be applied to Bayesian optimization.
- The choice of the Wasserstein distance because it's a metric, captures complex relationships between inputs, neighbourhood sizes and connectivity and provides geometrically meaningful distances.
- A Bayesian optimization method where both the kernel $K_{SE}(w, w') = e^{-\frac{\|w-w'\|^2}{2\sigma^2}}$ and the acquisition function operate in the Wasserstein space.
- A neural network which maps the results from the WST space back to the original search space $\mathcal{X}$.
- A measure of concentration around the global optimum in the space $\mathcal{W}$ as the ambiguity set built upon a notion of distance between histograms.
- Computational results which show that the Wasserstein algorithm outperforms both in terms of function evaluations and wall clock time, the basic BO algorithm and that the competitive edge increases substantially as the dimension of the search space increases.

1.3 Organization of the Paper

The contents of the paper are organized as follows. Section. 2 provides background knowledge in particular about Wasserstein distance in Sect. 2.1 and Bayesian optimization in Sect. 2.2. Section 3 establishes the Wasserstein based Bayesian optimization algorithms. Section 4 the computational results on a test function. Section 5 provides conclusions and perspectives.

2 Background

2.1 Wasserstein Distance

A The WST distance between continuous probability distributions is:

$$W_p(P^{(1)}, P^{(2)}) = \left(\inf_{\gamma \in \Gamma(P^{(1)}, P^{(2)})} \int_{X \times X} d(x^{(1)}, x^{(2)})^p d\gamma(x^{(1)}, x^{(2)}) \right)^{\frac{1}{p}} \tag{5}$$

where $d(x^{(1)}, x^{(2)})$ is also called ground distance (usually it is the Euclidean norm), $\Gamma(P^{(1)}, P^{(2)})$ the set of all joint distributions $\gamma(x^{(1)}, x^{(2)})$ whose marginals are respectively $P^{(1)}$ and $P^{(2)}$ is the set of all transportation plans and $p > 1$ is an index.

For one dimensional distributions WST can be written in an explicit form. Let $\widehat{P}^{(1)}$ and $\widehat{P}^{(2)}$ be the cumulative distribution for one-dimensional distributions $P^{(1)}$ and $P^{(2)}$ on the real line and $\left(\widehat{P}^{(1)}\right)^{-1}$ and $\left(\widehat{P}^{(2)}\right)^{-1}$ be their quantile functions. In this case:

$$W_p\left(P^{(1)}, P^{(2)}\right) = \left(\int_0^1 \left|\left(\widehat{P}^{(1)}\right)^{-1}\left(x^{(1)}\right) - \left(\widehat{P}^{(2)}\right)^{-1}\left(x^{(2)}\right)\right|^p dx\right)^{\frac{1}{p}} \tag{6}$$

Let's now consider the case of a discrete distribution P specified by a set of support points x_i with $i = 1, \ldots, m$ and their associated probabilities w_i such that $\sum_{i=1}^{m} w_i = 1$ with $w_i \geq 0$ and $x_i \in M$ for $i = 1, \ldots, m$. Usually, $M = \mathbb{R}^d$ is the d-dimensional Euclidean space with the l_p norm and x_i are called the support vectors. M can also be a symbolic set provided with a symbol-to-symbol similarity. P can also be written using the notation:

$$P(x) = \sum_{i=1}^{m} w_i \delta(x - x_i) \tag{7}$$

where $\delta(\cdot)$ is the Kronecker delta. The WST distance between two distributions $P^{(1)} = \left\{w_i^{(1)}, x_i^{(1)}\right\}$ with $i = 1, \ldots, m_1$ and $P^{(2)} = \left\{w_i^{(2)}, x_i^{(2)}\right\}$ with $i = 1, \ldots, m_2$ is the solution of the following linear program:

$$W\left(P^{(1)}, P^{(2)}\right) = \min_{\gamma_{ij} \in \mathbb{R}^+} \sum_{i \in I_1, j \in I_2} \gamma_{ij} d\left(x_i^{(1)}, x_j^{(2)}\right) \tag{8}$$

The cost of transport between $x_i^{(1)}$ and $x_j^{(2)}$, $d\left(x_i^{(1)}, x_j^{(2)}\right)$, is defined by the p-th power of the norm $\|x_i^{(1)}, x_j^{(2)}\|$ (usually the Euclidean distance). We define two index sets $I_1 = \{1, \ldots, m_1\}$ and I_2 likewise, such that:

$$\sum_{i \in I_1} \gamma_{ij} = w_j^{(2)}, \forall j \in I_2 \tag{9}$$

$$\sum_{j \in I_2} \gamma_{ij} = w_i^{(1)}, \forall i \in I_1 \tag{10}$$

Equations 9 and 10 represent the in-flow and out-flow constraint, respectively. The terms γ_{ij} are called matching weights between support points $x_i^{(1)}$ and $x_j^{(2)}$ or the optimal coupling for $P^{(1)}$ and $P^{(2)}$. Discrete optimal transport is a linear program and thus can be solved exactly in $\mathcal{O}(n^3 \log n)$ with interior point methods. In practice a version with entropic smoothing (Peyré and Cuturi 2019; Flamary et al., 2021) has proven more efficient. The discrete version of the WST distance, for $p = 1$, is usually called Earth Mover Distance (EMD). For instance, when measuring the distance between grey scale images, the histogram weights are given by the pixel values and the coordinates by the pixel positions. In the specific case of histograms, the entries γ_{ij} denote how much of the bin i has to be moved to bin j. The computation of EMD turns out to be the solution of a minimum cost flow problem on a bi-partite graph where the bins of $P^{(1)}$ are the source nodes and the bins of $P^{(2)}$ are the sinks while the edges between sources and sinks are the transportation costs. In the case of one-dimensional histograms, the computation of WST reduces to the comparison of two 1-dimensional histograms which can be performed by a simple sorting and the application of the following equation.

$$W_p\left(P^{(1)}, P^{(2)}\right) = \left(\frac{1}{n} \sum_{i}^{n} \left|w_i^{(1)*} - w_i^{(2)*}\right|^p\right)^{\frac{1}{p}} \tag{11}$$

where $x_i^{(1)*}$ and $x_i^{(2)*}$ are the sorted samples. To apply directly this formula the histograms have to be aligned. For each histogram, the cumulated relative frequency is computed, whose inverse, as in Eq. 6, is the quantile function $Q(s)$ which for a histogram is piecewise linear. Hence the integral, as in Eq. 6, can be computed as the sum of simple intervals.

In this paper, we embed input data as probability distributions in Wasserstein spaces. The Wasserstein distance can be analysed also from the embedding point of view. Wasserstein spaces are very large meaning that many spaces can embed into them with low distortion. The reverse direction, embedding WST spaces into others, is more difficult and relatively well studied only for discrete distributions. This is motivated by interest in more efficient algorithms by embedding into spaces with easily-computed metrics. In this paper we have used an instance of this approach embedding the WST distance for univariate histograms into the space of quantile functions, as shown in Eq. 6. A recent paper on the subject of embeddings and Wasserstein spaces is (Frogner et al. 2019) where the specific cases of graph embeddings and word embeddings are considered.

2.2 Bayesian Optimization

Bayesian Optimization is a statistically principled approach to adaptively and sequentially select points in which to evaluate the objective function with the aim of optimizing f with a small number of function evaluations. In order to be sample efficient, we need a way of extrapolating our belief about what f looks like at points not yet evaluated. In Bayesian Optimization this is enabled by a surrogate model which should also be able to quantify the uncertainty of its predictions in form of a posterior distribution over function values $f(x)$ at points x. Posteriors represent the belief a model has about the function values about points not yet observed. Therefore, the posterior is the distribution over the outputs conditioned on the data observed so far. When using the Gaussian process model the posterior is given explicitly as a multivariate distribution.

A Gaussian process (GP) is a probability distribution over functions denoted as $f(x) \sim GP(\mu(x), k(x, x'))$ where $\mu(x)$ is the mean function of the GP and $k(x, x')$ is the covariance function (aka kernel). Therefore, GP is as a collection of random variables, any finite number of which have a joint Gaussian distribution and $f(x)$ can be considered as a sample from a multivariate normal distribution (Frazier 2018; Archetti and Candelieri 2019).

Let denote with $X_{1:n} = \left\{x^{(i)}\right\}_{i=1,...,n}$ a set of n points in $\Omega \subset \mathfrak{R}^d$ and with $y_{1:n} = \left\{f\left(x^{(i)}\right) + \varepsilon\right\}_{i=1,...,n}$ the associated function values, possibly noisy with ε a zero-mean Gaussian noise $\varepsilon \sim \mathcal{N}(0, \lambda^2)$. Then the posterior predictive mean $\mu(x)$ and standard deviation $\sigma^2(x)$, conditioned on $X_{1:n}$ and $y_{1:n}$, are given by the following equations:

$$\mu(x) = k(x, X_{1:n})\left[K + \lambda^2 I\right]^{-1} y_{1:n} \tag{12}$$

$$\sigma^2(x) = k(x, x) - k(x, X_{1:n})\left[K + \lambda^2 I\right]^{-1} k(X_{1:n}, x) \tag{13}$$

where $k(x, X_{1:n}) = \left\{k\left(x, x^{(i)}\right)\right\}_{i=1,\dots,n}$ and $K \in \Re^{n \times n}$ with entries $K_{ij} = k\left(x^{(i)}, x^{(j)}\right)$.

The choice of the kernel should reflect prior beliefs over the structural properties of $f(x)$, specifically its smoothness. Almost every kernel has its own hyperparameters to tune – usually via Maximum Log-likelihood Estimation (MLE) or Maximum A Posteriori Probability (MAP) – for reducing the potential mismatches between prior smoothness assumptions and the observed data. Common kernels for GP regression are:

- Squared Exponential: $k_{SE}(x, x') = e^{-\frac{\|x-x'\|^2}{2\ell^2}}$
- Exponential: $k_{EXP}(x, x') = e^{-\frac{\|x-x'\|}{\ell}}$
- Power-exponential: $k_{PE}(x, x') = e^{-\frac{\|x-x'\|^p}{\ell^p}}$
- Matérn3/2: $k_{M3/2}(x, x') = \left(1 + \frac{\sqrt{3}\|x-x'\|}{\ell}\right) e^{-\frac{\sqrt{3}\|x-x'\|}{\ell}}$
- Matérn5/2: $k_{M5/2}(x, x') = \left[1 + \frac{\sqrt{5}\|x-x'\|}{\ell} + \frac{5}{3}\left(\frac{\|x-x'\|}{\ell}\right)^2\right] e^{-\frac{\sqrt{5}\|x-x'\|}{\ell}}$

The acquisition function is the mechanism behind the sample efficiency of Bayesian optimization method: it manages the balancing between exploration and exploitation and is an important concept also outside machine learning (Candelieri et al. 2021). It drives the search of the new evaluation points towards points with potential high values of objective function either because value of $\mu(x)$ is high or the uncertainty represented by $\sigma^2(x)$ is high (or both). A widely used acquisition function is the Expected Improvement (EI) given by the expected improvement on $f(x)$ with respect to the predictive distribution of the surrogate model. Another common acquisition is based the Confidence Bound concept. Lower Confidence Bound – LCB – and Upper Confidence Bound – UCB – in the minimization and maximization case, respectively. The next point to evaluate is given by the minimizer of:

$$LCB(x) = \mu(x) - k\sigma(x) \tag{14}$$

where $k \geq 0$ is the parameter to manage the exploration/exploitation trade-off.

3   The WST-BO Algorithm

3.1   BO in the Wasserstein Space

To train the GP model in the WST space we need the function evaluation dataset:

$$D_{1:n} = (\mathbf{X}_{1:n}, \mathbf{W}_{1:n}, \mathbf{y}_{1:n}) \tag{15}$$

where:

- $X_{1:n} = \{x^{(i)}\}_{i=1:n}$ are the locations queried
- $W_{1:n} = \{w^{(i)}\}_{i=1:n}$ are the associated histograms (in $\mathcal{W}$)
- $y_{1:n} = \{y^{(i)}\}_{i=1:n}$ are the observed function values.

The Gaussian Process is trained in the Wasserstein space and the updating formulas (Eqs. 12 and 13) become:

$$\mu(w) = k(w, W_{1:n})[K + \lambda^2 I]^{-1} y_{1:n} \tag{16}$$

$$\sigma^2(w) = k(w, w) - k(w, W_{1:n})[K + \lambda^2 I]^{-1} k(W_{1:n}, w) \tag{17}$$

where $K_{ij} = k_{\mathcal{W}}(w^{(i)}, w^{(j)})$ and $k(W_{1:n}, w)$ is a vector whose i-th component is $k_{\mathcal{W}}(w^{(i)}, w)$. The minimization of LCB$(\cdot) = \mu(\cdot) - \beta^{\frac{1}{2}}\sigma(\cdot)$ yields to $w^{(n+1)} = \text{argmin} LCB(w)$ used to augments the dataset and to update the kernel matrix and the GP as in Eqs. 16 and 17.

3.2 Mapping $\mathcal{W}$ into $\mathcal{X}$

We need to map back $w^{(n+1)}$ in the $\mathcal{X}$ space to obtain the value $x^{(n+1)}$ and consequently $y^{(n+1)}$. The mapping $\mathcal{W} \rightarrow \mathcal{X}$, is performed by a neural network: from $w^{(n+1)}$ we obtain $\theta^{(n+1)}$ an approximation of $(h_1(x^{(n+1)}), \ldots, h_M(x^{(n+1)}))$. Next, we learn a neural network-based regression model $\Phi : \Theta \rightarrow \mathcal{X}$. The number of layers of the neural network is 3 and in each layer the number of neurons is $\max(n, d)$. Using this model, we obtain the new point $\Phi(\theta^{(n+1)}) = x^{(n+1)}$ which yields $y^{(n+1)} = f(x^{(n+1)})$ which yields to $\tilde{w}^{(n+1)}$. Finally, we update both, the regression model Φ and the GP conditioned to this new observation $(x^{(n+1)}, \tilde{w}^{(n+1)}, y^{(n+1)})$.

4 Computational Results

The results are partial and preliminary. Only one function, Alpine 01 is reported (Al-Roomi 2015).

$$f(x) = \sum_{i=1}^{n} |x_i \sin(x_i) + 0.1 x_i| \tag{18}$$

The algorithms have been implemented using BoTorch (Balandat et al. 2020) a Python library for Bayesian Optimization part of the PyTorch ecosystem. BoTorch provides an easy-to-use interface for defining, managing, and running sequential experiments and a modular interface for composing Bayesian optimization primitives as probabilistic models, acquisition functions, and optimizers. BoTorch organizes the computations in batches of size q. The acquisition function used is Lower Confidence Bound (LCB).

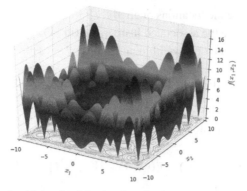

Fig. 1. Alpine 01 objective function in the case of $d = 2$.

All the following results refer to 5 optimization runs; in the figure the mean and standard deviation over the different runs are reported.

Dimension $d = 5$. The plain BO (RBF kernel in X) has a consistently better performance.

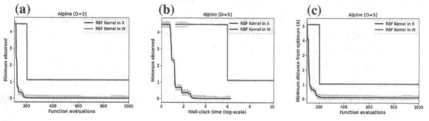

Fig. 2. (a) and (b) represent the minimum observed as a function respectively of function evaluations and wall-clock time; (c) minimum distance averaged over five simulations.

Dimension $d = 10$. The algorithms in the Wasserstein space perform significantly better confirming the superiority of the search in WST space at very low evaluation counts). WST kernel suffers in terms of wall clock time because its preliminary implementation in BoTorch has some significant inefficiencies which will be eliminated in the next version of WST Kernel.

Dimension $d = 50$. The advantage of the Wasserstein embedding grows with the number of dimensions.

4.1 Convergence

The ambiguity set is built upon a notion of distance between histograms. Let H^* the histogram associated to x^* and $f(x^*)$, the ambiguity set of the optimal solution in $\mathcal{W}$ is given by $\mathcal{W}(\varepsilon, H^*) = \{H : W(H^*, H) \leq \varepsilon\}$. The cardinality of $\mathcal{W}(\varepsilon, H^*)$ is a measure of concentration around the global optimum in the space $\mathcal{W}$ (for $\varepsilon = 0.05$ the cardinality is 77).

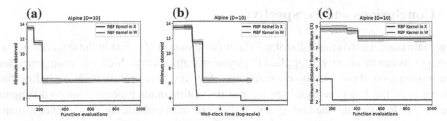

Fig. 3. (a) and (b) represent the minimum observed as a function respectively of function evaluations and wall-clock time; (c) minimum distance averaged over five simulations.

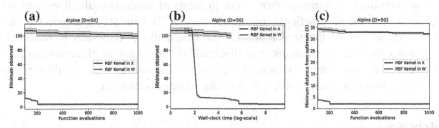

Fig. 4. (a) and (b) represent the minimum observed as a function respectively of function evaluations and wall-clock time; (c) minimum distance averaged over five simulations.

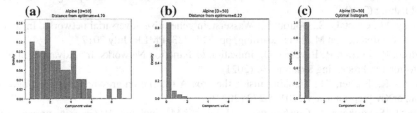

Fig. 5. Early iteration stage (a), late iteration stage (b), optimal histogram (c).

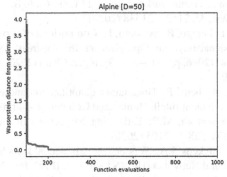

Fig. 6. The Wasserstein distance between the optimal histogram and the sequence $\left\{w^{(i)}\right\}_{i=1,\ldots,n}$.

5   Conclusions and Perspectives

The main conclusion is that a distributional representation of points in the search space as histograms can be effectively applied to Bayesian optimization. Metric learning encodes the information about the objective function for Bayesian optimization including the geometry of the data generated in the sequential optimization process. One could regard the bins as features and the histogram representation as embedding x in a lower dimensional feature space, the Wasserstein space. The Wasserstein distance has been chosen because it's a metric, captures complex relationships between inputs, neighbourhood sizes and connectivity and provides geometrically meaningful distances.

Computational experiments show, both in terms of function evaluations and wall clock time, how the new method outperform "vanilla" BO and its advantage increases with the dimension of the search space. The distributional approach is natural for simulation-optimization problems over discrete structures, sensor placement in physical and informational networks and stochastic vehicle routing. Also multi-objective problems fit into this scheme once a scalarizing strategy is adopted.

References

Al-Roomi, A.R.: Unconstrained Single-Objective Benchmark Functions Repository (2015)

Archetti, F., Candelieri, A.: Bayesian Optimization and Data Science. Springer International Publishing (2019)

Arjovsky, M., Chintala, S., Bottou, L.: Wasserstein generative adversarial networks. In: International Conference on Machine Learning, pp. 214–223. PMLR, July 2017

Astudillo, R., Frazier, P.: Bayesian Optimization of Function Networks. In: Advances in Neural Information Processing Systems, 34 (2021)

Astudillo, R., Frazier, P.I.: Thinking inside the box: A tutorial on grey-box Bayesian optimization. arXiv preprint arXiv:2201.00272 (2022)

Bachoc, F., Suvorikova, A., Ginsbourger, D., Loubes, J.M., Spokoiny, V.: Gaussian processes with multidimensional distribution inputs via optimal transport and Hilbertian embedding. Electron. J. Stat. **14**(2), 2742–2772 (2020)

Balandat, M., et al.: BoTorch: a framework for efficient Monte-Carlo Bayesian optimization. Adv. Neural. Inf. Process. Syst. **33**, 21524–21538 (2020)

Candelieri, A., Giordani, I., Perego, R., Archetti, F.: Composition of kernel and acquisition functions for High Dimensional Bayesian Optimization. In: Kotsireas, I.S., Pardalos, P.M. (eds.) LION 2020. LNCS, vol. 12096, pp. 316–323. Springer, Cham (2020). https://doi.org/10.1007/978-3-030-53552-0_29

Candelieri, A., Ponti, A., Archetti, F.: Uncertainty quantification and exploration–exploitation trade-off in humans. J. Ambient Intell. Humanized Comput., 1–34 (2021)

Cohen, S., Arbel, M., Deisenroth, M.P.: Estimating barycenters of measures in high dimensions. arXiv preprint arXiv:2007.07105 (2020)

De Plaen, H., Fanuel, M., Suykens, J.A.: Wasserstein exponential kernels. In: 2020 International Joint Conference on Neural Networks (IJCNN), pp. 1–6. IEEE, July 2020

Frazier, P.I.: Bayesian optimization. In: Recent Advances in Optimization and Modeling of Contemporary Problems, pp. 255–278. INFORMS (2018)

Frogner, C., Mirzazadeh, F., Solomon, J.: Learning embeddings into entropic wasserstein spaces. arXiv preprint arXiv:1905.03329 (2019)

Gangbo, W., McCann, R.J:. Shape recognition via Wasserstein distance. Quarterly of Applied Mathematics, 705–737 (2000)

Hein, M., & Bousquet, O.: Hilbertian metrics and positive definite kernels on probability measures. In: International Workshop on Artificial Intelligence and Statistics, pp. 136–143. PMLR, January 2005

Kandasamy, K., Neiswanger, W., Schneider, J., Poczos, B., Xing, E.P.: Neural architecture search with bayesian optimisation and optimal transport. Advances in neural information processing systems, 31 (2018)

Kandasamy, K., Schneider, J., Póczos, B.: High dimensional Bayesian optimisation and bandits via additive models. In: International conference on machine learning, pp. 295–304. PMLR, June 2015

Kantorovich, L.V.: On the translocation of masses. In: Dokl. Akad. Nauk. USSR (NS), vol. 37, pp. 199–201 (1942)

Kuhn, D., Esfahani, P.M., Nguyen, V.A., Shafieezadeh-Abadeh, S:. Wasserstein distributionally robust optimization: Theory and applications in machine learning. In: Operations research & management science in the age of analytics, pp. 130–166. Informs (2019)

Lau, T.T.K., Liu, H.: Wasserstein Distributionally Robust Optimization via Wasserstein Barycenters (2022)

Mohajerin Esfahani, P., Kuhn, D.: Data-driven distributionally robust optimization using the Wasserstein metric: Performance guarantees and tractable reformulations. Math. Programm. **171**(1), 115–166 (2018)

Monge, G.: Mémoire sur la théorie des déblais et des remblais. De l'Imprimerie Royale (1781)

Moriconi, R., Deisenroth, M. P., Sesh Kumar, K.S.: High-dimensional Bayesian optimization using low-dimensional feature spaces. Mach. Learn. **109**(9), 1925–1943 (2020a)

Moriconi, R., Kumar, K.S.S., Deisenroth, M.P.: High-dimensional Bayesian optimization with projections using quantile Gaussian processes. Optim. Lett. **14**(1), 51–64 (2019). https://doi.org/10.1007/s11590-019-01433-w

Nguyen, V.A., Shafieezadeh-Abadeh, S., Kuhn, D., Mohajerin Esfahani, P.: Bridging Bayesian and minimax mean square error estimation via Wasserstein distributionally robust optimization. Mathematics of Operations Research (2021)

Öcal, K., Grima, R., Sanguinetti, G.: Parameter estimation for biochemical reaction networks using Wasserstein distances. J. Phys. A: Math. Theor. **53**(3), 034002 (2019)

Peyré, G., Cuturi, M.: Computational optimal transport: With applications to data science. Found. Trends® Mach. Learn. **11**(5–6), 355–607 (2019)

Pflug, G.C., Pichler, A.: A distance for multistage stochastic optimization models. SIAM J. Optim. **22**(1), 1–23 (2012)

Ponti, A., Candelieri, A., Archetti, F.: A Wasserstein distance based multiobjective evolutionary algorithm for the risk aware optimization of sensor placement. Intell. Syst. Appl. **10**, 200047 (2021)

Ponti, A., Candelieri, A., Giordani, I., Archetti, F.: Probabilistic measures of edge criticality in graphs: a study in water distribution networks. Appl. Network Sci. **6**(1), 1–17 (2021). https://doi.org/10.1007/s41109-021-00427-x

Redko, I., Courty, N., Flamary, R., Tuia, D.: Optimal transport for multi-source domain adaptation under target shift. In: The 22nd International Conference on Artificial Intelligence and Statistics, pp. 849–858. PMLR, April 2019

Schieber, T.A., Carpi, L., Díaz-Guilera, A., Pardalos, P. M., Masoller, C., Ravetti, M.G.: Quantification of network structural dissimilarities. Nature Commun. 8(1), 1–10 (2017)

Smola, A.J., Schölkopf, B.: Learning with kernels, vol. 4. GMD-Forschungszentrum Informationstechnik

Sriperumbudur, B.K., Gretton, A., Fukumizu, K., Schölkopf, B., Lanckriet, G.R.: Hilbert space embeddings and metrics on probability measures. J. Mach. Learn. Res. **11**, 1517–1561 (2010)

Vayer, T., Chapel, L., Flamary, R., Tavenard, R., Courty, N.: Optimal transport for structured data with application on graphs. arXiv preprint arXiv:1805.09114 (2018)

Villani, C.: Optimal transport: old and new, vol. 338, p. 23. Springer, Berlin (2009)

Zhang, R., Golovin, D.: Random hypervolume scalarizations for provable multi-objective black box optimization. In: International Conference on Machine Learning, pp. 11096–11105. PMLR, November 2020

Network Vulnerability Analysis in Wasserstein Spaces

Andrea Ponti[1,4,5]([✉]), Antonio Irpino[3], Antonio Candelieri[1], Anna Bosio[2],
Ilaria Giordani[2,4], and Francesco Archetti[2,5]

[1] Department of Economics, Management and Statistics,
University of Milano-Bicocca, Milano, Italy
a.ponti5@campus.unimib.it
[2] Department of Computer Science, Systems and Communication,
University of Milano-Bicocca, Milano, Italy
[3] Dip. Matematica e Fisica Universitá della Campania "Luigi Vanvitelli" Caserta, Caserta, Italy
[4] Oaks S.R.L., Milano, Italy
[5] Consorzio Milano Ricerche, Milano, Italy

Abstract. The main contribution of this paper is the proposal of a new family
of vulnerability measures based on a probabilistic representation framework in
which the network and its components are modelled as discrete probability distributions. The resulting histograms are embedded in a space endowed with a metric
given by the Wasserstein distance. This representation enables the synthesis of a
set of discrete distributions through a barycenter and the clustering of distributions. We show that analyzing the networks as discrete probability distributions
in the Wasserstein space enables the definition of a new family of vulnerability
measures and the assessment of the criticality of each component. Computational
results on real-life networks confirm the validity of our basic assumption that distributional representation can capture the topological information embedded in
a network graph and yield more meaningful metrics than vulnerability measures
based on average values. The computation of the Wasserstein distance is equivalent
to the solution of a min-flow problem: its computational complexity has limited
its diffusion outside the imaging science community. To avoid this computational
bottleneck in this paper, we focus on a statistical approach that drastically reduces
the computational hurdles. This approach has been implemented in a software tool
HistDAWass. The linear complexity of this approach has also enabled the analysis
of large-scale networks.

Keywords: Vulnerability · Wasserstein distance · Network analysis

1 Introduction

There are many measures of network vulnerability: a recent paper (Freitas et al. 2021) is
a wide and updated survey that analyses eighteen such measures. The main contribution
of this paper is the proposal of a new family of vulnerability measures based on a probabilistic representation framework in which the network and its components are modeled

© Springer Nature Switzerland AG 2022
D. E. Simos et al. (Eds.): LION 2022, LNCS 13621, pp. 263–277, 2022.
https://doi.org/10.1007/978-3-031-24866-5_20

as discrete probability distributions. These distributions are embedded in a probabilistic space: in particular, we consider the discrete probability distribution The resulting histograms are embedded in a space endowed with a metric given by the Wasserstein (WST) distance. This paper also establishes a unifying mathematical and statistical framework that enables the computation of Wasserstein-based vulnerability measures and network clustering in the Wasserstein space. This distributional representation makes a better use of all the information hidden in the shortest paths in the network than the connectivity measures given by the average values. We use the Wasserstein distance among possible distances between discrete probability distributions due to its advantages over other distributional distances. It is close to natural perception, allows for flexible binning schemes and enables a synthetic representation of the shape of data through the barycenters. The barycenter is sensitive to the underlying geometry of the histograms. The Wasserstein barycenter, also called the Fréchet mean of distributions, can represent the mean variation of a set of histograms and offers a helpful synthesis of their structure. If we consider two nodes, the WST distance between their histograms captures our intuition of closeness between histograms. Therefore, the concept of barycenter enables clustering of distributions in a space whose metric is the Wasserstein distance. More simply, the barycenter in the space of distributions is the analog of the centroid when the clustering takes place in a Euclidean space. Considering barycenters as centroids, we can perform clustering in WST as the generalization of k-means in the space of discrete distributions endowed with the Wasserstein distance. The advantages of non-Euclidean distances in enabling new insights and a better statistical inference must be weighted against their computational costs. The computation of the Wasserstein distance is equivalent to a min-flow problem that typically has a cubical complexity. This paper focuses on a statistical approach (Irpino and Verde 2015), where the approximations are on the distributions themselves, which drastically reduces the computational hurdles for multivariate distributions. This approach has been implemented in a software tool, HistDAWass, whose basis is described in (Irpino and Verde, 2015), and has been used for the computations in this paper.

In this paper, we show that analyzing the networks as a discrete probability distribution in the Wasserstein space enables the definition of a set of new vulnerability measures. The WST distance between their barycenters gives the difference in vulnerability between two networks; If the two networks differ by one component (node or arc), their WSDT distance indicates the criticality of that component. This paper aims twofold: first to show how Wasserstein spaces enable a better representational flexibility than Euclidean spaces. Second to show how the employed computational method, with its low computational complexity, enables computational procedures which allow even large-scale networks to be analyzed in the new framework. This novel approach has been tested on a benchmark problem and a real-life large water distribution network, the assessment of their vulnerability and their clustering.

1.1 Related Works

An early proposal to measure network similarity as the distance distributions of the node-to-node distances is in (Schieber et al. 2017) which propose the Jensen-Shannon distance. Here we consider the Wasserstein distance due to its flexibility and natural interpretation.

Wasserstein Clustering

The basic computation of the optimal transport map between 2 discrete distributions involves solving a network flow problem whose computation is typically cubic in the sizes of the measure. In the case of simple ground costs, it is shown to be equivalent to a min-flow algorithm of quadratic computational complexity and, in specific cases, e.g., 1-dimensional, to linear. Computing the barycenter and clustering adds two further layers of computations. The problem has been investigated particularly for images (Bonneel et al. 2016). Flamary et al. (2021) have proposed a general package, "Python Optimal Transport," which offers several computational options. A seminal paper is (Applegate et al. 2011) where a method is proposed for clustering multidimensional distributions of patterns of mobile phone calls using earth mover distance, which reduces the cost to linear time. Building on flow representation, Atasu and Mittelholzer (2019) propose an approximate solution for Earth Mover Distance (a specific WST distance for discrete distributions) which attains linear complexity. A statistical approach for the data analysis in WST spaces is proposed in (Bigot 2020) and (Verdinelli and Wasserman 2019), which propose a modified Wasserstein distance for distribution clustering that inherits many properties of the Wasserstein distance at a lower computational cost. Another solution, which will be adopted in this paper and analyzed in Sect. 2, has been proposed in (Irpino and Romano 2007) and (Irpino and Verde 2015). This approach is extended by Balzanella and Irpino (2020) to the spatial prediction and dependence monitoring on georeferenced data streams. Yet another approach is in (Puccetti et al. 2020): their main contribution is the Iterative Swapping Algorithm (ISA) for computing barycenters, which has a quadratic time complexity and can also be applied to clustering problems and more complex optimization problems like the k-barycenter problem.

Wasserstein for Network Vulnerability

The literature on vulnerability and resilience in networks is extremely large. For general networks, an updated survey is in (Freitas et al. 2021). Here we limit to Water distribution Networks (WDN). The complex network analysis of water distribution systems has been introduced in (Yazdani and Jeffrey, 2011) and extended in (Yazdani and Jeffrey 2012) to weighted and directed network models. Graph-theoretic approaches have been the subject of several papers. Herrera et al. (2016) and Di Nardo et al. (2018) applied a graph-theoretic framework for assessing the resilience of sectorized water distribution networks (Candelieri et al. 2017) integrates network analysis and hydraulic simulation (Soldi et al. 2015). The same approach is extended in (Ulusoy et al. 2018) to a resilience measure based on a hydraulically informed measure of link criticality. Directly relevant to this paper is the new approach proposed in (Ponti et al. 2021a), which uses a representation of the network as a discrete probability distribution over the domain of node-to-node distances. As shown in (Ponti et al. 2021b) this also provides a link criticality index.

1.2 Our Contributions

- The distributional representation is shown to capture effectively the topological information embedded in network graphs and yield more meaningful metrics than centrality and vulnerability measures based on average values.
- The statistical approach in the computation of the Wasserstein distance has lower computational complexity than the optimal transport approach.
- The computation of barycenters in Wasserstein spaces enables the synthetic representation of the geometry of the data.
- Wasserstein-based clustering results in better quality and lower computational complexity than usual methods.
- The new family of vulnerability measures provides both aggregate evaluations (at network level) and indications of the criticality even of a single network element.

2 Wasserstein

2.1 Basic Definitions

The WST distance between continuous probability distributions is:

$$
W_p\left(P^{(1)}, P^{(2)}\right) = \left(\int_{X \times X} d\left(x^{(1)}, x^{(2)}\right)^p d\gamma\left(x^{(1)}, x^{(2)}\right)\right)^{\frac{1}{p}} \tag{1}
$$

where $d\left(x^{(1)}, x^{(2)}\right)$ is also called *ground distance* (usually it is the Euclidean norm), $\Gamma\left(P^{(1)}, P^{(2)}\right)$ denotes the set of all joint distributions $\gamma\left(x^{(1)}, x^{(2)}\right)$ whose marginals are respectively $P^{(1)}$ and $P^{(2)}$, and $p \geq 1$ is an index. When $p = 1$ and the probability distribution is discrete, the Wasserstein distance is also called the Earth Mover Distance (EMD). The EMD is the minimum energy cost of moving and transforming a pile of sand in the shape of $P^{(1)}$ to the shape of $P^{(2)}$. The cost is quantified by the amount of sand moved times the moving distance $d\left(x^{(1)}, x^{(2)}\right)$. The EMD then is the cost of the optimal transport plan. Some specific cases are very relevant in applications, where WST can be written in an explicit form. Let $\widehat{P}^{(1)}$ and $\widehat{P}^{(2)}$ be the cumulative distribution for one-dimensional distributions $P^{(1)}$ and $P^{(2)}$ on the real line and $\left(\widehat{P}^{(1)}\right)^{-1}$ and $\left(\widehat{P}^{(2)}\right)^{-1}$ be their quantile functions.

$$
W_p\left(P^{(1)}, P^{(2)}\right) = \left(\int_0^1 \left|\left(\widehat{P}^{(1)}\right)^{-1}(u) - \left(\widehat{P}^{(2)}\right)^{-1}(u)\right|^p du\right)^{\frac{1}{p}} \tag{2}
$$

2.2 The Space of Quantile Functions

As shown by Eq. (2), Wasserstein distance between one dimensional (1D) distributions relies on quantile functions, and, in the case of $p = 2$, it corresponds to a Euclidean distance between two non-decreasing functions with support $[0; 1]$. As shown by Dias and Brito (2015), being $\mathcal{F}(\mathbb{R}, \mathbb{R})$ a set of functions. $(\mathcal{F}, +, \cdot)$ is a vector space of functions equipped with the usual operations of addition $(f + g)(x) = f(x) + g(x)$, $\forall x \in \mathbb{R}$, and scalar product $(\lambda \cdot f)(x) = \lambda \cdot f(x)$, $\lambda \in \mathbb{R}$. Considering the case of the set $\mathcal{E} \subset \mathcal{F}$ of quantile functions such that $\mathcal{E}([0, 1], \mathbb{R})$, and with the non-negative first derivative, i.e., given $f \in \mathcal{E} : [0, 1] \to \mathbb{R}, f' \geq 0$, $\forall x \in (0, 1)$, the space $(\mathcal{E}, +, \cdot)$ is not a subspace of $(\mathcal{F}, +, \cdot)$ since it does not satisfy the definition of a vector space. In fact, if $\lambda < 0$ then $g(x) = \lambda \cdot f(x)$ in not a quantile function since $g' \leq 0$, $\forall x \in (0, 1)$. The last result impacts the addition operation too. For these reasons, $(\mathcal{E}, +, \cdot)$ is a semi-vector (or semi-linear space) space, i.e., only conical combinations of functions are possible.

Wasserstein Space

Even if $(\mathcal{E}, +, \cdot)$ is a semi-vector-space, (X, W_p), where W_p is the p-Wasserstein distance with $p \geq 1$ and $X \sim (\mathcal{E}^{-1})'$, , is a compact metric space and it is separable and complete (Bolley 2008). Especially, when x_0 is a 1D distribution function and $p = 2$, the $\uparrow_2$ Wasserstein distance, or the 2-Wasserstein distance, by means of convex analysis, the Agueh and Carlier (2011) provide conditions for the existence, uniqueness, characterization and regularity of Fréchet means of a set of distributions. No explicit solution to the Fréchet mean problem is available, in general, in the multivariate case. However, some approximate solutions have been recently proposed (Boissard et al. 2015; Cuturi and Doucet 2014).

2.3 The Wasserstein Distance for Discrete Distributions: The Optimal Transport Approach

Let us now consider the case of a discrete distribution P specified by a set of support points x_i with $i = 1, \ldots, m$ and their associated probabilities w_i such that $\sum_{i=1}^{m} w_i = 1$ with $w_i \geq 0$ and $x_i \in M$ for $i = 1, \ldots, m$. Usually, $M = R^d$ is the d-dimensional Euclidean space with the l_p norm and x_i are called the support vectors. M can also be a symbolic set provided with a symbol-to-symbol similarity. P can also be written using the notation:

$$P(x) = \sum_{i=1}^{m} w_i \delta(x - x_i) \tag{3}$$

where $\delta(\cdot)$ is the Kronecker delta. The WST distance between two distributions $P^{(1)} = \left\{ w_i^{(1)}, x_i^{(1)} \right\}$ with $i = 1, \ldots, m_1$ and $P^{(2)} = \left\{ w_i^{(2)}, x_i^{(2)} \right\}$ with $i = 1, \ldots, m_2$ is obtained by solving the following linear program:

$$W\left(P^{(1)}, P^{(2)}\right) = \sum_{i \in I_1, j \in I_2} \gamma_{ij} d\left(x_i^{(1)}, x_j^{(2)}\right) \tag{4}$$

The cost of transport between $x_i^{(1)}$ and $x_j^{(2)}$, $d\left(x_i^{(1)}, x_j^{(2)}\right)$, is defined by the p-th power of the norm $\|x_i^{(1)}, x_j^{(2)}\|$ (usually the Euclidean distance). We define two index sets $I_1 = \{1, \ldots, m_1\}$ and I_2 likewise, such that

$$\sum_{i \in I_1} \gamma_{ij} = w_j^{(2)}, \forall j \in I_2 \tag{5}$$

$$\sum_{j \in I_2} \gamma_{ij} = w_i^{(1)}, \forall i \in I_1 \tag{6}$$

Equations (5) and (6) represent the in-flow and out-flow constraints, respectively. The terms γ_{ij} are called matching weights between support points $x_i^{(1)}$ and $x_j^{(2)}$ or the optimal coupling for $P^{(1)}$ and $P^{(2)}$. . The discrete version of the WST distance is usually called Earth Mover Distance (EMD). For instance, when measuring the distance between greyscale images, the histogram weights are given by the pixel values and the coordinates by the pixel positions. Another way to look at the computation of the EMD is as a network flow problem. In the specific case of histograms, the entries γ_{ij} denote how much of the bin i has to be moved to bin j. The computation of EMD turns out to be the solution of a minimum cost flow problem on a bipartite graph where the bins of $P^{(1)}$ are the source nodes and the bins of $P^{(2)}$ are the sinks, while the edges between sources and sinks are the transportation costs. In the case of one-dimensional histograms, the computation of WST reduces to the comparison of two 1-dimensional histograms, which can be performed by a simple sorting and the application of Eq. (14)

$$W_p\left(P^{(1)}, P^{(2)}\right) = \left(\frac{1}{n} \sum_{i}^{n} \left| w_i^{(1)*} - w_i^{(2)*} \right|^p \right)^{\frac{1}{p}} \tag{7}$$

where $w_i^{(1)*}$ and $w_i^{(2)*}$ are the sorted samples. The major computational issue is the polynomial complexity of the linear programming solvers commonly used to compute WST. Starting from the consideration that the variables in w are more important than the matching weights, approximate solvers have been proposed, specifically, Sinkhorn solvers, which will be detailed later. Here it is just important to remark that they allow managing the trade-off between accuracy and computational cost through a regularization hyperparameter. Another approach is taken in (Ye et al. 2017) based on ADMM. Entropic regularization enables scalable computations, but large values of the regularization parameter could induce an undesirable smoothing effect, while low values not only reduce the scalability but might induce several numeric instabilities.

2.4 The Wasserstein Distance for Discrete Distributions: A Statistical Approach

In the case of one-dimensional pdf's Irpino and Romano (2007) and Irpino and Verde (2015) proposed how to compute in an efficient way $W_2\left(P^{(1)}, P^{(2)}\right)$. Irpino and Romano (2007) considered the computation of $W_2\left(P^{(1)}, P^{(2)}\right)$ when $P^{(1)}$ and $P^{(2)}$ are 1-D histograms also with different binning of the support. After a homogenization step, which allows for the description of the two histograms' cdf's through the same levels of probability, they show that the squared $W_2\left(P^{(1)}, P^{(2)}\right)$ can be obtained as a weighted sum of the squared distances of the bins' midpoints and half-widths in linear time with respect to the numbers of bins obtained by the homogenization step. They also proved that if $P^{(1)}$ has $n^{(1)}$ bins and $P^{(2)}$ has $n^{(2)}$ bins, after the homogenization step, the common number of bins $n^* \in \left(n^{(1)}, n^{(2)}\right), \ldots, \left(n^{(1)} + n^{(2)} - 1\right)$. Irpino and Verde (2015) proposed a set of basic statistics for 1-D-pdf-valued data where, assuming that all the pdf's have finite at least the first two moments, the squared $W_2\left(P^{(1)}, P^{(2)}\right)$ can be decomposed as follows:

$$W_2^2\left(P^{(1)}, P^{(2)}\right) = (\mu_1 - \mu_2)^2 + (\sigma_1 - \sigma_2)^2 + 2\sigma_1\sigma_2(1 - \rho_{12}),\qquad(8)$$

where:

- μ_1 and μ_2 are the expectations of the two *pdf*'s;
- σ_1 and σ_2 are the standard deviations of the two *pdf*'s;
- $\rho_{1,2}$ is the Pearson correlation between the quantile functions associated with the two *pdf*'s.

The decomposition, available only for 1-D-pdf-valued data, provides a framework from interpreting the variability of a set of distributions. The $W_2\left(P^{(1)}, P^{(2)}\right)$ is a metric that has been used for extending the k-means cost function to distribution-valued data (Verde and Irpino 2007) and it allows the definition of Fréchet means in terms of distributional-valued data. The decomposition is used to explain the within and between clusters sum of squares with respect to the three components, providing an effective tool for interpreting the clustering results.

2.5 Barycenter and Clustering

Consider a set of N discrete distributions, $P = \{P^{(1)}, \ldots, P^{(N)}\}$, with $P^{(k)} = \left\{\left(w_i^{(k)}, x_i^{(k)}\right) : i = 1, \ldots, m_k\right\}$ and $k = 1, \ldots, N$, then, the associated barycenter, denoted with $\underline{P} = \{(\underline{w}_1, x_1), \ldots, (\underline{w}_m, x_m)\}$, is computed as follows:

$$\underline{P} = \frac{1}{N}\sum_{k=1}^{N}\lambda_k W\left(P, P^{(k)}\right)\qquad(9)$$

where the values λ_k are used to weight the different contributions of each distribution in the computation. Without loss of generality, they can be set to $\lambda_k = \frac{1}{N}\forall k = 1, \ldots, N$.

The synthesis through a barycenter of a set of distributions have several advantages, among which: The Wasserstein barycenter, also called the Fréchet mean of distributions, and appears to be a noteworthy feature to represent the mean variation of a set of distributions and offers a useful synthesis of the structure of probability distributions, in particular:

- It is sensitive to the underlying geometry. Consider three distributions $P^{(1)} = \delta_0$, $P^{(2)} = \delta_\varepsilon$ and $P^{(3)} = \delta_{100}$. $W\left(P^{(1)}, P^{(2)}\right) \approx 0$, $W\left(P^{(1)}, P^{(3)}\right) \approx W\left(P^{(2)}, P^{(3)}\right) \approx 100$. The Total variation, Hellinger and Kullback-Leibler distances take the value 1; thus, they fail to capture our intuition that $P^{(1)}$ and $P^{(2)}$ are close to each other while they are far away from $P^{(3)}$.
- It is *shape-preserving*. Denote $P^{(1)}, \ldots, P^{(N)}$ and assume that each $P^{(j)}$ can be written as a location shift of any other $P^{(i)}$, with $i \neq j$. Suppose that each $P^{(j)}$ is defined as $P^{(j)} = N\left(\mu_j, \Sigma\right)$, then the barycenter has the closed form:

$$\underline{P} = N \left(\frac{1}{N} \sum_{j=1}^{N} \mu_j, \Sigma \right) \tag{10}$$

in contrast to the (Euclidean) average $\frac{1}{N} \sum_{j=1}^{N} P^{(j)}$.

Therefore, the concept of barycenter enables clustering among distributions in a space whose metric is the Wasserstein distance. More simply, the barycenter in the space of distributions is the analog of the centroid when the clustering takes place in a Euclidean space. The most common and well-known algorithm for clustering data in the Euclidean space is k-means. Since it is an iterative distance-based (aka representative-based) algorithm, it is easy to propose variants of k-means by simply changing the distance adopted to create clusters, such as the Manhattan distance (leading to k-medoids) or any kernel allowing for non-spherical clusters (i.e., kernel k-means). The crucial point is that only the distance is changed while the overall iterative two-step algorithm is maintained. This is also valid in the case of the WST k-means, where the Euclidean distance is replaced by WST and centroids are replaced by barycenters:

- **Step 1 – Assign.** Given the current k barycenters at iteration t, namely $\underline{P}_t^{(1)}, \ldots, \underline{P}_t^{(k)}$, clusters $C_t^{(1)}, \ldots, C_t^{(k)}$ are identified by assigning each one of the distributions $P^{(1)}, \ldots, P^{(N)}$ to the closest barycenter:

$$C_t^{(i)} = \left\{ P^{(j)} \in P : \underline{P}_t^{(i)} = W\left(Q, P^{(j)}\right) \right\}, \forall i = 1, \ldots, k \tag{11}$$

- **Step 2 – Optimize.** Given the updated composition of the clusters, update the barycenters:

$$\underline{P}_{t+1}^{(i)} = \frac{1}{\left| C_t^{(i)} \right|} \sum_{P \in C_t^{(i)}} W(Q, P) \tag{12}$$

As in k-means, a key point of WST k-means is the initialization of the barycenters. If all the distributions in P are defined on the same support, then they can be randomly initialized; otherwise, a possibility is to start from k distributions randomly chosen among those in P. Finally, termination of the iterative procedure occurs when the result of the assignment step does not change any longer, or a prefixed maximum number of iterations is achieved.

3   Distributional Representation of Networks

A histogram is a function m_k that counts the elements in a sample of n observations of a random variable that fall into each of k the disjoint categories (known as bins).

$$n = \sum_{k=1}^{K} m_k \tag{13}$$

To construct a histogram, the first step is to divide the support of the random variable into a number of intervals – and then compute the "weight" of the bin counting how many sampled values fall into each interval. The bins are usually specified as adjacent, consecutive, non-overlapping intervals and are usually of the same size. In this section a new analysis is performed in terms of node–node discrete distance distributions where the weights m_k are $P_i(k)$ the percentage of nodes which are connected to i at a distance k with each node $i = 1, \ldots, n$ of the graph $G(V, E)$.

$$P_i(k) = \frac{n_{i,k}}{n-1} \tag{14}$$

A first solution to aggregate the node to node distance distributions over the whole graph is given by the Euclidean average:

$$P_G(k) = \mu_k = \frac{1}{n} \sum_{i=1}^{n} \frac{n_{i,k}}{n-1} = \frac{1}{n} \sum_{i=1}^{n} P_i(k). \tag{15}$$

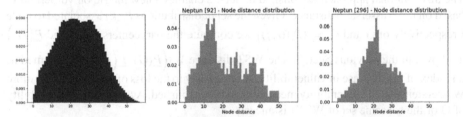

Fig. 1. The Euclidean average of the node-to-node distance of Neptun (left), for node 92 (center) and node 296 (right) of the Neptun network.

In this paper the Frechet mean of distributions (a.k.a. barycenter) will be introduced in the next section and discussed in Sect. 6. Figure 1 shows how the Euclidean mean is not shape preserving (from Sect. 6).

4   Vulnerability Measures

Given the graph $G(V, E)$ associated to the network, the removal of edge (i, j) yields $G' = G/(i, j)$. The question we address in this section is how does vulnerability change when edges are added or ablated from the network G. In particular, we assume the ablation of an edge, which corresponds to a failure in the network, either natural or intentional.

4.1 Efficiency-Based Vulnerability

Usual measures of network vulnerability are based on the concept of efficiency:

$$E(G) = \frac{1}{n(n-1)} \sum_{i,j \in V, i \neq j} \frac{1}{d_{ij}}, \tag{16}$$

where d_{ij} is the geodesic distance between nodes i and j.

The removal of edge (i,j) yields $G' = G/(i,j)$. Deleting an edge implies $E(G')$ smaller than $E(G)$. The increase of vulnerability induced by the removal of (i,j) is given by the relative loss of efficiency,

$$C_\Delta^E(i,j) = \frac{E(G) - E(G \setminus \{i,j\})}{E(G)} \tag{17}$$

The average network vulnerability is:

$$V_{MEAN}(G) = \frac{1}{n} \sum_{(i,j) \in E} C_\Delta^E(i,j) \tag{18}$$

The "worst case" network vulnerability is:

$$V_{MAX}(G) = \max_{(i,j) \in E} C_\Delta^E(i,j) \tag{19}$$

4.2 Wasserstein-Based Vulnerability

The distributional approach established in Sect. 2 enables a new metric on vulnerability based on the Wasserstein distance. Given the set of distributions $P_i(k)$ associated to node i respectively of G and $G' = G \setminus \{(i,j)\}$ we compute the barycenters $\underline{P}(G)$ and $\underline{P}(G')$ as shown in the previous Sect. 3. The WST distance $W_2\left(\underline{P}(G), \underline{P}(G')\right)$ is assumed as an index of the increase in vulnerability. Analogously to the loss of efficiency, two more Wasserstein-based vulnerability measures can be established. Average case vulnerability of G of the set same set of WST distances:

$$\frac{1}{|E|} \sum_{(i,j) \in E} W_2\left(\underline{P}(G), \underline{P}(G \setminus \{(i,j)\})\right) \tag{20}$$

For the "worst case" vulnerability:

$$\max_{(i,j) \in E} W_2\left(\underline{P}(G), \underline{P}(G \setminus \{(i,j)\})\right) \tag{21}$$

5   Data and Software Resources

5.1   Networks

Neptun is small WDN in Timisoara, Romania, more specifically it is a district metered area (DMA) of a large WDN, and it was a pilot area of the European project ICeWater (Fig. 2).

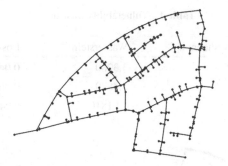

Fig. 2. The graph associated to Neptun WDN.

5.2 Software

We use the HistDAWass (Histogram Data Analysis using Wasserstein distance) package to analyse the 1D distributions extracted from the networks. HistDAWAss package contains statistical tools written in R, and it is open-source and freely available[1]. It was developed in the framework of Symbolic Data Analysis (SDA) (Bock and Diday 2000), a relatively new approach to the statistical analysis of multi-valued data, specifically for data described by univariate histograms. The methods and the basic statistics for histogram-valued data are based on the L2 Wasserstein metric between distributions, i.e., the Euclidean metric between quantile functions. The package extends several descriptive and exploratory techniques for multivariate data to datasets described by a table of univariate histograms. The core input of most of the methods is a table of histogram-valued data, namely, a table where the columns are histogram variables (as defined in the framework of SDA). The package contains tools for computing the basic statistics of histogram variables, such as Frechét means and Frechét variances, unsupervised classification techniques, such as hard and fuzzy k-means, hierarchical clustering and self-organizing maps, least-square regression techniques, and dimension reduction methods.

6 Computational Results

6.1 Vulnerability

The following table displays the differential vulnerability between networks G, G', G'' and G''' obtained removing e_1, e_2 and disconnecting the network by their joint removal. Jensen-Shannon is similar to the measure introduced in (Schieber et al. 2017), loss of efficiency is given by Eq. (13) (Table 1 and Fig. 3).

In the above heat maps the colour of each edge is given by the Wasserstein distance between the aggregate distributions of the original network and the one obtained deleting that edge. Left corresponds to Euclidean aggregation (Eq. 15) and right the Wasserstein aggregations i.e., the two barycenters (Eq. 9) (Figs. 4 and 5).

---

[1] https://github.com/Airpino/HistDAWass.https://cran.r-project.org/web/packages/HistDA Wass/index.html

Table 1. Vulnerability measures

Neptun	Jensen-Shannon	Wasserstein	Loss of efficiency
G, G'	0.2025	3.3280	0.0469
G, G''	0.2950	5.4870	0.0601
G, G'''	0.3286	12.1810	0.2432

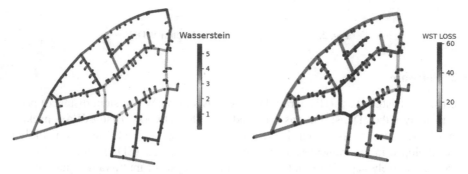

Fig. 3. Euclidean Heatmap (left), Wasserstein Barycenter heatmap (right).

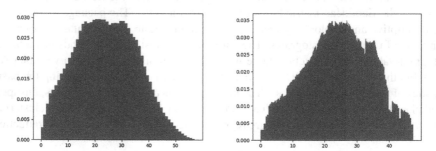

Fig. 4. Network mean histogram (left. Network barycenter histogram (right).

Again, we can observe, as already remarked in Sect. 3, that barycenters are shape-preserving.

6.2 Clustering

The following figure displays the clustering obtained using the k-means scheme but with the WST distance. The procedure is that described in Sect. 2.5. The number of clusters has been determined according to the Dunn-Index (Fig. 6).

$$\delta_{ij}(C_i, C_j) = \frac{1}{|C_i| \cdot |C_j|} \sum_{x \in C_i} \sum_{y \in C_j} W(x, y) \tag{22}$$

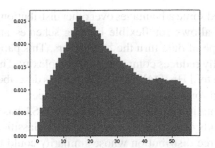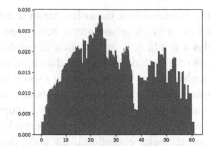

Fig. 5. Network mean histogram removing node 552 (left). Network barycenter histogram removing node 552 (right).

$$\Delta_i(C_i) = \frac{2}{|C_i| \cdot (|C_i| - 1)} \sum_{x \in C_i} \sum_{y \in C_i : x \neq y} W(x, y) \qquad (23)$$

$$DI = \frac{\min_{i,j \in \{1,2,3\}} \delta_{ij}(C_i, C_j)}{\max_{k \in \{1,2,3\}} \Delta_k(C_k)} \qquad (24)$$

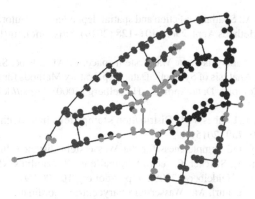

Fig. 6. The clustering results on Neptun WDN.

7 Conclusions and Perspectives

Modeling the network and its components as discrete probability distributions and embedding them in a probabilistic space has been shown to enable the methodological and operational definition of a new family of network vulnerability measures. Endowing this space with the Wasserstein metric also establishes a unifying mathematical and statistical framework that enables the computation of Wasserstein-based vulnerability measures and the barycentric synthesis of the network structure and clustering in the Wasserstein space.

The Wasserstein distance has demonstrated some advantages over other distributional distances. It is close to natural perception, allows for flexible binning schemes and enables a synthetic representation of the shape of data thru the barycenters. This paper focuses on a statistical approach that drastically reduces computational complexity. This approach has been implemented in a software tool HistDAWass whose basis is described in (Irpino and Verde 2015) and has been used for the computations in this paper.

Both modeling flexibility and low computational complexity make the WST-based framework naturally suitable for different kinds of analysis, such as network classification and temporal evolution using nodal degree distribution whose similarity could be effectively measured using WST.

References

Agueh, M., Carlier, G.: Barycenters in the Wasserstein space. SIAM J. Math. Anal. **43**(2), 904–924 (2011)

Applegate, D., Dasu, T., Krishnan, S., Urbanek, S.: Unsupervised clustering of multidimensional distributions using earth mover distance. In: Proceedings of the 17th ACM SIGKDD International Conference on Knowledge Discovery and Data Mining, pp. 636–644, August 2011

Atasu, K., Mittelholzer, T.: Linear-complexity data-parallel earth mover's distance approximations. In: International Conference on Machine Learning, pp. 364–373. PMLR, May 2019

Balzanella, A., Irpino, A.: Spatial prediction and spatial dependence monitoring on georeferenced data streams. Stat. Methods Appl. **29**(1), 101–128 (2019). https://doi.org/10.1007/s10260-019-00462-0

Bigot, J.: Statistical data analysis in the Wasserstein space. ESAIM: Proc. Surv. **68**, 1–19 (2020)

Bock, H.H., Diday, E.: Analysis of Symbolic Data: Exploratory Methods for Extracting Statistical Information from Complex Data. Springer, Heidelberg (2000). https://doi.org/10.1007/978-3-642-57155-8

Boissard, E., Le Gouic, T., Loubes, J.M.: Distribution's template estimate with Wasserstein metrics. Bernoulli **21**(2), 740–759 (2015)

Bolley, F.: Separability and completeness for the Wasserstein distance. In: Donati-Martin, C., Émery, M., Rouault, A., Stricker, C. (eds.) Séminaire de Probabilités XLI. LNM, vol. 1934, pp. 371–377. Springer, Heidelberg (2008). https://doi.org/10.1007/978-3-540-77913-1_17

Bonneel, N., Peyré, G., Cuturi, M.: Wasserstein barycentric coordinates: histogram regression using optimal transport. ACM Trans. Graph. **35**(4), 71–81 (2016)

Candelieri, A., Giordani, I., Archetti, F.: Supporting resilience management of water distribution networks through network analysis and hydraulic simulation. In: Proceedings of the 2017 21st International Conference on Control Systems and Computer Science (CSCS), pp. 599–605. IEEE (2017)

Cuturi, M., Doucet, A.: Fast computation of Wasserstein barycenters. In: International Conference on Machine Learning, pp. 685–693. PMLR, June 2014

Dias, S., Brito, P.: Linear regression model with histogram-valued variables. Stat. Anal. Data Min. **8**, 75–113 (2015)

Di Nardo, A., Giudicianni, C., Greco, R., Herrera, M., Santonastaso, G.F.: Applications of graph spectral techniques to water distribution network management. Water **10**, 45 (2018)

Flamary, R., et al.: Pot: python optimal transport. J. Mach. Learn. Res. **22**(78), 1–8 (2021)

Freitas, S., Yang, D., Kumar, S., Tong, H., Chau, D.H.: Graph Vulnerability and Robustness: A Survey. arXiv preprint arXiv:2105.00419 (2021)

Herrera, M., Abraham, E., Stoianov, I.: A graph-theoretic framework for assessing the resilience of sectorised water distribution networks. Water Resour. Manage. **30**(5), 1685–1699 (2016). https://doi.org/10.1007/s11269-016-1245-6

Ponti, A., Candelieri, A., Giordani, I., Archetti, F.: A novel graph-based vulnerability metric in urban network infrastructures: the case of water distribution networks. Water **13**(11), 1502 (2021)

Ponti, A., Candelieri, A., Giordani, I., Archetti, F.: Probabilistic measures of edge criticality in graphs: a study in water distribution networks. Appl. Network Sci. **6**(1), 1–17 (2021). https://doi.org/10.1007/s41109-021-00427-x

Puccetti, G., Rüschendorf, L., Vanduffel, S.: On the computation of Wasserstein barycenters. J. Multivar. Anal. **176**, 104581 (2020)

Schieber, T.A., Carpi, L., Díaz-Guilera, A., Pardalos, P.M., Masoller, C., Ravetti, M.G.: Quantification of network structural dissimilarities. Nat. Commun. **8**(1), 1–10 (2017)

Soldi, D., Candelieri, A., Archetti, F.: Resilience and vulnerability in urban water distribution networks through network theory and hydraulic simulation. Procedia Eng. **119**, 1259–1268 (2015)

Ulusoy, A.-J., Stoianov, I., Chazerain, A.: Hydraulically informed graph theoretic measure of link criticality for the resilience analysis of water distribution networks. Appl. Network Sci. **3**, 1–22 (2018)

Verdinelli, I., Wasserman, L.: Hybrid Wasserstein distance and fast distribution clustering. Electron. J. Stat. **13**(2), 5088–5119 (2019)

Yazdani, A., Jeffrey, P.: Complex network analysis of water distribution systems. Chaos: Interdiscipl. J. Nonlinear Sci. **21**, 016111 (2011)

Yazdani, A., Jeffrey, P.: Water distribution system vulnerability analysis using weighted and directed network models. Water Resour. Res. **48** (2012)

Ye, J., Wu, P., Wang, J.Z., Li, J.: Fast discrete distribution clustering using Wasserstein barycenter with sparse support. IEEE Trans. Signal Process. **65**(9), 2317–2332 (2017)

BERT Self-Learning Approach with Limited Labels for Document Classification

Carlos Eduardo de Lima Joaquim[1,2]([✉]) [iD] and Thiago de Paulo Faleiros[1] [iD]

[1] Departamento de Ciência da Computação, Universidade de Brasília,
Campus Universitário Darcy Ribeiro, 70910-900 Brasília, Brazil
carlos.joaquim@live.com, thiagodepaulo@unb.br
[2] Exército Brasileiro, Centro de Desenvolvimento de Sistemas, QGEx - Bloco G - 2o
Piso, 70630-901 Brasília, Brazil

Abstract. The remarkable production speed of documents and, consequently, the volume of unstructured data stored in the Brazilian Government facilities requires processes that enable the capacity of classifying documents. This requirement is compliant with the existing archival legislation. In this sense, Natural Language Processing (NLP) stands as an important asset related to document classification, considering the reality of current document production, where there is a considerable number of unlabeled documentary samples. The Self-Learning approach applied to the BERT fine-tuning step delivers a model capable of classifying a partially labeled set of data according to the Requirements Model for Computerized Document Management Systems (e-ARQ Brazil). The developed model was capable of reaching a human-level performance, outperforming Active Learning and BERT in a series of defined confidence levels.

Keywords: Self-learning · BERT · Natural language processing

1 Introduction

With the advent of the information age, where information and communication technologies became an essential asset, the potential use of applications exploring the possibilities of obtaining useful and timely knowledge became real. Institutions can potentialize its documentary collection value, transforming it into a valuable asset.

There is an appreciable amount of information being produced in a daily basis, with a significant part of this collection becoming records related to legal and historical matter. The legal value regards the value a document has to produce evidence before the law, and the historical value concerns documents related to institutional origin, rights and objectives, its organization and development [1].

© Springer Nature Switzerland AG 2022
D. E. Simos et al. (Eds.): LION 2022, LNCS 13621, pp. 278–291, 2022.
https://doi.org/10.1007/978-3-031-24866-5_21

The authors of [2] declare that large volumes of information available and stored make it difficult to access for the right information at the right time. Stating that this situation might lead to the information explosions, in accordance with [23]. Along the same line of thought, [21 as cited in 2] affirms that, at the tactical level, poor information quality compromises decision making.

Inner statistics from the Brazilian Army show that it is possible to apply this method to more than 22 million documents, evaluating and classifying them according to the current Federal regulations. The needed classification of this documentary mass is the first step towards the direction of delivering efficiency while following the present regulations. That massive amount of data would take too long to be processed and assessed, if considering the possibility of scrutiny carried out exclusively by human hands.

This documentary production started to increasingly grow after Computerized Document Management System's (CDMS) initiative took place. The Federal classification model, named e-ARQ, was the chosen model to be applied to the documents, given that this archetype is the standard reference to CDMS in Brazil. Thus, this intended study is related to the pressing need to properly classify documents produced by the Brazilian Army, allowing correct treatment and full compliance with the requirements established by the Government.

In the present scenario, the shortage of labeled samples shall be considered as a premise. With this information being known, one major question when classifying documents, while following the supervised learning paradigm, is the existing need of a substantial number of labeled samples, in order to adequately generalize the model and make predictions of unseen samples, the supervised learning approaches does not become suitable as a deemed solution.

As the foregoing restriction regarding the number of labeled samples emerges as a limitation in several scenarios, the opposite turns out to be true. While labeled data is expensive to obtain, unlabeled data is essentially free in comparison [14]. It can be seen that creating large datasets to certain supervised learning problems requires a great deal of human effort, pain and/or risk or financial expense [11,20]. This need for supervision poses a major challenge when we encounter critical scientific and societal problems where fine-grained labels are difficult to obtain [12].

In this line of thought, it can be seen in [20] that semi-supervised learning (SSL) provides a powerful framework for leveraging unlabeled data when labels are limited or expensive to obtain. The method can trace back to 1970s, and it attracts extensive attention since 1990s s [26 as cited in 13]

Considering that the size of modern real world datasets is ever-growing so that acquiring label information for them is extraordinarily difficult and costly [11,18], deep semi-supervised learning is becoming more and more popular [13].

It becomes an attractive approach towards addressing the lack of data, once, in contrast with supervised learning algorithms, SSL algorithms can improve their performance by also using unlabeled examples. Additionally SSL algorithms generally provide a way of learning about the structure of the data from the unlabeled examples, alleviating the need for labels [20].

There are various fields within semi-supervised learning of which self-learning is one [17]. Throughout this study, it was expected to find language model that would dexterously classify the partially labeled dataset set according to the e-ARQ, reaching human-level performance [25] in the classification results in a set of documents belonging to the Brazilian Army.

It was envisaged to expand the use of BERT [5], and replace the fully supervised fine-tuning stage for a self-learning method, termed BERT-SL, expecting that the process surrounding BERT downstream tasks will successfully achieve suitable scores when evaluated.

Furthermore, from an organizational perspective, it was expected that the results coming from this research fulfill the objective of allowing adequate support to properly classify documents, assisting document evaluation teams to do their job, according to what is determined by [6].

The remaining of this article is organized as follows: Sect. 2 presents the works related to what was developed in this article. Section 3 presents the methodology used in the development of the work, describing how the procedures and experiments were performed aiming the evaluation. Section 4 describes the results obtained, according to the established parameters. Finally, in Sect. 5, the conclusion was carried out, as well as opportunities for future work were pointed out.

2   Related Work

This research's related work derives from different areas, here named: text classification, limited labeled data, semi-supervised learning, and self-learning. As specified by [19], the problem of learning accurate text classifiers from limited numbers of labeled examples, by using unlabeled documents to augment the available labeled documents, was addressed by showing that the accuracy of learned text classifiers can be improved by augmenting a small number of labeled training documents with a large pool of unlabeled documents.

Their results showed EM to perform significantly better, mainly when there was little labeled data, though unlabeled data could throw off parameter estimation when one considered that the number of unlabeled documents was much greater than the number of labeled documents. The authors modulated the influence of the unlabeled data, in order to control the extent to which EM performs unsupervised clustering, by introducing a λ parameter into the likelihood equation, which decreased the contribution of unlabeled documents to parameter estimation.

Fragos, Belsis and Skourlas [7] focused on assessing the performance of two or more classifiers used in combination in the same classification task, classifying documents using two probabilistic approaches – Naïve Bayes and Maximum Entropy classification model – then combining the results of the two classifiers to improve the classification performance, using two merging operators, Max and Harmonic Mean.

The authors applied the X square test on the corpus and selected 2,000 higher ranked words for each category to be used in the maximum entropy model, and evaluated the classification performance of the classifiers using micro-Recall (μRe), micro-Precision (μPr) and micro-averaged F1 measure ($micro$-$F1$).

The Maximum Entropy model, presented by Fragos, Belsis and Skourlas [7], showed better performance than the Naïve Bayes classifier. Additionally, the two merging operators were used to combine results of the Naïve Bayes and SVM classifiers to improve performance, especially for the Recall rate. As results, it could be demonstrated that the merging operators do improve the performance, as indicated by the results for Micro-averaged F1 measure, that scored 0.90 and 0.91 for MaxC and HarmonicC operators respectively.

On the same pitch of [19], the use of self-learning and co-training is presented as a way to leverage the power of unlabeled data, together with labeled data, in [11]. The resulting work included the *TSentiment15*, an annotated Twitter dataset of 2015 comprising 228 million tweets without retweets and 275 million with retweets.

The authors evaluated the performance of self-learning and co-training and how it was affected by the amount of labeled data, the amount of unlabeled data and the confidence threshold, and, not only this, processed the available data as a batch and as a stream, showing that streaming achieved a comparable accuracy to the batch approach. The findings, in sentiment analysis, revealed that co-training performed better with limited labels, whereas self-training was best choice when significant amount of labeled data was available.

Iosifidis and Ntoutsi [11] used unigrams for self-learning, and unigram-bigrams and unigrams-SpecialF for co-training. In their experiments, although both self-learning and co-training benefited from more labeled data, when labeled data surpassed 40%, self-learning improved faster than co-training. When processing streaming, that revealed to be more efficient, history helped with the performance; notwithstanding the batch approach being better in terms of accuracy.

As avowed in [13], Li and Ye addressed issues related to generative model based schemes, that does not naturally work on discrete data. The authors bridged the idea of self-training and adversarial networks, to overcome their issues, by designing a Reinforcement Learning based Adversarial Networks for Semi-supervised Learning – RLANS – framework.

Howe, Khang and Chai [10] developed a comparative study on the performance of various machine learning("ML") approaches for classifying judgments into legal areas, using a novel dataset of 6,227 Singapore Supreme Court judgments and investigating how state-of-the-art NLP methods compared against traditional statistical models.

Dealing with small number of lengthy documents, the authors came to the conclusion that BERT models competed at disadvantage, because of the models' inability to be fine-tuned on longer input texts, having LSA based linSVM outperforming both word-embedding and language models.

While Howe, Khang and Chai [10] have found limitations regarding the text length, citing it as a disadvantage, Sun *et al.* [24] conducted exhaustive experiments to investigate different fine-tuning methods of BERT on text classification tasks and provided a general solution for BERT fine-tuning. This way they were able to reach new state-of-the-art results on eight widely-studied text classification datasets.

In [24], the problem related to the catastrophic forgetting was addressed as well. Considered a common problem in transfer learning, meaning that the pre-trained knowledge is erased while learning new knowledge [16 as cited in 24], it was managed by setting a lower learning rate, such as $2e - 5$ on BERT, so it could overcome the catastrophic forgetting problem. Aggressive learning rates, as $4e - 4$, lead the training set to fail to converge.

The authors were able to bring off experimental findings, reporting that the top layer of BERT showed to be more useful for text classification, the appropriate decreasing learning rate allows BERT overcome the catastrophic forgetting problem, whithin-task and in-domain further pre-training can significantly boost its performance, and the most important finding considered to this work, BERT can improve the task with small-size data.

Meng *et al.* [18] used the label name of each class to train classification models on unlabeled data, waiving the use of any labeled documents. Towards achieving their goals, they took advantage of pre-trained neural language models for document classification. In this abstraction only the label name of each class was provided to train a classifier on purely unlabeled data.

In the same field, in [8] BERT model is compared with a traditional machine learning NLP approach that trains machine learning algorithms in features retrieved by the Term Frequency - Inverse Document Frequency (TF-IDF) algorithm as a representative of traditional approaches. Experiments showed the superiority of BERT and its independence of features of the NLP problem such as the language of the text, adding empirical evidence to use BERT as a default technique in NLP problems [8].

3   Methodology

Considering that the main goal of this experiment is to assess the possible performance improvements that originate from the use of self-learning for downstream tasks, specifically text classification, the development of the methodology initially took place through the gathering of a specific dataset from a Military Organization (OM).

When it comes to the algorithm, initially a minimum set of labeled documents was trained in order to later classify the entire documents pertaining to the six chosen classes that encompass the following administrative actions, whose class id can be seen in Table 1: commendations, promotions, leaves, budget & finances, designations, and health. The process of labeling the unlabeled documents was based on the confidence level established as a threshold for the classification process.

Concerning the confidence level, the first experiments with classical BERT brought in f1-score 0.83 as one of the lowest results for specific classes, then a confidence level milestone was set having 0.82 as the basic confidence level for unlabeled samples' classification. Subsequently, the threshold possibilities were expanded and additional confidence levels were considered as an option, starting from 0.6, and reaching 0.95.

This procedure continued iteratively until there were no more documents remaining to be classified at a specific confidence level. Afterwards the results were compared to the results that stem from another classification methods, Active Learning, associated with Logistic Regression model applied on a TF-IDF vectorized corpus, and BERT itself.

3.1 Data

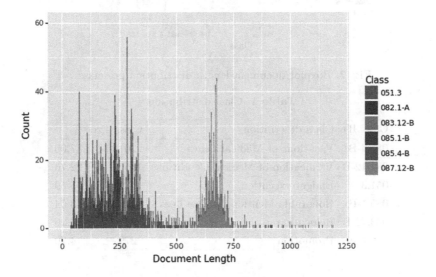

Fig. 1. Document length distribution

At the preprocessing stage, corrupted, drafts and non-processed documents were removed from the dataset, and all the corpus was converted to lowercase, along with manual preprocessing which comprehended: removing Portuguese stop words from the corpus using Natural Language Toolkit (NLTK) [3] library; Punctuation ablation; Numbers pertaining to itemization and object pronouns attached to Portuguese verbs removal; and conducting Lemmatization on the corpus using spaCy [9].

The corpus was then submitted to Ktrain [15] preprocessing methods. The resulting dataset had a 5,940 × 5,799 dimensionality, with unbalanced class distribution, dispersed over six distinct classes, as seen in Table 1.

After data transformation, the resulting document length distribution, and class distribution can be seen in Fig. 1, and Fig. 2, allowing one to perceive the final length distribution of the dataset, and the final length distribution per class.

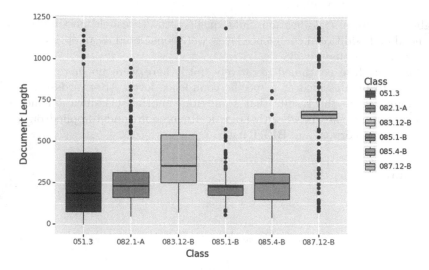

Fig. 2. Boxplot document length distribution per class.

Table 1. Class distribution

Class ID	Class description	Class size	%
085.4-B	Vacations or Medical Leave	2201	37.05
087.12-B	Verification of Medical Conditions	1352	22.76
051.3	Budget execution	926	15.59
085.1-B	Honorable Mention/Service Leave	810	13.64
083.12-B	Relocation	372	6.26
082.1-A	Promotion	279	4.70

In order to successfully establish a number of labeled documents, a classification tool, termed Document Classifier, was developed so the process of searching and classifying documents was conducted without further efforts regarding finding similar documents in the dataset.

3.2 BERTimbau

Throughout this work, finding an algorithm that could achieve state-of-the-art performance in Portuguese was a concern due to the existing linguistic bias in favor of languages that predominate on the areas where only major companies and research centers can afford training language models with billions of parameters on massive datasets [4].

The work of Souza *et al.* [22], BERTimbau, arouse as the current answer to this need, delivering state-of-the-art results for downstream natural language processing tasks in Portuguese language, and will be simply referred to as BERT throughout this study. The pre-trained BERT [5] based model, BERTimbau, was the model of choice, applied when running the fine-tuning step using the developed self-learning approach.

The model will be simply referred to as BERT throughout this study, having 12 layers, 768 hidden size, 12 attention heads, and 110M parameters. The maximum sentence length also observed BERT [5], following the $S = 512$ tokens limit. The pre-trained BERT [5] based model, BERTimbau, was the model of choice, applied when running the fine-tuning step using the developed self-learning approach.

3.3 Self-Learning

Bearing in mind that this research can find in self-learning a solid answer to the problem of limited number of labeled documents in a dataset, it is intended to assess the resulting performance of the algorithm by selecting a specific percentage of samples from the total amount of labeled documents, starting with 3% and increasingly growing up this number to 30%.

Table 2. Four group sample distribution

Dataset size	Training size	Validation size	Unlabeled size	Test size
200	100	50	50	5740
594	269	146	179	5346
1188	540	291	357	4752
1782	810	437	535	4158

As a means to better achieve the objectives of this research, the data was split in four sets. In this fashion, during this work one will observe, as exhibited in Table 2, the four sets being used, notably the training set, validation set, unlabeled set, and test set.

As a way of exploring the possibilities of the self-learning process, experiments using the validation set as the unlabeled set were carried out as well. In this configuration, the datasets do not include the unlabeled set, and the validation set suffered prejudice by being classified according to the established threshold, having its samples moved from the validation set to the training set.

Further experiments included model where line 13 of the Algorithm 1 was suppressed. Yet, the sample distribution in the experiments was as follows: the methods were equally exposed to the same number of samples, thus reflecting a factual scenario, which commonly occurs in everyday life.

Observing the aforementioned data organization, in each experiment, the selected data was submitted to the self-learning Algorithm 1, until there were no more unlabeled samples in the corresponding set. After the initial training procedure, the full subset of documents were then submitted to the trained model, without any labels.

Algorithm 1. BERT Self-Learning Approach Pseudocode

Input: $\mathcal{L}_{tr}$: labeled training set; $\mathcal{L}_v$: labeled validation set; $\mathcal{L}_{ts}$: labeled test set; $\mathcal{U}$: unlabeled set; δ: confidence threshold

Result: $\mathcal{T}$: labeled set; Φ_f: final classifier

1: $\mathcal{T} \leftarrow \mathcal{L}_{tr}, \mathcal{L}_v$
2: **while** *($\mathcal{U}$ is not empty)* **do**
3:    $\Phi \leftarrow$ new classifier
4:    $\Phi \leftarrow$ train new classifier on $\mathcal{T}$;
5:    **for** $i = 1$ **to** $|\mathcal{U}|$ **do**
6:       **if** *confidence of Φ.classify($\mathcal{U}_i$) $\geq \delta$* **then**
7:          $\mathcal{T} \leftarrow \mathcal{T} \cup \mathcal{U}_i$, where $\mathcal{U}_i$ is the i-th instance in $\mathcal{U}$
8:          Mark $\mathcal{U}_i$ as labeled;
9:       **end if**
10:   **end for**
11:   Update $\mathcal{U}$ by removing labeled instances;
12: **end while**
13: $\Phi_f \leftarrow$ train final classifier on $\mathcal{T}$
14: return $\mathcal{T}, \Phi_f$;

The resulting prediction for every sample, whose probabilistic classification was equal to or greater than the confidence level, allowed it to be incorporated, or not, in the labeled document selection of the training set for the next training session.

This process iteratively repeated its steps until all the documents in the $\mathcal{U}$ dataset were considered labeled – to expound, the whole dataset received a classification equal to or greater than the defined confidence level.

Downstream Task. The first experiment had as main objective nothing more than the discovery of hyperparameters, *i.e.*, learning rate, number of epochs, and batch size, that would be expected to best perform when classifying the available dataset, bringing off satisfying results. It was possible to find the benchmark after having thoroughly tested all possible combinations of the values of the hyperparameters defined in Table 3.

Table 3. Benchmark model hyperparameters

Learning rate	Batch size	Nº of Epochs
3e−5	2	4
4e−5	6	5
5e−5	8	6
6e−5	10	−
7e−5	−	−

Afterwards, series of self-learning experiments were conducted using the hyperparameters, and, then, more experiments were produced using classical BERT. The experiment was called BERT', being carried out having the data distributed, as presented in Table 4, and considered the most realistic scenario, once in real life there would be no more samples to feed classical BERT with.

Table 4. BERT' Data Distribution. In this distribution, as the unlabeled set was not used by the model, it was merged with the test set.

Dataset size	Training size	Validation size	Test size
200	100	50	5790
594	269	146	5525
1188	540	291	5109
1782	810	437	4693

4   Results and Discussion

In this section, the results of the experiments conducted using the self-learning approach during the fine-tuning step of BERT training are presented. The best hyperparameter experiment, where hyperparameter combinations would show the most prominent benchmark, registered that a 4.00E−5 learning rate, five epochs and batch size of two presented the best performance.

Table 5. Performance indicators (PI) of featured BERT-SL Classification Results. This table presents the best results for each dataset, outlined from the preceding experiments, against BERT' and Active Learning.

Dataset size	CL	F1-Score	PI	
			BERT'	AL
200	0.84	0.9771	0.0472	0.1330
594	0.82	0.9867	0.0165	0.0328
1188	0.90	0.9884	0.0146	0.0110
1782	0.90	0.9933	0.0121	0.0123

The ensuing experiments showed convincing results regarding the initial objective of this study, which was to find a self-learning method that could reach the human classification capacity.

As shown in Table 5, the use of self-learning also allowed the production of results comparable to BERT's, showing that it obtained excelling outcomes when

Table 6. Classical BERT classification results.

Dataset size	BERT' F1-score
200	0.9330
594	0.9708
1188	0.9743
1782	0.9814

compared to BERT, whose results are registered in Table 6, in several scenarios and confidence levels.

Despite of the highest score be related to the biggest dataset, the 1782-sample dataset, it is relevant to emphasize that the top gain came from the 200-sample dataset. The maximum gain coming from this experiment yielded a score 4.72% greater than BERT', and 13.30% beyond the active learning mark, as seen in Fig. 3, and detailed in Table 5.

The same Fig. 3 allows one to perceive that BERT-SL, when working with the 594-sample dataset, provided a gain of 1.65% over BERT', and 3.28% over the active learning method. When dealing with 1188-sample dataset, the proposed method was able to deliver scores 1.46% better than BERT', and 1.10% greater than the active learning method. The gain got lower as the dataset size increased, reaching a gain of 1.21% over BERT' for the 1782-sample dataset, and 1.23% over the active learning method for the same dataset.

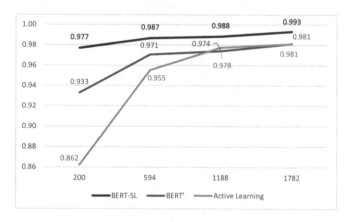

Fig. 3. BERT-SL overall performance, for every dataset, compared to BERT' and Active Learning experiments.

As presented in Fig. 4, throughout the undertook experiments, the language model resulting from the proposed method was able to outperform BERT' in every designed confidence level, surmounting the aforementioned method 100%

of the time, and surpassed the active learning process in every confidence level but 0.55 and 0.50, where BERT-SL was able to beat BERT' 87.50% of the time.

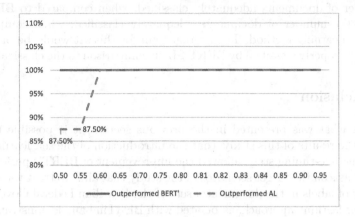

Fig. 4. Outperforming BERT' experiments grouped by confidence level.

Considering the complete testing universe, Fig. 5 delineates the results. It remains clear that BERT-SL excelled BERT' in every experiment, and surpassed the active learning in almost all of them, being the only exception the tests where the 1188-sample dataset served as the input, when 93,75% of the experiments surpassed the active learning method.

It has been successfully demonstrated that, even when more samples where fed to classical BERT, BERT-SL showed to be able to achieve better scores, mainly when it comes to labeled samples shortage. The experiment, in Table 5, showed that BERT-SL was capable of achieving up to 4.72% better performance than the classical BERT, and 13.30% than the Active Learning.

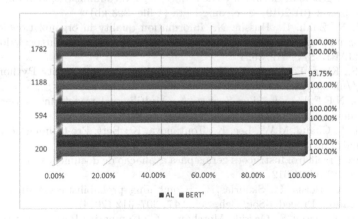

Fig. 5. Percentage of BERT-SL experiments that outperform BERT'.

As described above, it was observed that, for experiments involving a set of two hundred samples, BERT-SL obtained a superior performance capable of, considering the aforementioned 22 million documents, estimating in 1,039,272 the number of documents adequately classified, when compared to BERT'; in 2,926,811 the number of documents adequately classified, when compared to the Active Learning method. The average gain for this set would be of 314,524 documents properly classified by BERT-SL, in comparison to the classical BERT.

5   Conclusion

Observing what was presented in the previous section, it is possible to infer, based on the results of this study, that the introduction of the self-learning approach in the fine-tuning stage allowed the improvement of BERT's performance, with emphasis on the target scenarios of the study that aims to treat the problem of scarcity of labels in the documental sets of the Brazilian Federal Government.

The self-learning approach, associated with BERTimbau, demonstrated to be a promising method regarding NLP classification tasks, showing better results than the classical BERT when the same number of samples is available to both methods, surpassing the traditional method in every single experiment following this setup.

The method showed outstanding results associated with datasets whose labels reach only 3% of the samples, showing an increasing performance, when compared to classical BERT, as the number of available labeled samples decreases. It was therefore possible to achieve suitable results while carrying out the experiment, making it possible to apply the method to other datasets.

References

1. Nacional, A.: Gestão de documentos: curso de capacitação para os integrantes do sistema de gestão de documentos de arquivo siga, da administração pública federal. Course packet (01 2019), electronic Data (1 file: 993 kb)
2. Azemi, N., Zaidi, H., Hussin, N.: Information quality in organization for better decision-making. Int. J. Acad. Res. Bus. Soc. Sci. **7** (2018). https://doi.org/10.6007/IJARBSS/v7-i12/3624
3. Bird, S., Klein, E., Loper, E.: Natural Language Processing with Python (2009). https://nltk.org/book
4. Castro, N.F.F.d.S., da Silva Soares, A.: Multilingual transformer ensembles for portuguese natural language tasks (2020)
5. Devlin, J., Chang, M.W., Lee, K., Toutanova, K.: Bert: Pre-training of deep bidirectional transformers for language understanding (2019)
6. Exército Brasileiro: Instruções gerais para avaliação de documentos do exército (10 2019), eB10-IG-01.012
7. Fragos, K., Belsis, P., Skourlas, C.: Combining probabilistic classifiers for text classification. Procedia-Soc. Beh. Sci. **147**, 307–312 (2014)
8. González-Carvajal, S., Garrido-Merchán, E.C.: Comparing Bert against traditional machine learning text classification (2021)

9. Honnibal, M., Montani, I., Van Landeghem, S., Boyd, A.: spaCy: industrial-strength natural language processing in Python (2020). https://doi.org/10.5281/zenodo.1212303

10. Howe, J.S.T., Khang, L.H., Chai, I.E.: Legal area classification: a comparative study of text classifiers on Singapore supreme court judgments (2019)

11. Iosifidis, V., Ntoutsi, E.: Large scale sentiment learning with limited labels. In: Proceedings of the 23rd ACM SIGKDD International Conference on Knowledge Discovery and Data Mining. KDD 2017, New York, NY, USA, pp. 1823–1832. Association for Computing Machinery (2017). https://doi.org/10.1145/3097983.3098159, https://doi-org.ez54.periodicos.capes.gov.br/10.1145/3097983.3098159

12. Jean, N., Xie, S.M., Ermon, S.: Semi-supervised deep kernel learning: regression with unlabeled data by minimizing predictive variance (2019)

13. Li, Y., Ye, J.: Learning adversarial networks for semi-supervised text classification via policy gradient. In: Proceedings of the 24th ACM SIGKDD International Conference on Knowledge Discovery & Data Mining, pp. 1715–1723 (2018)

14. Liang, P.: Semi-supervised learning for natural language. Ph.D. thesis, Massachusetts Institute of Technology (2005)

15. Maiya, A.S.: ktrain: a low-code library for augmented machine learning. CoRR abs/2004.10703 (2020), https://arxiv.org/abs/2004.10703

16. McCloskey, M., Cohen, N.J.: Catastrophic interference in connectionist networks: The sequential learning problem. Psychol. Learn. Mot. **24**, 109–165 (1989)

17. McEntee, E.: Enhancing partially labelled data: self learning and word vectors in natural language processing (2019)

18. Meng, Y., et al.: Text classification using label names only: a language model self-training approach (2020)

19. Nigam, K., McCallum, A.K., Thrun, S., Mitchell, T.: Text classification from labeled and unlabeled documents using EM. Mach. Learn. **39**(2), 103–134 (2000)

20. Oliver, A., Odena, A., Raffel, C., Cubuk, E.D., Goodfellow, I.J.: Realistic evaluation of deep semi-supervised learning algorithms (2019)

21. Redman, T.C.: Improve data quality for competitive advantage. MIT Sloan Manage. Rev. **36**(2), 99 (1995)

22. Souza, F., Nogueira, R., Lotufo, R.: BERTimbau: pretrained BERT models for Brazilian Portuguese. In: 9th Brazilian Conference on Intelligent Systems, BRACIS, Rio Grande do Sul, Brazil, October 20–23 (2020, to appear)

23. Strong, D.M., Lee, Y.W., Wang, R.Y.: Data quality in context. Commun. ACM **40**(5), 103–110 (1997)

24. Sun, C., Qiu, X., Xu, Y., Huang, X.: How to fine-tune Bert for text classification? (2020)

25. Wolf, F., Poggio, T., Sinha, P.: Human document classification using bags of words, August 2006

26. Zhu, X.J.: Semi-supervised learning literature survey (2005). last modified on 19 July 2008

Autonomous Learning Rate Optimization for Deep Learning

Xiaomeng Dong[1,2]($\boxtimes$), Tao Tan[1], Michael Potter[1], Yun-Chan Tsai[1], Gaurav Kumar[1], V. Ratna Saripalli[1], and Theodore Trafalis[2]

[1] GE Healthcare, Chicago, USA
Xiaomeng.Dong@ge.com
[2] University of Oklahoma, Norman, USA

Abstract. A significant question in deep learning is: what should that learning rate be? The answer to this question is often tedious and time consuming to obtain, and a great deal of arcane knowledge has accumulated in recent years over how to pick and modify learning rates to achieve optimal training performance. Moreover, the long hours spent carefully crafting the perfect learning rate can be more demanding than optimizing network architecture itself. Advancing automated machine learning, we propose a new answer to the great learning rate question: the Autonomous Learning Rate Controller. Source code is available at https://github.com/fastestimator/ARC/tree/v1.0.

Keywords: Deep Learning · AutoML · Learning Rate · Optimization

1 Introduction

Learning Rate (LR) is one of the most important hyperparameters in deep learning training, a parameter everyone interacts with for all tasks. In order to ensure model performance and convergence speed, LR needs to be carefully chosen. Overly large LRs will cause divergence whereas small LRs train slowly and may get trapped in a bad local minima. As training schemes have evolved over time they have begun to move away from a single static LR towards scheduled LRs, as can be seen in a variety of state-of-the-art AI applications [2,23,30,36]. LR scheduling provides finer control of LRs by allowing different LRs to be used throughout the training. However, the extra flexibility comes with a cost: these schedules bring more parameters to tune. Given this tradeoff, there are broadly two ways of approaching LR scheduling within the AI community.

Experts with sufficient computing resources tend to hand-craft their own LR schedules, because a well-customized LR schedule can often lead to improvements over current state-of-the-art results. For example, entries in the Dawnbench [5] are known for using carefully tuned LR schedules to achieve world-record convergence speeds. However, such LR schedules come with significant drawbacks. First, these LR schedules are often specifically tailored to an exact problem configuration (architecture, dataset, optimizer, etc.) such that they do not generalize to other tasks. Moreover, creating these schedules tends to require a good deal of intuition, heuristics, and manual observation

© Springer Nature Switzerland AG 2022
D. E. Simos et al. (Eds.): LION 2022, LNCS 13621, pp. 292–305, 2022.
https://doi.org/10.1007/978-3-031-24866-5_22

of training trends. As a result, building a well-customized LR schedule often requires great expertise and significant computing resources.

In contrast, others favor existing task-independent LR schedules since they often provide decent performance gains with less tuning efforts. Some popular choices are cyclic cosine decay [24], exponential decay, and warmup [9]. While these LR schedules can be used across different tasks, they are not specially optimized for any of them. As a result, these schedules do not guarantee performance improvements over a consant LR. On top of that, many of these schedules still require significant tuning to work well. For example, in cyclic cosine decay, parameters such as l_{max}, l_{min}, T_0, and T_{multi} must all be tuned in order to function properly.

Recent advancements in AutoML on architecture search [21,29,38] and update rule search [1] have proved that it is possible to create automated systems that perform equal or better than human experts in designing deep learning algorithms. These successes have inspired us to tackle the LR scheduling problem. We aim to create a system that learns how to change LR effectively.

To that end we introduce ARC: an Autonomous LR Controller. It takes training signals as input and is able to intelligently adjust LRs in a real-time generalizable fashion. ARC overcomes the challenges faced by prior LR schedulers by encoding experiences over a variety of different training tasks, and by dynamically responding to each new training situation so that no manual parameter tuning is required.

ARC is also fully complementary to modern adaptive optimizers such as Adagrad [7] and Adam [17]. Adaptive optimizers compute updates using a combination of LR and 'adaptive' gradients. When gradients have inconsistent directions across steps, the scale of the adaptive gradient is reduced. Conversely, multiple updates in the same direction result in gradient upscaling. This is sometimes referred to as adaptive LR even though the LR term has not actually been modified. Our method is gradient agnostic and instead leverages information from various training signals to directly modify the optimizer LR. This allows it to detect patterns which are invisible to adaptive optimizers. Thus it can be used in tandem for even better results.

The key contributions of this work are:

1. An overall methodology for developing autonomous LR systems, including problem framing and dataset construction.
2. A comparison of ARC with popular LR schedules across multiple computer vision and language tasks.
3. An analysis of failure modes and unexpected behaviors from ARC, informing future directions for research.

2 Challenges and Constraints

Before delving into our methodology, we will first highlight some of the key challenges in developing an autonomous LR controller. These challenges explain many of our subsequent design decisions.

a) **Subjectivity.** Determining the superiority of one model over another (each trained with a different LR) is fraught with subjectivity. There are many different ways to measure model performance (training loss, validation loss, accuracy, etc.) and they may often be in conflict with one another.

b) **Cumulativeness.** Associating the current model performance with an LR decision at any particular step is challenging, since the current performance is the result of the cumulative effect of all previous LRs used during training.

c) **Randomness.** Randomness during training makes it difficult to compare two alternative LRs. Some common sources of randomness are dataset shuffling, data augmentation, and random network layers such as dropout. Any performance differences due to the choice of LR need to be large enough to overshadow these random effects.

d) **Scale.** Different deep learning tasks use different metrics to monitor training. The most task-independent of these are training loss, validation loss, and LR. Unfortunately, the magnitude of these values can still vary greatly between tasks. For example, categorical cross entropy for 1000-class classification is usually between 0 and 10, but a pixel-level cross entropy for segmentation can easily reach a scale of several thousand. Moreover, a reasonable LR for a given task can vary greatly, from 10^{-6} up to 10 or more.

e) **Footprint.** The purpose of having an automated LR controller is to achieve faster convergence and better results. Any solution must therefore have a sufficiently small footprint that using it does not adversely impact training speed and memory consumption.

3   Methods

3.1   Framing LR Control as a Learning Problem

We frame the development of ARC as a supervised learning problem: predicting the next LR given available training history. Due to challenge b), the model needs to observe the consequence of a specific LR for long enough to form a clear association between LR and performance. We therefore only modify the LR on a per-epoch basis.

Per challenges a) and d), as well as the desire to create a generalizable system, we cannot use any task-specific metrics. We also cannot rely on model parameters or gradient inspection since ARC would then become architecture dependent and would likely also fail challenge e). We therefore leverage only the historical training loss, validation loss, and LR as input features.

Due to challenges c) and d), rather than generating a continuous prediction of what new LR values should be, we instead pose this as a 3-class classification problem. Given the input features, should the LR: increase ($LR * 1.618$), remain the same ($LR * 1.0$), or decrease ($LR * 0.618$)? There is no theoretical basis for our choice of multiplicative factors here, but one increase followed by one decrease will leave you roughly where you started.

3.2   Generating the Dataset

Having framed the problem, we now need to generate a dataset on which we can train ARC. To do this we need data from real deep learning training tasks. For each task we used the following procedure to generate data:

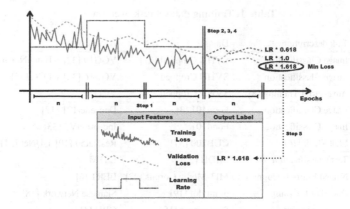

Fig. 1. Generating and Labeling a Data Point

1. Train n epochs with $LR = r$, then save the current state as checkpoint C
2. Reload C, train for n epochs with $LR = 1.618 * r$, then compute validation loss (l_+)
3. Reload C, train for n epochs with $LR = 1.0 * r$, then compute validation loss (l_1)
4. Reload C, train for n epochs with $LR = 0.618 * r$, then compute validation loss (l_-)
5. Note the LR which resulted in $min\{l_+, l_1, l_-\}$
6. Reload C, eliminate $max\{l_+, l_1, l_-\}$ and its corresponding LR, replace r with one of the two remaining LRs at random, and return to step 1

By executing steps 1 - 5 we can create one input/ground truth pair. The input features are the training loss, validation loss, and LR during step 1, concatenated with those same features from the previous $2n$ epochs of training. The label is the LR noted in step 5. This process is depicted in Fig. 1.

Steps 1 - 6 continue until training finishes. Assuming the total number of training epochs is N, then we get N/n data points from each training procedure. The random selection process in step 6 is used to explore a larger search space without causing the loss to diverge.

To address challenge d), for each instance of input data we apply z-score normalization to the training and validation losses. LR is normalized by dividing by its first value. All three features are then resized to length 300 using inter-nearest interpolation.

To help ensure generalization, we gathered 12 different computer vision and language tasks - each having a different configuration (dataset, architecture, initial LR, etc.) as shown in Table 1. Each of the 12 tasks were trained an average of 42 times. Each training randomly selected an optimizer from {Adam, SGD, RMSprop [34]}, an initial LR $r \in [\text{1e-2, 1e-5}]$, a value of $n \in [1, 10]$, and then trained for a total of $10n$ epochs. Thus approximately 5050 sample points were collected in total.

Table 1. Training dataset task overview

Task	Task description	Dataset	Architecture
1	Image Classification	SVHN Cropped [27]	VGG19 [32] + BatchNorm [13]
2	Image Classification	SVHN Cropped	VGG16 [32] + ECC [35]
3	Adversarial Training [8]	SVHN Cropped	VGG19
4	Image Classification	Food101 [3]	Densenet121 [12]
5	Image Classification	Food101	InceptionV3 [33]
6	Multi-Task [16]	CUB200 [37]	ResNet50 [10] + UNet [31]
7	Text Classification	IMDB [25]	LSTM
8	Named Entity Recognition	MIT Movie Corpus [22]	BERT [6]
9	One Shot Learning	omniglot [19]	Siamese Network [18]
10	Text Generation	Shakespear [15]	GRU [4]
11	Semantic Segmentation	montgomery [14]	UNet
12	Semantic Segmentation	CUB200	UNet

3.3 Correcting Ground Truth

Suppose that during step 5 of the data generation process you find that l_+, l_1, and l_- are 0.113, 0.112, and 0.111 respectively. Due to challenge c) it may not be appropriate to confidently claim that decreasing LR is the best course of action.

Luckily, there is one more datapoint we can use to reduce uncertainty. Suppose that during step 6 we choose to decrease the LR. Then the subsequent step 1 is repeating exactly the prior step 4. Let l_-^* be the validation loss at the end of step 1. If the relative order of l_+, l_1, and l_- is the same as the relative order of l_+, l_1, and l_-^*, then we consider our ground truth labeling to be correct (for example, if $l_-^* = 0.109$). On the other hand, if the relative ordering is different (for example, if $l_-^* = 0.115$), then random noise is playing a greater role than the LR in determining performance. In that case we take a conservative approach and modify the ground truth label to be 'constant LR'.

3.4 Building the Model

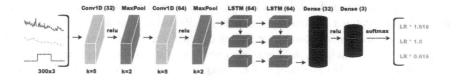

Fig. 2. Network architecture used in ARC

The ARC model architecture is shown in Fig. 2. It consists of three components: a feature extractor, an LSTM [11], and a dense classifier. The feature extractor consists of two 1D convolution layers, the LSTM of two stacked memory sequences, and the classifier of two densely connected layers. Considering challenge e), we chose layer parameters such that the total number of trainable model parameters is less than 80k. Compared to the millions of parameters which are common in current state-of-the-art models, this architecture should add relatively minimal overhead.

The model was trained with the corrected dataset from Sect. 3.3, split 70/30 between training and validation data. We used categorical cross entropy for the loss, and leveraged an Adam optimizer with the following parameters: $LR = 1e - 4$, $beta_1 = 0.9$, and $beta_2 = 0.999$. Training proceeded with a batch size of 128 for 300 epochs, with the best model being saved along the way. We define the 'best' model using a weighted accuracy, since predicting a constant LR is less problematic than calling for a change in the incorrect direction. The Reward/Penalty Matrix (RPM) for this metric is given in Table 2, with the corresponding metric computed from the Confusion Matrix (CM) per Eq. 1.

$$w_{acc} = \frac{\sum_i CM[i,i] * |RPM[i,i]|}{\sum_{i,j} CM[i,j] * |RPM[i,j]|} \tag{1}$$

Table 2. Reward/penalty matrix

Predict n Actual	Decrease	Constant	Increase
Decrease	+3	−1	−3
Constant	−1	+1	−1
Increase	−3	−1	+3

Once trained, the ARC model can be used to periodically adjust the LR for other models, as we will demonstrate in Sect. 4.

4 Experiments

In this section, we test how well ARC can guide training tasks on previously unseen datasets and architectures. Specifically, we deploy ARC on two computer vision tasks and one NLP task. For each task, we compare ARC against 3 standard LR schedules: Baseline LR (BLR) - in which LR is held constant, Cyclic Cosine Decay (CCD), and Exponential Decay (ED).

In order to gain a holistic view of the effectiveness of different schedulers, we use 3 different initial LRs for each task. Each training configuration is run 5 times in order to compute standard deviations. Scheduler performance is measured in two ways: best validation metric performance (e.g. highest accuracy), and fewest steps till convergence. We define convergence by running each of the 4 schedulers 5 times, finding the best metric scores for each of those 20 runs, and then picking the worst of those scores to be our convergence threshold. Schedulers which can reach that threshold the fastest are preferable.

4.1   Image Classification on CIFAR10

Table 3. Results for CIFAR10 Image Classification (best values in bold)

	Initial LR = 0.01		Initial LR = 0.001		Initial LR = 0.0001	
	Accuracy	Converge Step	Accuracy	Converge Step	Accuracy	Converge Step
BLR	90.16 ± 0.12	11020 ± 640	91.43 ± 0.15	10520 ± 621	88.83 ± 0.07	5340 ± 361
CCD	$\mathbf{92.20 \pm 0.12}$	8400 ± 228	92.88 ± 0.09	8040 ± 361	87.83 ± 0.31	7460 ± 196
ED	91.82 ± 0.28	$\mathbf{6800 \pm 580}$	92.43 ± 0.15	6900 ± 415	85.75 ± 0.28	9500 ± 1207
ARC	91.87 ± 0.36	7120 ± 671	$\mathbf{92.98 \pm 0.18}$	$\mathbf{5840 \pm 224}$	$\mathbf{89.87 \pm 0.58}$	$\mathbf{4440 \pm 258}$

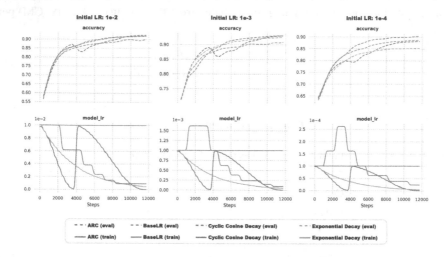

Fig. 3. Performance on CIFAR10

For our first experiment we trained a model to perform CIFAR10 image classification. We used the same architecture and preprocessing as proposed in [28]. We trained for 30 epochs (rather than 24 in the original implementation) using an Adam optimizer and a batch size of 128. Three different initial LRs were used: 1e−2, 1e−3, and 1e−4. For each initial LR, we compare the performance of BLR with the performance of ARC (invoked every 3 epochs), as well as CCD (using the settings proposed in [29] for CIFAR10), and ED (with a gamma of 0.9).

The results for all experiment runs are summarized in Table 3. Representative graphs of LR and accuracy for each method are given in Fig. 3.

When the initial LR is sufficiently large (1e−2 and 1e−3), all decaying LR schedulers outperform the baseline LR. From Fig. 3 (left and center), we can see that ARC also decided to decrease the LR. Amongst all LR schedulers, ARC perform the second best with the large initial LR (1e−2), and was the best performer with the medium initial LR (1e−3).

When the initial LR is small (1e−4), however, the drawback of statically decaying LR schedulers becomes evident: decaying an already small LR damages convergence. In this case, CCD and ED are beaten by the baseline LR. On the other hand, as shown in Fig. 3 (right), ARC is able to sense that the LR is too small and increasing it, achieving the best final accuracy and convergence speed.

4.2 Object Detection on MSCOCO

Table 4. Results for MSCOCO Object Detection (best values in bold)

	Initial LR = 0.01		Initial LR = 0.005		Initial LR = 0.001	
	mAP	Converge Step	mAP	Converge Step	mAP	Converge Step
BLR	16.3 ± 0.2	30380 ± 2905	16.3 ± 0.1	23140 ± 920	13.4 ± 0.1	16480 ± 519
CCD	16.8 ± 0.2	29600 ± 805	15.9 ± 0.1	28480 ± 1013	11.1 ± 0.1	24880 ± 637
ED	15.6 ± 0.1	36340 ± 5398	14.6 ± 0.1	37260 ± 4986	9.6 ± 0.1	38220 ± 4835
ARC	**17.1 ± 0.2**	**21440 ± 1384**	**16.5 ± 0.5**	**18820 ± 279**	**14.3 ± 0.2**	**12480 ± 979**

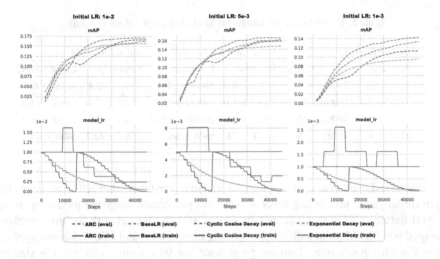

Fig. 4. Performance on MSCOCO

Our second and most time-consuming task is object detection using the MSCOCO dataset. We downscale the longest side of each image to 256 pixels in order to complete the trainings within a more reasonable computational budget. The RetinaNet [20] architecture was selected for this task. We used a batch size of 32 and trained for a total of 45000 steps, with validation every 1500 steps. We used a momentum optimizer with 0.9 for its momentum value, but kept all other parameters consistent with the official implementation. The configuration for our LR schedulers is the same as in Sect. 4.1, but with initial LRs of 0.01, 0.005, and 0.001 (we found that LRs outside of this range risked divergence or converged too slowly to be useful). For this task we use mean average precision (mAP) to benchmark model performance.

The results for all experiment runs are summarized in Table 4. Representative graphs of LR and mAP for each method are given in Fig. 4.

Interestingly, the largest LR (1e-2) we used was not large enough to allow ED to outperform the baseline LR. Unfortunately, larger initial LRs were found to lead to training divergence. This exposes a critical limitation of exponential LR decay: the rate of decay needs to be carefully tuned, otherwise LR can be either too large early on or too small later in training. On the other hand, both CCD and ARC outperform the baseline LR, with ARC achieving the best mAP and dramatically better convergence speed.

For the other two smaller LRs (5e-3 and 1e-3), both ED and CCD are worse than the baseline LR because they have no mechanism to raise the LR when doing so would be useful. In contrast, ARC can notice this deficiency and increase the LR accordingly - allowing it to achieve the best mAP and convergence scores across the board.

4.3 Language Modeling on PTB

Table 5. Results for PTB Language Modeling (best values in bold)

	Initial LR = 1.0		Initial LR = 0.1		Initial LR = 0.01	
	Perplexity	Converge Step	Perplexity	Converge Step	Perplexity	Converge Step
BLR	121.2 ± 3.8	7540 ± 3770	136.6 ± 0.3	8800 ± 63	314.5 ± 2.4	8640 ± 224
CCD	123.3 ± 3.9	11960 ± 6090	159.5 ± 1.0	16260 ± 206	450.3 ± 7.1	16160 ± 120
ED	122.9 ± 1.7	5900 ± 1909	201.4 ± 1.4	29000 ± 3592	603.9 ± 3.0	28400 ± 3769
ARC	$\mathbf{116.8 \pm 0.6}$	$\mathbf{4140 \pm 258}$	$\mathbf{124.7 \pm 2.6}$	$\mathbf{6840 \pm 80}$	$\mathbf{258.1 \pm 62.2}$	$\mathbf{6480 \pm 40}$

For our final task we move beyond computer vision to verify whether ARC can be useful in natural language processing tasks as well. We performed language modeling using the PTB dataset [26] with a vocabulary size of 10000. Our network for this problem leveraged 600 LSTM units with 300 embedding dimensions, and a 50% dropout applied before the final prediction. Training progressed for 98 epochs, with a batch size of 128 and a sequence length of 20. A Stochastic Gradient Descent (SGD) optimizer was selected, with initial LR values of 1.0, 0.1, and 0.01 (with 1.0 being the largest power of 10 we could find which did not lead to training divergence). Our CCD scheduler used $T_0 = 14$ and $T_{multi} = 2$ such that we could fit 3 LR cycles into the training window. The ED scheduler gamma value was set to 0.96, and ARC was invoked every 9 epochs. For this task we used perplexity to measure model performance (lower is better).

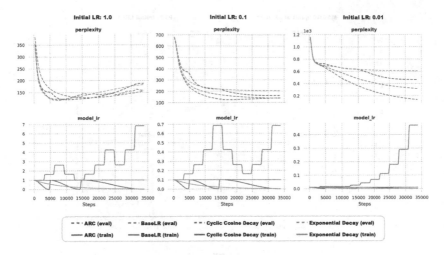

Fig. 5. Performance on PTB

The results for all experiment runs are summarized in Table 5. Representative graphs of LR and perplexity for each method are given in Fig. 5.

Surprisingly, even our largest initial LR did not allow CCD nor ED to outperform the baseline LR. The situation is similar to what we observed in Sect. 4.2 with ED. In contrast, ARC shows how LR changes can improve the training results significantly, especially in Fig. 5 (right) where the system continuously increased the LR throughout training, leading to the best final performance and convergence speeds.

5   Limitations and Unexpected Behaviors

As Sect. 4 demonstrates, ARC can be successfully deployed over a range of tasks, architectures, optimizers, and initial LRs. It does, however, have some limitations and failure modes which bear mentioning.

One limitation of ARC is that it requires a constant optimization objective. While this is often the case for real-world problem-solving tasks, it is not true of generative adversarial networks (GANs), where the loss of the generator is based on the performance of an ever-evolving discriminator. Thus ARC, while applicable to many problems, may not be appropriate for all genres of deep learning research.

A second limitation of the current ARC implementation is that it sometimes provides unreliable decisions if queried too frequently. While we found that once per epoch typically works, running every few epochs can be more reliable. In our experiments, we chose to evenly distribute 10 invocations throughout the training. A truly autonomous system ought to determine its own update frequency. We see eliminating this parameter as an area for future research.

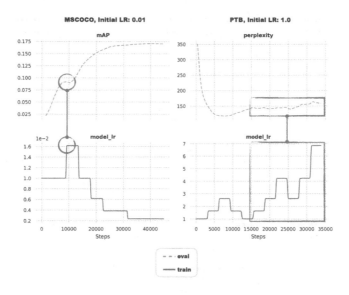

Fig. 6. Unexpected behaviors

As for failure modes, just like any other deep learning model, ARC can also make incorrect decisions. Figure 6 (left) shows an instance of object detection training where ARC incorrectly chose to raise the LR, leading to a small decrease in mAP. Nevertheless, ARC detected the problem and was then able to course correct and drop the LR again in order to get the training back on track. This bad behavior may have been due to the fact that less historical information is available early on during training.

Figure 6 (right) shows an interesting phenomena which may or may not be a failure mode. After achieving an optimal performance around step 7500, ARC started to drop the LR as might normally be expected to improve performance. However, after step 15000 it changed course and dramatically increased the LR, despite the fact that perplexity continued to get worse. Comparing this with what happened to the other schedulers for the same task configuration in Fig. 5 (left), however, ARC might actually be attempting to prevent the model from overfitting. This is something we hope to investigate more thoroughly in the future.

6   Conclusion

In this work we proposed an autonomous learning rate controller that can guide deep learning training to reach better results in less time. ARC overcomes several challenges in LR scheduling and is complementary to modern adaptive optimizers. We experimentally demonstrated its superiority to conventional schedulers across a variety of tasks, optimizers, and network architectures, as well as identifying several areas for future improvement. Not only that, ARC achieves its objectives without tangibly increasing the training budget, in sharp contrast to other AutoML paradigms. The true test of any automation system is not whether it can outperform any possible hand-crafted custom solution, but rather whether it can provide a high quality output with great efficiency. Given that, the fact that ARC actually does outperform popular scheduling methods while requiring no effort nor extra computation budget on the part of the end user makes it a valuable addition to the AutoML domain.

References

1. Bello, I., Zoph, B., Vasudevan, V., Le, Q.V.: Neural optimizer search with reinforcement learning. In: Precup, D., Teh, Y.W. (eds.) Proceedings of the 34th International Conference on Machine Learning, ICML 2017, Sydney, NSW, Australia, 6–11 August 2017. Proceedings of Machine Learning Research, vol. 70, pp. 459–468. PMLR (2017). https://proceedings.mlr.press/v70/bello17a.html
2. Bochkovskiy, A., Wang, C., Liao, H.M.: Yolov4: optimal speed and accuracy of object detection. CoRR abs/2004.10934 (2020). https://arxiv.org/abs/2004.10934
3. Bossard, L., Guillaumin, M., Van Gool, L.: Food-101 – mining discriminative components with random forests. In: Fleet, D., Pajdla, T., Schiele, B., Tuytelaars, T. (eds.) ECCV 2014. LNCS, vol. 8694, pp. 446–461. Springer, Cham (2014). https://doi.org/10.1007/978-3-319-10599-4_29
4. Cho, K., van Merrienboer, B., Bahdanau, D., Bengio, Y.: On the properties of neural machine translation: encoder-decoder approaches. In: Wu, D., Carpuat, M., Carreras, X., Vecchi, E.M. (eds.) Proceedings of SSST@EMNLP 2014, Eighth Workshop on Syntax, Semantics and Structure in Statistical Translation, Doha, Qatar, 25 October 2014. pp. 103–111. Association for Computational Linguistics (2014). https://doi.org/10.3115/v1/W14-4012, https://www.aclweb.org/anthology/W14-4012/
5. Coleman, C.A., et al.: Dawnbench: an end-to-end deep learning benchmark and competition, nIPS ML Systems Workshop (2017)
6. Devlin, J., Chang, M., Lee, K., Toutanova, K.: BERT: pre-training of deep bidirectional transformers for language understanding. In: Burstein, J., Doran, C., Solorio, T. (eds.) Proceedings of the 2019 Conference of the North American Chapter of the Association for Computational Linguistics: Human Language Technologies, NAACL-HLT 2019, Minneapolis, MN, USA, June 2–7, 2019, Volume 1 (Long and Short Papers), pp. 4171–4186. Association for Computational Linguistics (2019). https://doi.org/10.18653/v1/n19-1423
7. Duchi, J., Hazan, E., Singer, Y.: Adaptive subgradient methods for online learning and stochastic optimization. J. Mach. Learn. Res. 12(61), 2121–2159 (2011). https://jmlr.org/papers/v12/duchi11a.html
8. Goodfellow, I.J., Shlens, J., Szegedy, C.: Explaining and harnessing adversarial examples. In: Bengio, Y., LeCun, Y. (eds.) 3rd International Conference on Learning Representations, ICLR 2015, San Diego, CA, USA, May 7–9, 2015, Conference Track Proceedings (2015). https://arxiv.org/abs/1412.6572
9. Goyal, P., et al.: Accurate, large minibatch SGD: training imagenet in 1 hour. CoRR abs/1706.02677 (2017). https://arxiv.org/abs/1706.02677
10. He, K., Zhang, X., Ren, S., Sun, J.: Deep residual learning for image recognition. In: 2016 IEEE Conference on Computer Vision and Pattern Recognition, CVPR 2016, Las Vegas, NV, USA, 27–30 June 2016, pp. 770–778. IEEE Computer Society (2016). https://doi.org/10.1109/CVPR.2016.90
11. Hochreiter, S., Schmidhuber, J.: Long short-term memory. Neural Comput. 9, 1735–80 (1997). https://doi.org/10.1162/neco.1997.9.8.1735

12. Huang, G., Liu, Z., van der Maaten, L., Weinberger, K.Q.: Densely connected convolutional networks. In: 2017 IEEE Conference on Computer Vision and Pattern Recognition, CVPR 2017, Honolulu, HI, USA, 21–26 July 2017, pp. 2261–2269. IEEE Computer Society (2017). https://doi.org/10.1109/CVPR.2017.243

13. Ioffe, S., Szegedy, C.: Batch normalization: accelerating deep network training by reducing internal covariate shift. In: Bach, F.R., Blei, D.M. (eds.) Proceedings of the 32nd International Conference on Machine Learning, ICML 2015, Lille, France, 6–11 July 2015. JMLR Workshop and Conference Proceedings, vol. 37, pp. 448–456. JMLR.org (2015). https://proceedings.mlr.press/v37/ioffe15.html

14. Jaeger, S., Candemir, S., Antani, S., Wang, Y.X., Lu, P.X., Thoma, G.: Two public chest X-ray datasets for computer-aided screening of pulmonary diseases. Quant Imaging Med. Surg. 4(6), 475–477 (2014)

15. Karpathy, A.: The unreasonable effectiveness of recurrent neural networks. https://karpathy.github.io/2015/05/21/rnn-effectiveness/. Accessed 04 Nov 2020

16. Kendall, A., Gal, Y., Cipolla, R.: Multi-task learning using uncertainty to weigh losses for scene geometry and semantics. In: 2018 IEEE Conference on Computer Vision and Pattern Recognition, CVPR 2018, Salt Lake City, UT, USA, 18–22 June 2018, pp. 7482–7491. IEEE Computer Society (2018). https://doi.org/10.1109/CVPR.2018.00781, https://openaccess.thecvf.com/content_cvpr_2018/html/Kendall_Multi-Task_Learning_Using_CVPR_2018_paper.html

17. Kingma, D.P., Ba, J.: Adam: A method for stochastic optimization. In: Bengio, Y., LeCun, Y. (eds.) 3rd International Conference on Learning Representations, ICLR 2015, San Diego, CA, USA, 7–9 May 2015, Conference Track Proceedings (2015). https://arxiv.org/abs/1412.6980

18. Koch, G., Zemel, R., Salakhutdinov, R.: Siamese neural networks for one-shot image recognition (2015). iCML Deep Learning Workshop

19. Lake, B.M., Salakhutdinov, R., Tenenbaum, J.B.: Human-level concept learning through probabilistic program induction. Science 350(6266), 1332–1338 (2015)

20. Lin, T., Goyal, P., Girshick, R.B., He, K., Dollár, P.: Focal loss for dense object detection. In: IEEE International Conference on Computer Vision, ICCV 2017, Venice, Italy, 22–29 October 2017, pp. 2999–3007. IEEE Computer Society (2017). https://doi.org/10.1109/ICCV.2017.324

21. Liu, H., Simonyan, K., Yang, Y.: DARTS: differentiable architecture search. In: 7th International Conference on Learning Representations, ICLR 2019, New Orleans, LA, USA, 6–9 May 2019. OpenReview.net (2019). https://openreview.net/forum?id=S1eYHoC5FX

22. Liu, J., Cyphers, S., Pasupat, P., McGraw, I., Glass, J.R.: A conversational movie search system based on conditional random fields. In: INTERSPEECH 2012, 13th Annual Conference of the International Speech Communication Association, Portland, Oregon, USA, 9–13 September 2012, pp. 2454–2457. ISCA (2012). https://www.isca-speech.org/archive/interspeech_2012/i12_2454.html

23. Liu, Y., et al.: Roberta: a robustly optimized BERT pretraining approach. CoRR abs/1907.11692 (2019). https://arxiv.org/abs/1907.11692

24. Loshchilov, I., Hutter, F.: SGDR: stochastic gradient descent with warm restarts. In: 5th International Conference on Learning Representations, ICLR 2017, Toulon, France, 24–26 April 2017, Conference Track Proceedings. OpenReview.net (2017). https://openreview.net/forum?id=Skq89Scxx

25. Maas, A.L., Daly, R.E., Pham, P.T., Huang, D., Ng, A.Y., Potts, C.: Learning word vectors for sentiment analysis. In: Proceedings of the 49th Annual Meeting of the Association for Computational Linguistics: Human Language Technologies, pp. 142–150. Association for Computational Linguistics, Portland, Oregon, USA (June 2011). https://www.aclweb.org/anthology/P11-1015

26. Marcus, M.P.: Treebank-3 ldc99t42. web download. https://catalog.ldc.upenn.edu/LDC99T42. Accessed 04 Nov 2020
27. Netzer, Y., Wang, T., Coates, A., Bissacco, A., Wu, B., Ng, A.Y.: Reading digits in natural images with unsupervised feature learning. In: nIPS Deep Learning and Unsupervised Feature Learning Workshop (2011)
28. Page, D.C.: Cifar10-fast dawn benchmark implementation. https://github.com/davidcpage/cifar10-fast. Accessed 04 Nov 2020
29. Pham, H., Guan, M.Y., Zoph, B., Le, Q.V., Dean, J.: Efficient neural architecture search via parameter sharing. In: Dy, J.G., Krause, A. (eds.) Proceedings of the 35th International Conference on Machine Learning, ICML 2018, Stockholmsmässan, Stockholm, Sweden, July 10–15, 2018. Proceedings of Machine Learning Research, vol. 80, pp. 4092–4101. PMLR (2018). https://proceedings.mlr.press/v80/pham18a.html
30. Raffel, C., et al.: Exploring the limits of transfer learning with a unified text-to-text transformer. J. Mach. Learn. Res. **21**(140), 1–67 (2020). https://jmlr.org/papers/v21/20-074.html
31. Ronneberger, O., Fischer, P., Brox, T.: U-Net: convolutional networks for biomedical image segmentation. In: Navab, N., Hornegger, J., Wells, W.M., Frangi, A.F. (eds.) MICCAI 2015. LNCS, vol. 9351, pp. 234–241. Springer, Cham (2015). https://doi.org/10.1007/978-3-319-24574-4_28
32. Simonyan, K., Zisserman, A.: Very deep convolutional networks for large-scale image recognition. In: Bengio, Y., LeCun, Y. (eds.) 3rd International Conference on Learning Representations, ICLR 2015, San Diego, CA, USA, 7–9 May 2015, Conference Track Proceedings (2015). https://arxiv.org/abs/1409.1556
33. Szegedy, C., Vanhoucke, V., Ioffe, S., Shlens, J., Wojna, Z.: Rethinking the inception architecture for computer vision. In: 2016 IEEE Conference on Computer Vision and Pattern Recognition, CVPR 2016, Las Vegas, NV, USA, 27–30 June 2016, pp. 2818–2826. IEEE Computer Society (2016). https://doi.org/10.1109/CVPR.2016.308
34. Tieleman, T., Hinton, G.: Lecture 6.5-RmsProp: divide the gradient by a running average of its recent magnitude. In: COURSERA: Neural Networks for Machine Learning (2012)
35. Verma, G., Swami, A.: Error correcting output codes improve probability estimation and adversarial robustness of deep neural networks. In: Wallach, H.M., Larochelle, H., Beygelzimer, A., d'Alché-Buc, F., Fox, E.B., Garnett, R. (eds.) Advances in Neural Information Processing Systems 32: Annual Conference on Neural Information Processing Systems 2019, NeurIPS 2019, 8–14 December 2019, pp. 8643–8653. Canada, Vancouver, BC (2019)
36. Wang, C.Y., Mark Liao, H.Y., Wu, Y.H., Chen, P.Y., Hsieh, J.W., Yeh, I.H.: CSPNET: a new backbone that can enhance learning capability of CNN. In: Proceedings of the IEEE/CVF Conference on Computer Vision and Pattern Recognition Workshops, pp. 390–391 (2020)
37. Welinder, P., et al.: Caltech-UCSD Birds 200. Technical report CNS-TR-2010-001, California Institute of Technology (2010)
38. Zoph, B., Le, Q.V.: Neural architecture search with reinforcement learning. In: 5th International Conference on Learning Representations, ICLR 2017, Toulon, France, 24–26 April 2017, Conference Track Proceedings. OpenReview.net (2017). https://openreview.net/forum?id=r1Ue8Hcxg

Optimizing Data Augmentation Policy Through Random Unidimensional Search

Xiaomeng Dong[1,2]($\boxtimes$), Michael Potter[1], Gaurav Kumar[1], Yun-Chan Tsai[1], V. Ratna Saripalli[1], and Theodore Trafalis[2]

[1] GE Healthcare, San Ramon, USA
Xiaomeng.Dong@ge.com
[2] University of Oklahoma, Norman, USA

Abstract. It is no secret among deep learning researchers that finding the optimal data augmentation strategy during training can mean the difference between state-of-the-art performance and a run-of-the-mill result. To that end, the community has seen many efforts to automate the process of finding the perfect augmentation procedure for any task at hand. Unfortunately, even recent cutting-edge methods bring massive computational overhead, requiring as many as 100 full model trainings to settle on an ideal configuration. We show how to achieve equivalent performance using just 6 trainings with Random Unidimensional Augmentation. Source code is available at https://github.com/fastestimator/RUA.

Keywords: Deep learning · AutoML · Data augmentation

1 Introduction

Data augmentation is a widely used technique to improve deep-learning model performance. It is sometimes described as a "freebie" [1] because it can improve model performance metrics without incurring additional computational costs at inferencing time. Unfortunately, creating a good data augmentation strategy typically requires human expertise and domain knowledge [3], which is inconvenient during initial development as well as when transferring existing strategies between different tasks. In an effort to overcome these drawbacks, researchers have begun looking for an automated solution to data augmentation.

AutoAugment [2] and its variants (FastAA [13] and PBA [9]) automated the data augmentation process by introducing augmentation parameters which are then jointly optimized alongside the neural network parameters during training. While these methods do offer an automated solution to the problem, they also introduce massive search spaces which in turn significantly increase the time required to train a model. For example, AutoAugment uses Reinforcement Learning (RL) on a search space of size 10^{32}, which costs thousands of GPU hours to find a solution for a single task. Although later methods such as FastAA and PBA greatly improved the search and reduced computation requirements through data subsampling, they can still be undesirable due to the complexity of implementing joint optimization algorithms.

D. E. Simos et al. (Eds.): LION 2022, LNCS 13621, pp. 306–318, 2022.
https://doi.org/10.1007/978-3-031-24866-5_23

RandAugment [3] took a different approach by completely removing the policy optimization while achieving better results than prior methods. Unlike its predecessors which rely on applying RL to a search space of size 10^{32}, RandAugment uses only two global parameters, reducing the search space from 10^{32} to 10^2 so that a grid search can be a simple yet viable solution to the problem. As a result, RL is no longer needed for the policy search, making the method significantly easier to implement and more computationally feasible for practical usage.

Despite the significant complexity and efficiency enhancements made by RandAugment, there is still room for improvement. For example, the default setting of RandAugment uses a 10×10 grid search for the 10^2 search space. While it is technically possible to run any training task 100 times, the computational cost of doing so may still be prohibitive, especially on large-scale datasets.

To reduce costs, a sub-grid is often selected from the 10×10 grid for the actual search. Unfortunately, appropriate sub-grid selection is highly customized to specific problems. This re-introduces a requirement on human expertise and experience, which autonomous methods seek to avoid. For example, for Cifar100 [11] the proposed subgrid is $N \in \{1, 2\}, M \in \{2, 6, 10, 14\}$. For ImageNet [4], a ResNet50 model [8] uses the subgrid $N \in \{1, 2, 3\}, M \in \{5, 7, 9, 11, 13, 15\}$, whereas EfficientNet [16] on the same dataset searches $N \in \{2, 3\}$, and $M \in \{17, 25, 28, 31\}$. It is difficult to say what kind of intuition would allow someone to generate such sub-grids for previously unseen problems.

To address these problems, we propose Random Unidimentional Augmentation (RUA): a simpler yet more effective automated data augmentation workflow. The goal of RUA is to achieve the following two objectives:

1. Reduce the computational cost required to perform automated search, without sacrificing performance.
2. Eliminate the need for problem-specific human expertise in the process, enabling a fully automated workflow.

2   Methods

2.1   Dimensionality Reduction: 2D to 1D

There are 2 global parameters defined in the search space of RandAugment: M and N. M represents the global distortion magnitude which controls the intensity of all augmentation operations. N is the number of transformations to be applied in each training step. By default, M and N are both integers ranging from 1 to 10, with 10 giving the maximum augmentation effects.

Although the definitions of M and N are different, the end result of increasing their values is the same: more augmentation. If they could be merged into a single augmentation parameter, then the search space could be reduced by an order of magnitude. To check whether this might be possible, we ran RandAugment on a full 10×10 grid for two classification tasks. We used ResNet9 for Cifar10, and WRN-28-2 [19] for SVHN [14]. Their test accuracies are shown in Fig. 1.

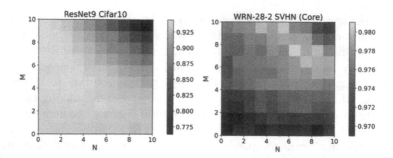

Fig. 1. Model accuracy as a function of M and N using RandAugment. Note the (accuracy) gradient as you traverse from the bottom left to the top right of each image.

The gradients in Fig. 1 show a diagonal trend from the bottom left to the top right. Although the optimal accuracy regions vary between the two problems, the fact that both exhibit an approximately diagonal gradient raises the possibility of traversing the two parameters simultaneously. We confirmed that this pattern is robust to variations in augmentation parameters, as well as across different architectures, tasks, and datasets. These results can be found later on in Fig. 3. We therefore introduce a single parameter $r \in [0, 1]$ such that $r = M/M_{max}$ and $r = N/N_{max}$. We then define our augmentation operation parameters directly in terms of r, eliminating the need to pick an explicit value for M_{max}. This parameterization can be found in Table 1. This formulation leaves N_{max} as the single open parameter in the method. While one could simply set $N_{max} = 10$ in the footsteps of RandAugment, it can also be set lower while still providing adequate gradient traversal. We defer further discussion of this to Sect. 2.4.

In situations where $r * N_{max}$ is not an integer, we apply $\lfloor r * N_{max} \rfloor$ augmentations, plus a final augmentation which executes with probability equal to the floating point remainder. For example, if $r * N_{max} = 3.14$, then 3 augmentations will be guaranteed, and a fourth will execute with 14% probability.

2.2 More Search with Less Computation

Another interesting observation one can make from Fig. 1 is that, traversing the diagonals of both Cifar10 and SVHN, accuracy first increases to a maximum and then decreases. In other words, there appears to be unimodality with respect to r. If we extract these diagonal terms and plot their relative accuracies against r (Fig. 2 top), the unimodal trend becomes more apparent.

The same trend can be observed in the RandAugment paper [3], reproduced here as Fig. 2 bottom. This demonstrates that the unimodal relationship persists across different network and dataset sizes.

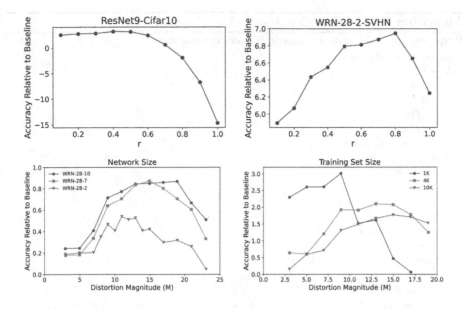

Fig. 2. RandAugment test accuracy as a function of r ($N_{max} = 10$).

In light of this unimodal property, we can leverage algorithms that are more efficient than grid search to explore a larger search space using less computation. One such algorithm is the golden-section search [10]. Golden-section search is a simple method that is widely used for finding the maximum or minimum of a unimodal function over a given interval. The pseudo code for golden-section search is given in Algorithm 1.

With golden-section search, every evaluation (after the first) of the search space will reduce the remaining search space by a constant factor of ≈ 0.618 (inverse golden ratio). As a result, we can search over 90% of the domain of r using only 6 evaluations. This makes it practical to search over the entire training dataset, without having to resort to subsampling like Fast AA or PBA. Note that this search space reduction does not require any human expertise or intervention, allowing the method to be used as an automated solution for a deep learning task.

2.3 RUA Augmentation Parameters

After our search space reduction, we are left with one parameter, r, which controls the global augmentation intensity. The exact manner of this control is given in Table 1 (right). A zero value of r means no augmentation, whereas a value of 1 achieves maximum augmentation.

This is a conceptual divergence from RandAugment, as 6 of their 14 transformations are not set up to scale this way. These transformations are marked with a "*" in Table 1 (left). For example, the transformation intensity of *Solarize* and

---

Algorithm 1. Golden Section Search (Max-Finding)

Require: Input function:f, range:$[a, b]$, max iterations:k

$\phi_1, \phi_2 \leftarrow \frac{\sqrt{5}-1}{2}, \frac{3-\sqrt{5}}{2}$

$h \leftarrow b - a$

$c \leftarrow a + \phi_2 * h$

$d \leftarrow a + \phi_1 * h$

$y_c \leftarrow f(c)$

$y_d \leftarrow f(d)$

for i from 1 to k **do**

  if $y_c > y_d$ **then**

    $b \leftarrow d$

    $d \leftarrow c$

    $y_d \leftarrow y_c$

    $h \leftarrow \phi_1 * h$

    $c \leftarrow a + \phi_2 * h$

    $y_c \leftarrow f(c)$

  else

    $a \leftarrow c$

    $c \leftarrow d$

    $y_c \leftarrow y_d$

    $h \leftarrow \phi_1 * h$

    $d \leftarrow a + \phi_1 * h$

    $y_d \leftarrow f(d)$

  end if

end for

if $y_c > y_d$ **then**

  return c

else

  return d

end if

Posterize are inversely correlated with r. Moreover, *Color*, *Contrast*, *Brightness*, and *Sharpness* are all 'shifted' in that they cause no augmentation when $r = 0.5$, whereas values closer to 0 or 1 lead to stronger alterations to the input.

In addition to aligning r with augmentation intensity, we also introduce non-deterministic parameter selection into our augmentations. For example, rather than rotating an image exactly ±30 degrees whenever the *Rotate* operation is applied, we instead draw from a random uniform distribution (U) to cover the augmentation space more thoroughly. The maximum intensity of certain augmentations are also increased to keep the expected intensity consistent in spite of the switch to uniform distributions. We justify each of these decisions with an ablation study in Sect. 3.2.

Table 1. Augmentations and their associated parameters. Augmentations marked with a "*" have non-zero impact at $r = 0$ under RA, but are zero-aligned under RUA.

Augmentations	RandAug (RA)	RUA
Identity	–	–
AutoContrast	–	–
Equalize	–	–
Rotate	degree $= \pm 30r$	degree $= U(-90r, 90r)$
Solarize*	threshold $= 256r$	threshold $= 256 - U(0, 256r)$
Posterize*	bit shift $= 8 - 4r$	bit shift $= U(0, 7r)$
Color*	factor $= 1.8r + 0.1$	factor $= 1 + U(-0.9r, 0.9r)$
Contrast*	factor $= 1.8r + 0.1$	factor $= 1 + U(-0.9r, 0.9r)$
Brightness*	factor $= 1.8r + 0.1$	factor $= 1 + U(-0.9r, 0.9r)$
Sharpness*	factor $= 1.8r + 0.1$	factor $= 1 + U(-0.9r, 0.9r)$
Shear-X	coef $= \pm 0.3r$	coef $= U(-0.5r, 0.5r)$
Shear-Y	coef $= \pm 0.3r$	coef $= U(-0.5r, 0.5r)$
Translate-X	coef $= \pm 100r$	coef $= U(-r, r) * width/3$
Translate-Y	coef $= \pm 100r$	coef $= U(-r, r) * height/3$

2.4 Selecting a Maximum N

One question which must be answered when applying RUA is what value to use as N_{max}. While one may be content to use 10, since that was the extent of the RandAugment search space, other numbers may well be equally valid. We ran a second grid search (Fig. 3) using our RUA augmentation parameters to verify that large values of N_{max} may not be necessary in order to achieve a good performance. Based on this search, we examined what outcomes a user would achieve if they ran RUA using different values of N_{max} ranging from 1 to 10. This sensitivity analysis is shown in Fig. 4. In our tests, setting $N_{max} > 5$ does not appear to provide any significant benefit, though small values like 1 or 2 can clearly be harmful, especially for ViT/Tiny ImageNet.

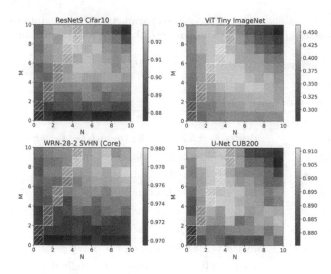

Fig. 3. Test performance as a function of M and N using RUA augmentation parameters. Note that the (accuracy/dice) gradients as you traverse from the bottom left to the top right of each image are similar to Fig. 1. Interestingly, this trend exists even in more recent attention-based architectures (ViT [6] on Tiny ImageNet [12]) and on segmentation tasks (U-Net [15] on CUB200 [17]). Cells which are candidates for our RUA search when $N_{max} = 5$ are hatched with white.

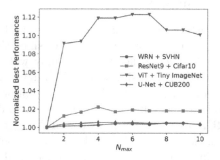

Fig. 4. The best performances along a diagonal path as a function of N_{max}.

Given that RUA is relatively insensitive to higher values of N_{max}, there are pragmatic reasons to choose values smaller than 10. Applying a large number of transformations during training can severely bottleneck the training speed. See Fig. 5 for an example. For our hardware, with any $N \geq 3$ the cpu-based preprocessing became rate limiting, especially once $N \geq 5$. This may be one reason why RandAugment never chose $N > 3$ in their sub-grid selections. With these factors in mind, we selected $N_{max} = 5$ for our final experiments.

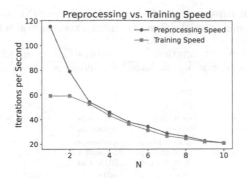

Fig. 5. Preprocessing and training speeds as a function of N. Measurements were taken on an AWS EC2 P3.2xlarge instance (8 core Intel Xeon CPU, NVIDIA Tesla V100 GPU). Training was conducted using the ResNet9 architecture on Cifar10.

3 Experiments

3.1 RUA Performance Assessment

In order to perform a direct comparison with previous works, we deploy RUA in the same training setting used by RandAugment on Cifar10, Cifar100, SVHN, and ImageNet. Details regarding the parameters used in each experiment are given in Table 2. There are a few things worth highlighting about our experimental parameters:

1. In order to be consistent with previous works, we also applied default augmentations before and after applying RUA augmentation on different tasks. For example, pad-and-crop, horizontal flip, and cutout [5] are used on the Cifar 10/100 datasets.
2. In Cifar10, RandAugment trained for 1800 epochs whereas the official implementation of PyramidNet [7] and ShakeDrop [18] trained for 300 epochs. We picked 900 epochs as a compromise between different official implementation settings.
3. In every dataset we hold out 5k training samples as evaluation data for selecting the best r. After selecting r, we put the hold-out set back into the training set and train again. We then record the test performance at the end of that final training.

Table 2. Experiment parameter details. Note that PyramidNet uses ShakeDrop regularization for consistency with the RandAugment experimental setup.

Dataset	CIFAR10	CIFAR10	CIFAR100	SVHN (Core)	ImageNet
Network	PyramidNet-272-200	Wide-ResNet-28-10	Wide-ResNet-28-10	Wide-ResNet-28-2	ResNet50
Epochs	900	200	200	200	180
Batch Size	128	128	128	128	4096
Image Preprocessing	mean-std-Normalize	mean-std-Normalize	mean-std-Normalize	Divide by 255	None
Augmentations	[pad-and-crop, horizontal flip, RUA, Cutout]	[pad-and-crop, horizontal flip, RUA, Cutout]	[pad-and-crop, horizontal flip, RUA, Cutout]	[RUA, Cutout]	[random resized crop, horizontal flip, RUA]
Optimizer	SGD	SGD	SGD	SGD	SGD
Weight Decay	1e−4	5e−4	5e−4	5e−4	1e−4
Initial LR	0.1	0.1	0.1	0.1	1.6
LR Schedule	Cosine Decay	Cosine Decay	Cosine Decay	Cosine Decay	×0.1 at epoch 60, 120, and 160
Momentum	0.9	0.9	0.9	0.9	0.9
N_{max}	5	5	5	5	5
Best r	0.867	0.6	0.733	0.8	0.666

Table 3. Experimental results for RUA compared with previous works. We report our average test accuracy over 10 independent runs (as in prior works). Best values in bold.

Methods	CIFAR10		CIFAR100	SVHN	ImageNet
	PyramidNet	WRN-28-10	WRN-28-10	WRN-28-2	ResNet50
Baseline	97.3	96.1	81.2	96.7	76.3
AA	**98.5**	**97.4**	82.9	98.0	77.6
Fast AA	98.3	97.3	82.7	–	77.6
PBA	**98.5**	**97.4**	83.3	–	–
RA	**98.5**	97.3	83.3	**98.3**	77.6
RUA	**98.5**	**97.4**	**83.6**	98.0	**77.7**

The final test results of RUA are shown in Table 3, where our performance scores are from an average of 10 independent runs. The results of previous methods including the baseline, AA, Fast AA, PBA, and RA are taken from previous work [3]. The best accuracies for each column are highlighted in bold. The search space and the number of iterations required by each method is shown in Table 4, with the best highlighted in bold.

Table 4. The search spaces of various auto-augmentation solutions. Fast AA and PBA search by training on subsampled datasets to improve search speed. Since reduced datasets can be equally applied to any of the above search methods, we directly compare iterations required by each search algorithm rather than dataset/hardware specific metrics. For each method, one iteration involves training the target model to convergence.

Methods	Search space order	Search iterations required
AA	10^{32}	15000
Fast AA	10^{32}	200
PBA	10^{61}	16
RA	10^{2}	100
RUA	10	6

As demonstrated in Table 3, RUA achieved equal or better test scores than previous state-of-the-art methods on 4 out of 5 tasks. For the Cifar10 tasks, we are equivalent to the best prior methods, with one-tailed t-test p-values of 0.0017 and 0.034. For Cifar100 and ImageNet our performance exceeds that of prior methods, with one-tailed t-test p-values of 0.002 and 0.039. On SVHN, despite being outperformed by RandAugment, RUA nonetheless achieved competitive performance on par with AutoAugment.

In addition to the performance, RUA also managed to reduce the search space by an order of magnitude and cut the training iteration requirements of the previous best method by more than 2x as shown in Table 4. This proved that the method could optimize the augmentation policy more efficiently than prior methods while achieving equivalent or better results.

3.2   Ablation Study

We conducted an ablation study on the various design decisions outlined in Sect. 2.3. The results of this study are given in Table 5. There are several noteworthy takeaways from these comparisons. First, making the "*" augmentations from Table 1 positively correlated with r is always beneficial. This can be seen through pairwise comparisons of rows 1 vs 5, 2 vs 6, 3 vs 7, and 4 vs 8. The second takeaway is that using a random distribution to draw the transformation arguments is always beneficial. This can be seen through pairwise comparisons of rows 1 vs 3, 2 vs 4, 5 vs 7, and 6 vs 8. Finally, increasing the maximum strength of augmentations (for example rotating ±90 rather than ±30) is always deleterious on its own (rows 1 vs 2 and 5 vs 6), but advantageous when paired with random sampling (rows 3 vs 4 and 7 vs 8). This is not particularly surprising since larger effects under deterministic sampling will consistently and seriously distort an image, whereas under uniform sampling they permit a larger exploration of the distortion space while still more often sampling less extreme distortions. All told, the best design was to apply all three modifications (row 8).

Table 5. An ablation study of the RUA design decisions from Sect. 2.3. A ResNet9 architecture was trained on Cifar10, with accuracies averaged over 10 independent runs. 'Aligned' indicates our modifications to the "*" transforms in Table 1, 'Random' indicates our use of a random uniform distribution, and 'Expanded' indicates the use of expanded augmentation parameters. Row 1 is thus analogous to running RandAugment using our dimensionality reduction and golden section search routine, and row 8 is the full RUA method.

	Aligned	Random	Expanded	Accuracy
1	0	0	0	0.916
2	0	0	1	0.912
3	0	1	0	0.917
4	0	1	1	0.920
5	1	0	0	0.917
6	1	0	1	0.915
7	1	1	0	0.920
8	1	1	1	0.922

4   Conclusion

In this work, we proposed Random Unidimensional Augmentation (RUA), an automated augmentation method providing several benefits relative to previous state-of-the-art algorithms. Our search space is one order of magnitude smaller than prior works, our transformations are more effective, and we leverage more efficient search algorithms. As a result of these improvements, RUA achieves equivalent results while requiring significantly less computation. We experimentally demonstrated RUAs strength on the same tasks used by previous works across various network architectures and datasets. Unlike previous methods, RUA does not rely on any problem-specific human expertise, making the method truly automated and thus fit for use in conjunction with larger autoML pipelines.

References

1. Bochkovskiy, A., Wang, C., Liao, H.M.: Yolov4: optimal speed and accuracy of object detection. CoRR abs/2004.10934 (2020). arxiv.org/abs/2004.10934
2. Cubuk, E.D., Zoph, B., Mané, D., Vasudevan, V., Le, Q.V.: Autoaugment: earning augmentation strategies from data. In: IEEE Conference on Computer Vision and Pattern Recognition, CVPR 2019, Long Beach, CA, USA, 16–20 June, 2019, pp. 113–123. Computer Vision Foundation/IEEE (2019). https://doi.org/10.1109/CVPR.2019.00020. http://openaccess.thecvf.com/content_CVPR_2019/html/Cubuk_AutoAugment_Learning_Augmentation_Strategies_From_Data_CVPR_2019_paper.html
3. Cubuk, E.D., Zoph, B., Shlens, J., Le, Q.: Randaugment: Practical automated data augmentation with a reduced search space. In: Larochelle, H., Ranzato, M., Hadsell, R., Balcan, M., Lin, H. (eds.) Advances in Neural Information Processing

Systems 33: Annual Conference on Neural Information Processing Systems 2020, NeurIPS 2020, 6–12 December 2020, virtual (2020). https://proceedings.neurips.cc/paper/2020/hash/d85b63ef0ccb114d0a3bb7b7d808028f-Abstract.html

4. Deng, J., Dong, W., Socher, R., Li, L., Li, K., Li, F.: Imagenet: A large-scale hierarchical image database. In: 2009 IEEE Computer Society Conference on Computer Vision and Pattern Recognition (CVPR 2009), 20–25 June 2009, Miami, Florida, USA, pp. 248–255. IEEE Computer Society (2009). https://doi.org/10.1109/CVPR.2009.5206848

5. Devries, T., Taylor, G.W.: Improved regularization of convolutional neural networks with cutout. CoRR abs/1708.04552 (2017). arxiv.org/abs/1708.04552

6. Dosovitskiy, A., et al.: An image is worth 16x16 words: Transformers for image recognition at scale. ICLR (2021)

7. Han, D., Kim, J., Kim, J.: Deep pyramidal residual networks. In: 2017 IEEE Conference on Computer Vision and Pattern Recognition, CVPR 2017, Honolulu, HI, USA, 21–26 July, 2017, pp. 6307–6315. IEEE Computer Society (2017). https://doi.org/10.1109/CVPR.2017.668

8. He, K., Zhang, X., Ren, S., Sun, J.: Deep residual learning for image recognition. In: 2016 IEEE Conference on Computer Vision and Pattern Recognition, CVPR 2016, Las Vegas, NV, USA, 27–30 June, 2016, pp. 770–778. IEEE Computer Society (2016). https://doi.org/10.1109/CVPR.2016.90

9. Ho, D., Liang, E., Chen, X., Stoica, I., Abbeel, P.: Population based augmentation: Efficient learning of augmentation policy schedules. In: Chaudhuri, K., Salakhutdinov, R. (eds.) Proceedings of the 36th International Conference on Machine Learning, ICML 2019, 9–15 June 2019, Long Beach, California, USA. Proceedings of Machine Learning Research, vol. 97, pp. 2731–2741. PMLR (2019). http://proceedings.mlr.press/v97/ho19b.html

10. Kiefer, J.: Sequential minimax search for a maximum. Proc. Am. Math. Soc. 4(3), 502–506 (1953). http://www.jstor.org/stable/2032161

11. Krizhevsky, A., Hinton, G.: Learning multiple layers of features from tiny images. Master's thesis, Department of Computer Science, University of Toronto (2009)

12. Li, F.F., Karpathy, A., Johnson, J.: Tiny imagenet visual recognition challenge (2016). https://www.kaggle.com/c/tiny-imagenet

13. Lim, S., Kim, I., Kim, T., Kim, C., Kim, S.: Fast autoaugment. In: Wallach, H.M., Larochelle, H., Beygelzimer, A., d'Alché-Buc, F., Fox, E.B., Garnett, R. (eds.) Advances in Neural Information Processing Systems 32: Annual Conference on Neural Information Processing Systems 2019, NeurIPS 2019, 8–14 December, 2019, Vancouver, BC, Canada, pp. 6662–6672 (2019). https://proceedings.neurips.cc/paper/2019/hash/6add07cf50424b14fdf649da87843d01-Abstract.html

14. Netzer, Y., Wang, T., Coates, A., Bissacco, A., Wu, B., Ng, A.Y.: Reading digits in natural images with unsupervised feature learning (2011). nIPS Deep Learning and Unsupervised Feature Learning Workshop

15. Ronneberger, O., P.Fischer, Brox, T.: U-net: Convolutional networks for biomedical image segmentation. In: Medical Image Computing and Computer-Assisted Intervention (MICCAI). LNCS, vol. 9351, pp. 234–241. Springer (2015). http://lmb.informatik.uni-freiburg.de/Publications/2015/RFB15a, (available on arXiv:1505.04597 [cs.CV])

16. Tan, M., Le, Q.V.: Efficientnet: Rethinking model scaling for convolutional neural networks. In: Chaudhuri, K., Salakhutdinov, R. (eds.) Proceedings of the 36th International Conference on Machine Learning, ICML 2019, 9–15 June 2019, Long Beach, California, USA. Proceedings of Machine Learning Research, vol. 97, pp. 6105–6114. PMLR (2019). http://proceedings.mlr.press/v97/tan19a.html

17. Welinder, P., et al.: Caltech-UCSD Birds 200. Technical report CNS-TR-2010-001, California Institute of Technology (2010)
18. Yamada, Y., Iwamura, M., Akiba, T., Kise, K.: Shakedrop regularization for deep residual learning. IEEE Access **7**, 186126–186136 (2019)
19. Zagoruyko, S., Komodakis, N.: Wide residual networks. In: Wilson, R.C., Hancock, E.R., Smith, W.A.P. (eds.) Proceedings of the British Machine Vision Conference 2016, BMVC 2016, York, UK, 19–22 September, 2016. BMVA Press (2016). http://www.bmva.org/bmvc/2016/papers/paper087/index.html

Evaluating Student Behaviour on the MathE Platform - Clustering Algorithms Approaches

Beatriz Flamia Azevedo[1,2]($^{\boxtimes}$) (iD), Ana Maria A. C. Rocha[2](iD),
Florbela P. Fernandes[1](iD), Maria F. Pacheco[1,3](iD), and Ana I. Pereira[1,2](iD)

[1] Research Centre in Digitalization and Intelligent Robotics (CeDRI),
Instituto Politécnico de Bragança, Bragança 5300-253, Portugal
{beatrizflamia,fflor,pacheco,apereira}@ipb.pt
[2] Algoritmi Research Centre, University of Minho, Campus Azurém,
Guimarães 4800-058, Portugal
arocha@dps.uminho.pt
[3] Center for Research and Development in Mathematics and Applications CIDMA,
University of Aveiro, Aveiro, Portugal

Abstract. The MathE platform is an online educational platform that aims to help students who struggle to learn college mathematics as well as students who wish to deepen their knowledge on subjects that rely on a strong mathematical background, at their own pace. The MathE platform is currently being used by a significant number of users, from all over the world, as a tool to support and engage students, ensuring new and creative ways to encourage them to improve their mathematical skills. This paper is addressed to evaluate the students' performance on the Linear Algebra topic, which is a specific topic of the MathE platform. In order to achieve this goal, four clustering algorithms were considered; three of them based on different bio-inspired techniques and the k-means algorithm. The results showed that most students choose to answer only basic level questions, and even within that subset, they make a lot of mistakes. When students take the risk of answering advanced questions, they make even more mistakes, which causes them to return to the basic level questions. Considering these results, it is now necessary to carry out an in-depth study to reorganize the available questions according to other levels of difficulty, and not just between basic and advanced levels as it is.

Keywords: Automatic clustering algorithms · Optimization · Bio-inspired methods · e-learning technology

This work has been supported by FCT Fundação para a Cincia e Tecnologia within the R&D Units Project Scope UIDB/00319/2020, UIDB/05757/2020, UIDP/05757/2020 and Erasmus Plus KA2 within the project 2021-1-PT01-KA220-HED-000023288. Beatriz Flamia Azevedo is supported by FCT Grant Reference SFRH/BD/07427/2021.

D. E. Simos et al. (Eds.): LION 2022, LNCS 13621, pp. 319–333, 2022.
https://doi.org/10.1007/978-3-031-24866-5_24

1   Introduction

In an era where the Internet and digital resources, in general, are forcing all teaching system levels to reinvent themselves, it becomes necessary and urgent to implement changes in teaching and learning processes [17]. Moreover, the COVID-19 pandemic showed how much investment in technological resources and literacy is still necessary in order to allow the strengthening of current educational systems and activities, contributing to increase the students' and teachers' interest in the subjects they are involved in. One way to do this is by applying digital educational technologies such as e-learning platforms.

In particular, Mathematics is considered a fundamental area for the construction of a sustainable knowledge economy, one of the great societal challenges of our time [4, 17]. However, this is one subject that students report most problems in learning and, therefore, it is essential to invest in different and engaging ways of teaching and learning mathematics. Today's students demand that their educational environments integrate the digital tools of the twenty-first century, adapting to their modern way of life and, in this context, the MathE learning environment can offer a valuable contribution to improve the students' confidence in their ability to learn mathematics.

MathE (mathe.pixel-online.org) is an e-learning platform where students from all over the world have free access to resources such as videos, exercises, training tests, and pedagogical materials covering several areas of mathematics taught in higher education courses. The MathE project offers an online tool for autonomous learning, available 24 hours per day, 7 days per week, where students can learn mathematics in an engaging way, more varied and more in line with the dynamics of the current generation of students than the traditional methods. MathE's purpose is to provide students and teachers with a new perspective on mathematical teaching and learning, relying on digital interactive technologies that enable autonomous study [4]. At its current stage, the platform is organized into three main sections, *Student's Assessment*, *MathE Library* and *Community of Practice*, in which fifteen mathematics topics are covered, among the ones that are in the classic core of graduate courses. A more detailed description of the sections and the covered topics can be found at [5].

In particular, the *Student's Assessment* section is composed of multiple-choice questions divided into topics, with two levels of difficulty—basic and advanced—among which the students can make their choice. The students can train and practice their skills in the *Self Need Assessment* (SNA) subsection. This subsection aims to provide the student with a self training assessment to test whether a certain topic that he/she has enrolled in is already properly understood. If a student or a teacher believe that the understanding of a given subject needs to be deepened, the student has the possibility of answering another training assessment to measure his/her degree of confidence in order to perform a final assessment. Each training assessment is be randomly generated from an assessments database composed of questions and their corresponding answers. In this way, the same student is able to answer different training assessments on the same topic. After the student submits a self-assessment test, the corresponding grade automatically appears, allowing self-assessment.

The MathE Platform is being improved, so that it becomes even more inter-active and gains intelligence for decision making. In this way, it is expected that in the near future the questions will be addressed to students in an autonomous way instead of in a randomized manner, as it currently is. One of the first nec-essary steps to achieve this is to recognize patterns in the data obtained so far. Thus, this work aims to evaluate the student's behavior when answering questions under the Linear Algebra topic of the SNA. Considering the obtained results it is expected to obtain information about the student's performance, that is, if they are getting the right answers or the wrong ones.

Currently, there are 99 teachers and 1161 students from different nationalities enrolled in the platform: Portuguese, Brazilian, Turk, Tunisian, Greek, German, Kazakh, Italian, Russian, Lithuanian, Irish, Spanish, Dutch, and Romanian. In this work, the performance of students using the Linear Algebra topic in the SNA section of the MathE platform will be evaluated. Linear Algebra is the most consulted and answered topic of the platform; this fact is not surprising, considering that Linear Algebra is a subject present in almost all curricula of higher education courses that include mathematics. For this reason, it was the topic chosen for the analysis herein described.

To perform the current research, the data collected over 3 years from students from different countries was analyzed by different clustering techniques in order to investigate the similarities and dissimilarities in the profiles of different groups of students in the topic Linear Algebra.

This paper is organized as follows: after the introduction, Sect. 2 presents an overview of clustering algorithms and also presents some recent work of bio-inspired clustering techniques. Section 3 introduces the clustering algorithms that will be applied in this work. The database composed of MathE student's performance in the MathE Self Need Assessment is described in Sect. 4. The results are presented and discussed in Sect. 5. Finally, the main conclusions and the future paths are described in Sect. 6.

2   An Overview on Clustering Algorithms and Related Works

Clustering is one of the most widely used methods for unsupervised learning and it is very useful in engineering, health sciences, humanities, economics, education, and in many other areas of knowledge that involve unlabeled datasets, i.e., sets of data where there is no defined association between input and output. Thus, clustering algorithms consist of performing the task of grouping a set of elements with similarities in the same group and dissimilarities in other groups [20].

A crucial step in clustering is to assess the member's proximity that composes a dataset and to partition the dataset into groups, considering the similarity and dissimilarity between a pair of elements. The partitioning method is one of the most common strategies used in clustering algorithms. This method provides a dataset partition into a pre-determined number of clusters, not known a priori. Each cluster is represented by its centroid vector, and the clustering process is carried out in an effort to iteractively optimize a criterion function and, at each execution step, all centroids are updated in an attempt to improve the quality of the final solution [16].

However, partitioning methods are known for their sensitivity to the initial position of the centroid, which may lead to weak solutions, getting stuck at the local optimum if the algorithm starts in a poor region of the problem space [16]. Moreover, the partitioning clustering algorithm heavily depends on the initial values of the cluster centers [8], which define the number of clustering partitions, as it is the number of groups that the dataset will be divided into.

Aiming to overcome these difficulties, the automatic clustering strategies that combine clustering and optimization techniques have helped to surpass these challenges, offering at the same time several improvements in clustering methods. The automatic clustering process consists of solving an optimization problem, aiming to minimize the similarity within a cluster and maximize the dissimilarity between clusters. Thus, most metaheuristic approaches are judged to fit well in the context of the new clustering paradigm [11].

In this context, several studies suggest using nature-inspired metaheuristics to select the optimal number of clusters and find a solution that maximizes the separation between different clusters and minimizes the distance between data points in the same cluster [18]. Eesa and Orman [8] present a bio-inspired Cuttlefish Algorithm (CFA) combined with the k-means algorithm for searching the best cluster centers that can minimize the clustering metrics and avoid getting stuck in local optima. Likewise, Singh [21] suggests using the Whale Optimization Algorithm (WOA) to improve the cluster exploration mechanism and solve the problem of local entrapment. Nemmich et al. [14] use Artificial Bees Colony Algorithm with a Memory Scheme to improve the k-means performance. So, in the approach presented in [14], a simple memory scheme is introduced to prevent visiting sites which are close to previously visited sites and to avoid visiting sites with the same fitness or worse. All of the enumerated approaches were tested on several benchmark datasets as well as, sometimes, on real-life problems, and the authors considered various statistical tests to justify the effectiveness of combining clustering algorithms and metaheuristics.

Nguyen and Kuo [15] present an automatic fuzzy clustering using a non-dominated sorting particle swarm optimization algorithm for categorical data. The method can identify the optimal number of clusters based on two objective functions that minimize the global compactness and fuzzy separation representing intra-cluster and inter-cluster distances. In its turn, [10] proposes a metaheuristic-based Possibilistic Multivariate Fuzzy Weighted c-means Algorithm (PMFWCM) for clustering mixed data (numerical and categorical). In this case, three metaheuristics, Genetic Algorithm (GA), Particle Swarm Optimization (PSO) and Sine Cosine algorithm (SCA) are used in different combinations with the PMFWCM for cluster analysis. Both authors claimed that the proposed algorithms work efficiently and determine the optimal number of cluster centers.

Another interesting approach is presented by Atabay et al. [2] which propose a clustering algorithm that integrates PSO and k-means algorithms. The sensitivity of the k-means algorithm to the initial choice of the centroids is solved by PSO integration. On the other hand, the ability to rapidly converge by transitioning the center of a cluster from the previous location to the average location of points belonging to that cluster in each iteration is used to accelerate convergence and improve the result of the PSO algorithm.

Considering what was described in the literature review, several approaches can be combined between bio-inspired optimization and clustering techniques, allowing to mitigate or eliminate some of the difficulties encountered by the methods using hybrid techniques. In this work, three bio-inspired metaheuristic approaches are considered, Genetic Algorithm (GA) [22], Particle Swarm optimization [9], and Differential Evolution (DE) [23], in order to find the optimum number of clusters to assess student performance from the MathE dataset. Besides, the results will be compared with k-means clustering.

3   Clustering Approaches

The cluster separation measure incorporates the fundamental features of some of the well-accepted similarity measures often applied to the cluster analysis problem and also satisfies certain heuristic criteria [1]. In this work, the Davies-Bouldin index (DB) [7] will be used as a clustering measure, that will define the number of cluster centroids, which is the number of groups that the dataset will be divided into.

3.1   Davies-Bouldin Index

Davies-Bouldin index (DB) is based on a ratio of intra-cluster and inter-cluster distances. It is used to validate cluster quality and also to determine the optimal number of clusters. Consider that cluster C have members $X_1, X_2, ..., X_m$. The goal is to define a general cluster separation measure, S_i and M_{ij}, which allows computing the average similarity of each cluster with its most similar cluster. The lower the average similarity, the better the clusters are separated and the better the clustering results. To better explain how to get the Davies-Bouldin index, four steps are considered [7].

In the first step, it is necessary to evaluate the average distance between each observation within the cluster and its centroid, that is the dispersion parameter S_i, also know as intra-cluster distance, given by Eq. (1),

$$S_i = \left\{ \frac{1}{T_i} \sum_{j=1}^{T_i} |X_j - A_i|^q \right\}^{\frac{1}{q}} \tag{1}$$

where, for a particular cluster i, T_i is the number of vectors (observations), A_i is its centroid and X_j is the jth (observation) vector.

The second step aims to evaluate the distance between the centroids A_i and A_j, given by Eq. (2), which is also known as inter-cluster distance. In this case,

a_{ki} is the kth component of the n-dimensional vector a_i, which is the centroid of cluster i, and N is the total number of clusters. It is worth mentioning that M_{ij} is the Minkowski metric of the centroids which characterize clusters i and j and $p = 2$ means the Euclidean distance.

$$M_{ij} = \left\{ \sum_{k=1}^{N} |a_k i - a_k j|^p \right\}^{\frac{1}{p}} = ||A_i - A_j||_p \qquad (2)$$

In the third step, the similarity between clusters, R_{ij}, is computed as the sum of two intra-cluster dispersions divided by the separation measure, given by Eq. (3), that is the within-to-between cluster distance ratio for the ith and jth clusters.

$$R_{ij} = \frac{S_i + S_j}{M_{ij}} \qquad (3)$$

Finally, the last step calculates the DB index, Eq. (4), that is, the average of the similarity measure of each cluster with the cluster most similar to it. R_i is the maximum of R_{ij} $i \neq j$, so, the maximum value of R_{ij} represents the worst-case within-to-between cluster ratio for cluster i. Thus, the optimal clustering solution has the smallest Davies-Bouldin index value.

$$DB = \frac{1}{N} \sum_{i=1}^{N} R_i \qquad (4)$$

Considering the definition of the DB index, a minimization problem can be defined, whose objective function is the DB index value. Thus, metaheuristics can be used in order to solve this problem as an evolutionary bio-inspired algorithm.

3.2 Evolutionary Bio-inspired Clustering Algorithms

The algorithms in the class of evolutionary computation start by randomly generating a set (population) of potential solutions. The population is represented by individuals arranged in the search space, which is the space where each variable can have values (some examples are $\mathbb{Z}^n$, $\mathbb{R}^n$, $\{0,1\}^n$, ...). The search space is delimited by the domain of the objective function, which ensures that all individuals are feasible solutions for the problem [22]. By iteratively applying the genetic operators like selection, crossover, and mutation (the most common ones), the population is being modified to obtain new feasible solutions. This process stochastically discards poor solutions and evolves more fit (better) solutions [6]. Due to the nature of these operators, which are based on Darwin's evolution principles (in which the most adapted individuals of a given population survive whereas the less adapted die to be replaced by their offspring [6, 22]), it is expected that the evolved solutions will become better generation by generation (iteration). Like any iterative process, the evolutionary algorithms require a stopping criterion to stop the search [22]. Some examples of stopping criteria are described in [3].

In this work, three bio-inspired evolutionary algorithms are used. Genetic Algorithms (GA) [22], which is based on the Darwinian principle of survival of the fittest and encoding of individuals; Differential Evolution (DE) [23], which are inspired by the theory of evolution using natural selection; and Particle Swarm Optimization (PSO) [9] that is an evolutionary algorithm, based on the behavior of birds flocking, or fish schooling. Figure 1 shows the GA, DE and PSO flowcharts.

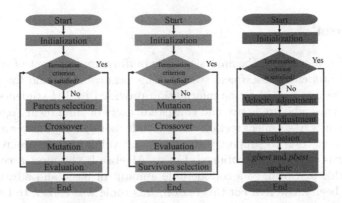

Fig. 1. The GA, DE and PSO flowcharts. Adapted from [13]

The main difference between the variants of the so-called automatic algorithms that will be used in this paper is the optimization process to define the DB-index, since each one of them employs a different bio-inspired optimization algorithm, that is GA, DE or PSO.

3.3 K-means Clustering Algorithm

The k-means partitioning clustering algorithm is one of the most well-known clustering algorithms, which requires a priori the definition of the number of clusters, being an example of an algorithm that is dependent on the initial solution, as mentioned in Sect. 2.

The k-means algorithm consists of trying to separate samples into groups of equal variance, minimizing a criterion known as the inertia or within-cluster sum-of-squares (WSS). As k-means is not an automatic clustering algorithm, it requires the definition of the initial parameter k, that represents the number of clusters division. The value of k can be specified by different techniques, such as Silhouette method, Davies-Boulding index, or Calinski Harabasz method [19]. Once this value is established, the k-means algorithm divides a set of X samples $X_1, X_2, ..., X_m$ into k disjoint clusters C, each described by the mean of the samples in the cluster, μ_i, also denoted as cluster "centroids". In this way, the k-means algorithm aims to choose centroids that minimize the inertia, or within-cluster sum-of-squares criterion, presented in Eq. (5) [1].

$$WSS = \sum_{i=0}^{m} \min ||X_j - A_i||^2, \text{ in which } \mu_i \in C \tag{5}$$

From these centers, a clustering is defined, grouping data points according to the center to which each point is assigned.

4 Dataset

This study is focused on the analysis of the performance of a set of students on the MathE Student's Assessment section. The data collected and the performed analysis take into consideration information provided by 134 students from different countries who are active and consistent users of the Linear Algebra topic of the Student's Assessment section. These students regularly answer and submit self-assessment tests to support their study and validate their progress on this topic. As was previously mentioned, Linear Algebra is the most accessed topic of the MathE platform, so a considerable amount of basic and advanced questions have been answered. For this reason, this topic was chosen to be analyzed through clustering algorithms.

In order to analyze the students' profile through clustering, the number of questions answered correctly and incorrectly for each student were evaluated, according to the basic and advanced levels. Then, the outlier students were identified through the Box plot method, and these students were removed out of the data set, leaving the information of 99 students for the analysis.

As previously mentioned, the questions available in MathE were divided into two levels of difficulty, basic and advanced. Hence, when a student selects a topic, he/she must also decide the difficult level of questions he/she wants to answer. After that the platform will provide a set of 7 random questions, available in the platform database, that belong from the chosen topic and level of difficult.

Over the 3 years of the platform's availability, 199 different questions were used, out of the 211 available in the platform's linear algebra database (142 basic and 69 advanced), being equal to 3696 the sum of times the available questions were used. Table 1 shows the number of correct and incorrect answers according to the question level. As can be seen, from the 3696 questions answered, 2919 were from the basic level and 777 from the advanced level, making a total of 1741 and 1955 correct and incorrect answers, respectively.

Table 2 presents the descriptive measures of the considered variables. The *Answers* column refers to the total number of basic or advanced questions that were answered correctly or incorrectly; *Min* and *Max* are the minimum and maximum values obtained in each variable; the column *No. Students* presents the number of students who answered a question correctly or incorrectly at basic or advanced levels. That is, out of the 99 students evaluated, 87 answered at least one basic question correctly and 88 answered at least one basic question

Table 1. Number of question answered according to the type of answer given

	Answer type		
Level	Correct	Incorrect	**Total**
Basic	1386	1533	2919
Advanced	355	422	777
Total	1741	1955	**3696**

incorrectly. On the other hand, only 26 correctly answered at least one advanced question and 23 incorrectly answered at least one advanced question.

Table 2. Descriptive measures

Variable	Answers	Min	Max	No. students
Correct basic	1386	0	25	87
Incorrect basic	1533	0	31	88
Correct advanced	355	0	7	26
Incorrect advanced	422	0	7	23

5   Results and Discussion

The MathE platform has the mission to offer a dynamic and compelling way of teaching and learning mathematics, relying on interactive digital technologies that enable autonomous study [5]. This work focuses on investigating students features, using clustering algorithm in order to recognize patterns in the students platform user's. In the future, these patterns will serve as a guidance to provide intelligence to the platform, making it capable of addressing questions in a personalized way according to each student's profile.

The information of the 99 students who used the Linear Algebra topic were considered in this analysis. The results obtained for the Linear Algebra topic - as was previously mentioned, the most widely chosen - can be inferred for the other less used topics of the platform.

Figure 2 shows the number of questions answered by each one of the 99 considered students, grouped by answered question level, ([a]-basic question and [b]-advanced questions). As already shown in Table 2 and better illustrated in Fig. 2, the range of answered basic questions varies from 0 to 35, while the advanced ones vary between 0 and 7. Hence, the figure offers a better perception of the profile of each individual student. It can be clearly seen that the students choose to answer more basic questions than advanced ones. However, even answering more basic than advanced questions, they end up making too many mistakes.

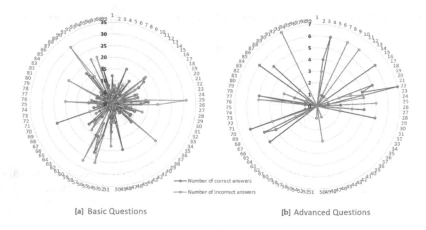

[a] Basic Questions   [b] Advanced Questions

Fig. 2. Student's Performance on basic and advanced questions

When a student selects a topic and a level on the Self-Need Assessment section, the MathE platform system provides the student with a subset of 7 questions randomly generated from the assessments database of the selected topic. Thus, when it sums up the number of questions answered by each student at each level, it is possible to evaluate, on average, how many tests these students used to answer. Thus, evaluating by question difficulty level, it can be deduced from Fig. 2(a) that the center of the figure, that is, with a range [0, 10], is comprised of students who answer 1 or 2 basic tests, which represent most of the students; within the range [11, 20], are the students answered more than 2 tests. Finally, above the range 20, are the students who answered at least 3 or more tests. Concerning the advanced questions (see Fig. 2(b)), we can say that students answer at most 2 tests of 7 advanced questions and do not return to this level afterwards.

Aiming to group the different profiles of students and analyzing the similarities and dissimilarities between students groups, the dataset was evaluated by clustering algorithms. Therefore, three automatic clustering algorithms were used to define the optimum number of clusters and establish their optimum position. Thence, three bio-inspired optimization strategies were considered, namely GA, PSO, and DE. More details about the algorithms' codification can be consulted at [24]. Moreover, the results of these three approaches were compared with k-means algorithm, which is an example of a non-automatic clustering algorithm [1].

For all bio-inspired algorithms, the common parameters used were: maximum number of clusters equal to 10; initial population equal to 100, maximum number of iterations equal to 250, which was also the stoppage criterion considered. For the GA, a rate of 0.8 was considered for selection and crossover, and 0.3 for a mutation. On the other hand, for PSO, the chosen rates were: global learning coefficient equal to 2, personal learning coefficient equal to 1.5, inertia weight equal to 1 and inertia weight damping equal to 0.99. Finally, for DE, the rates are equal to 0.2 for crossover and the scaling bound factor varies between $[0.2, 0.8]$. The results were obtained using an Intel(R) i5(R) CPU @1.60 GHz with 8 GB of RAM using Matlab software [12].

In order to perform the clustering analysis, 4 variables were defined, as presented below. Each of them describes the number of questions answered by a student according to the question level and the type of answer.

- **variable 1:** correct answers to basic questions
- **variable 2:** incorrect answers to basic questions
- **variable 3:** correct answers to advanced questions
- **variable 4:** incorrect answers to advanced questions

All clustering algorithms considered the four variables at the same time. The number of centroids and their positions defined by each algorithm are described on Table 3.

Table 3. Algorithms comparison

Algorithm	Centroid position				DB Index	Intra C. Dist.	Inter C. Dist.	Time (s)
	var1	var2	var3	var4				
Genetic Algorithm	C1	12.647	28.967	2.562e-06	1.087	0.612	10.359	30.128	59.217
	C2	4.461	3.531e-07	1.188	0.634		8.081		
Differential Evolution	C1	17.431	31	0.661	0	0.605	10.370	30.142	55.110
	C2	4.851	1.496	0.414	0		8.079		
PSO	C1	16.802	30.637	8.687e-08	0.374	0.598	10.989	31.382	42.821
	C2	3.866	2.0743	1.205	0.865		7.806		
k-means	C1	11.380	18.380	0.952	1.095	0.901	18.326	16.066	1.631
	C2	4.974	3.653	0.858	0.653		26.467		

Since in this paper, the automatic clustering algorithms are considered, it is up to the algorithm itself to define the optimal number of cluster division. In this case, the optimal value corresponds to the smallest DB-index, since the optimization algorithm goal is to minimize this parameter. Thence, from the results presented in Table 3, it is possible to observe that all algorithms pointed to 2 as the optimal cluster division number. That is, 2 cluster is the value that minimizes the DB-index for all considered bio-inspired clustering algorithms (GA, DE and PSO) and also by the Matlab function *evalclusters*, which was used to define the cluster number of the k- means.

From the results presented in Table 3 it can be said that the 3 evolutionary bio-inspired algorithms have similar behavior, both in the definition of the position of the centroids and also in the parameters value of DB index, intra-cluster distance and inter-cluster distance. However, the PSO bio-inspired clustering algorithm presented slightly better solutions, having obtained the lowest value of DB index and greater inter-clustering distance in less time than the GA and DE. PSO is one of the most famous bio-inspired algorithm due to its high exploration capacity, simplicity coding, and especially the high speed of convergence.

Such features were also evident from the results obtained in this work. Since a small size and low complexity dataset was considered, the similarity between the results of the bio-inspired algorithms is according to what was expected. As the complexity of the data increases, it is expected to find different amounts of clusters in each algorithm.

Regarding k-means, although it provides the solution in much less computational time than the other algorithms, the final solution is worst compared to the 3 bio-inspired algorithms in terms of DB index and also in relation to intra-cluster and inter-cluster distances. It is important to highlight that the DB index used in k-means was obtained by the Matlab function *evalclusters*, since k-means is not an example of automatic clustering, so it requires a specific technique to define the initial parameter k.

Due to the better performance of PSO algorithm its solution was chosen to be presented and analyzed. As it is no possible to represent a 4-dimensional graphic, Fig. 3 presents the clustering division, according to 3 to 3 variable combination.

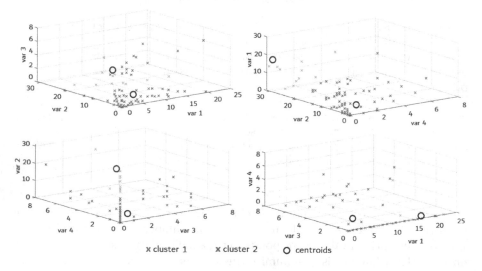

Fig. 3. PSO's bio-inspired clustering solution

Cluster 1, in blue, describes a group of students who answer more basic questions incorrectly. This cluster contains fewer elements, and it is slightly more compact than cluster 2. All students that belong to this cluster answered at least 18 basic questions incorrectly. However, it is not possible to establish an average value of basic questions answered by students in this cluster since the cluster elements are well scattered. Clearly, the characteristic of having at least 18 incorrect answers is the main point in establishing the division between cluster 1 and 2. Thus, it can be said that most basic questions were answered incorrectly. Moreover, most of the students in this cluster do not answer any advanced questions, and of those who do, only take one test of 7 questions in the SNA. Out of the answers, either they get 2 correctly and 5 incorrectly, or they answer all incorrectly.

Cluster 2, in red, has the group of students who made fewer mistakes in basic questions, that is, less than 15, but it is essential to consider that they answered fewer questions than students from cluster 1. In general, students from cluster 2 answer around 20 basic questions; they usually take 2 tests of 7 questions in the SNA, with half of these answers being correct. On the other hand, concerning the advanced questions, they usually answer 1 or at most 2 tests in the SNA, and of the questions answered, they usually provide 5 correct answers.

6 Conclusions

E-learning has already operated a transformation in higher education, and on-line platforms such as MathE are an opportunity to make learning more accessible, deepen student engagement and allow teachers to shift to a student-centered pedagogical model. This work aimed to evaluate the performance of a group of students who answered questions about Linear Algebra on the section Self Need Assessment of the MathE platform. For this purpose, data collected over 3 years was evaluated through different cluster techniques.

Through the performed analysis it can be concluded that most of the students who use the MathE platform, specifically on the Linear Algebra topic, have many difficulties in the subject, as they have a high error rate about the hit rate. Although the clustering algorithm separates the sample into two groups, it was not possible to establish a group of students whose performance was significantly better than the other's. Besides, the expressive number of incorrect answers indicates that it is urgent and mandatory to review the questions' difficulty level. However, it is also known that some teachers use the platform in the classroom to ascertain the level of the students at the beginning of the course. This may be the cause of the high number of questions incorrectly answered since many of them are answered before the students have contact with the concepts in the classroom.

Future research will focus on developing a more robust clustering analysis and new possibilities in combining bio-inspired algorithms. Besides, more of the topics covered by the MathE platform must be involved in the study as well as other students' features such as country and course information.

References

1. Arthur, D., Vassilvitskii, S.: K-means++: the advantages of careful seeding. In: Proceedings of the Eighteenth Annual ACM-SIAM Symposium on Discrete Algorithms, SODA 2007, pp. 1027–1035. Society for Industrial and Applied Mathematics, USA (2007). https://doi.org/10.1145/1283383.1283494
2. Atabay, H.A., Sheikhzadeh, M.J., Torshizi, M.: A clustering algorithm based on integration of k-means and pso. In: 1st Conference on Swarm Intelligence and Evolutionary Computation (CSIEC2016) - Higher Education Complex of Bam, pp. 59–63. Iran (2016). https://doi.org/10.1109/CSIEC.2016.7482110

3. Azevedo, B.F.: Study of Genetic Algorithms for Optimization Problems. Master's thesis, Instituto Politecnico de Braganca Escola Superior de Tecnologia e Gestao, Portugal, Braganca, Portugal (2020)
4. Azevedo, B.F., Amoura, Y., Kantayeva, G., Pacheco, M.F., Pereira, A.I., Fernandes, F.P.: Collaborative Learning Platform Using Learning Optimized Algorithms, vol. 1488. Springer (2021). https://doi.org/10.1007/978-3-030-91885-9-52
5. Azevedo, B.F., Pereira, A.I., Fernandes, F.P., Pacheco, M.F.: Mathematics learning and assessment using MathE platform: a case study. Educ. Inf. Technol. **27**(2), 1747–1769 (2021). https://doi.org/10.1007/s10639-021-10669-y
6. Bansal, J.C., Singh, P.K., Pal, N.R. (eds.): Evolutionary and Swarm Intelligence Algorithms. SCI, vol. 779. Springer, Cham (2019). https://doi.org/10.1007/978-3-319-91341-4
7. Davies, D.L., Bouldin, D.W.: A cluster separation measure. IEEE Trans. Pattern Anal. Mach. Intell. **PAMI-1**(2), 224–227 (1979). https://doi.org/10.1109/TPAMI.1979.4766909
8. Eesa, A.S., Orman, Z.: A new clustering method based on the bio-inspired cuttlefish optimization algorithm. Expert Syst.**37** (2020). https://doi.org/10.1111/exsy.12478
9. Kennedy, J., Eberhart, R.: Particle swarm optimization. In: Proceedings of ICNN'95 - International Conference on Neural Networks, vol. 4, pp. 1942–1948 (1995). https://doi.org/10.1109/ICNN.1995.488968
10. Kuo, R.J., Amornnikun, P., Nguyen, T.P.Q.: Metaheuristic-based possibilistic multivariate fuzzy weighted c-means algorithms for market segmentation. Appl. Soft Comput. J. **96** (2020). https://doi.org/10.1016/j.asoc.2020.106639
11. Kuo, R.J., Huang, Y.D., Lin, C.C., Wu, Y.H., Zulvia, F.E.: Automatic kernel clustering with bee colony optimization algorithm. Inf. Sci. **283**, 107–122 (2014). https://doi.org/10.1016/j.ins.2014.06.019
12. MATLAB: The mathworks inc (2019a). https://www.mathworks.com/products/matlab.html
13. Nakane, T., Bold, N., Sun, H., Lu, X., Akashi, T., Zhang, C.: Application of evolutionary and swarm optimization in computer vision: a literature survey. IPSJ Trans. Comput. Vis. Appl. **12**(1), 1–34 (2020). https://doi.org/10.1186/s41074-020-00065-9
14. Nemmich, M.A., Debbat, F., Slimane, M.: A data clustering approach using bees algorithm with a memory scheme. Lecture Notes Networks Syst. **50**, 261–270 (2019). https://doi.org/10.1007/978-3-319-98352-3-28
15. Nguyen, T.P.Q., Kuo, R.J.: Automatic fuzzy clustering using non-dominated sorting particle swarm optimization algorithm for categorical data. IEEE Access **7**, 99721–99734 (2019). https://doi.org/10.1109/ACCESS.2019.2927593
16. Pacifico, L.D.S., Ludermir, T.B.: An evaluation of k-means as a local search operator in hybrid memetic group search optimization for data clustering. Nat. Comput. **20**(3), 611–636 (2020). https://doi.org/10.1007/s11047-020-09809-z
17. Pedró, F., Subosa, M., Rivas, A., Valverde, P.: Artificial intelligence in education: challenges and opportunities for sustainable development (2019), uNESCO DOC Digital Library - Available online at https://unesdoc.unesco.org/ark:/48223/pf0000366994. Accessed May 2021
18. Qaddoura, R., Faris, H., Aljarah, I.: An efficient evolutionary algorithm with a nearest neighbor search technique for clustering analysis. J. Ambient. Intell. Humaniz. Comput. **12**(8), 8387–8412 (2020). https://doi.org/10.1007/s12652-020-02570-2

19. Saitta, S., Raphael, B., Smith, I.F.C.: A comprehensive validity index for clustering. Intell. Data Anal. **12**(6), 529–548 (2008). https://doi.org/10.3233/IDA-2008-12602

20. Shalev-Shwartz, S., Ben-David, S.: Understanding Machine Learning: From Theory To Algorithms. Cambridge University Press (2014)

21. Singh, T.: A novel data clustering approach based on whale optimization algorithm. Expert Syst. **38**(3) (2021). https://doi.org/10.1111/exsy.12657

22. Sivanandam, S.N., Deepa, S.N.: Introduction to Genetic Algorithms. Springer, 1 edn. (2008). https://doi.org/10.1007/978-3-540-73190-0

23. Storn, R., Price, K.: Differential evolution-a simple and efficient heuristic for global optimization over continuous spaces. J. Global Optim. **11**(4), 341–359 (1997). https://doi.org/10.1023/A:1008202821328

24. Yapiz: Evolutionary clustering and automatic clustering (2022). https://www.mathworks.com/matlabcentral/fileexchange/52865-evolutionary-clustering-and-automatic-clustering. Accessed 2 Feb 2022

Unsupervised Training for Neural TSP Solver

Elīza Gaile[1]([✉]), Andis Draguns[2], Emīls Ozoliņš[2], and Kārlis Freivalds[3]

[1] Faculty of Computing, University of Latvia, Riga, Latvia
eliiza.gaile@gmail.com
[2] Institute of Mathematics and Computer Science at University of Latvia, Riga, Latvia
andis.draguns@lumii.lv
[3] Institute of Electronics and Computer Science, Riga, Latvia
karlis.freivalds@edi.lv

Abstract. There has been a growing number of machine learning methods for approximately solving the travelling salesman problem. However, these methods often require solved instances for training or use complex reinforcement learning approaches that need a large amount of tuning. To avoid these problems, we introduce a novel unsupervised learning approach. We use a relaxation of an integer linear program for TSP to construct a loss function that does not require correct instance labels. With variable discretization, its minimum coincides with the optimal or near-optimal solution. Furthermore, this loss function is differentiable and thus can be used to train neural networks directly. We use our loss function with a Graph Neural Network and design controlled experiments on both Euclidean and asymmetric TSP. Our approach has the advantage over supervised learning of not requiring large labelled datasets. In addition, the performance of our approach surpasses reinforcement learning for asymmetric TSP and is comparable to reinforcement learning for Euclidean instances. Our approach is also more stable and easier to train than reinforcement learning.

1 Introduction

Traveling salesman problem (TSP) is a well-known combinatorial optimization problem that searches for the optimal way to traverse a graph while visiting each node exactly once. TSP and its variants have broad practical applications, e.g. in electronics and logistics [16]. Considering that TSP is an NP-hard problem, many analytical methods and heuristics have been handcrafted to solve this problem as optimally and efficiently as possible.

Neural networks have become powerful tools to solve various tasks and have shown encouraging results also for TSP. Training neural networks to solve combinatorial optimization tasks such as TSP presents distinct challenges for all learning paradigms - supervised (SL), unsupervised (UL), and reinforcement

This research is funded by the Latvian Council of Science, project lzp-2021/1-0479.

D. E. Simos et al. (Eds.): LION 2022, LNCS 13621, pp. 334–346, 2022.
https://doi.org/10.1007/978-3-031-24866-5_25

learning (RL). Recently, both supervised and reinforcement learning has been widely used to solve TSP, however, both of them have disadvantages.

While SL can perform better than other learning paradigms on fixed-size graphs (e.g. [11]), supervised learning requires optimally labeled TSP examples that are time consuming to produce even for moderately sized instances; besides that supervised learning cannot process multiple correct solutions. Although reinforcement learning does not have these problems, RL systems are often complex, unstable, and less sample efficient than other learning paradigms. Besides that, RL models are not well-suited for non-autoregressive models since that would require a vast action space with $O(n^2)$ continuous values. But autoregressive RL models rely on decoders that use node embeddings instead of adjacency matrix embeddings. This means that to solve asymmetric TSP using RL, the full adjacency matrix representation must be encoded in node embeddings. It can be done but requires powerful and complex encoders [14].

We propose to train Neural models for TSP in an unsupervised way through minimization of a differentiable loss. By exchanging supervised loss function with this unsupervised loss, it is possible to eliminate the need for labeled training data without using RL and hence avoid RL disadvantages as well. The loss function is constructed so that its global minimum corresponds to the optimal solution of relaxed TSP; we use variable discretization to obtain integer solutions. Since loss function is non-convex, direct minimization for a given TSP instance usually ends up a sub-optimal local minimum, but when used for training a neural network, the trained network manages to find close-to-optimal solutions.

The proposed loss function does not rely on labels so it can be applied on large instances while relieving us from the need to create a pre-solved dataset and does not have problems with multiple solutions. In addition, our unsupervised approach performs similarly to reinforcement learning when solving Euclidean instances and our approach works on asymmetric TSP which reinforcement learning struggles with.

2   Related Work

Over the last few years, many new machine learning approaches for solving TSP have been proposed. These approaches can be divided into two directions - hybrid methods where machine learning assists established classical heuristics and end-to-end neural solvers where ML model outputs solutions directly from the input.

Most of the notable end-to-end advances in TSP are based on sequence-to-sequence models [2,18], attention models [4,13,17] or graph neural networks [11,12,14]. Earlier works focused more on supervised learning (mostly based on Pointer Networks [18]), however after that more and more RL approaches emerged (e.g. [2,11,13,14]) and it was noted that reinforcement learning is generally more suitable for TSP than SL [9,10].

To our knowledge, there are no notable works on neural end-to-end TSP solvers with unsupervised learning despite UL also having the advantage of not requiring labeled data. There is also very little work on non-Euclidean TSP variants; most of the models used for solving TSP use only node coordinates without adjacency matrices as input and are only applied to planar TSP variants (e.g. [4, 11, 13, 18]).

3   Unsupervised TSP

To train a neural network in an unsupervised manner, an unsupervised loss function is needed. We construct a differentiable function $\mathcal{L} : ([0, 1]^{n \times n}, \mathbb{R}^{n \times n}) \to \mathbb{R}$, where n - number of nodes in a TSP instance. The input of this function is two matrices - matrix X which tells if each edge is a part of the proposed optimal tour (0 - not a part of the tour, 1 - part of the tour) and matrix C which is the adjacency matrix of the instance and contains weights of each edge. The output of $\mathcal{L}$ is an abstract numerical evaluation of how close the proposed tour is to the optimal solution; the smaller the output, the better.

3.1   Unsupervised Loss

Our approach to obtain such a loss function is based on a relaxation of an integer program for TSP that was first used by Dantzig, Fulkerson, and Johnson [1]. If the vertices are numbered from 1 to n, the TSP problem can be formulated as follows:

$$\min \sum_{i=1}^{n} \sum_{j \neq i, j=1}^{n} c_{ij} \cdot x_{ij}: \tag{1}$$

$$0 \leq x_{ij} \leq 1$$

$$\sum_{i=1, i \neq j}^{n} x_{ij} = 1 \qquad j = 1, \ldots, n;$$

$$\sum_{j=1, j \neq i}^{n} x_{ij} = 1 \qquad i = 1, \ldots, n;$$

$$\sum_{i \in Q} \sum_{j \neq i, j \in Q} x_{ij} \leq |Q| - 1 \qquad \forall Q \subsetneq \{1, \ldots, n\}, |Q| \geq 2$$

The last constraint guarantees that the solution has no tours smaller than the whole graph and therefore is called the subtour constraint. The subtour constraint checks $O(2^n)$ node subsets. To avoid this exponential growth, the subtour constraint can be replaced by a heuristic algorithm that looks only at a part of subsets (the ones which are more likely to violate the constraint) and can therefore run in polynomial time. If the chosen heuristic is re-applied many times and each time the found constraint violations are corrected, a solution with no violations will be achieved.

For the heuristic, the parametric connectivity [1] is used. First, this heuristic replaces the previous subtour contraint with

$$\text{cut}(Q, V - Q) \geq 1 \qquad \forall Q \subset \{1, \ldots, n\}, |Q| \geq 2, \qquad (2)$$

where V - the set of all vertices and $\text{cut}(S, T) = \sum_{i \in S} \sum_{j \in T} x_{ij}$. Similarly to DFJ, this constraint also ensures that there are no subtours in the TSP solution.

Second, the heuristic imagines the proposed solution X as a separate graph with edge weights corresponding to x_{ij} values (these are values that determine whether edge is the solution). The heuristic calculates and then orders this graph's subsets by a parameter ϵ, which is the maximum edge weight such that for any two vertices in the subset, they are connected by a path in with all edges of weight at least ϵ. Assuming that the first three constraints of DFJ are satisfied, subsets of the graph with the largest ϵ have the smallest cuts (and the largest expected values to violate subtour constraint), therefore n subsets with the largest ϵ parameters are chosen to be checked.

By combining the relaxation of the integer program and the parametric connectivity heuristic, the loss function for TSP with solution matrix X and adjacency matrix C is obtained. We evaluate the length of the tour and how much each of the DFJ constraints are violated (0 corresponds to satisfied constraints):

$$\mathcal{L} = \alpha \cdot \sum_{i=1}^{n} \sum_{j=1, j \neq i}^{n} c_{ij} \cdot x_{ij} \qquad (3)$$

$$+ \beta \cdot \left[\sum_{j=1}^{n} \left(1 - \sum_{i=1, i \neq j}^{n} x_{ij} \right)^2 + \sum_{i=1}^{n} \left(1 - \sum_{j=1, j \neq i}^{n} x_{ij} \right)^2 \right]$$

$$+ \gamma \cdot \sum_{Q \in S} (1 - \text{cut}(Q, V - Q))^2$$

where S - node subsets chosen by the parametric connectivity heuristic that violate the subtour constraint; α, β, γ - scalars for scaling.

When the summands of the loss function are scaled appropriately, the global minimum of the function thus created satisfies all constraints and has the shortest tour, and can therefore express the quality of a proposed TSP solution without optimal labels. For this paper, we experimentally found that values $\alpha = 5$, $\beta = 1$, $\gamma = 1$ work well.

Additionally, the proposed loss function $\mathcal{L}$ is differentiable w. r. t. solution X and can therefore be used to train a neural network directly. It is not differentiable w. r. t. to the choice of node subsets S and if we wanted to perfectly evaluate the proposed solution, we should look at all its subsets. However, the use of a neural network helps to get closer results without checking all subsets of a particular instance and avoids exponential count of such subsets.

3.2   Variable Discretization

To achieve a differentiable loss function, a relaxed TSP problem is used, and the optimal solution of such a relaxation does not always correspond to integer solutions [7]. To ensure that the solutions provided by minimization of the unsupervised loss function are integer or can be easily converted to integer solutions we use the Gumbel-Softmax technique [8,15]. We add Gumbel noise of a certain magnitude to logits and then apply softmax to get x_{ij}. To understand why adding noise helps to obtain integer solutions, consider how noise affects the x_{ij}. When logits are near zero ($x_{ij} = 0.5$) the addition of noise will produce large fluctuations in x_{ij} and consequently large and random loss. But when logits are large (x_{ij} close to integer points) softmax is saturated and the impact of noise becomes negligible. Therefore to minimize the expected loss, the network will be encouraged to produce integer or near-integer solutions. This principle has been described in more detail in other works [5].

Since we use greedy search to read out the obtained solution we are not required to fully discretize the solution and noise addition is performed only to improve the solution quality. We experimentally found out that in Euclidean TSP case the best results are obtained without noise but in Asymmetric TSP noise of magnitude 0.1 works well.

3.3   Implementation

Implementation of the first three summands of $\mathcal{L}$ is straightforward. Our approach to efficiently implement the last summand and the parametric connectivity heuristic is based on a known method [1]. To find the n subsets with the largest ϵ parameters and therefore the smallest expected cuts, we start with an empty graph $X_0 = (\{1, ..., n\}, \varnothing)$. By adding edges of the proposed solution X to X_0 in a descending edge weight order, components of X_0 are subsets of X whose ϵ is at least the value of the last edge added. Hence every time the addition of an edge changes components of X_0, the new component corresponds to the subset with the next largest ϵ. This way we can bypass the explicit calculation of ϵ.

For the implementation of this algorithm, only the list of components of X_0 is needed. At first, each vertex is in its own component. If the endpoints of the edge with the next largest weight belong to different components, they are merged together by changing the component list; then the cut of the vertex subset of the new component is checked. If the cut sum value in either direction is smaller than one, this new component is insufficiently connected to the rest of the graph and forms a subtour. This is repeated until there are only two components left.

To make our method easily usable for both symmetric and asymmetric graphs, all graphs are implemented as directed; when ordering edge weights and adding edges to X_0, a sum of both direction weights is used. However, when observing cuts, each direction is checked separately as this can provide more information.

It has been shown that parametric connectivity heuristic can be implemented with complexity $O(n^2 \alpha(n^2))$ [1], α - the inverse of the Ackermann function. However, since the parametric connectivity heuristic is used only when training the

Algorithm 1. Parametric connectivity

Input: Adjacency matrix X, where weight of each edge describes if that edge is in the optimal tour; number of vertices n

1: Initialize empty set of cuts that violate subtour constraint S
2: Initialize array of component indexes for each vertex $comp$; $comp \leftarrow [1..n]$
3: Initialize counter i; $i \leftarrow 0$

4: Add all edges to E_{decr}
5: Sort E_{decr} in decreasing order of edge weight

6: **for** $edge$ in E_{decr} **do**
$\qquad\qquad\qquad\qquad\qquad\qquad\qquad$ ▷ Find components of both $edge$ endpoints
7: $\qquad c \leftarrow comp[edge.\text{endpoint}_1]$
8: $\qquad c_0 \leftarrow comp[edge.\text{endpoint}_2]$

9: $\qquad$ **if** $c\: != c_0$ **then**
10: $\qquad\qquad i \leftarrow i + 1$ $\qquad\qquad\qquad\qquad\qquad\qquad$ ▷ Merge components together
11: $\qquad\qquad$ **for** $v \leftarrow 1$ **to** n **do**
12: $\qquad\qquad\qquad$ **if** $comp[v]$ is c_0 **then**
13: $\qquad\qquad\qquad\qquad comp[v] \leftarrow c$

14: $\qquad\qquad$ Initialize cut values for each direction cut_{in}, cut_{out}
15: $\qquad\qquad cut_{in}, cut_{out} \leftarrow 0$
16: $\qquad\qquad$ **for** $v_1 \leftarrow 1$ **to** n **do**
17: $\qquad\qquad\qquad$ **for** $v_2 \leftarrow 1$ **to** n **do**
$\qquad\qquad\qquad\qquad\qquad\qquad\qquad\qquad$ ▷ Find values of cuts in both directions
18: $\qquad\qquad\qquad\qquad c_1 \leftarrow comp[v_1]$
19: $\qquad\qquad\qquad\qquad c_2 \leftarrow comp[v_2]$
20: $\qquad\qquad\qquad\qquad$ **if** $c_1 == c$ and $c_2\: != c$ **then**
21: $\qquad\qquad\qquad\qquad\qquad cut_{in} \leftarrow cut_{in} + X[v_1, v_2]$
22: $\qquad\qquad\qquad\qquad$ **if** $c_2 == c$ and $c_1\: != c$ **then**
23: $\qquad\qquad\qquad\qquad\qquad cut_{out} \leftarrow cut_{out} + X[v_2, v_1]$

$\qquad\qquad\qquad$ ▷ Add a subset of vertices (cut) for each violated constraint
24: $\qquad\qquad$ **for** sum in $[c_{in}, c_{out}]$ **do**
25: $\qquad\qquad\qquad$ **if** $sum < 1$ **then**
26: $\qquad\qquad\qquad\qquad$ Initialize empty set Q
27: $\qquad\qquad\qquad\qquad$ **for** $v \leftarrow 1$ **to** n **do**
28: $\qquad\qquad\qquad\qquad\qquad$ **if** $comp[v] == c$ **then**
29: $\qquad\qquad\qquad\qquad\qquad\qquad Q \leftarrow Q \cup \{v\}$
30: $\qquad\qquad\qquad\qquad S \leftarrow S \cup \{Q\}$
31: $\qquad\qquad$ **if** $i = n - 2$ **then return** S

model, a more advanced implementation is not necessarily needed. Our implementation of this algorithm is simpler and the complexity of our approach is $O(n^3)$ (see Algorithm 1).

4   Neural Model

Our model is based on Joshi et al. [9] and follows the same pipeline: the input is a graph represented by its adjacency matrix C and we train our model to output matrix $X \in [0,1]^{n \times n}$. This matrix shows which edges belong to the optimal tour and its values can be viewed as probabilities. We use a graph neural network that uses both edges and nodes, and each edge and node of the graph is embedded as d-dimensional vector. Lastly, we use our differentiable loss function $\mathcal{L}$ to train the network.

To get the full predicted tours from X (e.g. to use for evaluation), we use greedy search. Greedy search finds the complete tour by starting from a random node and traversing along the heaviest edge which is available until a Hamiltonian cycle is formed.

We also need neural models with supervised learning and reinforcement learning to compare our approach to different learning paradigms. To ensure fairness, we use the same SL and RL models as Joshi et al. [9] used in his work.

4.1   Graph Neural Network

Our graph neural network (GNN) consists of several layers and each layer ℓ consists of message passing and updating of node and edge embeddings. For initial node embeddings $h_{ij}^{\ell=0}$ and edge embeddings $e_{ij}^{\ell=0}$ we use d-dimensional linear projections of node coordinates and normalized edge weights respectively:

$$h_{ij}^{\ell=0} = node_{ij} \cdot W_1 + b_1 \tag{4}$$

$$norm\_c_{ij} = \frac{c_{ij}}{\sqrt{\frac{1}{n} \sum_k^n \sum_l^n (c_{kl})^2}} \tag{5}$$

$$e_{ij}^{\ell=0} = norm\_c_{ij} \cdot W_2 + b_2 \tag{6}$$

To update edge and node features message passing is used. Two types of messages are computed from each edge (outgoing from vertex and incoming to vertex) using simple MLP networks, after that vertices gather and process all messages from their adjacent edges and the vertex itself:

$$out\_state_{ij} = \frac{\sum_{k=1}^n (\text{MLP}_1(e_{ik}^\ell))}{\sqrt{n}} \tag{7}$$

$$in\_state_{ij} = \frac{\sum_{k=1}^n (\text{MLP}_2(e_{kj}^\ell))}{\sqrt{n}} \tag{8}$$

$$vertex\_state_{ij} = [in\_state, out\_state, h_{ij}^\ell] \tag{9}$$

Next a new edge embedding candidate for each edge is obtained from the processed messages of adjacent vertices, and the embedding of each edge is updated by combining the old embedding with the new candidate ($tile_{ij}$ is used for ease of implementation):

$$tile_{ij} = \left.\begin{bmatrix} vertex_{ij} \\ \vdots \\ vertex_{ij} \end{bmatrix}\right\} n \text{ times} \tag{10}$$

$$candidate_{ij} = \text{MLP}_3([e_{ij}^{\ell}, tile_{ij}, tile_{ij}^T]) \tag{11}$$

$$e_{ij}^{\ell+1} = e_{ij}^{\ell} \cdot \sigma(a \cdot \mathbf{A}^{\ell}) + \mathbf{B}^{\ell} \cdot candidate \tag{12}$$

Lastly, the updating of node embeddings h_{ij}^{ℓ} is done by using MLP network on information available to vertices:

$$h_{ij}^{\ell+1} = \text{MLP}_4(h_{ij}^{\ell}) \tag{13}$$

Each layer contains multilayer perceptrons MLP_1, MLP_2, MLP_3, MLP_4, each with 3 layers (including input and output layer); learnable parameters $\mathbf{W}_1$, $\mathbf{W}_2$, $\mathbf{A}$, $\mathbf{B} \in \mathbb{R}^d$ and b_1, $b_2 \in \mathbb{R}$ as well as a scalar value a (we experimentally determined $a = 10$ to work well).

To decode edge embeddings from the last layer of GNN and get probabilities for each edge to belong to optimal tour, we first use two layer MLP to get logits from embeddings; after that we get probability matrix X from logits via softmax over each edge.

If a symmetrical TSP variant is being tackled (e.g. Euclidean TSP), we make logits symmetrical before softmax by taking the mean of logits in each edge direction.

5   Evaluation

We carry out several experiments to compare our unsupervised approach to both supervised and reinforcement approaches.

All datasets are generated randomly. For symmetric graphs, we choose points in a unit square and get adjacency matrices as Euclidean distances between those points. For asymmetric graphs, random adjacency matrices are generated with each edge weight ranging from 0 to 1. Correct solution tours (required for evaluation and supervised learning) are computed using *Concorde* solver [3] and *Gurobi* optimizer [6] for symmetric and asymmetric cases respectively.

We explore all methods on fixed-size graphs of 20 and 50 vertices on Euclidean TSP and on asymmetric TSP. For unsupervised and reinforcement learning training 128000 examples are randomly generated in each of the 100 epochs; for supervised learning, a larger set of 1280000 samples and their solutions are generated beforehand. For evaluation, 1280 samples and their solutions are generated of each type, i.e. TSP and ATSP on respective graph sizes.

To fairly compare between different paradigms, supervised and unsupervised models differ only in the loss function. Comparison with reinforcement learning is not as straightforward considering that the RL model is auto-regressive and builds the solution step by step as opposed to SL and UL models, which are non-autoregressive and produce the solution in one shot. To ensure the comparison is as fair as possible, for RL we use the corresponding encoder described in Joshi et al. [9]. Unfortunately, this means that for asymmetric TSP this encoder has to embed the adjacency matrix into node embeddings. Nonetheless, Table 1 contains a summary of the main hyperparameters used for comparing SL, RL, and our UL method. We also follow the experimental setup of Joshi et al. [9], but some parameters have been adjusted for hardware limitations.

Table 1. Training parameters for SL, UL and RL models.

Parameter	SL	UL	RL
Epochs	1	100	100
Epoch size	12800000	128000	128000
Batch size ($n = 20, 50$)	128, 32	128, 32	128, 32
Encoder layers ($n = 20, 50$)	16, 8	16, 8	16, 8
Number of parameters	354562	354562	379072
Learning rate	10^{-4}	10^{-4}	10^{-4}
Embedding and hidden dimensions	64	64	64

The output of the model is the probabilities of edges to belong to the correct tour; hence, the greedy search method is used to get a valid tour prediction. Results are compared using optimality gap, i.e. the average percentage ratio between the predicted tour and the correct one. We also look at inference time (1280 samples) and inspect the consistency of validation results throughout training to observe any unstable behaviours.

Tables 2 and 3 shows results of solving Euclidean and asymmetric TSP using different learning paradigms. We evaluate all methods on fixed-size graphs of 20 and 50 vertices.

Table 2. Optimality gap and inference time of TSP using SL, RL and UL

Method	TSP20		TSP50	
	Opt. Gap	Time	Opt. Gap	Time
SL	0.219	2.914	4.870	8.141
RL	2.752	3.163	7.954	8.881
UL	1.289	2.852	11.419	8.468

In all of the experiments supervised learning shows superior results to other learning paradigms, which we explain by our experiments being of fixed-sized graphs and supervised learning having all the information of training instances. This also coincides with the current literature [10].

When comparing unsupervised learning with reinforcement learning, we can observe that results on Euclidean instances are ambiguous, as UL performs better

Table 3. Optimality gap and inference time of ATSP using SL, RL and UL

Method	ATSP20		ATSP50	
	Opt. Gap	Time	Opt. Gap	Time
SL	17.640	3.225	83.377	9.598
RL	534.820	3.488	1439.005	9.208
UL	20.560	3.446	32.699	9.392

on 20 vertices, but RL surpasses UL when looking at instances with 50 vertices. However, results on asymmetric TSP are much more certain, where RL behaves very poorly. This suggests that the encoder used is not powerful enough to efficiently embed the adjacency matrix into node embeddings.

When comparing results between TSP and ATSP, we can see that performance for asymmetric instances is noticeably worse as it is a generally harder problem. However, UL achieves similar results to SL for asymmetric TSP, which may indicate the adaptiveness of the unsupervised approach.

Inference time results between learning paradigms are very similar and no noteworthy differences can be seen.

Figure 1 shows training behaviors of each of the learning paradigms in all experiments carried out (validation done with greedy search). It can be seen in Euclidean experiments (Fig. 1a, Fig. 1b) that reinforcement learning has several big fluctuations in the training process; we do not experience this with other learning paradigms. This type of unstable behaviour is a relatively more common behaviour for RL in general and leads to large amount of steps to train the neural network properly. The asymmetric training graphs (Fig. 1c, Fig. 1d) show small or no improvement in RL training over time, indicating that the encoder used is not suitable for asymmetric TSP.

To better see how unsupervised loss work with neural networks, we carried out an experiment to compare straightforward minimization of the loss function and its usage in our model. The minimization of the function was done using Adam optimizer (*learning rate* = 0.01) and we let it run on each instance of the evaluation datasets for 15000 steps which were empirically determined to be enough for most edges to be almost discrete. To get proper tours from the output, we use greedy search.

Results in this experiments for Euclidean TSP and asymmetric TSP can be seen in Table 4 and Table 5. We tracked both the optimality gap for each method as well as inference time for 1280 instances. It should be noted that time spent on loss minimization directly depends on optimization steps and could be reduced by possibly sacrificing the quality of the solution. For better comprehension of the experiment, we added average results of a random tour and also of a tour found with greedy search on the adjacency matrix.

As expected, we can see that the minimization of the loss function returns better results than just greedy search. When the loss function is used together with a neural network, the results are even better. This can be explained by

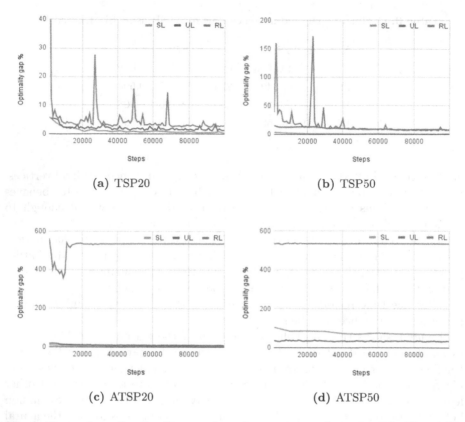

Fig. 1. Comparison of optimality gap throughout training when using RL, SL and UL

the fact that the loss function has many local minimums in which the optimizer can get trapped, but a neural network helps to overcome this. If we look at the inference times, we can see that individual optimization is very slow and is not practical for widespread use.

Table 4. Comparison of minimization of loss function and loss function used in neural network for TSP

Method	TSP20		TSP50	
	Opt. Gap	Time	Opt. Gap	Time
Random	186.959	0.007	371.908	0.011
Greedy search	17.620	0.169	22.801	1.220
Loss function	6.736	25181.865	16.560	54313.828
Neural network	1.289	2.852	11.419	8.468

Table 5. Comparison of minimization of loss function and loss function used in neural network for ATSP

Method	ATSP20		ATSP50	
	Opt. Gap	Time	Opt. Gap	Time
Random	556.233	0.007	1492.072	0.0011
Greedy search	91.194	0.169	145.298	1.220
Loss function	26.413	24448.681	47.762	54002.372
Neural network	20.560	3.446	32.699	9.392

6 Conclusions

We introduce a novel unsupervised learning approach for solving the TSP problem with neural networks. The basis of our unsupervised method is a new differentiable loss function that works on both Euclidean and asymmetric TSP. Unsupervised learning has the advantage over supervised learning of not needing large correctly labeled datasets. Our method performs similarly to reinforcement learning for Euclidean graphs with 20 and 50 vertices and outperforms reinforcement learning when looking at the stability of training or asymmetric graphs.

The loss function is constructed in a way to be easily modified with extra constraints. The addition of constraints can be done by expressing the constraint as a differentiable polynomial and adding it to the loss function. Considering that routing problems are very widespread and often have unique limitations, the addition of constraints is very relevant and may be very useful. This work does not explore this possibility further but in the future, we want to examine our work on TSP variants with additional constraints.

References

1. Applegate, D., Bixby, R., Chvátal, V., Cook, W.: Implementing the dantzig-fulkerson-johnson algorithm for large traveling salesman problems. Math. Program. **97**, 91–153 (2003)
2. Bello, I., Pham, H., Le, Q.V., Norouzi, M., Bengio, S.: Neural combinatorial optimization with reinforcement learning. ArXiv abs/1611.09940 (2017)
3. Concorde TSP Solver (2003). http://www.math.uwaterloo.ca/tsp/concorde/
4. Deudon, M., Cournut, P., Lacoste, A., Adulyasak, Y., Rousseau, L.M.: Learning heuristics for the tsp by policy gradient. In: CPAIOR (2018)
5. Frey, B.: Continuous sigmoidal belief networks trained using slice sampling. In: NIPS (1996)
6. Gurobi Optimization (2021). https://www.gurobi.com
7. Hougardy, S.: On the integrality ratio of the subtour lp for euclidean tsp. Oper. Res. Lett. **42**(8), 495–499 (2014)

8. Jang, E., Gu, S., Poole, B.: Categorical reparameterization with gumbel-softmax. In: 5th International Conference on Learning Representations, ICLR 2017, Toulon, France, 24–26 April, 2017, Conference Track Proceedings. OpenReview.net (2017). https://openreview.net/forum?id=rkE3y85ee

9. Joshi, C.K., Cappart, Q., Rousseau, L.M., Laurent, T., Bresson, X.: Learning tsp requires rethinking generalization. arXiv preprint arXiv:2006.07054 (2020)

10. Joshi, C.K., Laurent, T., Bresson, X.: On learning paradigms for the travelling salesman problem. ArXiv abs/1910.07210 (2019)

11. Joshi, C.K., Laurent, T., Bresson, X.: An efficient graph convolutional network technique for the travelling salesman problem. arXiv preprint arXiv:1906.01227 (2019)

12. Khalil, E.B., Dai, H., Zhang, Y., Dilkina, B., Song, L.: Learning combinatorial optimization algorithms over graphs. In: NIPS (2017)

13. Kool, W., Hoof, H.V., Welling, M.: Attention, learn to solve routing problems! In: ICLR (2019)

14. Kwon, Y.D., Choo, J., Yoon, I., Park, M., Park, D., Gwon, Y.: Matrix encoding networks for neural combinatorial optimization. ArXiv abs/2106.11113 (2021)

15. Maddison, C.J., Mnih, A., Teh, Y.W.: The concrete distribution: a continuous relaxation of discrete random variables. In: 5th International Conference on Learning Representations, ICLR 2017, Toulon, France, April 24–26, 2017, Conference Track Proceedings. OpenReview.net (2017). https://openreview.net/forum?id=S1jE5L5gl

16. Matai, R., Singh, S., Mittal, M.L.: Traveling salesman problem: an overview of applications, formulations, and solution approaches (2010)

17. Nazari, M., Oroojlooy, A., Snyder, L., Takác, M.: Reinforcement learning for solving the vehicle routing problem. In: NeurIPS (2018)

18. Vinyals, O., Fortunato, M., Jaitly, N.: Pointer networks. In: NIPS (2015)

Comparing Surrogate Models for Tuning Optimization Algorithms

Gustavo Delazeri[1], Marcus Ritt[1][✉], and Marcelo de Souza[2]

[1] Instituto de Informática, Universidade Federal do Rio Grande so Sul, Porto Alegre, Brazil
{gustavo.delazeri,marcus.ritt}@inf.ufrgs.br
[2] Departamento de Engenharia de Software, Universidade do Estado de Santa Catarina, Ibirama, Brazil
marcelo.desouza@udesc.br

Abstract. Tuning an algorithm requires to evaluate it under different configurations on several problem instances. Such evaluations are costly. A way to reduce the configuration time when developing tuners is to use surrogate models, which map configuration-instance pairs to the approximate algorithm performance and thus allow to replace algorithm runs by fast calls to the model. Most applications of surrogate models found in the literature focus on predicting algorithm running time; much less effort has been devoted to predicting the quality of solutions of optimization algorithms. In this paper, we present a comparative study of surrogate models for predicting solution quality. We evaluate several surrogate models from the literature, including random forests, gradient boosting methods, and neural networks, and compare ways of handling different classes of parameters, data imputation strategies, and codification of instances. We demonstrate for two heuristic algorithms that the best models can accurately reproduce effects observed when tuning with the ground truth. Our code is available (https://github.com/gutodelazeri/oracle).

Keywords: Automatic algorithm configuration · Surrogate models · Optimization algorithms

1 Introduction

Selecting the best algorithm among a set of candidates for solving a problem is a common challenge, since often several algorithms of complementary strengths are available, or algorithms are parameterized. Following the seminal work of [36] algorithm selection can be described as a mapping from a problem space $\mathcal{P}$ to an algorithm space $\mathcal{A}$ that maximizes some performance measure $m(p, a)$ for $p \in \mathcal{P}, a \in \mathcal{A}$. Here we are interested in selecting a single algorithm that has the best expected value with respect to some distribution over the problem space, a common approach in automatic algorithm configuration or tuning [24,25,32].

© Springer Nature Switzerland AG 2022
D. E. Simos et al. (Eds.): LION 2022, LNCS 13621, pp. 347–360, 2022.
https://doi.org/10.1007/978-3-031-24866-5_26

Often the algorithm space consists of a family of parameterized algorithms a_θ, with parameter settings θ from some parameter space $\Theta = \times_{i \in [n]} \Theta_i$, where Θ_i is the domain of parameter $i \in [n]$. Thus the problem of automatic algorithm configuration is to find the best configuration

$$\theta^* = \mathrm{argmax}_{\theta \in \Theta} E_{\mathcal{P}}[m(p, a_\theta)] \qquad \text{(AAC)}$$

for some distribution over the problems $\mathcal{P}$. For stochastic algorithms the problem space $\mathcal{P}$ can be understood as composed of instance-seed pairs together with an appropriate distribution. Parameters are often real or integer, but can also be ordinal, categorical, or conditional (i.e. only effective for specific settings of enabling parameters).

Solving problem (AAC) even approximately is hard, in particular because obtaining the measure $m(p, a)$ requires running an algorithm on a problem instance, which can take considerable time, often minutes or hours. As a consequence, the (second-level) problem of selecting among optimizers for problem (AAC) is even harder. On a family of parameterized optimizers this is also known as the problem of *hyperparameter tuning*. It can be made more tractable by providing a surrogate function $\hat{m}$ for measure m for some representative first-level configuration problems. Such surrogate functions are required to be:

1. fast to evaluate;
2. accurate in the sense that predicted values allow to correctly rank the performance of algorithms on instances;
3. handle all parameter classes, including numerical, categorical, and conditional parameters;
4. have a high fidelity in the sense that hyperparameter optimization on surrogates correctly predicts the behaviour under ground truth m.

Note that requirement 4 goes beyond simple accuracy, since we require the model to represent the ground truth on the level of observable effects, as opposed to just taking an accurate model as the ground truth for comparing hyperparameter optimizers. Clearly, as a function of the accuracy, this will be limited to effects of a minimal effect size. This requirement should allow to quickly assess the performance of new optimizers during design and test, and speed up benchmarking of optimizers.

In this paper we compare several representative models from the literature and evaluate their accuracy and ability to reproduce effects observable when comparing tuners. The paper is organized as follows. In the remainder of this section we discuss related work. We then introduce in Sect. 2 a selection of models for surrogate functions. Section 3 presents an experimental comparison of these surrogate models for tuning (hyper-)optimization algorithms. We conclude in Sect. 4.

1.1 Related Work

Approaches to surrogate functions make different assumptions on the domain and co-domain of the function to be modeled, on access to additional information such as derivatives, or bounds, and use different surrogate models. [17], for example, assumes a continuous, compact parameter space, a continuous function that is expensive to evaluate, with no additional information, and a model which is a weighted sum of radial basis functions centered at the samples plus a polynomial term. This is a form of *black-box* optimization, and many other approaches to black-box optimization assume real-valued parameters (e.g. [7,18,30]), some of which can be extended to handle integer parameters [8], or categorical and boolean parameters [1,4,41]. Approaches also differ in assumptions on the cost of evaluation. Expensive functions can take hours to evaluate, and thus the number of evaluations is typically limited to a few hundred, while for less expensive functions thousands of evaluations are common [38]. Other additional information that can be exploited in so-called *gray-box* optimization are objective functions of a known form (e.g. a sum of squares) the possibility of obtaining faster, less accurate samples in *multi-fidelity* objective functions [2], or assumptions on continuity (e.g. the knowledge of the Lipschitz constant [33]). Typical surrogate models include linear regression, kernel-based techniques (e.g. the radial basis functions mentioned above), Gaussian processes, (gradient-boosted) regression trees, random forests, and neural networks [9,31,39].

Most of the above approaches are concerned with optimization, and often follow some sequential model-based strategy, that successively acquires a promising sample (e.g. of best objective function value or highest information gain), updates the model, and returns the best sample when the computational budget is spent. In contrast, in this paper we are concerned with static surrogate models that represent typical tuning landscape well, as outlined in the introduction (requirements 1–4).

A collection of 110 hyperparameter optimization benchmarks, with 2 to 26 continuous, integer, categorical, or ordinal parameters has been introduced by [15]. It focuses on ease-of-use, reproducibility and benchmarks with multiple fidelity levels. All except six benchmarks which use random forests as a surrogate model are provided in a form of lookup tables. Closer to our work, [5] introduce a benchmark consisting of four expensive real-world problems (e.g. wind farm layout) that take from 2 to 60 seconds per evaluation. Two of these problems are continuous, two have mostly categorical variables. Besides comparing tuners, they compare mean errors of seven surrogate models on the best 1K of random samples, as well as samples collected during tuning on two problems. Experiments show that random forests [34] and gradient boosted regression trees (via XGBoost [6]) are among the best models, and these models tend to be more accurate when trained on random samples.

[11] have compared eight models to construct surrogate benchmarks for hyperparameter tuning, namely linear and ridge regression, two forms of support vector machines, Gaussian processes, random forests, and gradient boosted trees. Models have been built from 2K to 20K samples collected during the execution of four tuners on nine datasets. The datasets have 3 to 36 categorical and continuous parameters, with a dimension in [4, 82] after one-hot encoding of categorical variables. Results show gradient boosted trees and random forests to have best accuracy, as measured by root mean square error (RMSE) and Spearman's rank

correlation, followed by Gaussian processes, with random forests giving the most "similar" tuned values. Even with similar values, however, a consistent ranking of tuners on ground truth and surrogate models is not guaranteed. [14] extends the former work to algorithm configuration by handling conditional variables by imputation of default or midpoint values, imputation of censored data, and explicitly handling of randomized algorithms. Quantile random forests serve as a surrogate model. There are 11 scenarios with about 10–300 parameters and 30–300 instance features, with nine scenarios predicting runtime. Experiments show a reasonable reproduction of the rank of tuners over the course of the tuning process.

Different from most of the above approaches, here we are interested in models for parameter spaces (requirement 3) when instances features are not available, and on approximating objective function values as opposed to running time, which is the focus of most of the literature on models for algorithm tuning.

2   Surrogate Models for Optimization Algorithms

In this section we present the selected procedures for data pre-processing and surrogate models. Concerning data pre-processing, all selected models are able to handle real and integer parameters, so we have to define only how to handle categorical and conditional variables to attend requirement 3. For categorical variables, if the underlying surrogate model is not already able to handle them directly, we test three encoding strategies: one-hot, binary, and by-index encoding. In all three a variable with n categories is first mapped to an integer $i \in [n]$. Then, for one-hot encoding i is mapped to e_i, where $e_i = (0, \ldots, 1, \ldots, 0)$ is the ith unit vector, i.e. to n new binary variables; for binary encoding i is mapped to its base-2 representation $(i)_2$, leading to $\lceil \log_2 n \rceil$ new binary variables; for index encoding, i is used directly. The categorical variables include the instances. Since the instances may have a considerable influence on the predicted variable, and the number of instances is limited, we study two further strategies for this case: instance-wise models, where we build an individual model for each instance, and a combined model for all instances where we treat the instance as a normal categorical variable. For conditional variables, we impute values, and test three strategies: imputing a fixed, chosen from the parameter's domain uniformly at random, imputing the default value, and imputing random values. We have excluded out-of-domain, quantile, or mean imputation since previous work found no significant differences to default imputation, which is similar to our strategy.

From previous work it is also clear that tree-based models are often competitive [12,27]. Thus we include random forests and gradient boosted trees. We have chosen two widely used implementations, random forest run [3] and ranger [44]. [16] have studied models for tabular data over real, integer and categorical features, and have found two deep neural network architectures, namely ResNet and FT-Transformer, as well as gradient boosted trees to be among the best models. Therefore we include these two neural network architectures, as well as two implementations of gradient boosted trees, namely CatBoost [35] and

XGBoost [6]. We did not include Gaussian process models in our study, since previous work indicate that random forest perform better with large parameter spaces with both numerical and categorical parameters [11].

The ResNet consists of a number of residual blocks, each block consisting of two layers: a linear layer followed by a rectified linear unit (ReLU), and a second linear layer. The residual blocks are followed by a final ReLU and a linear layer that maps to the regression variable [19]. The FT-Transformer consists of a number of transformer blocks, each consisting of the typical multi-head attention with query-key-value layers followed by a linear layer [42]). Categorical variables are handled as proposed by [16], who propose a specific embedding for each possible value. The number of layers of both architectures, as well as several other parameters are tuned following [16], as explained below in the experimental section.

Finally we include an interpolated tabular model, that uses Shephard's method of inverse distance weighting [37] with a definable cutoff after considering the k closest samples. Closest samples are found by using the ball tree structure from scikit-learn. For distance weighting we use the heterogeneous Euclidean-overlap metric of [43]: for numerical parameters, the distance is defined as difference in parameter values, normalized to $[0, 1]$; for categorical parameters the distance is 0 when parameter values are equal, and 1 otherwise. Undefined categorical values always have distance 1 to all other values.

All models are trained to learn to predict objective function values from the ground truth data, i.e. our approach can be classified as pointwise learning-to-rank (as opposed to learn to rank pairs or larger subsets of inputs directly).

3   Experimental Results

The main goal of the experiments is to compare the performance of the selected models for modeling optimization landscapes. We use two measures to judge performance. First, we measure the accuracy of the models in ranking pairs of parameter settings. This is done by comparing pairs of samples from the test set using the predicted and ground truth values. The accuracy is then defined as the number of correctly predicted orders over all pairs. Accuracy values a are in $[0, 1]$ and are related to Kendall's tau [29] by $\tau = 2a - 1$.

We first report on the effects of instance representation and pre-processsing in Sect. 3.2. In Sect. 3.3 we analyze accuracy as measured by the relative root mean squared error and Kendal's τ. Finally, we measure the performance by the fidelity with which the different models are able to reproduce effects that can be observed on the ground truth data. For this experiment we have selected three parameters of irace and compare it on different settings. These experiments are explained in Sect. 3.4.

3.1   Methodology

We have done the experiments using the tuner irace [32]. The irace configurator applies iterated racing, where in each iteration candidate parameter settings are run on a number of instances, statistically significantly worse configurations are discarded, and new configurations are sampled according to a distribution of good parameter values, which evolves with each candidate parameter setting. These distributions are represented by their marginal distributions on each parameter and converge to the best settings. If not otherwise specified we run irace with default parameters, and run all tests with three fixed budgets. We have chosen a maximum budget of 3 K evaluations, the smallest possible budget for irace of 780 evaluations, and an intermediate budget of 1890 evaluations. Therefore the results do not depend on the concrete computing platform or the running time. When running times are mentioned they have been obtained on a PC with a 12-core AMD Ryzen 9 3900X processor with 32 of main memory, running Ubuntu Linux 20.04. All models run in a server that responds queries on parameter settings by irace, generated by a wrapper that replaces the algorithm to be tuned. Our code is available at https://github.com/gutodelazeri/oracle.

Configuration Scenarios. We evaluate all surrogate models on two configuration scenarios, namely ACOTSP and LKH. Both are heuristic algorithms for the symmetric traveling salesperson problem (TSP). We use 10 Random Uniform Euclidean TSP instances with 2000 cities, generated using the *portgen* instance generator from the 8th DIMACS Implementation Challenge [28]. Both ACOTSP and LKH scenarios are part of the AClib benchmark library for algorithm configuration [26]. The ACOTSP scenario implements ant colony optimization algorithms for TSP [10]. We use ACOTSP version 1.03, available at [40]. It has 4 real, 4 integer, and 3 categorical parameters; 5 of these parameters are conditional. The LKH scenario concerns the Lin-Kernighan-Helsgaun algorithm for TSP [20–22]. We use LKH version 2.0.9 [23]. It has 12 integer and 9 categorical parameters, all of them unconditional. In both scenarios we use no additional instance features, since we are only interested in building surrogate models for the selected instances. We test, however, the case of a separate model for each instance, see below, which arguably corresponds to the best possible set of features for the selected instances.

Models. We summarize the selected models in Table 3.1. All models use their default parameters, with the exception of the neural networks, which are tuned using the process proposed by [16], and PyRFR, which uses the configuration found by [13].

Model	Instances	Categoricals	Imputation
Interpolation	split, merged	model-defined	model-defined
CatBoost	split, merged	model-defined	model-defined
XGBoost	split, merged	binary, one-hot, index	fixed, default, random
skranger	split, merged	binary, one-hot, index	fixed, default, random
PyRFR	split, merged	binary, one-hot, index	fixed, default, random
FT-Transformer	merged	model-defined	model-defined
ResNet	merged	model-defined	model-defined

Datasets. For each of the above scenarios we created a dataset by sampling 5 K configurations uniformly at random for each instance, for a total of 50 K samples with their corresponding (ground truth) objective function value. In both scenarios objective function values were obtained by fixing the seed of the pseudo-random generator to 1, i.e. we limit ourselves here to studying a deterministic version of both algorithms. The cost metric for both scenarios is the objective function value obtained after a running time limit of 10 s. Samples for ACOTSP have been obtained with a single trial (`tries=1`), those for LKH with a single run (`RUNS=1`).

In the experiments, these datasets are split into a training set, a test set, and a validation set. We use three sizes of the training set, namely 300, 3 K, and 30 K samples. Training sets are generated to be laminar, i.e. smaller training sets are contained in larger ones. The remaining samples are evenly split among the test and the validation set. The test set is used for tuning the architecture of the neural networks (FT-Transformers and ResNets). We study all models with these three sizes, except for the models based on neural networks which require a higher number of samples and therefore are trained only with a training set of size 30 K. To evaluate accuracy in Sects. 3.2 and 3.3 we select 5 K samples from the test and validation set.

For handling categorical variables we study three ways of encoding them, as mentioned above: binary encoding, one-hot encoding, and encoding by the index. These encodings are applied to all models which cannot handle categorical data directly, namely gradient-boosted trees as implemented by XGBoost, and random forests as implemented in skranger and PyRFR. skranger has support for declaring variables to be categorical. However, these variables have to be numerical. For this reason, we test in skranger all encodings, but declare the encoding variables as categorical. We set handling of categorical values in skranger to consider all partitions (`respect_categorical_features="partition"`), since the number of values in the dataset is small, and testing all partitions will give the best possible splits.

For handling different instances we test two strategies: training a surrogate model for each instance individually, or training a single surrogate model for all instances. These strategies are called "split" and "merged", respectively, in the experiments below. For "merged" models the instance is added as an additional categorical variable to the dataset, and is subject to the above transformations of categorical variables, when applicable.

In summary, the XGBoost model and the random forests (skranger, PyRFR) require imputation of missing values and encoding of conditionals; CatBoost, and the neural networks can handle all parameters as given. Furthermore, the interpolation baseline also does not need any pre-processing.

3.2 Effects of Instance Representation and Pre-processing

In a first experiment we analyze the influence of the representation of the instances, the parameter imputation for conditional variables, and handling of categorical variables. Table 1 shows for the three different sample sizes the average relative root mean squared error (RRMSE) R and the average Kendall's τ for the different strategies. Averages apply only to the models for which different strategies have been applied (see Table 3.1). In the comparisons below we report p-values of sign tests comparing errors and τ values over all models.

Table 1. Relative root mean squared error R and Kendall's τ for different sample sizes N, and different representation, imputation, and encoding strategies.

N	Instance				Imputation						Encoding					
	Merged		Split		Default		Fixed		Random		One hot		Binary		Index	
	R	τ	R	τ	R	τ	R	τ	R	τ	R	τ	R	τ	R	τ
300	0.26	0.48	0.27	0.46	0.14	0.50	0.14	0.47	0.14	0.51	0.26	0.47	0.27	0.45	0.26	0.48
3000	0.15	0.63	0.16	0.61	0.08	0.67	0.09	0.67	0.09	0.63	0.15	0.64	0.16	0.61	0.15	0.62
30000	0.07	0.78	0.08	0.76	0.04	0.78	0.04	0.78	0.05	0.75	0.07	0.79	0.08	0.75	0.07	0.78

We can observe that errors decrease and Kendall's τ consistently increases with increasing sample sizes. For 30 K samples all RRMSE are below 0.1, and the value of Kendall's τ above 0.75. Turning to representation of instances, we find that using a single model (strategy "merged") is better than strategy "split" which models each instance individually (all $p < 0.001$). Note that since we have 10 instances, each individual model is built with only 10 % of the samples. We next look at the imputation strategy. In this case averages are only over the ACOTSP scenario, since scenario LKH does not have conditional variables. Here differences are only significant for larger sample sizes, and indicate that imputing fixed or default values tends to be better than imputing random values for 3 K samples (all $p < 0.004$) and for 30 K samples (all $p < 0.001$), but not significantly different from each other. Finally we turn to the encoding. Here we find that binary and index encoding are not significantly different (all $p > 0.02$), but for larger sample size one-hot encoding performs better than the other strategies (all $p < 0.001$ for 3 K and 30 K samples).

Based on these results, we focus in the remaining experiments on a single model, and fix imputation and encoding strategies, where they apply, to "fixed" and "one-hot", and compare the resulting seven models.

3.3   Accuracy of the Surrogate Models

Table 2 shows RRMSE and Kendall's τ for the seven selected models, both scenarios and all three samples sizes, except for the neural networks, which have been trained only with 30 K samples. Again we can see that RRMSE decreases and Kendall's τ in agreement increases with the sample size. We also can see that the error for scenario LKH is considerably higher than for ACOTSP. A sample size of 300 seems inadequate for a good prediction, and increasing the number of samples to 3 K and then 30 K quality improves considerably, in particular for scenario LKH, although it remains harder to predict, with RRMSE values of about

0.1. The baseline model using interpolation shows worse performance, although it is not far off from the other models. Overall both two gradient-boosted trees and the random forest pyRFR work well, and the Transformer-based neural network has the best performance.

Table 2. Relative root mean squared error R and Kendall's τ for different sample sizes N, and both scenarios for all seven selected models.

Scen.	Model	300		3 K		30 K	
		R	τ	R	τ	R	τ
ACOTSP	Interpolation	0.14	0.57	0.10	0.67	0.07	0.78
ACOTSP	skranger	0.14	0.49	0.09	0.68	0.05	0.79
ACOTSP	PyRFR	0.13	0.50	0.08	0.71	0.04	0.81
ACOTSP	XGBoost	0.13	0.54	0.06	0.72	0.03	0.79
ACOTSP	CatBoost	0.12	0.66	0.07	0.74	0.03	0.82
ACOTSP	ResNet	–	–	–	–	0.02	0.80
ACOTSP	FT-Transformer	–	–	–	–	0.02	0.84
LKH	Interpolation	0.45	0.15	0.36	0.30	0.19	0.71
LKH	skranger	0.45	0.36	0.29	0.52	0.12	0.77
LKH	PyRFR	0.44	0.45	0.23	0.59	0.09	0.83
LKH	XGBoost	0.43	0.48	0.24	0.66	0.09	0.80
LKH	CatBoost	0.45	0.44	0.20	0.65	0.10	0.78
LKH	ResNet	–	–	–	–	0.10	0.74
LKH	FT-Transformer	–	–	–	–	0.09	0.80

3.4 Agreement in Reproduction of Effects

In this section we assess how well the surrogate models can replicate the observable effects caused by changing the default configuration of irace. We have selected the three parameters shown in Table 3 and set them to the two listed levels, with all remaining parameters kept at their defaults. Then we compare the ground truth (i.e. irace executing the algorithms) to the results obtained using the surrogate models trained with 30 K samples. For the comparison we ran irace with the three budget values 780, 1890, and 3000 as explained above, and replicate each run 10 times with different seeds for irace. The best obtained configuration is then evaluated on all 10 instances.

To compare the models, we follow [13] and introduce scores that reflect the concordance of effects in surrogate models with the ground truth. For each of the three budgets and two scenarios, we compare the objective values found by the best configuration for a high and low level of one of the three parameters. The comparison is based on a Mack-Skillings test with a confidence level $\alpha = 0.05$ and Bonferroni correction for multiple tests. In each case we attribute a score

Table 3. Selected parameters of irace with a brief description, and the tested levels.

Parameter	Description	Levels
firstTest	No. of instances evaluated before first statistical test	2, 10
elitistLimit	Max. no. of statistical tests without elimination	1, 5
confidence	Confidence level of statistical test	0.2, 0.95

Table 4. Overall scores for all seven models, three effects, and two scenarios.

Model	Confidence		ElitistLimit		firstTest	
	ACOTSP	LKH	ACOTSP	LKH	ACOTSP	LKH
CatBoost	0.17	0.00	0.00	0.00	0.00	0.33
FT-Transformer	0.17	0.00	0.00	0.00	0.00	0.33
Interpolation	0.17	0.00	0.00	0.00	0.33	0.33
PyRFR	0.17	0.00	0.00	0.00	0.00	0.33
XGBoost	0.33	0.33	0.00	0.00	0.00	0.33
resnet	0.33	0.17	0.00	0.00	0.00	0.33
skranger	0.33	0.17	0.00	0.00	0.00	0.33

that is 0, if the low level is significantly higher, 0.5 if both levels are statistically not different, and 1 if the high level is significantly higher. To compare a model to the ground truth, we report the average absolute distance of the model's scores to the scores of the ground truth. In this model, a score of 0 corresponds to complete agreement with the ground truth, and a score of 1 to complete disagreement.

Table 4 shows the results. We can see that there is overall a very good agreement of models and ground truth, with scores never higher than 0.33, with one exception. Models of best accuracy, as reported in the previous section, also have best scores, with two exceptions: the interpolation baseline has also a comparable score, while XGBoost is slightly worse on parameter "confidence". A closer look at the ground truth shows that for parameter "elitistLimit" the effects are not statistically significant. This explains the good scores for all models, which also find no effect, but limits the scope of the conclusions. In contrast, setting parameter "firstTest" to the high level is in four of six cases statistically different, and the models are able to reproduce this effect mostly.

We finally have a look at evaluation times, to see to what extent requirement 1 is satisfied. Average evaluation times per call to the models ranges from 1 ms to about 160 ms with two neural networks being the slowest to evaluate (without using a GPU). Therefore, in our experiments speedups range from 50 to 10000. Since the evaluation times are independent from algorithm execution times and grow only slowly with the number of samples, clearly speedups will grow with algorithm execution times.

4   Conclusions

In this paper we have compared surrogate models for the tuning of optimization algorithms, including several strategies for handling categorical and conditional variables, if the underlying model cannot represent them directly, and two ways for handling instances, namely by a single model or by per-instance models. We find that a one-hot encoding with a fixed value imputation and a combined model work best. Among the models one based on random forests (pyRFR) and two gradient-boosted trees (XGBoost, CatBoost) work well. A neural network (FT-Transformer) has the overall best performance. This also holds when evaluating the agreement of the surrogate models to the ground truth with the effects of changing tuner parameters, although XGBoost performs worse in this setting. Overall we can confirm that models that have been found to work well in the literature for surrogate models for hyperparameter optimization and execution time, also work well for objective function values, and that neural networks maybe an interesting alternative, which confirms findings of [16] with regard to tabular data. In future work, we plan to extend the scope of this study to more scenarios and a broader selection of tuners and models.

Acknowledgments. This research has been supported by Coordenação de Aperfeiçoamento de Pessoal de Nível Superior – Brasil (CAPES) – Finance Code 001. M. de Souza acknowledges the support of the Santa Catarina State University, Brasil. M. Ritt acknowledges the support of CNPq, Brasil (grant 437859/2018-5) and Google Research Latin America (grant 25111).

References

1. Akiba, T., Sano, S., Yanase, T., Ohta, T., Koyama, M.: Optuna: a nextgeneration hyperparameter optimization framework. In: Proceedings of the 25rd ACM SIGKDD International Conference on Knowledge Discovery and Data Mining (2019)
2. Astudillo, R., Frazier, P.I.: Thinking inside the box: a tutorial on grey-box bayesian optimization. In: Proceedings of the 2021 Winter Simulation Conference, December 2021
3. AutoML. RFR: A extensible C++ library for random forests with Python bindings, December 2022. https://github.com/automl/random_forest_run
4. Bergstra, J., Yamins, D., Cox, D.: Making a science of model search: hyperparameter optimization in hundreds of dimensions for vision architectures. In: Dasgupta, S., McAllester, D. (eds.) Proceedings of the 30th International Conference on Machine Learning, vol. 28. Proceedings of Machine Learning Research 1. Atlanta, Georgia, USA: PMLR, June 2013, pp. 115–123. https://proceedings.mlr.press/v28/bergstra13.html
5. Bliek, L., Guijt, A., Karlsson, R., Verwer, S., de Weerdt, M.: EXPObench: benchmarking surrogate-based optimisation algorithms on expensive black-box functions. In: CoRR abs/2106.04618 (2021). arXiv: 2106.04618
6. Chen, T., Guestrin, C.: XGBoost: a scalable tree boosting system. In: Proceedings of the 22nd ACM SIGKDD International Conference on Knowledge Discovery and Data Mining, KDD 2016, pp. 785–794. ACM, San Francisco (2016). ISBN: 978-1-4503-4232-2. https://doi.org/10.1145/2939672.2939785

7. Claesen, M., Simm, J., Popovic, D., Moreau, Y., Moor, B.D.: Easy hyperparameter search using optunity. In: CoRR abs/1412.1114 (2014). arXiv: 1412.1114
8. Costa, A., Nannicini, G.: RBFOpt: an open-source library for black-box optimization with costly function evaluations. Math. Program. Comput. **10**(4), 597–629 (2018). https://doi.org/10.1007/s12532-018-0144-7
9. Cowen-Rivers, A., Lyu, W., Wang, Z., Tutunov, R., Jianye, H., Wang, J., Ammar, H.: HEBO: heteroscedastic evolutionary bayesian optimisation, December 2020. https://valohaichirpprod.blob.core.windows.net/papers/huawei.pdf
10. Dorigo, M., Stützle, T.: Ant Colony Optimization. MIT Press, Cambridge (2004)
11. Eggensperger, K., Hutter, F., Hoos, H.H., Leyton-Brown, K.: Efficient benchmarking of hyperparameter optimizers via surrogates. In: Bonet, B., Koenig, S. (eds.) Proceedings of the Twenty-Ninth AAAI Conference on Artificial Intelligence, 25–30 January 2015, pp. 1114–1120. AAAI Press, Austin (2015). http://www.aaai.org/ocs/index.php/AAAI/AAAI15/paper/view/9993
12. Eggensperger, K., Hutter, F., Hoos, H.H., Leyton-Brown, K.: Efficient benchmarking of hyperparameter optimizers via surrogates. In: Bonet, B., Koenig, S. (eds.) Proceedings of the Twenty-Ninth AAAI Conference on Artificial Intelligence, 25–30 January, 2015, pp. 1114–1120. AAAI Press, Austin (2015). http://ceur-ws.org/Vol-1201/paper-06.pdf
13. Eggensperger, K., Lindauer, M., Hoos, H.H., Hutter, F., Leyton-Brown, K.: Efficient benchmarking of algorithm configuration procedures via model-based surrogates. In: CoRR abs/1703.10342 (2017). arXiv: 1703.10342
14. Eggensperger, K., Lindauer, M., Hoos, H.H., Hutter, F., Leyton-Brown, K.: Efficient benchmarking of algorithm configurators via model-based surrogates. In: Mach. Learn. **107**(1), 15–41 (2018). https://doi.org/10.1007/s10994-017-5683-z
15. Eggensperger, K., et al.: HPOBench: a collection of reproducible multi-fidelity benchmark problems for HPO. In: Thirty-fifth Conference on Neural Information Processing Systems Datasets and Benchmarks Track (Round 2) (2021)
16. Gorishniy, Y., Rubachev, I., Khrulkov, V., Babenko, A.: Revisiting deep learning models for tabular data. In: CoRR abs/2106.11959 (2021). arXiv: 2106.11959
17. Gutmann, H.-M.: A radial basis function method for global optimization. In: J. Global Optim. **19**(3), 201–227 (2001). issn: 0925–5001. https://doi.org/10.1023/A:1011255519438
18. Hansen, N., Ostermeier, A.: Completely derandomized self-adaptation in evolution strategies. Evolutionary Comput. **9**(2), 159–195 (2001). https://doi.org/10.1162/106365601750190398
19. He, K., Zhang, X., Ren, S., Sun, J.: Deep residual learning for image recognition. In: 2016 IEEE Conference on Computer Vision and Pattern Recognition (CVPR). IEEE, June 2016. https://doi.org/10.1109/cvpr.2016.90
20. Helsgaun, K.: An effective implementation of the lin-kernighan traveling salesman heuristic. Eur. J. Oper. Res. **126**, 106–130 (2000)
21. Tinós, R., Helsgaun, K., Whitley, D.: Efficient recombination in the Lin-Kernighan-Helsgaun traveling salesman heuristic. In: Auger, A., Fonseca, C.M., Lourenço, N., Machado, P., Paquete, L., Whitley, D. (eds.) PPSN 2018. LNCS, vol. 11101, pp. 95–107. Springer, Cham (2018). https://doi.org/10.1007/978-3-319-99253-2_8
22. Helsgaun, K.: General k-opt Submoves for the Lin-Kernighan TSP Heuristic. Math. Programm. Comput. **1**(2-3), 119–163 (2009)
23. Helsgaun, K.: Source Code of the Lin-Kernighan-Helsgaun Traveling Salesman Heuristic (2018). http://webhotel4.ruc.dk/~keld/research/LKH

24. Hutter, F., Hoos, H.H., Leyton-Brown, K.: Sequential model-based optimization for general algorithm configuration. In: Coello, C.A.C. (ed.) LION 2011. LNCS, vol. 6683, pp. 507–523. Springer, Heidelberg (2011). https://doi.org/10.1007/978-3-642-25566-3_40

25. Hutter, F., Hoos, H.H., Leyton-Brown, K., Stützle, T.: ParamILS: an automatic algorithm configuration framework. J. Artif. Intell. Res. **36**, 267–306 (2009). https://doi.org/10.1613/jair.2861

26. Hutter, F., López-Ibáñez, M., Fawcett, C., Lindauer, M., Hoos, H.H., Leyton-Brown, K., Stützle, T.: AClib: a benchmark library for algorithm configuration. In: Pardalos, P.M., Resende, M.G.C., Vogiatzis, C., Walteros, J.L. (eds.) LION 2014. LNCS, vol. 8426, pp. 36–40. Springer, Cham (2014). https://doi.org/10.1007/978-3-319-09584-4_4

27. Hutter, F., Xu, L., Hoos, H.H., Leyton-Brown, K.: Algorithm runtime prediction: Methods & evaluation. Artif. Intell. **206**, 79–111 (2014). https://doi.org/10.1016%2Fj.artint.2013.10.003. https://doi.org/10.1016/j.artint.2013.10.003

28. Johnson, D.S., McGeoch, L.A., Rego, C., Glover, F.: 8th DIMACS Implementation Challenge: The Traveling Salesman Problem (2001). http://dimacs.rutgers.edu/archive/Challenges/TSP

29. Kendall, M.G.: A new measure of rank correlation. Biometrika **30**(1-2), 81–93 (1938). https://doi.org/10.1093/biomet/30.1-2.81

30. Klein, A., Dai, Z., Hutter, F., Lawrence, N., González, J.: Meta-surrogate benchmarking for hyperparameter optimization. In: Proceedings of the 33rd International Conference on Neural Information Processing Systems. Curran Associates Inc., Red Hook (2019)

31. Lindauer, M., et al.: SMAC3: a versatile bayesian optimization package for hyperparameter optimization. In: CoRR (2021). arXiv: 2109.09831 [cs.LG]

32. López-Ibáñez, M., Dubois-Lacoste, J., Pérez Cáceres, L., Stützle, T., Birattari, M.: The irace package: iterated racing for automatic algorithm configuration. Oper. Res. Perspect. **3**, 43–58 (2016). https://doi.org/10.1016/j.orp.2016.09.002

33. Malherbe, C., Vayatis, N.: Global optimization of lipschitz functions. In: Proceedings of the 34th International Conference on Machine Learning - Volume 70, ICML 2017. Sydney, NSW, Australia: JMLR.org, pp. 2314–2323 (2017)

34. Pedregosa, F., et al.: Scikit-learn: machine learning in python. J. Mach. Learn. Res. **12**, 2825–2830 (2011)

35. Prokhorenkova, L., Gusev, G., Vorobev, A., Dorogush, A.V., Gulin, A.: CatBoost: unbiased boosting with categorical features. In: Proceedings of the 32nd International Conference on Neural Information Processing Systems. NIPS 2018, pp. 6639–6649. Curran Associates Inc., Montréal (2018)

36. Rice, J.R.: The algorithm selection problem. Adv. Comput. **15**, 65–118 (1976). https://doi.org/10.1016/S0065-2458(08)60520-3

37. Shepard, D.: A two-dimensional interpolation function for irregularlyspaced data. In: Proceedings of the 1968 23rd ACM National Conference. ACM Press (1968). https://doi.org/10.1145/800186.810616

38. Škvorc, U., Eftimov, T., Korošec, P.: GECCO black-box optimization competitions. In: Proceedings of the Genetic and Evolutionary Computation Conference Companion. ACM, July 2019. https://doi.org/10.1145/3319619.3321996

39. Springenberg, J.T., Klein, A., Falkner, S., Hutter, F.: Bayesian optimization with robust bayesian neural networks. In: Lee, D., Sugiyama, M., Luxburg, U., Guyon, I., Garnett, R. (eds.) Advances in Neural Information Processing Systems, vol. 12. Curran Associates Inc. (2016). https://proceedings.neurips.cc/paper/2016/file/a96d3afec184766bfeca7a9f989fc7e7-Paper.pdf

40. Stützle, T.: ACOTSP: a software package of various ant colony optimization algorithms applied to the symmetric traveling salesman problem (2002). http://www.aco-metaheuristic.org/aco-code
41. Turner, R., et al.: Black-Box Optimization for Machine Learning (2020). https://github.com/rdturnermtl/bbo_challenge_starter_kit
42. Vaswani, A., et al.: Attention is all you need. In: Guyon, I., et al. (eds.) Advances in Neural Information Processing Systems, vol. 30. Curran Associates Inc. (2017). https://proceedings.neurips.cc/paper/2017/file/3f5ee243547dee91fbd053c1c4a845aa-Paper.pdf
43. Wilson, D.R., Martinez, T.R.: Improved heterogeneous distance functions. J. Artif. Intell. Res. 6, January 1997. https://doi.org/10.1613/jair.346
44. Wright, M.N., Ziegler, A.: Ranger: a fast implementation of random forests for high dimensional data in C++ and R. J. Stat. Softw. 77(1)(2017). https://doi.org/10.18637/jss.v077.i01

Search and Score-Based Waterfall Auction Optimization

Dan Halbersberg[✉], Matan Halevi, and Moshe Salhov

Playtika Ltd., Hahoshlim St. 8, Herzliya, Israel
{danh,matanh,moshesa}@playtika.com

Abstract. Online advertising is a major source of income for many online companies. One common approach is to sell online advertisements via waterfall auctions, through which a publisher makes sequential price offers to ad networks. The publisher controls the order and prices of the waterfall in an attempt to maximize his revenue. In this work, we propose a methodology to learn a waterfall strategy from historical data by wisely searching in the space of possible waterfalls and selecting the one leading to the highest revenues. The contribution of this work is twofold; First, we propose a novel method to estimate the valuation distribution of each user, with respect to each ad network. Second, we utilize the valuation matrix to score our candidate waterfalls as part of a procedure that iteratively searches in local neighborhoods. Our framework guarantees that the waterfall revenue improves between iterations ultimately converging into a local optimum. Real-world demonstrations are provided to show that the proposed method improves the total revenue of real-world waterfalls, as compared to manual expert optimization. Finally, the code and the data are available here.

Keywords: Auction optimization · Real-time bidding · Search and score · Waterfall

1 Introduction

Online advertisements serve as a major source of income, for many online companies [15]. Whenever a user surfs on a website or utilizes a mobile app, an advertisement real-estate, known as *ad-slots* are allocated. Each ad-slot is populated by a relevant advertisement. The slot owner is called a *publisher* and the advertisement owner is called a *supplier*. The publisher sells ad-slots to suppliers via *ad networks* such as Facebook, Google, etc. Interactions between publishers and suppliers take place through real-time bidding auctions.

The publisher's goal is to maximize the selling price of each ad-slot. There are several approaches to choosing a particular ad network for a given slot. One common approach is known as the waterfall strategy [2,6,7,11,14,18]. A waterfall is a list of instances, where each *instance* belongs to a specific ad network and is associated to a specific price [14]. For each ad-slot, the ad networks

© Springer Nature Switzerland AG 2022
D. E. Simos et al. (Eds.): LION 2022, LNCS 13621, pp. 361–378, 2022.
https://doi.org/10.1007/978-3-031-24866-5_27

are sequentially approached, according to a pre-configured list of instances. Each ad network can accept or reject to buy the slot for the given pre-defined instance price. If the ad network rejects the price, the slot is offered to the next instance in line until an ad network accepts the terms. Since the waterfall is predefined by the publisher, optimally determine the strategy, i.e., the order and pricing configuration of all instances in the waterfall, remains a significant challenge [1,3].

To empower the user experience, this bidding process must be completed in real-time in order to empower user experience [17,19]. Therefore, it is important to find the best strategy, such that the timing of the last approached instance will not breach the time constraint. Additionally, the supplier limits the number of instances, to minimize the auction overhead [3].

Many publishers decide on the ordering of ad networks based on human experience, trial and error, or other similar inefficient methods [3]. Besides for the fact that these methods cannot guarantee an optimal strategy, it is challenging to scale them for large publishers who need to manage many waterfalls for different platforms, operation systems, countries, etc. In recent years, growing attention has focused on automating and optimizing this process, in order to increase the revenue. An online learning algorithm was proposed by [14] to solve this problem, and other reinforcement methods were proposed [1–3,18] as well. These methods focus on predicting the ad network's pricing strategy. In many real-world cases, the waterfall strategy is repeatedly operating over the same users and over several online sales events. This data stream can be accumulated and utilized to further improve any waterfall strategy.

In this study, we propose a novel approach that utilizes user-accumulated data that is measured during multiple visits to the publisher's web site or app. More explicitly, we suggest to use this information to estimate the perceived valuation of each user, by each ad network. These personalized valuations inform our hypothesis regarding auction events via simulations. The ability to simulate the actual effect of the bidding process allows us to design an efficient local search strategy for designing a locally optimal waterfall strategy. Applying the optimal waterfall strategy allows the publisher to not only gain more profits from the auction, but also to automate the process and adapt to temporal changes in user valuation, which reduces the overhead from marketing experts.

The rest of the paper is organized as follows. Section 2 provides the relevant background and surveys related works. In Sect. 3, we present the proposed framework and detail its characterizations. Furthermore, Sect. 4 demonstrates the application of the suggested method using both simulated and real-world data. Finally, in Sect. 5 we provide our conclusions.

2   Background and Related Works

A waterfall, W, is an ordered list of r instances. The ith instance W_i, where $1 \leq i \leq r$ is associated to a specific ad network for a given price, $p_i \in \{0, ..., M\}$, where M is the maximal price allowed by the ad network [14]. Thus, each ad

Instance Id	Ad network	Price per 1,000 impressions
1	⊙ AdMob	100
2	f facebook	90
3	◁ unity	70
4	⊙ AdMob	50
5	Vungle	40
6	hyprMX	35
7	f facebook	35
8	◁ unity	20
9	f facebook	10
10	⊙ AdMob	7
11	hyprMX	5
12	◁ unity	2

Fig. 1. An example of a waterfall with twelve instances of five ad networks.

network in the waterfall may have several instances. For example, in Fig. 1, the ad network $AdMob$ has three instances 1, 4, and 10 for $100, $50, and $7, respectively. The waterfall strategy is not tailored to a specific user valuation, but rather is designed to be optimal for the entire user population (i.e., all users will run through the same waterfall).

Most of the works in this field are based on concepts pertaining to reinforcement learning [3]. One such algorithm [14] is a multi-armed bandits algorithm. The idea of the algorithm is to learn the valuation distribution of each ad network. For that, the publisher adaptively chooses waterfall strategy, receives feedback (accept/reject), and evaluates the performance using a regret function. However, this online algorithm has several restrictive assumptions, such as: each ad network has a single instance and that the valuations are unique per ad network, yet equal for all users. Similar assumptions were also made by other researchers [3]. Recently, [2] proposed to utilize the ad-request information to learn a model that predicts the probability of an ad network to buy the ad-slot for the given price. The outputs are then fed into a Monte-Carlo algorithm, which optimizes the state-action values, accordingly. However, proposing a policy for each ad-request (even in the case that they are grouped by their commonality) is practically impossible.

The waterfall optimization can be defined as a local search problem, where the task is to maximize the total waterfall strategy revenue over all users. Search and score (S&S) is a heuristic method that belongs to a family of local search algorithms [12]. Methods following this heuristic aim to solve computationally hard optimization problems [4,12]. One such method is known as hill climbing [16]; an iterative algorithm is initialized by an arbitrary solution and then attempts to find a better solution candidate by making a local change to the best solution so far. The best solution candidate is adopted by the algorithm, which is then followed by a similar evaluation of any incremental change to that selected solution, in the next iteration. The algorithm terminates once the improvement is negligible.

A major limitation of the hill climbing procedure is that it can converge into a local minimum. This limitation is heavily dependent on the search starting point. To overcome this limitation, another heuristic search known as the Monte Carlo Tree Search (MCTS) was proposed [10]. As opposed to the hill climbing procedure, the MCTS searches for the most promising next solution candidate in a decision-making problem, combining the precision of tree search with the generality of random simulation. Algorithms following the MCTS approach adopt, in each iteration, the change with the greatest potential with respect to future iterations. One relevant implementation of a hill climbing procedure, which inspired our proposed S&S–based waterfall optimization algorithm, is the well-known K2 algorithm [9]. The K2 is a heuristic search algorithm for learning the structure of a Bayesian network that best fits the data. The algorithm starts with a random graph, and considers all local neighbor graphs, at each iteration. A neighbor graph is defined as an equal graph with a single change that can be: edge addition, deletion or reversal. By likening the waterfall W to a serial graph, we can equate a node in the graph to an instance in the waterfall and edges to the waterfall order. Using the estimated valuation matrix B, a procedure, similar to that of the K2, can be designed over the space of all valid waterfall graphs. In the following section, we will describe this local search procedure in detail.

3   Proposed Method

To utilize the fact that users are frequently visiting the publisher's website or app, we propose a two-stage framework; First, estimate valuation matrix, $B \in \mathcal{R}^{U \times K}$ from historical data, where U is the number of distinct users and K is the number of ad networks. B holds the perceived value of each user by each ad network (Sect. 3.1). Second, search for the optimal waterfall by simulating auction events utilizing B, which can approximate the revenue effect of order and pricing changes in the waterfall. More explicitly, in the second stage, our proposed framework uses the valuation matrix in an iterative manner (Sect. 3.2). In this way, we are able to define the problem as a local search problem over the space of valid waterfalls.

The main contributions of the proposed method are: 1) modeling user pricing per ad network, by learning each user's Beta distribution parameters, given their respective historical pricing data; 2) minimizing the information requirement by explicitly utilizing sales events, while implicitly utilizing rejected bid information from the sales data. The two contributions rely on the fact that the advertisement process is ongoing and most user ad-slots were sold several times in the past.

3.1   Estimate the Valuation Matrix

The valuation matrix (B) is a key component in our S&S–based waterfall optimization algorithm as the entire search procedure depends on B to simulate

---

Algorithm 1. Estimate the valuation matrix

1: **Input:** Dataset D
2: **Output:** Valuation matrix B
3:
4: **for** each User u and ad network k **do**
5:     $V_{u,k}$ = collect all past sell events of u to k from D
6:     **if** $V_{u,k} \neq \emptyset$ **then**
7:         $B_{u,k} = beta.fit(V_{u,k})$
8:     **else if** u was sold to at least a ad networks $z \neq k$ **then**   ▷ Imputation needed
9:         $V_{u,k}$ = collect all past sell events of u to $\forall z \neq k$
10:         $B_{u,k} = beta.fit(V_{u,k})$
11:     **else**
12:         $B_{u,k} = B_k$ ▷ Use global parameters of k

---

auction events. If B is wrong, the results would be misleading. One simple approach to estimating B is to take the average sell price of each user, per ad network. However, the main limitations of this metric are: 1) it will generate a deterministic value that is less suitable for simulation purposes; and 2) in many cases, the average is not a good representative of the user valuation. To overcome these limitations, one can suggest to replace the simple average with a normal distribution estimation, and then during the search phase, sample from this distribution. Nevertheless, normal distribution does not necessarily fit the user valuation distribution, and therefore, we propose to use the Beta distribution, which allows for a more flexible representation of the user's valuation. This distribution was also found appropriate by other researchers [8,14]. Algorithm 1 loops over each user ($u, \quad 1 \leq u \leq U$) and collects the user data (vector $V_{u,k}$) per ad network ($k, \quad 1 \leq k \leq K$) for the beta estimation ($B_{u,k}$). This is our training data. However, if no data is available, i.e., the user was never sold to that particular ad network, the algorithm tries two estimation methods; 1) if the user was sold in the past to at least a (in our experiments we use $a = 3$) other ad networks, it uses their data for the estimation; or 2) if (sufficient) other ad networks data are not available, it uses the global beta distribution parameters of the specific ad network in question. The output of the Algorithm is the valuation matrix $B_{u,k}$ of dimension $U \times K$, where U is the number of distinct users and K is the number of ad networks. In our case, the main advantage of the Beta distribution over, for example, the simple mean value, is that it generalizes well to a stochastic process, allowing us to represent different types of users. For example, some users could have a Poisson-like distribution while others can be normal, exponential, etc. In general, the beta distribution is appropriate when the true probability distribution is unknown [13].

Another method that can be used to estimate the valuation matrix is via a classifier (e.g., CATboost, Neural Network, etc.). Following [3], we propose to utilize historical data to predict the valuation of each user, and more explicitly, to train classifiers per ad network. To achieve this goal, one can use information from past auction events to correlate between the dependent variable

---

Algorithm 2. Run users in a waterfall

1: **Input:** Valuation matrix B, Waterfall W
2: **Output:** q'
3:
4: **Initialization:**
5: **for** each instance W_i in the waterfall W **do**
6:     $q_i' = 0$
7: **Start:**
8: **for** each User u **do**
9:     $i = 0$ ▷ Go over the waterfall from top to bottom
10:     **while** $i \leq r$ **do**
11:         $k = $ ad network of W_i
12:         $P_{u,k} = $ Sample a value from $B_{u,k}$
13:         **if** $P_{u,k} > p_i$ **then**
14:             $q_i' + = 1$
15:             Break
16:         $i + = 1$

Algorithm 3. Evaluate the valuation matrix

1: **Input:** Valuation matrix B, Waterfall W
2: **Output:** Similarity score
3:
4: $q' = $ Run users in the waterfall based on their valuation matrix ▷ Call Algorithm 2
5: $Score = 0$
6: **for** each instance W_i in the waterfall W **do**
7:     $Score + = \frac{|q_i' - q_i|}{q_i} \times W e_i$

(auction price) and other available independent variables such as: time, geography, demographic, device information, in-app activities, etc. Once such classifiers are trained, we may use them to fill-in valuation matrix, $B_{u,k}$.

Evaluate the Valuation Matrix Estimation in Terms of Accuracy: It is crucial to validate the capability of the estimated Beta distributions in B to accurately generate pricing predictions, before moving forward to search for the best fitted waterfall, using that matrix. The B-based pricing predictions ability is a key component in our framework. Large prediction errors will result in a misleading waterfall strategy that will produce reduced revenues, when run over the actual user population.

To evaluate B-based prediction accuracy, we propose to generate sale predictions for validation data that is accumulated similarly to the training data. Each sale event is an advertisement (sometimes called an *impression*) and has corresponding sale pricing. Acceptable B will predict the number of impressions with good accuracy, as compared to the given validation impressions data. The performance evaluation process is detailed in Algorithms 2 and 3. The goal of Algorithm 2 is to predict how each ad network instance in a predefined waterfall strategy would behave, given B. Let $q \in \mathcal{Z}^{+r}$ be the measured vector of the actual outcomes of a waterfall strategy in the form of the number of impressions per ad network instance and q_i an impression assignment of a specific instance. The output of Algorithm 2 is a vector, q', which holds the number of predicted impressions of each instance in the waterfall. Algorithm 3 takes W and its corresponding impressions, q, as input. Additionally, Algorithm 3 compares q and q' computed in Algorithm 2, based on Eq. 1.

$$Score = \sum_{i=1}^{r} \frac{|q'_i - q_i|}{q_i} \times We_i, \tag{1}$$

where q'_i and q_i are the simulated and real number of impressions in the ith instance, respectively, r is the length of the waterfall, and We_i is the weighted revenue of the ith instance: $We_i = \frac{Revenue(W_i)}{\sum Revenue(W_i)}$. If this score is low enough, we say that B represents the true value of the users, as perceived by the different ad networks.

3.2 The Search Procedure

First, we define the score function that is used to compare waterfalls, as part of the S&S procedure:

$$Revenue(W) = \sum_{i=1}^{r} q_i \times p_i. \tag{2}$$

The revenue of each candidate waterfall is calculated using Algorithm 2. That is, once the algorithm runs all the users through the candidate waterfall, the number of impressions of each instance, q'_i, are updated. Figure 2 shows an example output of Algorithm 2, where the last column is q'. Using Eq. 2, we can sum up the multiplication of the price and the number of impressions to get the total revenue of the candidate waterfall.

Next, our search procedure considers all local neighbors (i.e., candidate waterfalls) in each iteration, where a local neighbor is defined as the current waterfall, except for a single change that can be: 1) instance addition, 2) instance removal,

Instance Id	Ad network	Price	Impressions
1	AdMob	90	2,013
2	facebook	82	4,226
3	unity	66	4,101
4	AdMob	45	3,505
5	Vungle	25	2,009
6	facebook	5	5,107

Fig. 2. An example of a waterfall with six instances and their associate daily number of impressions for a given date. The total daily revenue of the waterfall is $1,032 since the prices in the waterfall are for batches of 1,000 users.

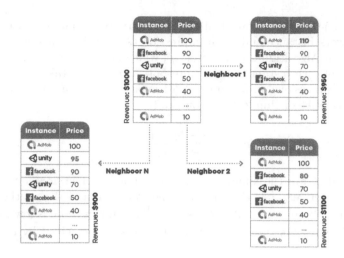

Fig. 3. An illustration of a single iteration in the S&S-based waterfall optimization algorithm. In each neighbor waterfall the changed price is marked in red. (Color figure online)

or 3) a changed instance price (increase or decrease). To narrow-down the search space, we restrict the prices to be discrete (however, one can choose other quantiles such as 50 cents, 10 cents, etc.). By scoring each neighbor waterfall, the algorithm can evaluate how the waterfall would be affected by each specific incremental change. In each iteration, the algorithm adopts the change leading to the highest improvement in revenue. Figure 3 illustrates an example of a waterfall with a revenue of $1,000 and three of its neighbor waterfalls. The second neighbor has the highest revenue, and therefore, is selected as the incremental change to the next iteration.

Following, we present our proposed S&S–based waterfall auction optimization algorithm (Algorithm 4). The algorithm takes (lines 1–5) an initial waterfall (W^0), a valuation matrix (B), the maximal number of iterations (Max_iter) and a threshold (ϵ) for the stopping condition as inputs. The initial waterfall can be an empty/random waterfall, or an existing/human expert waterfall. After initialization (lines 9–13) is complete, the algorithm iterates, taking into consideration all local changes (line 17) and adopting the one with highest revenue (line 22), for each iteration. Finally, it terminates once the maximal number of iterations is reached or the difference in revenue between two successive iterations is determined to be lower than ϵ (line 26).

Our waterfall auction optimization algorithm is based on the heuristic hill climbing method [12] that can neither guarantee reaching the global optimum nor converging to a local or the global optimum in a reasonable time frame. Since our proposed algorithms adopt a neighbor waterfall only if its revenue is strictly above the aforesaid threshold, then no cycles can exist within the search

---

Algorithm 4. Search and score procedure

1: **Input:** Initial waterfall W^0 ▷ This could be the current existing waterfall
2:         Valuation matrix B
3:         Max_iter
4:         ϵ
5: **Output:** Waterfall W
6:
7: **Initialization:**
8: q' = Run users in the waterfall W^0 based on their valuation matrix B ▷ Call
   Algorithm 2
9: Optimal_revenue = Revenue(W^0)
10: Convergence = True
11: $W = W^0$
12: Iter = 0
13: **Start:**
14: **while** Convergence **do**
15:     Iter += 1
16:     Prev_revenue = Optimal_revenue
17:     neighbors = list of neighbors of W
18:     **for** each neighbor g in neighbors **do**
19:         Run the users (B) in g ▷ Call Algorithm 2
20:         Curr_revenue = Revenue(g)
21:         **if** Curr_revenue > Prev_revenue **then**
22:             Prev_revenue = Curr_revenue
23:             $W = g$
24:     **if** Iter==Max_iter **or** Prev_revenue - Optimal_revenue < ϵ **then**
25:         Convergence = False
26:         Optimal_revenue = Prev_revenue

---

procedure, and thus, the algorithms must converge. The convergence rate, however, is problem-dependent [9]. Luckily, our target function is the total revenue of the waterfall strategy and the majority of the revenue comes from the higher section of the waterfall, where the prices are relatively high. Since the algorithm selects the change leading to the highest improvement in revenue for each iteration, it will first optimize the higher part of the waterfall, enabling the best solution to be reached relatively quickly. Therefore, one can restrict the number of iterations, as we show in our empirical evaluation in Sect. 4.

The Monte Carlo Tree Search (MCTS): Another search procedure we chose to apply as part of our waterfall auction optimization algorithm is a MCTS-like procedure. The goal of this algorithm is to expand the search space in order to avoid a local maximum, due to the non-convex nature of our score function. Algorithm 5 (see Appendix A) trades between accuracy and complexity, as it evaluates more candidate waterfalls at the expense of a higher run-time. At each iteration, the algorithm will adopt the neighbor waterfall with the greatest revenue potential; not necessarily the one with the current highest revenue. All other components of the MCTS–based waterfall optimization algorithm are the same as the regular S&S (i.e., Algorithms 1–3).

Complexity: The computational complexity of our proposed methods is composed of two elements, corresponding to the two folds of our framework. The first, which dictates the complexity, is the estimations of B per user and ad network. In total, the algorithm has to estimate $U \times K$ Beta distributions parameters, each from x samples per user (past sell events), which depends on the data time-period (x is monotonic with the data duration). The second, less dominant element of our framework is the hill climbing search. In this stage, the algorithm is bounded by $O(U \times r)$ for each candidate waterfall, since in the worst case, each user runs through the entire waterfall. The number of candidate waterfalls at each of the it iterations is bounded by $O(3r + MK)$, and thus, the second element of our framework is bounded by $O[it \times U \times r \times (3r + MK)]$.

The number of possible waterfalls is $(M \times K)^r$, where M is the number of unique prices (M is also the maximal price in our discrete case). Although the number of possible iterations is bounded by the number of possible waterfalls, in practice, it is restricted to a value between $10 - 50$. This is because the optimization is mainly affected by changes in the higher section of the waterfall (as described in Sect. 3.2), and thus, we achieve most of the improvement in revenue at an early stage of the search process.

Moreover, both parts of the computations (i.e., estimating B and the hill climbing search) could easily be parallelized: 1) estimating B could be parallelized by distributing the data by users, and; 2) as part of the search procedure, one could evaluate at each iteration the candidate waterfalls in parallel and by that reduce the algorithm run-time to $O(it \times U \times r)$. Also, note that it is small and that $O(U \times r)$ refers to the worst case scenario. In practice, assuming, for example, that the impressions are distributed uniformly across the waterfall, this bound is reduced by half, as Eq. 3 shows regarding the number of requests to ad networks:

$$\#\_of\_requests = \sum_{i=1}^{r} i \times \frac{U}{r} = \frac{r(r+1)}{2} \times \frac{U}{r} = \frac{U(r+1)}{2}. \tag{3}$$

Finally, in terms of the data we need to store, our proposed algorithms require only the successful events. This is a huge advantage over other reinforcement learning algorithms (e.g., [2,14]), which require both accepted and rejected auction events. This allows us to reduce the volume of stored data by over 90%.

4   Empirical Evaluation

In this section, we report on our experiments with synthetic and real-world waterfalls. With respect to the latter, we experimented with four different waterfalls linked to different countries, to increase the variability of our results. To maintain confidentially, we will refer to them as $Waterfall_A - Waterfall_D$. We compared the two variations of the search procedure; S&S– and MCTS–based waterfall optimization algorithms, to a human expert optimization based on total revenue and computational complexity. The waterfalls and their associated data can all be found online in our supplementary materials. The human expert optimization is actually the most common approach in the industry, and therefore, it can be used as a solid baseline.

4.1   Synthetic Data

In this section, we report on our experiment with synthetic data. The motivation for this experiment was twofold: 1) to show that our proposed S&S–based algorithm may converge to the optimal solution regardless of the initialization, using a toy example, and; 2) to show that the simple S&S–based algorithm does not fall behind the MCTS–based algorithm, in terms of accuracy, while it convergences significantly faster. To demonstrate these advantages, we selected four ad networks and synthetically sampled 4×10^5 users, where the beta distribution parameters per ad network were: $Beta(\alpha = 1, \beta = 6), Beta(2, 6), Beta(10, 5)$, and $Beta(6, 1)$. We selected these parameters so that the distributions would substantially overlap each other (as can be seen in Fig. 4). In addition, we initialized the algorithms with five different waterfalls, i.e., different in the order and the prices of the instances, to show that the convergence was not random.

Figure 5a shows the learning curve of the two search algorithms for an empty waterfall initialization. The optimal solution ($835.4) was calculated using an exhaustive search over all possible solutions (total of 30 discrete prices and 4 ad networks, generating $4! \times 30^4$ candidate waterfalls). It can be seen that both S&S– and MCTS–based waterfall optimization algorithms converge to a solution that is close to the optimal one ($830.9 and $835.4 for the S&S and MCTS, respectively). Although the S&S–based algorithm did not achieve as optimal a solution as the MCTS–based algorithm, their revenue is close enough ($\frac{830.9}{835.4} = 0.995$). In addition, the fact that they both have a similar number of iterations (≈ 40) is deceiving, as the number of examined neighbors is significantly lower for the S&S–based algorithm (as can be seen in Fig. 5b). The S&S–based algorithm simulated only 200 waterfalls, as opposed to the over 4,000 waterfalls simulated by the MCTS–based algorithm. Therefore, concerning the trade off accuracy-runtime, this experiment demonstrates the S&S–based algorithm might be superior to the MCTS–based algorithm, with respect to both accuracy and complexity.

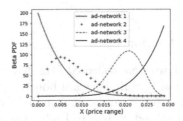

Fig. 4. The beta distributions used to sample the synthetic valuations.

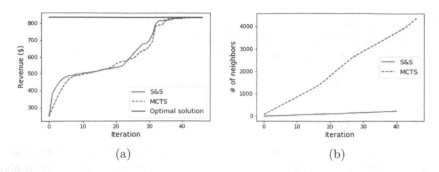

(a)   (b)

Fig. 5. a) The learning curve measured in revenue ($) for the S&S– (solid red line) and MCTS–based (dashed blue line) algorithms comparing to the optimal solution that was found using exhaustive search and b) The cumulative number of neighbors (candidate waterfalls) examined by the S&S– (solid red line) and MCTS–based (dashed blue line) algorithms, with respect to the learning curve in Fig. 5 (Color figure online)a.

Table 1. The revenue($) achieved by the two search algorithms for five initializations, where the optimal solution is $835.4

	Init_1	Init_2	Init_3	Init_4	Init_5
S&S	833.9	830.9	822.5	829.5	827.9
MCTS	835.4	835.4	835.4	831.4	832.4
Difference (%)	0.2	0.5	1.6	0.2	0.5

Table 1 shows the results of the S&S– and MCTS–based waterfall optimization algorithms for five initializations: 1) the true order, but with different prices, 2) empty waterfalls, 3) all prices being equal to the average valuation, and 4–5) in the opposite order and with different prices. It can be seen that, except for the last two initializations, the MCTS–based algorithm learned the optimal waterfall, while the S&S–based algorithm did not recover the optimal waterfall for any of the initializations. Nevertheless, the S&S–based algorithm is only 0.6% lower than the MCTS–based algorithm on average. That said, it is ≈ 20 quicker, as measured by the number of neighbors evaluated during the learning. Therefore, we can conclude that for small non-complex waterfall optimization problems, the S&S–based algorithm can almost achieve the global optimum and it is monotonic with the MCTS–based algorithm, but faster in several orders of magnitudes.

4.2   The Real-World Auction Data

The data used in our analyses were collected from four different real waterfalls. The "raw-data" used in our experiments are hourly aggregated, per user and ad network. The given data were randomly sampled from a cohort of 60 days which include 355, 905 users and 27, 615, 614 advertisements (an average of 78 advertisements per user). The data processing is described in detail, in Table 4 in Appendix B.

Next, we describe the experiment methodology. The given data includes 60 days of waterfall sales and strategy measurements. For each validation day number d, we use its past 30 days as training data. The result is training set X_t and a corresponding validation data X_v. Given the data train-validation split we used Algorithm 1 on X_t to learn the beta distributions in B. Given B, we predicted the outcome of applying the waterfall strategy W on samples from B using Algorithm 2. Furthermore, we generated the corresponding impressions vector q'. This vector was compared to the given impressions vector q of the waterfall from day d using Eq. 1. Notice, that q' was estimated based on data ended in day $d - 1$, while q refers to day d. This is to prevent over-fit and to demonstrate the prediction capability of B.

Since a sale event can take place only if the user valuation is larger than the instance price, we need to increase the values sampled from B by a small constant ϵ. To enable this to occur, we define a coefficient vector ζ that includes one coefficient per ad network. We learn the ζ that minimize Eq. 1. However, optimizing all coefficients at once is practically unfeasible, since a change in one coefficient requires to re-predict the waterfall outcome and re-evaluate B, based on Eq. 1. Therefore, we suggest to learn the coefficients in a coordinate descent methodology [5]. That is, randomly selecting an ad network and optimizing its coefficient using a grid search. Then, move to the next ad network and continue the optimization in a round-robin manner, until the improvement in Eq. 1 is negligible. This procedure allows to loop over the ad networks several times. Finally, after B is estimated, run Algorithms 4 and 5 to find the converged waterfall. We use the current real waterfalls as the initial waterfall inputs for the learning algorithms. This enable us to exploit the human expert knowledge.

Table 2 shows the valuation matrix fitness to each waterfall, as measured by Eq. 1. The values in the table are all positive, but are not bounded i.e., $[0 - \infty]$. There is a trade-off between an accurate valuation matrix and a generalized one. One could define the valuation matrix as the true absolute sales prices of each user that will generate q, but this will result in over-fit that will mislead the waterfall strategy over new data. Thus, the proposed valuation matrix, is a probabilistic estimation of longer time period, which will generalize well, and allows the algorithm to explore and exploit the data. Table 2 demonstrates that, for $Waterfall_A$, the valuation matrix is the most accurate with a mean error of 0.29 and a corresponding std of 0.06. This is the result of a larger number of ad sales per user in $Waterfall_A$, as compared to the other waterfalls.

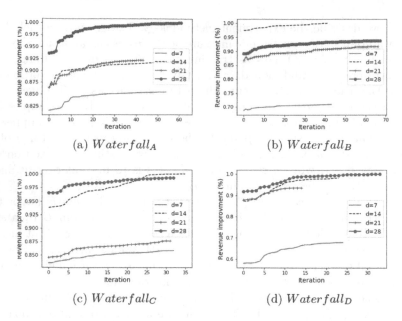

(a) $Waterfall_A$ (b) $Waterfall_B$

(c) $Waterfall_C$ (d) $Waterfall_D$

Fig. 6. The learning curve, measured as the improvement in revenue, for the four real-world waterfalls ($A - D$). Each color represents a point in time (d) to which the algorithm was applied. (Color figure online)

Table 2. The valuation matrix fitness for four waterfalls and four time-points (d) measured by mean absolute error (low is better).

	$d = 7$	$d = 14$	$d = 21$	$d = 28$	Mean (std)
$Waterfall_A$	0.30	0.37	0.27	0.23	0.29 (0.06)
$Waterfall_B$	0.62	0.78	0.70	0.66	0.69 (0.07)
$Waterfall_C$	0.46	0.54	0.48	0.54	0.51 (0.04)
$Waterfall_D$	0.74	0.61	0.80	0.74	0.72 (0.08)

We will now present the results of our proposed search-based waterfall optimization algorithms. Figure 6 shows the learning curves of the S&S–based waterfall optimization algorithm for the four waterfalls in four time points during the experiment. In general, the algorithm converges after 10–20 iterations. Also, it can be seen that $Waterfall_A$ (Fig. 6a) has the fastest learning curve, as compared to the other waterfalls. This is due to the origin of the users in that particular waterfall. Furthermore, it can bee seen that the results for $d = 7$ (red solid line) are always the lowest. This is due to the environment's characteristics in that period of time. Conversely, for $d = 28$ (blue dot line) the results are always the highest (except for $Waterfall_B$).

Table 3 summarizes the results of the two search–based waterfall optimization algorithms, as compared to the human–based performances for the four real-world waterfalls. The human expert had manually inspected the original waterfalls at each time point, according to their number of impressions, prices, and ad network capacity, and following, recommendations for the changed waterfalls were evaluated in terms of total revenue (recall that this is the most common approach in the industry). The table reveals that human experts had the lowest improvement in revenue, while the MCTS had the highest average improvement rate. Although the MCTS was superior to the S&S-based waterfall optimization algorithm for three out of four real waterfalls, this came at the expense of a much higher run-time: 20–25 times higher. The run-time results could be seen in the last two columns of Table 3. Note that we did not parallelize the algorithms since a run-time of $\approx$ 1 hour is reasonable for an off-line algorithm that usually runs once a day (for a more complex system that could be scaled).

Table 3. The revenue improvement (%) and run-time complexity (measured in hours), as compared to the baseline (current revenue) for the four waterfalls, as well as to the two search-based optimization algorithms and human expert.

	Human expert (%)	S&S-based (%)	MCTS-based (%)	S&S-based (hours)	MCTS-based (hours)
$Waterfall_A$	0.6	6	5	1.2	27.0
$Waterfall_B$	0.6	4	4	1.8	47.4
$Waterfall_C$	0.1	4	5	0.8	17.1
$Waterfall_D$	0.2	11	12	0.5	11.6

Finally, we experimented our algorithms with empty and a random waterfall initializations. First, we revealed that the empty initialization yield the worst results. The main reason is that the algorithm terminates too fast, and thus, converges to a poor local maximum. Second, we found that the random initialization was also inferior to the human expert initialization, which make sense, since the human expert knowledge is valuable.

5   Summary

In this study, we suggest a framework to learn about waterfalls from historical data. The settings in our proposed framework are: offline learning, multiple instances per ad network, unknown number of instances, and discrete prices. To the best of our knowledge, this is the first attempt to tackle the problem from the user's perspective. The main advantages of the proposed method are: it utilizes expert knowledge as an initial waterfall for the S&S procedure, and; the proposed method does not requires an online search step. In many cases, online measurements are not feasible. Finally, our method uses only the successful events, while implicitly utilizing rejected bid information. This allows to significantly reduce the volume of the stored data by 90%.

Future research should focus on 1) improving the valuation matrix estimation. Two proposals in this direction could be to use a more general distribution, such as Dirichlet, or to incorporate the rejected events in the valuation estimation process; 2) to dispense with the discrete price assumption, and; 3) to investigate other algorithms for solving this optimization problem so as to find the globally optimal waterfall strategy. This is crucial, since the proposed search-based waterfall optimization algorithms can only guarantee sub-optimal solutions.

6   Appendix

Appendix A - The Pseudo-code for the Monte Carlo Tree Search Algorithm

In this Appendix, we present the pseudo-code for the MCTS proposed algorithm. As opposed to the S&S-based algorithm, here, the algorithm will adopt the neighbor waterfall with the greatest revenue potential at every iteration; not necessarily the one with the current highest revenue. From an algorithm perspective, the difference is that there are two *for loops* in lines 18 and 20.

Algorithm 5. Monte Carlo tree search procedure

1: **Input:** Initial waterfall W^0 ▷ This could be the current existing waterfall
2:        Valuation matrix B
3:        Max_iter
4:        ϵ
5: **Output:** Waterfall W
6:
7: **Initialization:**
8: $q' =$ Run users in the waterfall W^0 based on their valuation matrix B ▷ Call Algorithm 2
9: Optimal_revenue = Revenue(W^0)
10: Convergence = True
11: $W = W^0$
12: Iter = 0
13: **Start:**
14: **while** Convergence **do**
15:     Iter += 1
16:     Prev_revenue = Optimal_revenue
17:     neighbors = list of neighbors of W
18:     **for** each neighbor g in neighbors **do**
19:         grand_neighbors = list of neighbors of g
20:         **for** each neighbor gg in grand_neighbors **do**
21:             Run the users (B) in gg ▷ Call Algorithm 2
22:             Curr_revenue = Revenue(gg)
23:             **if** Curr_revenue > Prev_revenue **then**
24:                 Prev_revenue = Curr_revenue
25:                 $W = g$
26:     **if** Iter==Max_iter **or** Prev_revenue - Optimal_revenue < ϵ **then**
27:         Convergence = False
28:         Optimal_revenue = Prev_revenue

Appendix B - An Example for the Data Processing Flow

Table 4 describes the data processing from raw-data (Table 4a) into a valuation matrix (Table 4c), using Algorithm 1. Table 4b is the output of row 5 in Algorithm 1. For example, rows 1 and 4 that are marked in red-bold in Table 4a are the raw data of user '4421AB3' and 'G' with a single impression each. These two rows are converted to a vector with (at least) the two entries '[0.02, 0.19]' that are marked with a red-bold box in Table 4b, before the beta distribution parameters, $Beta(\alpha = 0.93, \beta = 10.99)$, are estimated as marked in red-bold in Table 4c.

Table 4. Raw-data samples from the auction dataset ($Waterfall_A$) and their processed output.

Date	Hour	ad network	User id	Impressions	Revenue
01/01/2021	19	G	4421AB3	1	0.020
01/01/2021	18	F	345ADB	2	0.022
01/01/2021	21	U	12345AS	1	0.015
01/01/2021	19	G	4421AB3	1	0.019
01/01/2021	18	F	345ADB	2	0.018
01/01/2021	22	U	4421AB3	1	0.015
01/01/2021	20	G	12345AS	3	0.057
...					

(a) An example of the raw-data

User id	G	F	...
4421AB3	[0.020,0.019,...]	[...]	
345ADB	[...]	[0.011,0.011,0.009,0.009,...]	
12345AS	[0.019,0.019,0.019,...]	[...]	
...			

(b) Vectorization of Table 4a

User id	G	F	...
4421AB3	($\alpha = 0.93, \beta = 10.99$)	($\alpha = 1.06, \beta = 0.25$)	
345ADB	($\alpha = 0.69, \beta = 6.95$)	($\alpha = 0.51, \beta = 0.67$)	
12345AS	($\alpha = 0.36, \beta = 0.50$)	($\alpha = 1.64, \beta = 0.89$)	
...			

(c) The valuation matrix representation of Table 4b of after Beta estimations

References

1. Afshar, R.R., Rhuggenaath, J., Zhang, Y., Kaymak, U.: A reward shaping approach for reserve price optimization using deep reinforcement learning. In: Proceedings of the 9th International Joint Conference on Neural Networks (2021)
2. Afshar, R.R., Zhang, Y., Firat, M., Kaymak, U.: A decision support method to increase the revenue of ad publishers in waterfall strategy. In: IEEE Conference on Computational Intelligence for Financial Engineering and Economics, pp. 1–8. IEEE (2019)
3. Afshar, R.R., Zhang, Y., Firat, M., Kaymak, U.: A reinforcement learning method to select ad networks in waterfall strategy. In: Proceedings of the 11th International Conference on Agents and Artificial Intelligence (2019)
4. Battiti, R., Brunato, M., Mascia, F.: Reactive search and intelligent optimization, vol. 45. Springer Science & Business Media (2008)
5. Dimitri, P.: Bertsekas and Athena Scientific. Convex optimization algorithms, Athena Scientific Belmont (2015)
6. Busch, O.: The programmatic advertising principle. In: Busch, O. (ed.) Programmatic Advertising. MP, pp. 3–15. Springer, Cham (2016). https://doi.org/10.1007/978-3-319-25023-6_1
7. Chakraborty, T., Even-Dar, E., Guha, S., Mansour, Y., Muthukrishnan, S.: Approximation schemes for sequential posted pricing in multi-unit auctions. In: Saberi, A. (ed.) WINE 2010. LNCS, vol. 6484, pp. 158–169. Springer, Heidelberg (2010). https://doi.org/10.1007/978-3-642-17572-5_13
8. Chou, P.-W., Maturana, D., Scherer, S.: Improving stochastic policy gradients in continuous control with deep reinforcement learning using the beta distribution. In: Proceedings of the 34th International Conference on Machine Learning, pp. 834–843. PMLR (2017)
9. Cooper, G., Herskovits, E.: A bayesian method for the induction of probabilistic networks from data. Mach. Learn. 9(4), 309–347 (1992)
10. Coulom, R.: Efficient selectivity and backup operators in Monte-Carlo tree search. In: van den Herik, H.J., Ciancarini, P., Donkers, H.H.L.M.J. (eds.) CG 2006. LNCS, vol. 4630, pp. 72–83. Springer, Heidelberg (2007). https://doi.org/10.1007/978-3-540-75538-8_7
11. Despotakis, S., Ravi, R., Sayedi, A.: First-price auctions in online display advertising. J. Marketing Res. 58(5), 888–907 (2021)
12. Hoos, H., Stützle, T.: Stochastic local search: Foundations and applications. Elsevier (2004)
13. Johnson, N., Kotz, S., Balakrishnan, N.: Continuous univariate distributions, vol. 289. John Wiley and sons (1995)
14. Kveton, B., Mahdian, S., Muthukrishnan, S., Wen, Z., Xian, Y.: Waterfall bandits: Learning to sell ads online. arXiv preprint:1904.09404 (2019)
15. Muthukrishnan, S.: Ad exchanges: research issues. In: Leonardi, S. (ed.) WINE 2009. LNCS, vol. 5929, pp. 1–12. Springer, Heidelberg (2009). https://doi.org/10.1007/978-3-642-10841-9_1
16. Russell, S., Norvig, P.: Artificial intelligence: A modern approach. Prentice Hall (2002)
17. Ting, M., Grislain, N.: Maximizing net income of the auction waterfall with an abort decision tree. arXiv preprint arXiv:1809.01245 (2018)
18. Wang, J., Zhang, W., Yuan, S.: Display advertising with real-time bidding (RTB) and behavioural targeting. arXiv preprint arXiv:1610.03013, 2016
19. Zhang, W., Yuan, S., Wang, J.: Optimal real-time bidding for display advertising. In: Proceedings of the 20th ACM SIGKDD International Conference on Knowledge Discovery and Data Mining, pp. 1077–1086 (2014)

Survey on KNN Methods in Data Science

Panos K. Syriopoulos$^{(\boxtimes)}$, Sotiris B. Kotsiantis⬤, and Michael N. Vrahatis⬤

Computational Intelligence Laboratory, Department of Mathematics,
University of Patras, 26110 Patras, Greece
{p.syriopoulos,sotos,vrahatis}@math.upatras.gr

Abstract. The k-nearest neighbors (KNN) algorithm remains a useful and widely applied approach. In the recent years, we have seen many advances in KNN methods, but few research works give a holistic account of all aspects of KNN and the progress made. This paper is a brief survey on modern KNN methods and their role in data science. Furthermore, we survey: the challenges, how they are approached in the literature, the impact of the distance metric, several KNN variations, as well as query methods.

Keywords: KNN methods · Data science · Instance based learning

1 Introduction

KNN belongs to the family of instance-based learning algorithms, a concept explained by Aha et al. [2]. Simply put, the training instances are stored in memory without explicitly learning a model. The training instances (referred to as "knowledge", training set, or simply dataset) will only be processed in the prediction phase. For each new data instance, a query is made to the knowledge base and, only then, the knowledge returned by the query is processed to produce a prediction. In the simplest case of KNN, the query returns the k nearest data points to the new instance (based on some distance or similarity metric). The intuition is that similar examples are good predictors of unseen examples. Apart from the simplicity of the algorithm, KNN classification enjoys a theoretical guarantee: the probability of error is bounded above by twice the Bayes probability of error (Loizou and Maybank [33]).

In KNN classification, the output is a class membership and is usually determined by the majority class of the instances returned. In KNN regression, the output is determined by the average of neighboring data points. With little to no training time, KNN is a useful tool for "off-the-bat" analysis of data sets. It is versatile, easy to implement and makes no assumptions on the data. An early review of KNN classification is given by Cunningham and Delany [10]. The core issues were: (a) how to determine the appropriate similarity/distance metric, (b) how to determine the user-defined parameter k, (c) how to address computational complexity issues in large datasets. Many works have addressed these issues since then. This survey gives an overview of modern developments

© Springer Nature Switzerland AG 2022
D. E. Simos et al. (Eds.): LION 2022, LNCS 13621, pp. 379–393, 2022.
https://doi.org/10.1007/978-3-031-24866-5_28

with respect to these core issues but also intends to exhibit the utility of KNN in data science in general.

The structure of this paper is as follows: Sects. 2–4 are concerned with the high level concepts of KNN algorithms. Specifically, Sect. 2 is a recap of the challenges related to KNN, and provides descriptions of how certain algorithms tackle these issues. Section 3 is focused on the choice of distance metric, metric learning and the role of feature space transformation. Section 4 describes several recent KNN variations that we find interesting. The rest of the chapters are as follows: Sect. 5 is concerned with prototype selection and generation, and feature selection methods. Section 6 is dedicated to matching issues related to hashing algorithms, partition trees, and graph techniques. A lot of work related to scaling KNN for big data is also presented in Sect. 6. Finally, a synopsis and concluding summary are presented in Sect. 7.

2   Challenges

KNN is a simple and non parametric algorithm (does not make any assumption on the distribution of the training instances). However, proper tuning of the hyper-parameter k is of crucial importance. Figure 1(a) is an example where different parameter values result in different classification outcome. If $k = 3$ the unseen observation (the 'x' mark on the figure) is classified as blue, whereas if $k = 6$ it is classified as orange. Figure 1(b) demonstrates how the decision boundary might look like for a particular choice of k. A comprehensive tutorial is given by Cunningham and Delany [11].

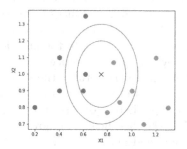 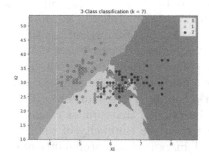

(a) k-nearst neighbors for k=3 and k=6.   (b) Example decision boundary plot.

Fig. 1. In plot (a) the unseen observation is classified as blue for $k = 3$ while it is classified as orange for $k = 6$. Plot (b) exhibits the decision boundary formed in a 3-class example. (Color figure online)

The factors that might affect KNN performance include:

1. **Outliers and Noise:** Outliers have a higher chance of misclassification and noise makes the determination of a suitable k value a more challenging task.

2. **Overlapping class regions:** The region defining a class, or fragments of a class, may overlap with the region of another class. When classes intertwine, the classification decision may be prone to error.
3. **Class imbalance:** Some classes in the data set may contain significantly less observations than other classes. This biases the prediction in favor of the over-represented classes, especially when there is a high degree of overlap.

A recent survey on the challenges related to KNN, together with experimental results, has been given by Zhang in [66]. The current chapter is complementary by offering insightful descriptions of the challenges and of how they are addressed, together with key results regarding the use of KNN algorithms in general.

The weakness of KNN in overlapping regions can be described in terms of the statistical properties of the classes. Consider two Gaussian distributions with different means and variances, and overlapping density functions. The Gaussian with the smaller variance tends to dominate the decision of KNN, as samples tend to form dense clusters in the region around the mean. In this example, Tang and He [49] demonstrate that Bayesian estimation outperforms regular KNN. To tackle this challenge they propose *Extended-NN* (ENN) which looks at data points that consider the unseen example to be in their k-neighborhood. The method iteratively assigns the unseen observation to classes and calculates a class-wise coherence statistic. Total coherence is calculated by the average number of correct inclusion in the k-neighborhoods of data points (of a given class) that consider the unseen observation to be their k-neighbor, and summing over all the classes. The assignment with largest total coherence wins. ENN is shown to outperform regular KNN consistently. Nonetheless, their paper demonstrates that the incorporation of Bayesian estimation may yield promising results.

Class imbalance phenomena have been widely studied in the machine learning literature (for example He and Garcia [20], Fernández et al. [14]). Traditional pre-processing techniques include re-sampling and cost-sensitive learning. Unfortunately, re-sampling techniques can fail to improve the accuracy of KNN due to the usually sparse nature of minority instances in space. As a response, researchers devise KNN specific approaches, many of which revolve around formulating more appropriate decision rules.

Ando proposes in [3] *Weighted Class-wise Nearest Neighbors* (WCNN). He estimates and compares conditional probabilities of class membership given majority or minority class. The estimates are based on the distance of the unseen observation from: (a) the kth nearest majority observation, (b) the kth nearest minority observation. The intuition is that the distance from the kth minority class observation is generally larger. WCNN biases the decision towards the minority class. The decision function is linear in the k-distances and the weights are optimised for non-linear performance loss in a convex setting. Furthermore, by construction, WCNN can work with distance data. To reduce the perplexity of the distance features, WCNN is paired with a hierarchical clustering algorithm that considers class labels. The major sub-components are regarded as supplementary labels to facilitate optimisation. The key advantages of WCNN are that (a) data-balancing preprocessing algorithms may be avoided, (b) it works directly with pairwise distance data.

Zhang et al. [70] proposed the *K-Rare Nearest Neighbors* (KRNN) where k is variable so as to encompass at least some examples of the under-represented classes, i.e. the total number of observations in a query's neighborhood is variable. Given an unknown instance, the probability of it belonging to a minority class is modeled with a binomial distribution. The confidence interval for the proportion of minority observations in the data set is calculated, and a modified confidence interval is used for the proportion of minority observations in the neighborhood of the query. Additionally, two modified Laplace estimates for the posterior class probability are formed, and the relative position of the confidence intervals determines which Laplace estimate is used in the decision function. The confidence levels for the global and local confidence intervals are hyperparameters of the algorithm. Experiments indicate a performance improvement over WCNN on the datasets used.

When it comes to noise we can distinguish between noise in the attributes and noise in the prediction variable. When attributes are numeric, noise manifests itself in the statistical properties of the classes. For categorical variables, there could be errors in the attributes and/or misclassified data points. All types of noise deteriorate the performance of KNN. Assuming that correctly classified points dominate their respective regions in feature space, sensitivity to noise can be reduced by choosing a larger k value. Early noise reduction techniques involved disregarding misclassified points in the training set. However, the sensitivity of KNN's performance to noise has motivated KNN-based anomaly detection techniques. The general idea is scoring points based on their similarity to their neighbors, for example Bandaragod et al. [7], Pang et al. [38]. We have not come across a paper that considers correlations between attribute values and noise in a KNN framework.

Apart from competitive performance, KNN based anomaly detection algorithms demonstrate a gravity defiant learning curve. The general understanding is that, for most learning algorithms, the error rate decreases with the size of the dataset, i.e. the more data the better. Contrary to the conventional wisdom, anomaly detection techniques based on KNN reach peak performance when the training dataset is small. Ting et al. [50] use a computational geometry argument to find closed form expressions for the lower and upper bounds of the *area under the receiver operating characteristic curve* for the 1-NN anomaly detection scheme. It is shown that the expected bounds depend on both the size of the dataset and the proportion of anomalies, and that the bounds reach an optimal value at a finite dataset size. Experimental results show that other KNN-based anomaly detectors demonstrate the same behavior. This result consolidates the role of KNN methods in data science as it provides a solid theoretical ground for the use of KNN ensembles in anomaly detection, which would be especially useful in the context of big data. More specifically, KNN detectors can work in parallel on subsamples of the dataset, and achieve maximal expected accuracy. Triguero et al. [52] advocate the use of KNN methods as means of creating smart data out of big data, the main tools being KNN based noise reduction methods, and missing value imputators. Note that noise reduction methods should be applied prior to imputators to reduce bias.

The final issue of this chapter concerns the choice of the hyperparameter k. To the best of our knowledge the most straightforward approach seems to be the best in terms of accuracy. Most notably, García-Pedraja et al. [17] propose the assignment of a k value to each observation in the knowledge base, using ten-fold cross-validation. The values are chosen by considering both the local best k and the global best k in order to avoid large deviations in the values assigned to neighboring points. New instances, then, inherit the k value of their nearest neighbor. Experiments indicate improved accuracy over regular KNN. Approaches based on direct (convex) optimization are also present, and can also improve accuracy. Zhang et al. [67] propose a sparse reconstruction of the dataset from itself through multiplication with a weight matrix. Non-negativity is enforced to ensure that points are reconstructed from points with correlated features and l_1 regularization is added to the cost function to ensure sparsity. The number of non-zero entries indicates the suitable k value for each observation. An extra term is added to ensure that points are reconstructed with points with similar features (i.e. are close in the feature space). Their work is extended in Zhang et al. [68] where a decision tree, named k^*-tree, is constructed. The k^*-tree can be searched efficiently and results in faster query times (by storing relevant points in the leafs) for a small loss of accuracy.

3 Choice of Distance Metric

In general, data instances with d features are considered as points within an d-dimensional feature space. Since the prediction is determined by the nearest neighbors, the nature of KNN is such that the relative position of instances is more significant than their absolute positions. Ideally, the distance metric should minimize the distance between similarly classified instances, while maximizing distance between instances of different classes.

Abu Alfeilat et al. [1] tested a large number of distance metrics on real world datasets and found that the choice of metric significantly affects classification performance. Specifically, there is no optimal distance metric that is suitable for all datasets, and metrics from the same families showed similar classification results. Additionally, some distance metrics were found to be more tolerant to noise than others. It could be the case that some application domains favor certain metrics over others. Hu et al. [23] conducts similar experiments for the medical domain and find that the Chi-square distance function performs the best.

While cross validation is an option for finding a suitable distance metric, another approach is distance metric learning. These methods involve applying a (linear or nonlinear) transformation to the feature space in order to maximise classification accuracy. Xing et al. [63] was the first to propose a convex optimization approach. They considered the Mahalanobis distance defined by

$d_A(x, y) = (x - y)^\top A(x - y)$, where $A = W^\top W$ is a positive semi-definite matrix $A \succeq 0$ (here W corresponds to the space transformation). The intuition is: find the optimal matrix A that minimizes distances between similarly labeled points and maximizes distances between points in different classes. To see this, let $y_{ij} = 1$ if the i'th and j'th points in the knowledge base belong to the same class (i.e. $y_i = y_j$), and zero otherwise. The objective function can be written as:

$$\text{maximize} \quad \sum_{i,j} (1 - y_{ij}) \sqrt{d_A(x_i, x_j)},$$

$$\text{subject to:} \quad \sum_{i,j} y_{ij} d_A(x_i, x_j) \leqslant 1, \tag{1}$$

$$A \succeq 0.$$

The objective function aims to maximize the distance between points belonging to different classes while the first condition forces points belonging to the same class to remain close to each other. In a similar approach Shalev-Shwartz et al. [42] learns a Mahalanobis metric online and tries to enforce a scalar threshold b such that points in the same class are at most $b - 1$ distance apart while points in different classes are at least $b + 1$ distance apart. Both these approaches improve the performance of KNN and have the advantage that the optimization problems are convex. However there is a key observation relevant to KNN that exposes a weakness of linear space transformations. Points of the same class may cluster in different locations in feature space and may exhibit different patterns in their relevant positions with points from other classes. Goldberger et al. [18] tries to circumvent this problem by introducing the novel idea of *Neighborhood Component Analysis* (NCA). NCA randomly assigns neighborhoods to points and maximizes the leave-one-out probability that the KNN label is correct. The probability that point j is in the neighborhood of point i is inversely related to their distance in the transformed space. Inspired by the idea that only the distances of neighbors are relevant in the objective function of Eq. (1), Weinberger et al. [59] formulated a convex optimization problem similar to that of Eq. (1) that sums over neighbors (instead of over all pairs of points) and whose resulting transformation is known as the *Large Margin Nearest Neighbor* (LMNN).

It is clear that the choice of metric is relevant in KNN applications and that the methods of this chapter alleviate the adverse effects of data peculiarities. While linear transformations do improve the accuracy of KNN, their weakness comes from the fact that they are global, yet, the distributions of classes differ in different regions of the feature space. As a result, and by extension, all literature on space transformations is relevant for improving KNN classification accuracy.

4   Variations of KNN

Here we briefly recite a selection of recent KNN methods that we find interesting in dealing with the challenges in Sect. 2. Although this chapter provides a less than complete account, we believe that the our references include a satisfactory number of readings.

The state of the art seems to revolve around meaningful compositions of existing methods and ideas. The works cited in this paragraph have been used in conjunction with each other to produce better KNN algorithms. Pascal and Yoshua [54] devised a variant of KNN based on k-local hyperplanes. The intuition is that classes lie on non-linear manifolds on the feature space. The core idea is to linearly approximate the manifold locally, and assign unseen observations based on their distance from the (approximate) manifold of each class. Particularly, suppose a set of class labels $\mathcal{C} = \{1, 2, \ldots, N_C\}$ where N_C is the number of classes. For each unseen data point, locate the k nearest neighbors from each class (for a total of kN_C nearest neighbors). Say $\mathcal{N}_{C_i} = \{x_1, x_2, \ldots, x_k\}$ are the k closest points with class label i. The local approximation of the corresponding manifold is a linear combination of the points in $\mathcal{N}_{C_i}$. The coefficients of the linear combinations are optimised to minimize the distance to the unseen observation for each i in $\mathcal{C}$, and the point is assigned to the class of the closest linear combination. Another important "building block" is the family of fuzzy KNN algorithms. We direct the reader to Derrac et al. [12] for a taxonomy and experimental analysis. Meanwhile, several researchers have proposed the use of different k-values for different regions of the feature space, i.e. Wettschereck and Dietterich [60], Wang et al. [56], Garcia et al. [17]. Among others, Garcia et al. [17] proposed a tenfold cross validation scheme for the assignment of different values of k to each point in the knowledge base. Unclassified examples would inherit the k value of their closest neighbor. Furthermore, several researchers adopted weighting schemes to weight the labels of the k-nearest neighbors (usually according to distance), i.e. Dudani [13], Liu and Chawla [32], Gou et al. [19].

Recently, researchers combine two or several of the known approaches, together with their own innovation, and produce algorithms that are less vulnerable to noise and outliers, class imbalance, and class region distribution. Susan and Kumar [47] applied a linear transformation (such as LMNN or NCA, see Sect. 3) to the feature space and adopted a new decision rule. They split the k nearest neighbors in two clusters. One consists of the neighbors that are closest to the farthest neighbor, and the other consists of the neighbors that are closest to the closest neighbor. The latter cluster, then, determines the final classification. The rule is resilient to outliers, noise, and class imbalances because the feature space transformation has expanded distances between instances of different classes. Yu et al. [65] combined k-local hyperplane distances with a fuzzy relative transform decision rule to tackle class imbalances. Zhang et al. [69] employed the locality preserving projection (by He and Niyogi [22]) to reconstruct the test sample from the training sample, resulting in a weighting of nearest neighbors. These are but a few examples of successful combinations of ideas.

5   Feature Selection and Data Reduction

This chapter is concerned with data reduction techniques. These methods revolve around disregarding irrelevant and/or redundant dimensions of a dataset, or, disregarding unnecessary data points. These methods are valuable in practice as they may result in greater accuracy, reduced runtimes, and reduced memory requirements.

The accuracy of distance based algorithms can be severely degraded with high-dimensional data due to the curse of dimensionality. Feature selection (FS) techniques aim at reducing the dimensionality of the data, thus, improving the accuracy of the learning algorithm. The goal is to identify a small subset of features that maximizes a measure of accuracy. A broad categorization of FS methods includes filter-based methods, wrapper-based methods, embedded methods, and hybrid methods. For more information we refer the reader to Li et al. [30]. Lets consider a general strategy for FS that includes (a) a feature subset selector, (b) a feature subset evaluator. Usually, in wrapper-based methods, a machine learning model is re-trained and tested with the dataset projected on the current feature subset (in order to evaluate its performance). In these cases, KNN offers a significant speedup when the distance function is calculated recursively. For example, the d-dimensional Euclidean distance satisfies:

$$d_E(x, x')^2 = \sum_{i=1}^{d-1}(x_i - x_i')^2 + (x_d - x_d')^2.$$

Wang et al. take the aforementioned approach in [55]. Many approaches in the literature select KNN for subset evaluation, e.g. Tahir et al. [48] propose a Tabu search strategy for subset generation. A challenge in the FS literature is instability. In many cases the number of features far exceeds the number of observations. As a result, the features selected by certain algorithms (most notably the random forest approach) highly depend on the initial data sample. Ensemble KNN wrapper methods are believed to be able to tackle instability issues (Li et al. [31], Park and Kim [39]). In the context of filter-based methods, a comparative study by Rogati and Yang [41] showed that KNN methods were amongst the top three performers at year 2001, indicating that KNN can benefit greatly from FS methods.

In a work more intimately related to KNN, Xiao and Chaovalitwongse [62] showed that the FS problem can be cast as a convex optimisation problem if the decision is based on the distance to the centroids of each class. The idea is to learn a Mahalanobis matrix (similarly to what is shown in Sect. 3) and add a l_1 regularization term in the loss function. l_1 regularization promotes sparsity for the Mahalanobis matrix, which effectively nullifies certain dimensions of the data.

Data reduction methods attempt to reduce the number of training instances. The idea is to select, or artificially generate prototypes that faithfully represent the given target concept. Ideas include retaining instances that are close to the decision boundary, or, retaining instances near the centers of the class clusters. Wrapper-based subset search and evaluation strategies are also present. A complete taxonomy, comparison, and extensive experimentation is given in Garcia et al. [16]. A valuable insight is that efficient query methods, such as approximate nearest neighbors (described in the next chapter), can compete in run-time performance with regular KNN even when the initial dataset is reduced in size. Another observation is that the time complexity of these algorithms were generally $O(n^2 d^2)$ or higher, where n is the number of data points and d the number of features, together with high storage requirements. Arnaiz-González et al. [5] propose a linear complexity algorithm using locality sensitivity hashing (LSH). Even though the data reduction rates and the resulting classification accuracy were not among the top performers, it is a solution for extremely large datasets, and their paper provides descriptions of many other data reduction algorithms. Additionally, Triguero et al. [51] provide a distributed solution using the MapReduce framework. Recent publications, among others, include prototype selection for imbalanced datasets: Sisodia and Sisodia [45], prototype selection with local feature weighting: Zhang et al. [71], prototype selection for KNN regression: Song et al. [46].

6 Nearest Neighbor Matching Algorithms

It is important to note that when working with high dimensional features, there is no known exact nearest-neighbor search algorithm with acceptable efficiency. To enhance the speed of queries, most practical applications settle for approximate search. The neighbors returned by approximate search techniques may not be the k-nearest, but they are typically close to the k-nearest neighbors. Regardless of whether KNN is used as a classifier, or as a tool for aforementioned purposes, this chapter aims to show that KNN can be a scalable solution.

When the data set is large, $O(dn)$ query time-complexity renders KNN algorithms intractable for certain applications. Several solutions have been studied. When the number of training instances is large, special data structures (partitioning trees, neighboring graph techniques) can enhance query speeds significantly. The main disadvantage of such methods is how they scale with the dimensionality of the data. Techniques such as hashing counteract this problem.

A simple and popular hashing method is the locality sensitive hashing (LSH), Indyk and Motwani [25]. The idea is to hash the data so that the probability that hashes coincide is much higher for points that are close together. Formally, locality sensitivity is defined by four parameters, (r_1, r_2, p_1, p_2). Given some distance function d, a family of hash functions $\mathcal{H}$ is $(r1, r2, p1, p2)$-sensitive if for any two points p, q in the data set the following conditions are fulfilled:

1. if $p \in B(q, r_1)$, then $\Pr_{h \in \mathcal{H}}[h(p) = h(q)] \geqslant p_1$,

2. if $p \notin B(q, r_2)$, then $\Pr_{h \in \mathcal{H}}[h(p) = h(q)] \leqslant p_2$,

where $B(p, r)$ denotes the hypersphere centered at point p with radius r. For this to be meaningful, it is essential that probabilities p_1, p_2 satisfy $p_1 > p_2$, and $r_1 < r_2$. The gap between p_1 and p_2 can be amplified by concatenating several hash functions. Practically, a collection of hashes split the data points into several partitions. To process a query, brute force search is applied to elements of the partitions with corresponding hash values. LSH is grounded in the theory of random projections. In the simplest case, points are projected on random lines passing through the origin. These lines are then discretized into small line segments each with a corresponding id. Thus, a table of hash codes is created, each entry having pointers to the corresponding data points. This allows for the creation of data structures that can be searched efficiently. A generalization named *Density Densitive Hashing* (DSH) also exploits the distribution of the data [28].

Hashing methods result in fast and effective queries. Query times achieved are sublinear. In general, the quality of the hash functions determine the quality of the method (for details see Muja and Lowe [36]). In fact, the space partitions produced by random projections had been widely studied in the average case. In the last decade a lot of work has been dedicated in studying worst case scenarios. To illustrate what is meant by average and worst case scenario, consider the random projection method described in the previous paragraph. If a dataset consists of points sparsely distributed around the origin, the probability of collision of far away points is small. On the contrary, if there is a dense cluster of points in the dataset (and away from the origin), these points are likely to collide with far away points. For this reason, data-dependent hashing methods have been developed. These methods aim for data optimal (approximate) nearest neighbors search, for details see He et al. [21], Xu et al. [64], Iwamura et al. [26], Andoni and Razenshteyn [4]. In the recent years, the literature on hashing methods has grown and we direct the reader to the survey by Wang et al. [57].

In the family of partitioning trees, the kd-tree (Friedman et al. [15]), has been one of the best known neighbor matching algorithms with logarithmic time complexity, but it scales poorly with the number of dimensions (and in fact is comparable to exhaustive search when the number of dimensions is high, see Indyk [24]). Several authors have improved kd-trees in order to speed up KNN search. To name a few, Beygelzimer et al. [8] proposed cover-trees, Silpa-Anan and Hartley [44] proposed optimised kd-trees. Nister and Stewenius [37] proposed the vocabulary tree, which uses hierarchical k-means. In a comparison by Muja and Lowe [35] it was shown that the multiple randomized trees are the most effective for high dimensional data.

Jegou et al. [27] proposed a product quantization approach in which the feature space is decomposed into low dimensional subspaces in which the data points are represented by compact codes. Babenko and Lempitsky [6] proposed the inverted multi-index, obtained by replacing the standard quantization with product quantization. A more in-depth analysis is given in review papers by Vasuki and Vanathi [53], Wu and Yu [61].

Nearest neighbor graph methods build graphs where vertices are data points or subsets of data points. In this case the query is an effective exploration of the graph. Empirical results place graph methods in the current state of the art for query methods. We direct the reader to recent surveys by Wang et al. [58] and Shimomura et al. [43].

Finally, a lot of work has been done for decentralized framework solutions. Chatzimilioudis et al. [9] developed Spitfire, a high performance distributed algorithm. Gieseke et al. [40] presented a GPU based algorithm for *kd*-trees. Kim et al. [29] propose parallel KNN using MapReduce. Maillo et al. [34] also provided a solution for exact *k*-nearest neighbor classification based on Spark.

7   Synopsis and Concluding Remarks

We have studied a variety of aspects related to KNN. Challenges related to: (a) noise and outliers, (b) overlapping class regions, and (c) class imbalanced data have been analyzed through key works in the literature. The role of KNN in data pre-processing, including de-noising and missing value imputations has been presented. The choice of hyperparameter k has also been analyzed. The impact of the distance metric used has been explored, together with metric-learning techniques. The merit of KNN in feature selection algorithms has been discussed, together with the effect of such methods on KNN's accuracy. We further explored data reduction techniques. Moreover, query methods, together with mentions of parallel and distributed solutions have been mentioned.

We would like to point out here that KNN is a non parametric, instance based algorithm that makes no assumptions about the underlying data distribution. This trait is crucial due to the fact that real world data rarely obey typical theoretical assumptions. KNN is easy to implement and its basic principle is easy to understand. Some challenges regarding the choice of the right distance metric have been overcome with data driven approaches like metric learning. Localization of the k-value, and different decision rules (subspace distances, fuzzy criteria, feature weighting etc.) have also increased classification accuracy.

The state-of-the-art seems to revolve around the meaningful composition of ideas that make the algorithm more resilient to class imbalances, noise, and outliers. The two major drawbacks are the storage requirements and the query run-time complexity. Methods of prototype selection and generation can reduce the number of training instances required without hindering classification accuracy greatly. Methods of feature selection can reduce the dimensionality of the data by discarding irrelevant/redundant features. KNN based methods have been developed for noise reduction, outlier detection, and missing value imputation among other applications. Theoretical results indicate that KNN anomaly detectors demonstrate a gravity defiant learning curve. These issues enable the usage of KNN anomaly detectors ensembles in the context of large datasets.

Approximate nearest neighbor methods significantly reduce query run-times, allowing for applications on data intensive domains. Hashing, quantization, neighboring graph techniques, and indexing methods all contribute to the diverse literature that enables the use of KNN in large data domains. GPU based distributed algorithms and adaptations for data streams are a testament to KNN's utility as a classification method or as part of larger machine learning models.

References

1. Alfeilat, H.A., et al.: Effects of distance measure choice on K-nearest neighbor classifier performance: a review. Big Data, **7** (2019)
2. Aha, D.W., Kibler, D., Albert, M.K.: Instance-based learning algorithms. Mach. Learn. (1991)
3. Ando, S.: Classifying imbalanced data in distance-based feature space. Knowl. Inf. Syst. **46** (2016)
4. Andoni, A., Razenshteyn, I.: Optimal data-dependent hashing for approximate near neighbors. In: Proceedings of the Forty-Seventh Annual ACM Symposium on Theory of Computing, STOC 2015, pp. 793–801. Association for Computing Machinery, New York, NY (2015). ISBN 9781450335362
5. Arnaiz-González, Á., Díez-Pastor, J.-F., Rodríguez, J.J., García-Osorio, C.: Instance selection of linear complexity for big data. Knowl.-Based Syst. **107**, 83–95 (2016)
6. Babenko, A., Lempitsky, V.: The inverted multi-index. IEEE Trans. Pattern Anal. Mach. Intell. **37**(6), 1247–1260 (2014)
7. Bandaragoda, T.R., Ting, K.M., Albrecht, D., Liu, F.T., Wells, J.R.: Efficient anomaly detection by isolation using nearest neighbour ensemble. In: 2014 IEEE International Conference on Data Mining Workshop, pp. 698–705. IEEE (2014)
8. Beygelzimer, A., Kakade, S., Langford, J.: Cover trees for nearest neighbor. In: Proceedings of the 23rd International Conference on Machine Learning, pp. 97–104 (2006)
9. Chatzimilioudis, G., Costa, C., Zeinalipour-Yazti, D., Lee, W.-C., Pitoura, E.: Distributed in-memory processing of all k nearest neighbor queries. IEEE Trans. Knowl. Data Eng. **28**(4), 925–938 (2015)
10. Cunningham, P., Delany, S.: k-nearest neighbour classifiers. Mult Classif. Syst. **54**, 04 (2007)
11. Cunningham, P., Delany, S.J.: k-nearest neighbour classifiers - a tutorial. ACM Comput. Surv. (CSUR) **54**(6), 1–25 (2021)
12. Derrac, J., García, S., Herrera, F.: Fuzzy nearest neighbor algorithms: taxonomy, experimental analysis and prospects. Inf. Sci. **260**, 98–119 (2014)
13. Dudani, S.A.: The distance-weighted k-nearest-neighbor rule. IEEE Trans. Syst. Man Cybern. **SMC-6**(4), 325–327 (1976)
14. Fernández, A., del Río, S., Chawla, N.V., Herrera, F.: An insight into imbalanced big data classification: outcomes and challenges. Complex Intell. Syst. **3**(2), 105–120 (2017)
15. Friedman, J.H., Bentley, J.L., Finkel, R.A.: An algorithm for finding best matches in logarithmic expected time. ACM Trans. Math. Softw. (TOMS) **3**(3), 209–226 (1977)
16. Garcia, S., Derrac, J., Cano, J., Herrera, F.: Prototype selection for nearest neighbor classification: taxonomy and empirical study. IEEE Trans. Pattern Anal. Mach. Intell. **34**(3), 417–435 (2012)
17. García-Pedrajas, N., Romero del Castillo, J.A., Cerruela-García, G.: A proposal for local k values for k -nearest neighbor rule. IEEE Trans. Neural Netw. Learn. Syst. **28**(2), 470–475 (2017)
18. Goldberger, J., Hinton, G.E., Roweis, S., Salakhutdinov, R.R.: Neighbourhood components analysis. In: Saul, L., Weiss, Y., Bottou, L. (eds.) Advances in Neural Information Processing Systems, vol. 17. MIT Press, Cambridge (2004)

19. Gou, J., Du, L., Zhang, Y., Xiong, T.: A new distance-weighted k-nearest neighbor classifier. J. Inf. Comput. Sci. **9**, 1429–1436 (2012)
20. He, H., Garcia, E.A.: Learning from imbalanced data. IEEE Trans. Knowl. Data Eng. (2009)
21. He, J., Liu, W., Chang, S.-F.: Scalable similarity search with optimized kernel hashing. In: Proceedings of the 16th ACM SIGKDD International Conference on Knowledge Discovery and Data Mining, pp. 1129–1138 (2010)
22. He, X., Niyogi, P.: Locality preserving projections. In: Thrun, S., Saul, L., Schölkopf, B. (eds.) Advances in Neural Information Processing Systems, vol. 16. MIT Press, Cambridge (2003)
23. Hu, L.-Y., Huang, M.-W., Ke, S.-W., Tsai, C.-F.: The distance function effect on k-nearest neighbor classification for medical datasets. Springerplus **5**(1), 1–9 (2016). https://doi.org/10.1186/s40064-016-2941-7
24. Indyk, P.: Nearest neighbors in high-dimensional spaces (2004)
25. Indyk, P., Motwani, R.: Approximate nearest neighbors: towards removing the curse of dimensionality. In: Conference Proceedings of the Annual ACM Symposium on Theory of Computing, pp. 604–613, October 2000
26. Iwamura, M., Sato, T., Kise, K.: What is the most efficient way to select nearest neighbor candidates for fast approximate nearest neighbor search? In: Proceedings of the IEEE International Conference on Computer Vision, pp. 3535–3542 (2013)
27. Jegou, H., Douze, M., Schmid, C.: Product quantization for nearest neighbor search. IEEE Trans. Pattern Anal. Mach. Intell. **33**(1), 117–128 (2010)
28. Jin, Z., Li, C., Lin, Y., Cai, D.: Density sensitive hashing. IEEE Trans. Cybern. **44**(8), 1362–1371 (2013)
29. Kim, W., Kim, Y., Shim, K.: Parallel computation of k-nearest neighbor joins using mapreduce. In: 2016 IEEE International Conference on Big Data (Big Data), pp. 696–705. IEEE (2016)
30. Li, J., et al.: Feature selection: a data perspective. ACM Comput. Surv. (CSUR) **50**(6), 1–45 (2017)
31. Li, S., Harner, E.J., Adjeroh, D.A.: Random KNN feature selection - a fast and stable alternative to random forests. BMC Bioinform. **12**(1), 1–11 (2011)
32. Liu, W., Chawla, S.: Class confidence weighted kNN algorithms for imbalanced data sets. In: Huang, J.Z., Cao, L., Srivastava, J. (eds.) PAKDD 2011. LNCS (LNAI), vol. 6635, pp. 345–356. Springer, Heidelberg (2011). ISBN 978-3-642-20847-8. https://doi.org/10.1007/978-3-642-20847-8_29
33. Loizou, G., Maybank, S.J.: The nearest neighbor and the Bayes error rates. IEEE Trans. Pattern Anal. Mach. Intell. **PAMI-9**(2), 254–262 (1987)
34. Maillo, J., Ramírez, S., Triguero, I., Herrera, F.: KNN-IS: an iterative spark-based design of the k-nearest neighbors classifier for big data. Knowl.-Based Syst. **117**, 3–15 (2017)
35. Muja, M., Lowe, D.G.: Fast approximate nearest neighbors with automatic algorithm configuration. VISAPP (1), 2 (331–340), 2 (2009)
36. Muja, M., Lowe, D.G.: Scalable nearest neighbor algorithms for high dimensional data. IEEE Trans. Pattern Anal. Mach. Intell. **36**(11), 2227–2240 (2014)
37. Nister, D., Stewenius, H.: Scalable recognition with a vocabulary tree. In: 2006 IEEE Computer Society Conference on Computer Vision and Pattern Recognition (CVPR 2006), vol. 2, pp. 2161–2168. IEEE (2006)
38. Pang, G., Ting, K.M., Albrecht, D.: LeSiNN: detecting anomalies by identifying least similar nearest neighbours. In: 2015 IEEE International Conference on Data Mining Workshop (ICDMW), pp. 623–630. IEEE (2015)

39. Park, C.H., Kim, S.B.: Sequential random k-nearest neighbor feature selection for high-dimensional data. Expert Syst. Appl. **42**(5), 2336–2342 (2015)
40. Patwary, M.M.A., et al.: Panda: extreme scale parallel k-nearest neighbor on distributed architectures. In: 2016 IEEE International Parallel and Distributed Processing Symposium (IPDPS), pp. 494–503. IEEE (2016)
41. Rogati, M., Yang, Y.: High-performing feature selection for text classification. In: Proceedings of the Eleventh International Conference on Information and Knowledge Management, pp. 659–661 (2002)
42. Shalev-Shwartz, S., Singer, Y., Ng, A.Y.: Online and batch learning of pseudometrics. In: Proceedings of the Twenty-First International Conference on Machine Learning, ICML 2004, pp. 94. Association for Computing Machinery, New York (2004)
43. Shimomura, L.C., Oyamada, R.S., Vieira, M.R., Kaster, D.S.: A survey on graph-based methods for similarity searches in metric spaces. Inf. Syst. **95**, 101507 (2021)
44. Silpa-Anan, C., Hartley, R.: Optimised KD-trees for fast image descriptor matching. In: 2008 IEEE Conference on Computer Vision and Pattern Recognition, pp. 1–8. IEEE (2008)
45. Sisodia, D., Sisodia, D.S.: Quad division prototype selection-based k-nearest neighbor classifier for click fraud detection from highly skewed user click dataset. Int. J. Eng. Sci. Technol. **28**, 101011 (2022)
46. Song, Y., Liang, J., Lu, J., Zhao, X.: An efficient instance selection algorithm for k nearest neighbor regression. Neurocomputing **251**, 26–34 (2017)
47. Susan, S., Kumar, A.: DST-ML-EkNN: data space transformation with metric learning and elite k-nearest neighbor cluster formation for classification of imbalanced datasets. In: Chiplunkar, N.N., Fukao, T. (eds.) Advances in Artificial Intelligence and Data Engineering. AISC, vol. 1133, pp. 319–328. Springer, Singapore (2021). https://doi.org/10.1007/978-981-15-3514-7_26
48. Tahir, M.A., Bouridane, A., Kurugollu, F.: Simultaneous feature selection and feature weighting using hybrid Tabu search/K-nearest neighbor classifier. Pattern Recogn. Lett. **28**(4), 438–446 (2007)
49. Tang, B., He, H.: ENN: extended nearest neighbor method for pattern recognition [research frontier]. IEEE Comput. Intell. Mag. **10**(3), 52–60 (2015)
50. Ting, K.M., Washio, T., Wells, J.R., Aryal, S.: Defying the gravity of learning curve: a characteristic of nearest neighbour anomaly detectors. Mach. Learn. **106**(1), 55–91 (2017)
51. Triguero, I., Peralta, D., Bacardit, J., García, S., Herrera, F.: MRPR: a mapreduce solution for prototype reduction in big data classification. Neurocomputing **150**, 331–345 (2015)
52. Triguero, I., García-Gil, D., Maillo, J., Luengo, J., García, S., Herrera, F.: Transforming big data into smart data: an insight on the use of the k-nearest neighbors algorithm to obtain quality data. WIREs Data Min. Knowl. Discov. **9**(2) (2019)
53. Vasuki, A., Vanathi, P.: A review of vector quantization techniques. IEEE Potentials **25**(4), 39–47 (2006)
54. Vincent, P., Bengio, Y.: K-local hyperplane and convex distance nearest neighbor algorithms. In: Dietterich, T., Becker, S., Ghahramani, Z. (eds.) Advances in Neural Information Processing Systems, vol. 14. MIT Press, Cambridge (2001)
55. Wang, A., An, N., Chen, G., Li, L., Alterovitz, G.: Accelerating wrapper-based feature selection with k-nearest-neighbor. Knowl.-Based Syst. **83**, 81–91 (2015)
56. Wang, J., Neskovic, P., Cooper, L.N.: Neighborhood size selection in the k-nearest-neighbor rule using statistical confidence. Pattern Recogn. **39**(3), 417–423 (2006)

57. Wang, J., Zhang, T., Song, J., Sebe, N., Shen, H.T.: A survey on learning to hash. IEEE Trans. Pattern Anal. Mach. Intell. **40**(4), 769–790 (2018)
58. Wang, M., Xu, X., Yue, Q., Wang, Y.: A comprehensive survey and experimental comparison of graph-based approximate nearest neighbor search. arXiv preprint arXiv:2101.12631 (2021)
59. Weinberger, K., Blitzer, J., Saul, L.: Distance metric learning for large margin nearest neighbor classification, January 2005
60. Wettschereck, D., Dietterich, T.: Locally adaptive nearest neighbor algorithms. In: Cowan, J., Tesauro, G., Alspector, J. (eds.) Advances in Neural Information Processing Systems, vol. 6. Morgan-Kaufmann, Burlington (1993)
61. Wu, Z., Yu, J.: Vector quantization: a review. Front. Inf. Technol. Electron. Eng. **20**(4), 507–524 (2019). https://doi.org/10.1631/FITEE.1700833
62. Xiao, C., Chaovalitwongse, W.A.: Optimization models for feature selection of decomposed nearest neighbor. IEEE Trans. Syst. Man Cybern. Syst. **46**(2), 177–184 (2016)
63. Xing, E., Jordan, M., Russell, S.J., Ng, A.: Distance metric learning with application to clustering with side-information. In: Becker, S., Thrun, S., Obermayer, K. (eds.) Advances in Neural Information Processing Systems, vol. 15. MIT Press, Cambridge (2002)
64. Xu, H., Wang, J., Li, Z., Zeng, G., Li, S., Yu, N.: Complementary hashing for approximate nearest neighbor search. In: 2011 International Conference on Computer Vision, pp. 1631–1638 (2011)
65. Yu, Z., Chen, H., Liu, J., You, J., Leung, H., Han, G.: Hybrid k-nearest neighbor classifier. IEEE Trans. Cybern. **46**(6), 1263–1275 (2016)
66. Zhang, S.: Challenges in KNN classification. IEEE Trans. Knowl. Data Eng. 1 (2021)
67. Zhang, S., Li, X., Zong, M., Zhu, X., Cheng, D.: Learning k for KNN classification. ACM Trans. Intell. Syst. Technol. (TIST) **8**(3), 1–19 (2017)
68. Zhang, S., Li, X., Zong, M., Zhu, X., Wang, R.: Efficient KNN classification with different numbers of nearest neighbors. IEEE Trans. Neural Netw. Learn. Syst. **29**(5), 1774–1785 (2017)
69. Zhang, S., Cheng, D., Deng, Z., Zong, M., Deng, X.: A novel KNN algorithm with data-driven k parameter computation. Pattern Recogn. Lett. **109**, 44–54 (2018). Special Issue on Pattern Discovery Multi-Source Data (PDMSD)
70. Zhang, X., Li, Y., Kotagiri, R., Wu, L., Tari, Z., Cheriet, M.: KRNN: k rare-class nearest neighbour classification. Pattern Recogn. **62**, 33–44 (2017)
71. Zhang, X., Xiao, H., Gao, R., Zhang, H., Wang, Y.: K nearest neighbors rule combining prototype selection and local feature weighting for classification. Knowl.-Based Syst. **243**, 108451 (2022)

Constrained Shortest Path
and Hierarchical Structures

Adil Erzin[1,2]($\boxtimes$) (iD), Roman Plotnikov[1,2] (iD), and Ilya Ladygin[3]

[1] Sobolev Institute of Mathematics, SB RAS, Novosibirsk 630090, Russia
{adilerzin,prv}@math.nsc.ru
[2] St. Petersburg State University, St. Petersburg 199034, Russia
[3] Novosibirsk State University, Novosibirsk 630090, Russia

Abstract. The Constrained Shortest Path (CSP) problem is as follows. An n-vertex graph is given, two weights are assigned to each edge: "cost" and "length". It is required to find a min-cost bounded-length path between a given pair of vertices. The problem is NP-hard even when the lengths of all edges are the same. Therefore, various *approximation* algorithms have been proposed in the literature for it. The constraint on path length can be accounted for by considering one aggregated edge *weight* equals to the linear combination of the cost and length. By varying the value of the Lagrange multiplier in the linear combination, a feasible solution delivers a minimum to the objective function with new weights. At the same time, as usually, the Dijkstra's algorithm or its modifications are used to construct a shortest path with the current weights of the edges. However, in the large graphs, this approach may turn out to be time-consuming. In this paper, we propose to search a solution, not in the original graph but in the specially constructed hierarchical structures (HS). We show that the shortest path in the HS is constructed with $O(m)$-time complexity, where m is the number of edges/arcs of the graph, and the approximate solution in the case of integer costs and lengths is found with $O(m \log n)$-time complexity. In result of a priori analysis of the algorithm its accuracy estimation turned out to depend on the parameters of the problem and can be significant. Therefore, to evaluate the algorithm's effectiveness, we conducted a numerical experiment on the graph of roads of megalopolis and randomly constructed metric unit-disk graphs (UDGs). The numerical experiment results show that in the HS, solution is built 10–100 times faster than in the methods which use Dijkstra's like algorithm to build a min-weight path in the *original graph*.

Keywords: Constrained shortest path · Hierarchical structures · Polynomial algorithms · Complexity · Simulation

---

The research was supported by the Russian Science Foundation (grant No. 19-71-10012 "Multi-agent systems development for automatic remote control of traffic flows in congested urban road networks").

D. E. Simos et al. (Eds.): LION 2022, LNCS 13621, pp. 394–410, 2022.
https://doi.org/10.1007/978-3-031-24866-5_29

1 Introduction

In the modern communication networks, to meet the Quality of Service (QoS) requirements, it is necessary to take into account more than one characteristic of each element [9,12,14,20]. In this paper we are considering the following problem. Given a weighted digraph $G = (V, A)$, where V is the set of vertices ($|V| = n$), A is the set of arcs ($|A| = m$). Two non-negative values (*length* and *cost*) are assigned to each arc. It is required to find a min-cost bounded-length path between a given pair of vertices s and t (s-t path). In the literature, this problem is mentioned as Constrained Shortest Path (CSP) problem. CSP is NP-hard, both in the general graphs [4] and in the acyclic networks [20]. Exact exponential [11,13,21] and approximation polynomial algorithms [6,8,12,14,15,18–20,22] are proposed to solve it.

The exact Constrained Bellman-Ford (CBF) algorithm proposed in [21] has exponential complexity, but it is faster than brute force on average. The main idea behind this algorithm is to systematically search for the least cost paths while monotonically increasing the length. First, the algorithm finds a min-cost s-t path. Next, for each vertex u, a list of min-length paths from s to u is created. Then, a vertex is selected that lies on the s-t path with minimal cost, the list of which contains the path that satisfies the constraint. The algorithm then explores the neighbors of this vertex using breadth-first search [3], and (if necessary) adds new paths to the lists of neighbors. This process continues until the length constraint is met and there is a path for further exploration.

Another exact algorithm is the Pulse algorithm proposed in [13]. Its essence is to apply an impulse from the vertex s to the neighboring vertices, then from all neighboring to the next neighbors, etc. Each time, the following characteristics of the partial path are stored in memory: the vertices passed, the value of the objective function, and the current length. When the impulse reaches the vertex t, then the constructed path along with all the characteristics is stored. In this way, all possible paths can be found, including an optimal one. The difference between this algorithm and the full enumeration lies in the special strategies for cutting off partial paths ("pulses"). In the paper, these strategies are *dominance*, *bounds* and *infeasibility*. The essence of the dominance strategy is to remember the best paths in terms of the cost and length, bounds strategies is the systematic pruning of paths with the worse objective function than the paths already found. Infeasibility allows to cut off pulses that are unpromising in length at an early stage (this is achieved by calculating the shortest distance from t to each other vertex).

Hassin in [6] proposed two ε-approximation algorithms for the case of positive arc weights with time complexity $O((\frac{mn}{\varepsilon}+1)\log\log B)$ and $O(\frac{mn^2}{\varepsilon}\log\frac{n}{\varepsilon})$, where B is an upper bound on the path cost. The first algorithm uses the upper and lower bounds (UB and LB respectively). At the start of the program, they are given the values $LB = 1$, UP is the sum of $(n-1)$ largest arc costs. Then, using a special testing procedure, the estimates are systematically improved, and, using the results obtained, new arc costs are set in the form $c'(u,v) = \lfloor\frac{c(u,v)(n-1)}{\varepsilon LB}\rfloor$ $\forall(u,v) \in E$, which allows us to obtain the required path. Orda [14] and Lorenz et al. [12] modified ε-approximation algorithms to scale better in hierarchical networks.

A special place among the approximation algorithms is occupied by backward-forward heuristic (BFH). First, for each vertex $u \in V$, two $u - t$ paths are searched: min-cost path and min-length path. This can be done, for example, with the Reverse-Dijkstra algorithm [2]. Then, starting from the vertex s, a modification of Dijkstra's algorithm is applied, in which an additional condition is used to relax the arc, using the previously found paths (arc relaxation has the same meaning as in the usual Dijkstra algorithm). Examples of algorithms using similar approaches have been given by Reeves and Salama [16], Sun and Langendorfer [17]. A similar algorithm for a multi-constrained problem was proposed by Ishida [7].

For large road networks, Wang et al. [19] developed the constrained labeling algorithm COLA. It is based on two special properties that are characteristic for large road networks. First, road networks are usually (roughly) planar, which makes it possible to effectively divide the graph into several subgraphs with the special boundary vertices, between which it is required to find a path inside each subgraph. Secondly, often in the solutions of CSP problems on road networks there are a small number of landmarks [5] – the vertices that are present in valid paths much more often than others. According to experiments, an algorithm that takes into account these properties copes with mainland-sized road networks many times better than other algorithms.

In the generalization of the CSP problem – Multi-Constrained Path (MCP) problem – each arc has more than two parameters and it is required to find a path that satisfies each constraint with respect to the corresponding parameter. A summary and comparison of algorithms that solve MCP can be found in [11].

1.1   Our Contribution

Our approach to find an approximate solution to the CSP is based on the Lagrange Relaxation Aggregated Cost (LARAC) algorithm developed in [8] and summarized in [22]. In this approach, the Lagrange multiplier $\alpha > 0$ is introduced, and instead of the cost a_{ij} and length b_{ij} of the arc (i, j), one aggregated weight $c_{ij} = a_{ij} + \alpha b_{ij}$ is used. For a fixed value of α, a path $P(\alpha)$ of minimal weight $c(\alpha)$ is constructed of the cost $a(\alpha)$ and length $b(\alpha)$. If the length of the path exceeds the allowable value, then the value of α increases. Otherwise, it decreases.

To reduce the complexity, we build the special hierarchical structures (HS), in which the copies of the same vertex can be located at the several neighboring levels, and the arcs connect the vertices of the neighboring levels. However, in the HS, the sink t is incident to the arcs from all adjacent vertices, regardless of the level of their location. Further, using the heuristic considerations, the additional arcs are added to the HS instead of some paths in the original metric graph.

We have shown that the shortest path in the HS is constructed with $O(m)$-time complexity, where m is the number of arcs/edges in the original graph. If the graph is sparse, then this is a big gain compared to the Dijksra's algorithm and its modifications in the original graph. Obviously, not all arcs of the original graph are included in the HS, so the found path may differ from the shortest one. To compare the running time and the accuracy of our approach, a numerical experiment was carried out. The simulation shows that the construction of the shortest path in HS is several times faster than in the original graph. At the same time, in the HS, the solutions close to the optimal ones are constructed.

The rest of the paper is organized as follows. In the next section, we present a formulation of the CSP. In the third section, we present the procedures for constructing the hierarchical structures. Section 4 is devoted to the description of algorithm A_α, which ideologically coincides with the LARAC [8, 22] and builds an approximate solution to the problem. This section also provides estimates for the running time and accuracy of the A_α. Section 5 describes the numerical experiment, as well as the results of the simulation. The last section concludes the paper.

2   Problem Formulation

Let a mixed graph $G = (V, A)$, $|V| = n$, $|A| = m$, be given, whose arcs/edges we will call the *arcs* for definiteness. To each arc $(i, j) \in A$ two non-negative numbers: *cost* a_{ij} and *length* b_{ij} are assigned. We assume that the graph does not contain a pair of vertices $i, j \in V$ linked by a simple path P_{ij} from i to j in which all internal vertices (that is, not coinciding with i and j) have degree equal to 2. If such path exists, then instead of it we add one arc (i, j), the cost of which is equal to the sum of the costs $a_{ij} = \sum\limits_{(p,q) \in P_{ij}} a_{pq}$, and the length is equal to the sum of the lengths $b_{ij} = \sum\limits_{(p,q) \in P_{ij}} b_{pq}$ of the arcs included in it. It is required to find a path from the vertex $s \in V$ to the vertex $t \in V$ (s-t path) of the minimal cost and the length no more than $\beta > 0$. If Π is a set of simple s-t paths, then it is required to find a path $P \in \Pi$, which is the solution to the following problem.

$$\sum_{(i,j) \in P} a_{ij} \to \min_{P \in \Pi}; \tag{1}$$

$$\sum_{(i,j) \in P} b_{ij} \leq \beta. \tag{2}$$

The problem (1) is polynomially solvable, but the problem (1)–(2) is NP-hard even if the lengths of the arcs are equal [4].

3   Hierarchical Structures

First, let us consider an acyclic digraph (Fig. 1a) with one non-negative weight assigned to each arc. If the vertices s and t are known, then in this case the HS is constructed without loss of arcs as follows. We place the vertex s to the level 0. Then any vertex i falls into the level $l \geq 1$ if there is a path from s to i consisting

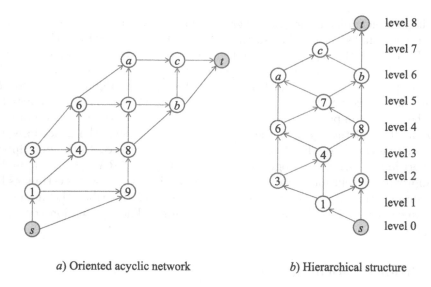

a) Oriented acyclic network   *b*) Hierarchical structure

Fig. 1. HS for acyclic digraph

of l arcs, but there is no $s - i$ path consisting of $l + 1$ or more arcs (Fig. 1*b*). As a result, the destination vertex t gets to some last level L. If there was an arc (i, j) in the original graph G, then the same arc exists in the HS. In this case, it is obvious that the vertex j is on a level with greater number compared to the level number of the location of the vertex i. Moreover, in the process of building a HS, we can simultaneously build the shortest paths to each vertex (see the red arcs in Fig. 1). To do this, we consider in turn the vertices of the levels $1, \ldots, L$. Among the arcs entering the vertex i, which is at the level l, we choose one that belongs to the shortest path going from s to i. This is easy to implement by storing the length of the shortest path to every vertex adjacent to i that is on a level less than l. As a result, the shortest s-t path will be constructed with $O(m)$-time complexity.

If graph G is arbitrary (not acyclic directed), then the placement of the vertices at the levels of the HS is ambiguous. In this case, we may build a k-HS, where k is a positive integer constant. In the k-HS, k copies of each vertex i are located at the levels $l, l+1, \ldots, l+k-1$, where l is the minimal number of edges in the path from s to i (see example of k-HS in Fig. 2, where $k = 2$). In the k-HS, the arcs link only the vertices of neighboring levels, except for the vertex t, which is connected with all adjacent vertices, regardless of their placement level (Fig. 2*b*).

In the example in Fig. 2*b* the images of the vertex 4 are located at the levels 2 and 3. The arc $(1, 4)$ enters the vertex 4 at the level 2, and the arc $(3, 4)$ enters the vertex 4 at the level 3. Some vertices in the HS may turn out to be dead ends – no arcs go out of them. In Fig. 2*b* such vertices are 5 at the level 3 and 7 at the level 5.

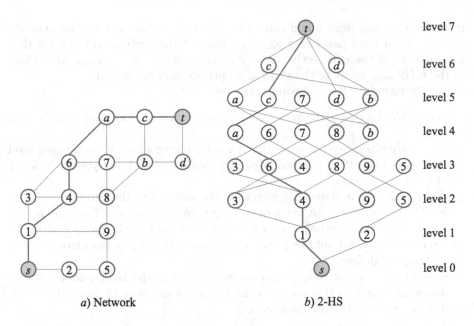

a) Network   *b*) 2-HS

Fig. 2. 2-HS for the mixed graph

If it is required to build the paths of minimal weight to all vertices in the HS, then this can be done during the construction of a k-HS, similarly to the method described above. For this, the vertices of the levels $1, \ldots, L$ are considered in turn. For any vertex i of the l-th level, an incoming arc (p, i) is chosen such that $c_{pi} + d_p = \min_q(c_{qi} + d_q)$, where the vertex q is at the level $l - 1$, and d_q is the minimal weight of the $s - q$ path (it was found earlier). In the example in Fig. 2b the arcs included in the shortest paths are red. Since $k = const$, the time complexity of finding the min-cost paths is still equal to $O(m)$.

In the 1-HS, each vertex i goes to the certain level which number equals to the minimal number of arcs in the $s - i$ path. As a result, all s-t paths in the 1-HS consist of the minimal number of arcs. In the k-HS, $k > 1$, the number of arcs in the s-t paths, as well as the number of paths, is greater, which makes it possible to find a path better than in the 1-HS. In Fig. 2 the shortest path is indicated by bold red edges.

3.1   Algorithm k-HSp_{max}

Since not all promising paths fall into the k-HS, then when the nodes are the points in the plane (metric graph), we add some arcs to the k-HS that connect vertices of the non-adjacent levels based on the following heuristic. For each vertex $v \in V$, we choose a *perspective* arc $a(v)$ outgoing from v in the direction of the sink t, defined by equation $a(v) = \arg\max_{(ij)\in A} |\vec{ij}| \cos(\angle(\vec{ij}, \vec{it})/c_{ij}$ if it is greater than 0. For each vertex $v \in V$ and any integer $p \in [1, n-1]$, a path (if it exists) outgoing from v and consisting of p perspective arcs can be uniquely defined. For a fixed parameter p_{max}, we connect the vertices in the k-HS that are the ends of the perspective paths consisting of $p = 2, \ldots, p_{max}$ arcs. Thus, in the k-HSp_{max} no more than np_{max} additional arcs are added.

Let us introduce the following notations:

- $vertexMinLayers$ – an array that for each vertex stores its minimal level in HS;
- $vertexMaxLayers$ – an array that for each vertex stores its maximum level in HS (so, in k-HS, $0 \le vertexMaxLayers[i] - vertexMinLayers[i] \le k - 1$ $\forall i \in \{0, \ldots, n-1\}$);
- $\vec{ij}$ – a vector on a plane that goes from the vertex i to the vertex j.
- $bestOutgoingArc[i]$ - an index of a perspective arc outgoing from the node i (or -1, if it does not exist). For each vertex $v \in V$ and any integer $p \in [1, n-1]$, a path (if it exists) outgoing from v and consisting of p perspective arcs can be uniquely defined.
- $vertexInfo$ – a matching between a pair of vertex and level, and a pair of a min-weight path to this vertex in the HS and a sequence of arcs in this path from a vertex of the previous level.

The pseudocode of the algorithm k-HSp_{max} presented in Algorithm 1. First, for each vertex $i \in \{0, \ldots, n-1\}$ its perspective arc $bestOutgoingArc[i]$ is calculated. Then the $vertexMaxLayers[i]$ and $vertexMinLayers[i]$ are set to -1 for all $i \in V \setminus \{s\}$, and they set to 0 for $i = s$. The queue Q is initially formed by the only pair $(s, 0)$ and $vertexInfo[(s, 0)] = (0, \emptyset)$. The main phase of the algorithm consists of the breadth-first search using the queue Q. For each considered pair of vertex and level (i, l) the weight of a best found subpath w_i^l and its part from the vertex of a previous level P_i^l are known, and on each iteration the direct descendants of a vertex and the ends of the chains constructed by perspective arcs (we call this neighborhood $N(i)$) are considered. For each vertex $j \in N(i)$ a path from i to j with the weight w_{ij} is known, as well as $vertexMinLayers[j]$ and $vertexMaxLayers[j]$. A Q or $vertexInfo$ may be changed only in one of the three following cases:

1. $vertexMinLayers[j] = vertexMaxLayers[j] = -1$. In this case we set $vertexMinLayers[j] = vertexMaxLayers[j] = l + 1$, $vertexInfo[j, l + 1] = (w_i^l + w_{ij}, P_{ij})$ and push the pair $(j, l + 1)$ to the queue Q.
2. $(vertexMinLayers[j] \le l + 1) \wedge (vertexMaxLayers[j] \ge l + 1)$. In this case the value of w_j^{l+1} is known. If $w_j^{l+1} > w_i^l + w_{ij}$, then we set vertexInfo$[j, l + 1] = (w_i^l + w_{ij}, P_{ij})$.
3. $(vertexMaxLayers[j] - vertexMinLayers[j] < k - 1) \wedge (vertexMaxLayers[j] = l) \wedge (maxLayer = -1 \vee l < maxLayer)$. In this case, as well as in the case 1, we set $vertexMaxLayers[j] = l + 1$, $vertexInfo[j, l + 1] = (w_i^l + w_{ij}, P_{ij})$ and push the pair $(j, l + 1)$ to the queue Q.

Once all predecessors of the target vertex are considered, the number of maximal level is known. After that all vertices on the level that exceeds the maximal level number are not added to the queue. During the procedure, the s.c. best child of a target is found. Starting from this vertex and the corresponding level, the sought for path is recovered using a backtracking procedure.

Algorithm 1. Heuristic algorithm k-HSp_{max} of shortest path construction using the HS and perspective arcs merging

---

1: *Input*: $G = (V, A)$ is an original graph, p_{max} – a maximum number of perspective arcs in a path, k – a maximal number of levels that may contain copies of one vertex; Q – an empty queue of pairs of two integers

2: *Output*: P – an s-t path in G;

3: **for** $i = 1, \ldots, n$ **do**

4:     Find *bestOutgoingArc*[i];

5:     *vertexMaxLayers*[i] = *vertexMinLayers*[i] = -1;

6: **end for**

7: *vertexMaxLayers*[s] = *vertexMinLayers*[s] = 0;

8: Push $(s, 0)$ to Q, set *vertexInfo*[$(s, 0)$] = $(0, \emptyset)$;

9: Set *bestTargetChild* = $(-1, -1)$, *maxLayer* = -1;

10: **while** Q is not empty **do**

11:     $(i, l) = Q.Pop()$;

12:     **for all** $j \in N(i)$ **do**

13:         Calculate w_{ij} and P_{ij};

14:         **if** $j = t$ **then**

15:             Update *bestTagretChild* and *maxLayer*;

16:         **end if**

17:         **if** *vertexMinLayers*[j] = *vertexMaxLayers*[j] = -1 **then**

18:             Set *vertexMinLayers*[j] = *vertexMaxLayers*[j] = $l + 1$;

19:             Set *vertexInfo*[$(j, l + 1)$] = $(w_i^l + w_{ij}, P_{ij})$;

20:             $Q.Push(j, l + 1)$;

21:         **else if** (*vertexMinLayers*[j] $\leq l + 1$) $\wedge$ (*vertexMaxLayers*[j] $\geq l + 1$) $\wedge$ $(w_j^{l+1} > w_i^l + w_{ij})$ **then**

22:             $w_j^{l+1} = w_i^l + w_{ij}$;

23:             *vertexInfo*[$(j, l + 1)$] = $(w_i^l + w_{ij}, P_{ij})$;

24:         **else if** (*vertexMaxLayers*[j] - *vertexMinLayers*[j] $< k - 1$) $\wedge$ (*vertexMaxLayers*[j] = l) $\wedge$ (*maxLayer* = -1 $\vee$ $l <$ *maxLayer*) **then**

25:             Set *vertexMaxLayers*[j] = $l + 1$;

26:             *vertexInfo*[$(j, l + 1)$] = $(w_i^l + w_{ij}, P_{ij})$;

27:             $Q.Push(j, l + 1)$;

28:         **end if**

29:     **end for**

30: **end while**

31: Starting from the *bestTargetChild*, using the data stored in the *vertexInfo*, recover the s-t path

---

4   Algorithm A_α

The algorithm presented below essentially coincides with LARAC [8,22], but we describe it in the following interpretation convenient for us. Instead of two characteristics of each arc $(i,j) \in A$: cost a_{ij} and length b_{ij}, we introduce one aggregated characteristic equal to $c_{ij}(\alpha) = a_{ij} + \alpha b_{ij}$, $\alpha \geq 0$, which we call the *weight* of the arc. Let us denote by $P(\alpha)$ a min-weight s-t path when the weights of the arcs are equal to $c_{ij}(\alpha)$, $(i,j) \in A$. Its cost is $a(\alpha)$ and its length is $b(\alpha)$. If the path $P(0)$ is feasible, i.e. the inequality (2) $b(0) \leq \beta$ is satisfied, then this is the optimal path. Otherwise, the value of α should be increased until we find the minimal $\alpha = \alpha^*$ for which the s-t path $P(\alpha^*)$ is feasible. To find α^* one can apply a dichotomy algorithm (the pseudo code is presented in Algorithm 2). The authors in [8,22] use an alternative way to change the α values.

Algorithm 2. Algorithm A_α

1: *Input*: $G = (V, A)$ is an original graph, $\alpha_{max}, \varepsilon > 0$
2: *Output*: s-t path P in G;
3:   $\alpha_l = 0; \alpha_r = \alpha_{max}$;
4: **if** $P(\alpha_r)$ is not feasible **then**
5:     Solution does not exist. **stop algorithm**
6: **end if**
7: **if** $P(\alpha_l)$ is feasible **then**
8:     **return** $P(\alpha_l)$
9: **end if**
10: **while** $\alpha_r - \alpha_l > \varepsilon$ **do**
11:   $\alpha_m = (\alpha_l + \alpha_r)/2$
12:   **if** $P(\alpha_m)$ is feasible **then**
13:     $\alpha_r = \alpha_m$;
14:   **else**
15:     $\alpha_l = \alpha_m$;
16:   **end if**
17: **end while**
18: **return** $P(\alpha_l)$;

Of course, $P(\alpha^*)$ is not always an optimal solution to the problem (1)–(2) (see Fig. 3a). However, in the case of the integer costs and lengths of the arcs, we can estimate the accuracy of the resulting solution, as well as the number of steps to find the α^*. Indeed, an arbitrary s-t path on a plane with a horizontal coordinate axis α and a vertical axis y is characterized by the straight line $y = b\alpha + a$. The length b of the path determines the slope of this line, and the cost a of the path determines the point of intersection of the line with vertical axis. The entire set of the s-t paths forms a minorant of straight lines whose slope decreases with increasing α. It is required to find the minimal $\alpha = \alpha^*$ for which the minorant is determined by the straight line $y = b\alpha^* + a$ with $b \leq \beta$. The path $P(\alpha^*)$ corresponding to this line is an approximate solution to the problem (1)–(2).

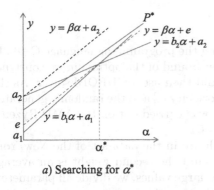
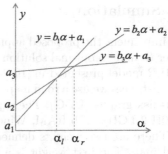

a) Searching for α^* b) Lower estimate per step in α

Fig. 3. Representation of the s-t paths by the straight lines on the plane $(\alpha, 0, y)$

Let us assume that α^* is determined by the intersection of the lines $y = b_1\alpha + a_1$ and $y = b_2\alpha + a_2$, $b_2 < \beta$, $b_1 > \beta$. From the equality $b_1\alpha^* + a_1 = b_2\alpha^* + a_2$, we get that $\alpha^* = \frac{a_2 - a_1}{b_1 - b_2}$. If $a_{ij} \le a_{max}$ and $b_{ij} \le b_{max}$, then $\alpha^* \le a_{max}n$.

Assume that the line corresponding to the optimal path passes above the minorant at the point α^* (green line in Fig. 3a). Then the parameter e of the line $y = \beta\alpha + e$ passing through the point of intersection of the lines $y = b_1\alpha + a_1$ and $y = b_2\alpha + a_2$ (dashed green line in Fig. 3a), is the lower bound for the optimum. We have $\beta\alpha^* + e = b_2\alpha^* + a_2$, whence $e = a_2 - (\beta - b_2)\alpha^*$. Therefore, taking into account the integer parameters, the ratio is

$$\varepsilon \le \frac{a_2}{e} = \frac{a_2}{a_2 - (\beta - b_2)\frac{a_2 - a_1}{b_1 - b_2}} \le \frac{1}{1 - \frac{\beta - a_2}{b_1 - b_2}} \le \frac{1}{1 - \frac{\beta - 1}{\beta}} \le \beta.$$

As mentioned above, the dichotomy method can be used to find α^*. Let us estimate the number of iterations of the method for integer parameters. An upper bound for the value of α^* was obtained above. Let us find a lower bound for the difference of neighboring α. To do this, we take three lines $y = b_1\alpha + a_1$, $y = b_2\alpha + a_2$, $y = b_3\alpha + a_3$, $b_1 > b_2 > b_3$, $a_1 < a_2 < a_3$, which form two neighboring break points α_l and α_r, $\alpha_l < \alpha_r$, of the minorant. We have $b_1\alpha_l + a_1 = b_2\alpha_l + a_2$ and $b_3\alpha_r + a_3 = b_2\alpha_r + a_2$. Consequently,

$$\alpha_r - \alpha_l = \frac{a_3 - a_2}{b_2 - b_3} - \frac{a_2 - a_1}{b_1 - b_2} \ge \frac{1}{(b_2 - b_3)(b_1 - b_2)} > \frac{1}{b_1 b_2} \ge \frac{1}{b_{max}^2 n^2}.$$

If K is the maximum number of iterations in the dichotomy method, then $a_{max}n/2^K \le 1/b_{max}^2 n^2$. Hence $K = O(\log n)$.

If, for each value of α, Dijkstra's algorithm is used to find the min-weight path, then the complexity of constructing an approximate solution to the problem (1)-(2) is $O(n^2 \log n)$. If we look for the min-weight path in the HS, then the complexity of obtaining an approximate solution is $O(m \log n)$.

The resulting guaranteed accuracy is rough. Therefore, in the next section we describe a numerical experiment in which the running time and the accuracy of the solution are compared.

5   Simulation

We implement the proposed approach using the programming language C++. In order to find an optimal solution or a lower bound of the optimum, we construct an ILP model presented in [10] for CSP and then use a GUROBI solver. As the test instances we use a road map of New York city [1] and the randomly generated unit-disk graphs (UDG). The experiment was carried out on the AMD Ryzen 5 3550H 2.1 GHz 8 Gb RAM, Windows 10 × 64.

There are two weights defined for each arc in the data set of the New York road map. The first weight is a distance, and the second weight is an average traveling time between the nodes. To avoid large values, we divide all parameters by 100 and left only the integer parts of the resulting numbers.

UDGs are constructed in the following way. Firstly, a set of nodes is randomly uniformly spread over a square region. Then, each two nodes are connected by two oppositely directed arcs iff the distance between them is not more than a predefined value r. After that, two weights are assigned to each arc. The first weight equals to the distance, and the second weight equals to the distance multiplied by noise factor – a random real uniquely generated for each arc and uniformly distributed in the segment [1, 3].

Practically, actual running time spent to find one-weight shortest path (SP) depends on the proximity between the source s and target t. That is why we consider separately instances when the distance between s and t is small (25 % of the graph diameter), medium (50%), and large (75%). For each graph instance and each variant of the distance between source and target, we generate 10 random problem instances.

We test different variants of HS based heuristic in order to find a best combination of its parameters. To be precise, for each $k = 1, 2, 3$ and $p_{max} = 1, 2, 3$ we run k-HSp_{max} on each test instance and compare their performance with Dijkstra's algorithm (it is called Dij below). Also, we use each heuristic that solves SP problem in the LARAC based algorithm that approximately solves CSP. Note that for the LARAC based approach we use the rules of updating α from [8]. To denote these algorithms the prefix A_ is used.

In Fig. 4 the results on the New York map are presented. Here and in the next figures, the average values among launching the algorithms on 10 random instances are presented, and the vertical intervals stand for the standard deviations. On the one hand, as it is seen in Fig. 4a, the HS based algorithms bring significant performance error, but, on the other hand, it noticeably decreases with growth of k and p_{max}, and, according to Fig. 4b, these heuristics spend less time than Dijkstra's algorithm. Of course, the running time also increases with growth of k and p_{max}, so the moderate values of these parameters may be chosen to achieve less quality degradation with significant speedup.

The results of application of these algorithms to the LARAC based approach for CSP are presented in Fig. 5. The average path lengths are presented in Fig. 5a, the average ratio values that was obtained on the cases when GUROBI found optimal solution are presented in Fig. 5b, and Fig. 5c shows the average running time. Note that GUROBI failed to find solution to the large-size instances,

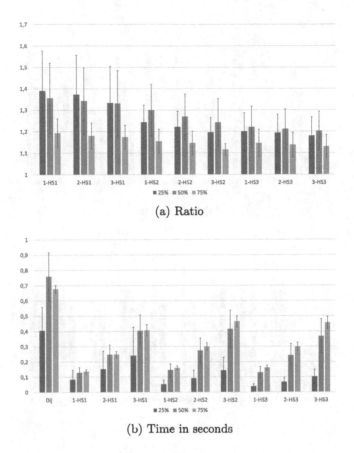

(a) Ratio

(b) Time in seconds

Fig. 4. Algorithms results for the SP problem on the New York map. Average values and standard deviations.

when the distance between s and t is 75% of the metric graph diameter. Here, again, one can observe that using Dijkstra's algorithm allows to get more precise solution on average but HS based heuristics allow to find approximate solution 10–100 times faster.

We also test all algorithms on the UDGs that were constructed as described above. The points for the graph generation are randomly spread on the unit-side square. Graph density depends on two parameters: n – the number of vertices, and r – the *disk* radius that defines connectivity between each pair of vertices. There are three UDG variants tested: (1) $n = 10000$ and $r = 0.1$, (2) $n = 10000$ and $r = 0.2$, and (3) $n = 100000$ and $r = 0.025$. All tested heuristics construct a *near optimal solution*: the ratio is at most 1.002. Therefore, it is worth comparing only a running time. Figure 6 presents the average running time for solving the SP problem, and in Fig. 7 the average running time for solving the CSP problem are shown. It can be noticed that for the both problems in the UDG the usage of HS based heuristics instead of Dijkstra's algorithm is justified since they construct almost optimal solution an order of magnitude faster.

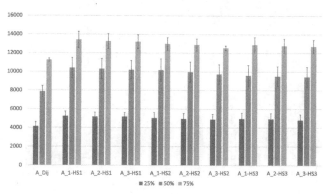

(a) Path length

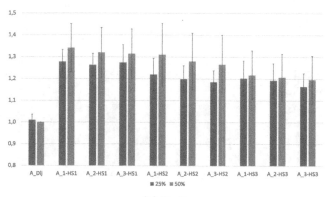

(b) Ratio

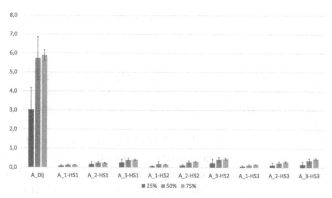

(c) Time in seconds

Fig. 5. Algorithms results for the CSP problem on the New York map. Average values and standard deviations.

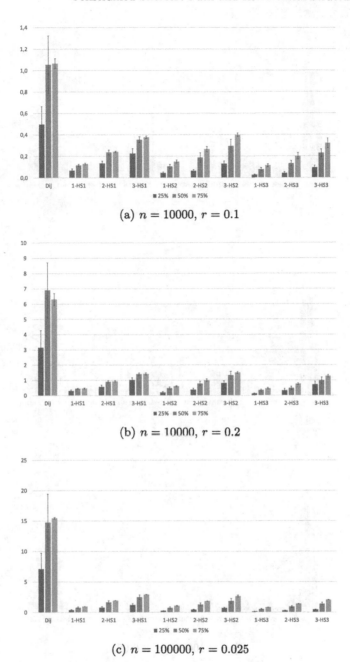

(a) $n = 10000$, $r = 0.1$

(b) $n = 10000$, $r = 0.2$

(c) $n = 100000$, $r = 0.025$

Fig. 6. Time in seconds of solving SP problem on the UDG. Average values and standard deviations.

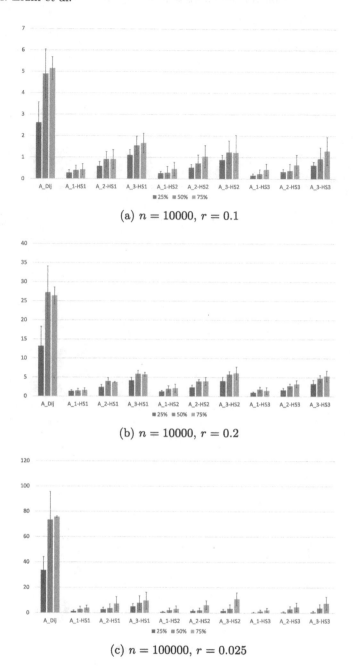

(a) $n = 10000$, $r = 0.1$

(b) $n = 10000$, $r = 0.2$

(c) $n = 100000$, $r = 0.025$

Fig. 7. Time in seconds of solving CSP problem on the UDG. Average values and standard deviations.

6   Conclusion

This paper considers NP-hard Constrained Shortest Path (CSP) problem, when to each arc of the given n-nodes graph two characteristics are assigned: *cost* and *length*, and it is required to find a min-cost bounded-length path between the given pair of nodes. The constraint on a path length is accounted for by considering one aggregated edge *weight* equals to the linear combination of the cost and length using a Lagrange multiplier as in [8]. By varying the multiplier value, a feasible solution delivers a minimum to the function with new weights. Then we are looking for the solution, not in the original graph but in the specially constructed hierarchical structures (HS). We show that the shortest path in the HSs is constructed with $O(m)$-time complexity, where m is the number of edges/arcs in the graph, and the approximate solution in the case of integer costs and lengths of the edges is found with $O(m \log n)$-time complexity. An a priori estimate of the accuracy turned out to depend on the parameters of the problem and can be significant. Therefore, to evaluate the algorithm's effectiveness, we conducted a numerical experiment on the graph of roads of the megalopolis and randomly constructed unit-disk graphs (UDGs). The simulation shows that in the HSs, a solution is built 10–100 times faster than by the methods using Dijkstra like algorithms in the *original graph*.

References

1. 9th DIMACS Implementation Challenge. http://www.dis.uniroma1.it/challenge9/download.shtml
2. Ahuja, R.K., et al.: Network Flows: Theory, Algorithms, and Applications. Prentice Hall Inc, London (1993)
3. Cormen, T.H., et al.: Introduction to Algorithms. The MIT Press, Cambridge (2000)
4. Garey, M.S., Johnson, D.S.: Computers and Intractability: Guide to the Theory of NP-Completeness. (Ed by, W.H. Freeman), New York (1979)
5. Goldberg, A.V., Chris, H.: Computing the shortest path: a search meets graph theory. In: SODA 2005 (2005)
6. Hassin, R.: Approximation schemes for the restricted shortest path problem. Math. Oper. Res. **17**(1), 36–42 (1992)
7. Ishida, K., et al.: A delay-constrained least-cost path routing protocol and the synthesis method. In: Proceedings of the 5th International Conference on Real-Time Computing Systems and Applications, pp. 58–65. IEEE (1998)
8. Jüttner, A., et al.: Lagrange relaxation based method for the QoS routing problem. IEEE INFOCOM **2001**, 859–868 (2001)
9. Koster, A.M.C., Muñoz, X. (eds.): Graphs and Algorithms in Communication Networks. Springer, Heidelberg (2014). https://doi.org/10.1007/978-3-642-02250-0
10. Handler, G., Zang, I.: A dual algorithm for the constrained shortest path problem. Networks **10**, 293–310 (1980)
11. Kuipers, F.A., et al.: An overview of constraint-based path selection algorithms for QoS routing. IEEE Commun. Mag. **40**(12), 50–55 (2002)
12. Lorenz, D.H., et al.: Efficient QoS partition and routing of unicast and multicast. In: Proceedings of IWQoS 2000, pp. 75–83 (2000)

13. Lozano, L., Medaglia, A.L.: On an exact method for the constrained shortest path problem. Comput. Oper. Res. **40**, 378–384 (2013)
14. Orda, A.: Routing with end-to-end QoS guarantees in broadband networks. IEEE/ACM Trans. Netw. **7**(3), 365–374 (1999)
15. Pugliese, L.D.P., et al.: The resource constrained shortest path problem with uncertain data: a robust formulation and optimal solution approach. Comput. Oper. Res. **107**, 140–155 (2019)
16. Reeves, D.S., Salama, H.F.: A distributed algorithm for delay-constrained unicast routing. IEEE/ACM Trans. Netw. **8**(2), 239–250 (2000)
17. Sun, Q., Langendorfer, H.: A new distributed routing algorithm for supporting delay-sensitive applications. Comput. Commun. **21**, 572–578 (1998)
18. Wang, H., et al.: A bio-inspired method for the constrained shortest path problem. Sci. World J. **2014**, 271280 (2014)
19. Wang, S., et al.: Effective indexing for approximate constrained shortest path queries on large road networks. Proc. VLDB Endow. **10**(2), 61–72 (2016)
20. Wang, Z., Crowcroft, J.: Quality-of-service routing for supporting multimedia applications. IEEE Sel. Areas Commun. **14**(7), 1228–1234 (1996)
21. Widyono R.: The design and evaluation of routing algorithms for real-time channels. Technical report TR-94-024, University of California at Berkeley & International Computer Science Institute (1994)
22. Xiao, Y., et al.: The constrained shortest path problem: algorithmic approaches and an algebraic study with generalization. AKCE J. Graphs. Combin. **2**(2), 63–86 (2005)

Investigation of Graph Neural Networks for Instance Segmentation of Industrial Point Cloud Data

Sandeep Jalui[1]([⊠])(iD) and Evangelia Agapaki[2]([⊠])(iD)

[1] Electrical and Computer Engineering, Herbert Wertheim College of Engineering, University of Florida, Gainesville, FL 32611, USA
sjalui@ufl.edu
[2] Construction Management, M.E. Rinker, Sr. School of Construction Management, University of Florida, Gainesville, FL 32603, USA
agapakie@ufl.edu
https://www.evagapaki.com/

Abstract. The concept of Digital Twins (DTs) was introduced 15 years ago. There have been many research methodologies and software to implement DTs in different industries including manufacturing, construction, product design, and other fields. DTs help quantify operational risks, improve production time, assist predictive maintenance, enable real-time remote monitoring and thus better financial decision-making. The generation of DTs for existing industrial sites necessitates the use of laser scanners for the acquisition of point cloud data that capture the existing (as-is) conditions. Currently, human modelers manually segment point cloud data by overlaying 3D CAD models on top of the laser scans or validating laser scanned point clouds with 2D documentation and drawings. Our previous work has achieved effective point cloud processing with techniques such as instance segmentation and class segmentation of the collected and registered industrial point cloud data. Instance Segmentation is an important method of clearly partitioning each object to a human-understandable point cluster in complex laser scanned data, creating a Geometric Digital Twin of Industrial Facilities. The industrial point cloud data consists of pipes, valves, cylinders, and various other combinations of geometric shapes. Segmenting such data is a difficult task as the data is too complex to visualize and understand. In our previous work, CLOI-NET, which is the state-of-the-art architecture for instance segmentation of industrial point clouds, achieves instance segmentation with average accuracy of 70% using graph connectivity algorithms. This proved that there is scope for more accurate instance segmentation of complex industrial point cloud data with a focus on identifying topological connectivity between components of the point cloud (e.g., a piping network). Also, there have been many research methods, for instance segmentation in city-scale and indoor environments classifying cars, people, buildings, trees, roads, using different types of neural networks with satisfactory performance having average accuracy upto 85%. In this paper, we discuss the best algorithms/networks like Graphical Neural Networks

© Springer Nature Switzerland AG 2022
D. E. Simos et al. (Eds.): LION 2022, LNCS 13621, pp. 411–428, 2022.
https://doi.org/10.1007/978-3-031-24866-5_30

and 3D-CNNs and how they can be used to perform instance segmentation of industrial data, which will eventually lead to a better version of DT implementation specifically for industrial point cloud data.

Keywords: Digital Twin · Instance segmentation · Graphical Neural Network · Geometric deep learning · Point clouds · Computer vision

1   Introduction

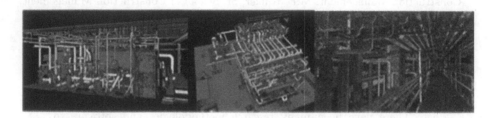

Fig. 1. Class segmentation of industrial objects (Agapaki, 2020) [3]

Maintaining a safe environment in industrial facilities is critical for improving work processes efficiently as well as economic aspects of their operation and management. It has been estimated that due to poor maintenance of equipment, the downtime costs are around $50 billion per year with 44% of equipment failing due to aging processes [1]. Also, the Chartered Institute of Buildings in the U.K. estimates that refurbishing and retrofitting 93% of the industrial equipment will be one of the major goals by 2050 [2]. These problems can be resolved by the generation and maintenance of up-to-date Digital Twins (DTs). However, the cost of generating DTs is so high that it outweighs the benefits. This is because manual modeling of labor cost and data collection using laser scanners is expensive.

Thus, there is a need to develop solutions which need less labor-intensive industrial modeling tools and will help increase the productivity of industrial assets. Such problems can be solved by generating DTs with the help of Artificial Intelligence and Neural Networks. In our previous work as shown in Fig. 1, we developed and implemented a CLOI-NET [3] framework, which aims to simplify the above process as discussed. The CLOI-NET algorithms were the first framework to generate geometric digital twins from terrestrial laser scanned industrial data. This framework follows three steps: (a) Semantic or class Segmentation (b) Instance Segmentation, and (c) Fitting to other geometric representations. Semantic segmentation helps us to identify different classes or objects in a point cloud while instance segmentation is an extension of semantic where it differentiates each instance of the semantic category.

The class segmentation problem is solved, and 70% of the total labor hours can be saved for this task [3]. While instance segmentation has been developed using graph connectivity algorithms in our previous work [4], the problem remains due to low accuracy as well as missing relationships between instances. The output of instance segmentation is important as it can provide sufficient geometric information to generate DTs by fitting 3D representations (e.g., meshes, parametric shapes) to the instance point clusters. In our previous research [4], we used a Breadth First Search algorithm [5] to implement instance segmentation. The implementation of this algorithm on industrial point cloud data revealed low accuracy due to high data complexity, however there is room for improvement if we apply more advanced models tailored to industrial environments. To give some examples, the highest recall value of segmented cylinders is 61.3% and lowest is 48.5% for 50% IoU on four industrial laser scanned, point cloud datasets. Similarly, the average precision for I-beams was 59.1% and average recall was 64.2%. Also, a petrochemical plant suffers most from over-segmentation due to sparse point density of the data. There are many research organizations which are described below who have worked on instance segmentation using different combinations of neural networks and they have successfully implemented those on real world datasets of indoor or outdoor scenes with sufficient accuracy scores.

The segmentation output in computer vision problems is usually evaluated using the mean Intersection over Union (mIoU) metric. The IoU is the ratio between the area of overlap and the area of union between the ground truth and the predicted instance labels. AP is the area under the Precision-Recall Curve(AUC-PR) evaluated at some % of IoU threshold. AP50 is evaluated at 50% IoU threshold and AP75 at 75% IoU threshold. mAP is the average of different AP values calculated for respective classes [6]. Models such as Soft-Group [7], HAIS [8], SSTNET [9], Dyco3D [12] have achieved some significant results with respect to instance segmentation of point clouds. On the ScanNet v2 dataset, the AP50 score of softgroup is 76.1%, followed by 69.9% and 69.8% of HAIS and SSTNET respectively. The SSTNet [9] model has achieved an average recall value of 73.4% and mean average precision (mAP) 54.1% on an indoor point cloud dataset of office spaces (S3DIS), while on ScanNet(v2) dataset, mAP@0.5 is 69.8% and mAP is 50.6%. Another state-of-the-art instance segmentation network (OccuSeg [10]) achieved mAP@0.5 of 63.4% and 47.1% on the ScanNET(v2) and the SceneNN dataset respectively. Also, networks such as Res-UNet-R/H [13] are successful in 3D instance segmentation of mitochondria cells from human and rat samples.

While these networks solve instance segmentation, there are some robust networks who have achieved panoptic segmentation. Panoptic segmentation unifies both class and instance segmentation methods. Table 1 below shows some of the state-of-the-art panoptic segmentation networks on semantic Kitti datasets with their respective mean Intersection over Union (mIoU) scores. As per our study, we found that GP-S3Net [19] and EfficienctLPS [20] gives the best state-of-the-art output on panoptic segmentation tasks in terms of mIoU score. Semantic Kitti [22,23] dataset is based on real world objects such as roads, buildings, cars, people, trees and many others. Apart from the techniques mentioned, we also found that Graphical Neural Networks [14,16] have huge scope of being applied to 3-D pointcloud datasets performing both semantic and instance segmentation. Thomas N. Kipf and Max Welling have explained the fundamentals of Graph Convolutional Networks (GCN) in their prior work [25]. In this paper, we discuss some of the existing techniques, which have been tested on indoor buildings and city datasets, how we can leverage these techniques and update our own state-of-the-art architecture CLOI-NET to give satisfactory performance on Industrial Point Clouds.

Table 1. Panoptic segmentation models [24]

Approach	mIoU (%)
GP-S3Net	70.8
EfficientLPS	61.4
DS-Net	61.6
Panoster	59.9
Panoptic4D (single scan)	61.3
MOPT	52.6
Panoptic RangeNet	50.9

2   Background on Point Cloud Instance Segmentation Methods

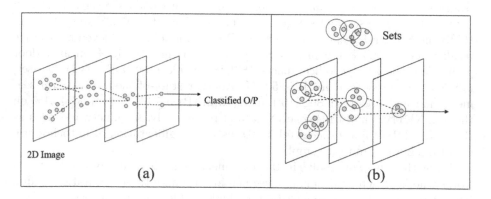

Fig. 2. Convolutional neural networks and PointNET-based methods

As we can see in Fig. 2 (a), Convolutional Neural Networks (CNNs) can be applied on the 2-D image that segment objects because they are powerful algorithm for image processing, and require a regular grid as input. However, point clouds are typically sparse and not evenly distributed on a regular grid. If point clouds were to be placed on a regular 2D grid, they will generate an uneven number of points, thus there is no benefit in applying convolutions as it will give less accurate output [27]. Another approach is using PointNETs which are shown in the Fig. 2 (b). PointNET [41] is one of the best architectures to process an unordered set of points for semantic feature extraction, because of its simplicity (MLP layers) and invariance in permutations of input points. This is the reason the development of CLOI-NET uses PointNET algorithms as the backbone network for semantic segmentation. PointNET uses a hierarchical neural network, named as PointNET++ [28], which processes a set of points sampled in a metric space in a hierarchical manner. Using a distance metric, the set of points are partitioned into overlapping local regions. Then, by applying MLP layers, local features are extracted capturing fine geometric features from small neighborhoods. Such features are grouped further until PointNET++ obtains the features of the whole point set. This technique is quite accurate, but does not allow the aggregation of local information which restricts the performance of the algorithm [16].

However, there has been a follow-up work on PointNET [41] such as DGCNN [26] which constructs graphs, thereby improving the accuracy performance but at the cost of memory consumption and runtime. Another approach is by using grouping based instance segmentation, which is demonstrated by SoftGroup algorithm. This method has a bottom-up pipeline that generates per-point predictions such as semantic maps, geometric shifts and then aggregates points into instances segmentation [7]. SoftGroup is a useful method to investigate compared to our previous efforts on CLOI-NET for pure improvements of the instance segmentation task. Out of all the methods discussed above, we researched that if point clouds are considered as graphical data by defining nodes and edges [29] and training a classification algorithm using Graphical Neural Networks will be a huge advancement with expected accurate predictions and will be the first benchmark for instance segmentation of industrial point clouds.

2.1   Geometric Deep Learning Methods

One of the advantages of geometric deep learning methods is that they directly process 3D points. SSTNet [9] learns point-wise semantic and instance-level features separately and then it efficiently aggregates these features as superpoints via point-wise pooling. The important part in this algorithm is semantic superpoint tree (SST), where superpoints are its tree leaves. SSTNET achieved 67.8% AP@50, being the state-of-the-art instance segmentation network on real-world indoor laser scanned data. The limitation of this method is that topological relationships between instances cannot be predicted using this method. OccuSeg [10] takes the 3D geometry model data as input, and produces point-wise predictions of instance level semantic information. While PointGroup [11] is a bottom-up 3D instance segmentation framework.

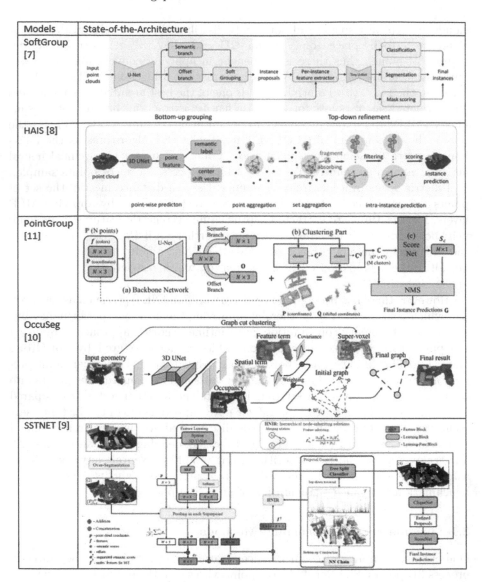

Fig. 3. SOTA literature review based on Geometric Deep Learning method

Based on the semantic predictions for the original coordinate space and shifted coordinate space, this network is trained to group points into different clusters and ScoreNet is used for prediction. The architectural model of HAIS [8] and SoftGroup [7] is built in a similar way as PointGroup. However, Softgroup [7] has achieved the best results by surpassing the strongest previous method by a significant margin of +6.2% on the ScanNet v2 hidden test set and +6.8% on S3DIS Area 5 of AP_50. The main limitation of such methods is that they are suitable for small datasets not for large datasets due to their computational complexity. RandLA-Net [17], KPConv [18] and PointNet++ [28] are point based methods. Figure 3 shows different architectures based on 3D Geometric Deep Learning Methods.

2.2 Projection Based Methods

Projection based methods transform input point clouds in 2D and use CNNs to segment objects in 2D images. These networks have been widely applied in autonomous driving. The projection can be either the topdown Bird-Eye-View or spherical Range-View as shown in Fig. 4 [35–39]. The computation speed of these models is very high but accuracy is low because of information loss during projections. One of the high-performing networks is EfficientLPS [20]. This model is built on the EfficientNet family of networks and thus the name EfficientLPS. EfficientNet [21] is a scaling method that uniformly scales depth, width, and resolution to increase the performance for convolutional neural networks. EfficientLPS architecture consists of projection, backbone, semantic head, instance head and post-processing modules. Using a scan unfolding technique [32], 3D point cloud data is projected into 2D data. It is then fed as an input into a backbone module consisting of a Proximity Convolution Module, an EfficientNet encoder, 2-way FPN and Range Encoder Network (REN) in parallel. The REN is implemented parallel to the encoder to encode range channel of the projection and fuse it with the 2-way FPN outputs to calculate range-aware multi-scale features and then outputs are fed as an input to the semantic and instance head.

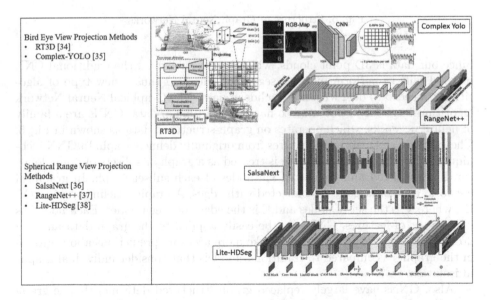

Fig. 4. SOTA literature review based on Projection Based Methods

The logits from the heads are combined in the panoptic fusion model and using the K-NN algorithm the output is projected back to 3D. The mIoU score for SemanticKITTI test set is 61.4% and for validation set is 64.9% which is quite low as compared to instance segmentation networks. A limitation of projection-based methods is that they cannot be easily applied on densely populated scenes due to information loss and many object occlusions.

2.3   Graph Neural Networks

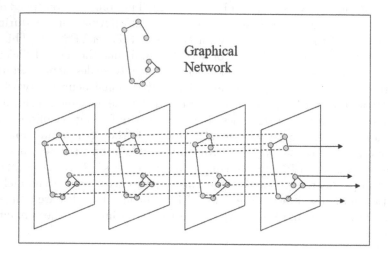

Fig. 5. Structure of Graph Neural Networks

Since our data is 3D point clouds, which is non-euclidean, the traditional CNN method fails to perform instance segmentation. And thus a new type of algorithm is needed to deal with non-euclidean data. The Graphical Neural Network algorithm has a lot of potential to handle such type of data. GNNs are a family of neural networks, which operates on graph-structured data as shown in Fig. 5. They extract and utilize the features from originally defined graph. In GNN techniques the entire point cloud data is treated as a graph or subsets of pointclouds can be considered and connected via nodes of each subset graph. In graphical network, each data point is connected with edges. A graph is defined as G = (V, E), where V is the set of nodes and E is the edges between nodes. Each node has its own set of features. GNNs can be easily applied to the graph data and can do segmentation tasks. GNNs can make more accurate predictions about groups in the graph, as compared to traditional models that consider individual graphs in isolation.

Also, GNNs have largely replaced graph-structured data models like graph kernels and random-walk methods because of their inherent flexibility to model the underlying systems better. GCN, GAT, GraphSAGE, GIN are some of the types of GNNs based on message-passing forms in GNN architectures [33]. However, there are many challenges in graphical neural network modelling. The graphs do not have consistent structure, therefore prediction becomes a challenging task. The node order equivalency is another challenge. Even though the graph structure might look the same but node labelling will be different after every iterations. As compared to images, where pixels have absolute fixed positions. If the nodes are permuted in some way, the outcome representation of the nodes should also be permuted in the same way after applying graph algorithm. The GNN models applied on the data should be scalable. In our point cloud dataset, we cannot consider every point as a node in the graph, otherwise it will be computationally expensive. Rather, we consider one point from the whole object as node and connect with other nodes through edges. GNNs can solve multiple problems like Node Classification which classifies individual nodes, Graph Classification which classifies entire graphs, Node Clustering which groups similar nodes, Link Prediction helps in finding missing links and Influence Maximization identifies individual nodes. Figure 6 shows the different type of models discussed above [33].

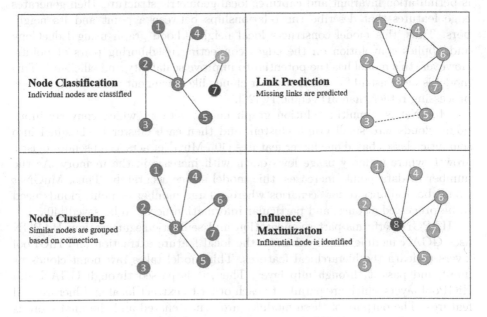

Fig. 6. Types of GNN solutions

Weijing and Rajkumar [14] have worked on developing the Point-GNN algorithm which performs the object detection on pointclouds data. Their approach is composed of three steps: a) Graph generation, b) GNN layers of T iterations, c) bounding box generation based on the classification and localization output of the GNN. The point clouds are represented as $P = \{p_1,...,p_N\}$, where $p_i = (x_i, y_i, z_i, s_i)$ is a point with both 3D coordinates and the state value $s_i \in R_k$ a k-length vector that represents the laser intensity or the features encoding the surrounding objects. S_i preserves the information within the original point cloud. Using the point cloud P, a graph is constructed G = (P, E) where P is considered as vertices and connects to its neighbour points within a fixed radius r and $E = \{(p_i, p_j) \mid ||x_i - x_j||_2 < r\}$ [15].

Here, voxels [30] are also used to reduce the density of point clouds that reduces the computational time. These graphs are then passed through multiple multi-layer perceptron (MLP) layers and then aggregated through a max function. In the third step, using non-maximum suppression (NMS), multiple bounding boxes of the same object are merged into one and confidence score is assigned. The highest Average Precision (AP) on the KITTI dataset is 88.33%, 51.92% and 78.6% on car, pedestrian and cyclist predictions. As shown in Fig. 7 which shows various types of approaches for segmentation algorithms using graphical neural networks.

DGCNN model is inspired by PointNet [41] and convolution operations which performs point cloud classification, point cloud part segmentation and semantic segmentation. It integrates EdgeConv into the PointNet architecture. EdgeConv is permutation invariant and captures local geometric structure, then generates edge features that describe the relationships between a point and its neighbors. Thus, this model constructs local neighborhood graph using EdgeConv and applies convolution on the edges connecting neighboring pairs of points. However, this model has the potential to improve scalability and efficiency. This model is more useful for abstract point clouds like document retrieval and image processing rather than 3D geometry [26].

MuGNet is a multi-resolution graph neural network which converts input point clouds into small graph clusters and then each cluster is classified into semantic classes based on cluster features [40]. MuGNet acts as a pseudo-logistic growth where memory usage lows down with increase in the memory. As the number of data points increases, this model scales accurately. Thus, MuGNet would be appropriate for scenarios where a large number of point clouds need to be processed at once, and particular information needs to be saved [40].

HAPGN performs part segmentation and semantic segmentation. HAPGN uses GGAN module which enhances the local feature extraction and HiGPool layers to learn the hierarchical features. This model takes raw point clouds as input and pass it through mlp layer. Then, it is passed through GGAN and HiGPool layers which are parallel to each other to extract local and hierarchical features. The outputs of these modules are concatenated and the final score is computed through an MLP layer [42]. However, this algorithm needs a lot of improvement when it comes to unbalanced large scale point cloud datasets. The accuracy scores for all these models are shown in Table 2.

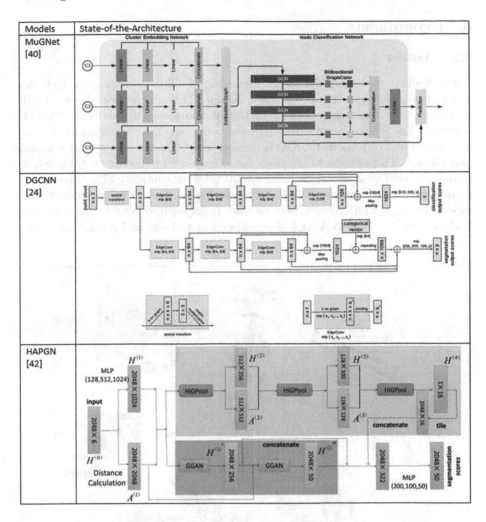

Fig. 7. SOTA literature review based on Graph Neural Network

Table 2. Performance of Graph Models on S3DIS dataset

Models	Overall accuracy (%)	mIoU (%)
MuGNet	88.5	69.8
DGCNN	85	59.2
HAPGN	85.8	62.9

3   Experiments

3.1   Dataset

Figure 8 shows the CLOI [3] dataset consisting of different shapes like Cylinder, Valve, Flange, Angle, Ibeam, Channel, Elbow. The hrs mentioned refer to the modelling time taken when performing the modelling task manually in Revit [31] software. The % mentioned is the frequency of appearance in the entire dataset. The maximum objects in the CLOI dataset are Cylinder followed by channels and elbows. The objects which does not comes under the above category, we have annotated that as other which constitutes approximately 27%. The dataset is unbalanced due to nature of the target regions and needs to considered while modelling algorithm. Also, CLOI dataset is segregated into 6 different Area folders, where Area_1 to Area_5 is considered as training and Area_6 as testing.

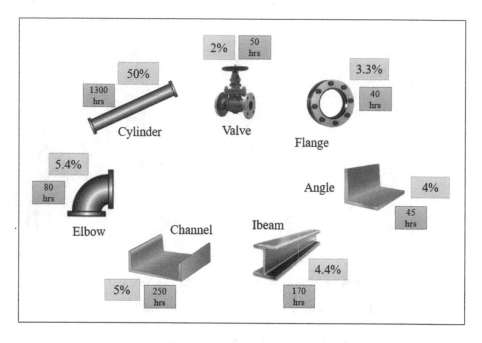

Fig. 8. CLOI dataset

3.2   Implementation

The above CLOI dataset was used for training with SoftGroup algorithm to check the behavior of geometric deep learning methods. As SoftGroup has achieved the highest benchmark on S3DIS dataset for Instance Segmentation, this algorithm was first and becomes the basis for implementing neural network based algorithm as compared to graph connectivity algorithm. This experiment was tested with different combination of learning rate, epochs, GPU configuration and different parameters of the dataset. This experiment is run in 4 stages:- pre-training, training, inference and visualization. We found the respective mIoU scores for different classes with 30 epochs and learning rate 0.004 and is mentioned in the Table 3. The accuracy score was around 62.3% with mean IoU 27.2%. The running time for each training codes takes approximately 6 hrs for 30 epochs with 1 GPU node and 200 GB memory.

Table 3. SoftGroup implementation on CLOI dataset

Objects	Cylinder	Valve	Channel	Ibeam	Elbow	Angle	Flange	Other
mIoU(%)	50.7	49.1	5.0	43.4	13.5	0.7	0.0	55.0

4 Discussion

Based on the investigation of all algorithms discussed in this paper. We implemented the SoftGroup algorithm on CLOI dataset and there are some challenges which need to be addressed. There is room for lot of improvements for this experiment. The computational time and resources needs to be reduced. Also, the overall accuracy and mIoU scores needs to be increased for better instance segmentation. Also, the new algorithm should be able to train accurately on unbalanced datasets. Our review of the existing literature indicates that graph neural network architectures are appropriate for the inference of the links between segmented instances (edges of the graph). Our proposed architecture is that we can combine the SoftGroup with Graphical Neural Network to overcome the challenges. Our hypothesis is that with given the instance point clusters as input, we predict the edges of the graph. The input of this process will be a graph with a set of nodes $\{i_{s1}, i_{s2}, ..., i_{sn}\}$ and edges $\{(i_{s1}, i_{s2}), (i_{s2}, i_{s3}), ...\}$, where edges are defined for each pair of nodes that are relatively in close proximity to each other. This is considered as the "ground truth graph" for the inference of topological relationships in a piping network.

A piping network consists of pipes, valves, elbows and flanges. After the input graph is defined, the next step is to identify feature vectors, $f = \{f_1, f_2, ..., f_n\}$, where N is number of features of each instance node. Each feature vector will be obtained by using a geometric deep learning algorithm such as SoftGroup for classifying the node and extracting the feature vector of its last network layer. In addition to the extracted feature vector, another set of object-level geometric features needs to be obtained. These features are a set of orientation vectors, radius (for the case of pipes), center, starting and end point coordinates of each instance node. Then, a graph attention network will be deployed similar to the one proposed in CurvaNet architecture [34], where the input of the network is the feature vector of each instance, and the outputs are link edges between the instances. The outline of our suggested methodology is presented in the Fig. 9.

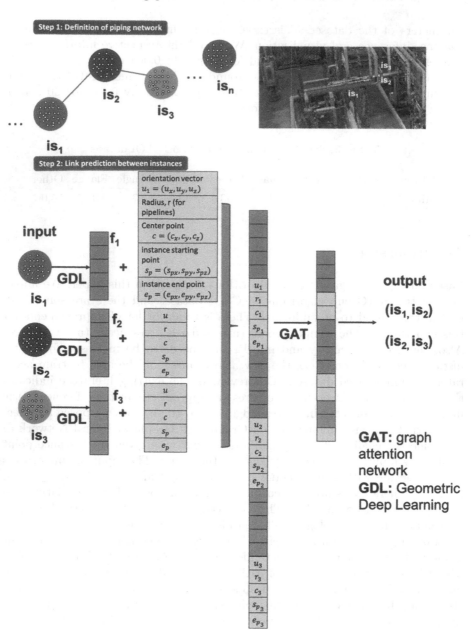

Fig. 9. Discussion

Pytorch Geometric [43] library can be used to built the proposed architecture. This library is built on PyTorch specifically for designing graph neural network based methods. However, there are still some open-ended research questions (RQs), which we intend to answer while implementing this methodology:

- RQ1: Which is the most appropriate GDL instance segmentation architecture to extract instance feature vectors? (Current research)
- RQ2: Which are other node definitions that can be considered for the piping network graph (e.g., partial point cloud instances and missing edges)?
- RQ3: Which additional features can be added to the geometric feature vector (i.e., curvature, normals)

5   Conclusions

The major challenge we are facing is dealing with huge and complex point cloud data without sacrificing instance segmentation accuracy. By investigating various types of 3D segmentation networks based on geometric techniques, projection and graphical based methods, we implemented a SoftGroup algorithm on CLOI dataset and found the results not so satisfactory but it was the first benchmark if we consider neural network based models as compared to a previous method which was only breadth first search algorithm. Also, while researching various graphical-based neural networks implemented on 3D point clouds data shows the average accuracy of 85% and thus there can be a high likelihood of success with using graphical techniques and potentially we will be able to implement and deploy a high-performing instance segmentation network specifically for industrial point clouds. Implementing a successful instance segmentation method, which will resolve issues related to the faster implementation of DTs in the manufacturing industry, can be of substantial importance.

Acknowledgments. We thank the Florida Space Institute (NASA) for sponsoring this research. We gratefully acknowledge the collaboration of all academic and industrial project partners. Any opinions, findings and conclusions or recommendations expressed in this material are those of the authors and do not necessarily reflect the views of the institutes mentioned above.

References

1. Thomas, D.S., Thomas, D.S.: The costs and benefits of advanced maintenance in manufacturing. US Department of Commerce, National Institute of Standards and Technology, Gaithersburg, MD, USA (2018)
2. Edwards, J., Townsend, A.: Buildings under refurbishment and retrofit. Carbon Action 2050 (2011)
3. Agapaki, E., Brilakis, I.: CLOI-NET: class segmentation of industrial facilities point cloud datasets. Adv. Eng. Inform. **45**, 101121 (2020)
4. Agapaki, E., Brilakis, I.: Instance segmentation of industrial point cloud data. J. Comput. Civ. Eng. **35**(6), 04021022 (2021)

5. Hsu, C.-M., Lian, F.-L., Ting, J.-A., Liang, J.-A., Chen, B.-C.: Road detection based on bread-first search in urban traffic scenes. In: 2011 8th Asian Control Conference (ASCC), pp. 1393–1397. IEEE (2011)
6. https://towardsdatascience.com/on-object-detection-metrics-with-worked-example-216f173ed31e
7. VuVu, T., Kim, K., Luu, T.M., Nguyen, T., Yoo, C.D.: SoftGroup for 3D instance segmentation on point clouds. In: Proceedings of the IEEE/CVF Conference on Computer Vision and Pattern Recognition, pp. 2708–2717 (2022)
8. Chen, S., Fang, J., Zhang, Q., Liu, W., Wang, X.: Hierarchical aggregation for 3D instance segmentation. In: Proceedings of the IEEE/CVF International Conference on Computer Vision, pp. 15467–15476 (2021)
9. Liang, Z., Li, Z., Xu, S., Tan, M., Jia, K.: Instance segmentation in 3D scenes using semantic superpoint tree networks. In: Proceedings of the IEEE/CVF International Conference on Computer Vision, pp. 2783–2792 (2021)
10. Han, L., Zheng, T., Xu, L., Fang, L.: OccuSeg: occupancy-aware 3D instance segmentation. In: Proceedings of the IEEE/CVF Conference on Computer Vision and Pattern Recognition, pp. 2940–2949 (2020)
11. Jiang, L., Zhao, H., Shi, S., Liu, S., Fu, C.W., Jia, J.: Pointgroup: dual-set point grouping for 3d instance segmentation. In: Proceedings of the IEEE/CVF Conference on Computer Vision and Pattern Recognition, pp. 4867–4876 (2020)
12. He, T., Shen, C., Hengel, A.V.: Dynamic convolution for 3D point cloud instance segmentation. arXiv preprint arXiv:2107.08392, 18 July 2021
13. Li, M., Chen, C., Liu, X., Huang, W., Zhang, Y., Xiong, Z.: Advanced deep networks for 3D mitochondria instance segmentation. arXiv preprint arXiv:2104.07961 (2021)
14. Shi, W., Rajkumar, R.: Point-GNN: graph neural network for 3D object detection in a point cloud. In: Proceedings of the IEEE/CVF Conference on Computer Vision and Pattern Recognition, pp. 1711–1719 (2020)
15. Bentley, J.L., Stanat, D.F., Hollins Williams Jr., E.: The complexity of finding fixed-radius near neighbors. Inf. Process. Lett. 6(6), 209–212 (1977)
16. Tailor, S.A., de Jong, R., Azevedo, T., Mattina, M., Maji, P.: Towards efficient point cloud graph neural networks through architectural simplification. In: Proceedings of the IEEE/CVF International Conference on Computer Vision, pp. 2095–2104 (2021)
17. Hu, Q., et al.: Randla-net: efficient semantic segmentation of large-scale point clouds. In: Proceedings of the IEEE/CVF Conference on Computer Vision and Pattern Recognition, pp. 1110–811117 (2020)
18. Thomas, H., Qi, C.R., Deschaud, J.E., Marcotegui, B., Goulette, F., Guibas, L.J.: KpConv: flexible and deformable convolution for point clouds. In: Proceedings of the IEEE/CVF International Conference on Computer Vision, pp. 6411–6420 (2019)
19. Razani, R., Cheng, R., Li, E., Taghavi, E., Ren, Y., Bingbing, L.: GP-S3NET: graph-based panoptic sparse semantic segmentation network. In: Proceedings of the IEEE/CVF International Conference on Computer Vision, pp. 16076–16085 (2021)
20. Sirohi, K., Mohan, R., Büscher, D., Burgard, W., Valada, A.: EfficientLPS: efficient lidar panoptic segmentation. IEEE Trans. Robot. (2021)
21. Tan, M., Le, Q.: EfficientNet: rethinking model scaling for convolutional neural networks. In: International Conference on Machine Learning. PMLR (2019)

22. Behley, J., et al.: Semantickitti: a dataset for semantic scene understanding of lidar sequences. In: Proceedings of the IEEE/CVF International Conference on Computer Vision (2019)
23. Geiger, A., Lenz, P., Urtasun, R.: Are we ready for autonomous driving? The KITTI vision benchmark suite. In: 2012 IEEE Conference on Computer Vision and Pattern Recognition. IEEE (2012)
24. http://www.semantic-kitti.org/tasks.html#panseg
25. Kipf, T.N., Welling, M.: Semi-supervised classification with graph convolutional networks. arXiv preprint arXiv:1609.02907, 9 Sept 2016
26. Wang, Y., Sun, Y., Liu, Z., Sarma, S.E., Bronstein, M.M., Solomon, J.M.: Dynamic graph CNN for learning on point clouds. ACM Trans. Graph. (TOG). **38**(5), 1–2 (2019)
27. Su, H., Maji, S., Kalogerakis, E., Learned-Miller, E.: Multi-view convolutional neural networks for 3D shape recognition. In: Proceedings of the IEEE International Conference on Computer Vision, pp. 945–953 (2015)
28. Qi, C.R., Yi, L., Su, H., Guibas, L.J.: PointNet++: deep hierarchical feature learning on point sets in a metric space. In: Guyon, I., et al. (eds.) Advances in Neural Information Processing Systems, vol. 30, Curran Associates Inc. (2017)
29. Scarselli, F., Gori, M., Tsoi, A.C., Hagenbuchner, M., Monfardini, G.: The graph neural network model. IEEE Trans. Neural Netw. **20**(1), 61–80 (2008)
30. Meng, H.-Y., Gao, L., Lai, Y.-K., Manocha, D.: VV-net: voxel VAE net with group convolutions for point cloud segmentation. In: Proceedings of the IEEE/CVF International Conference on Computer Vision, pp. 8500–8508 (2019)
31. https://www.autodesk.com/products/revit/overview?term=1-YEAR&tab=subscription&plc=RVT
32. Triess, L.T., Peter, D., Rist, C.B., Zöllner, J.M., Scan-based semantic segmentation of lidar point clouds: an experimental study. In: 2020 IEEE Intelligent Vehicles Symposium (IV), pp. 1116–1121. IEEE (2020)
33. Daigavane, A., Ravindran, B., Aggarwal, G.: Understanding convolutions on graphs. Distill **6**(9), e32 (2021)
34. He, W., Jiang, Z., Zhang, C., Sainju, A.M.: CurvaNet: geometric deep learning based on directional curvature for 3D shape analysis. In: Proceedings of the 26th ACM SIGKDD International Conference on Knowledge Discovery and Data Mining, pp. 2214–2224 (2020)
35. Zeng, Y., et al.: RT3D: real-time 3-D vehicle detection in lidar point cloud for autonomous driving. IEEE Robot. Autom. Lett. **3**(4), 3434–40 (2018)
36. Simon, M., Milz, S., Amende, K., Gross, H.-M.: Complex-YOLO: an Euler-region-proposal for real-time 3D object detection on point clouds. In: Leal-Taixó, L., Roth, S. (eds.) ECCV 2018. LNCS, vol. 11129, pp. 197–209. Springer, Cham (2019). https://doi.org/10.1007/978-3-030-11009-3_11
37. Cortinhal, T., Tzelepis, G., Aksoy, E.E.: Salsanext: fast semantic segmentation of lidar point clouds for autonomous driving. **3**(7). arXiv preprint arXiv:2003.03653 (2020)
38. Milioto, A., Vizzo, I., Behley, J., Stachniss, C.: RangeNet++: fast and accurate LiDAR semantic segmentation. In: IEEE/RSJ International Conference on Intelligent Robots and Systems (IROS), pp. 4213–4220 (2019). https://doi.org/10.1109/IROS40897.2019.8967762
39. Razani, R., Cheng, R., Taghavi, E., Bingbing, L.: Lite-HDSeg: LiDAR semantic segmentation using lite harmonic dense convolutions. In: IEEE International Conference on Robotics and Automation (ICRA), pp. 9550–9556 (2021). https://doi.org/10.1109/ICRA48506.2021.9561171

40. Xie, L., Furuhata, T., Shimada, K.: MuGNet: multi-resolution graph neural network for segmenting large-scale pointclouds. In: Conference on Robot Learning. PMLR (2021)

41. Qi, C.R., et al.: Pointnet: deep learning on point sets for 3D classification and segmentation. In: Proceedings of the IEEE Conference on Computer Vision and Pattern Recognition (2017)

42. Chen, C., et al.: Hapgn: hierarchical attentive pooling graph network for point cloud segmentation. IEEE Trans. Multimedia **23**, 2335–2346 (2020)

43. Fey, M., Lenssen, J.E.: Fast graph representation learning with PyTorch Geometric. arXiv preprint arXiv:1903.02428 (2019)

Fitness Landscape Ruggedness Impact on PSO in Dealing with Three Variants of the Travelling Salesman Problem

Abtin Nourmohammadzadeh[1]([✉])[ID], Malek Sarhani[1,2][ID], and Stefan Voß[1][ID]

[1] Institute of Information Systems, University of Hamburg, Hamburg, Germany
{abtin.nourmohammadzadeh,stefan.voss}@uni-hamburg.de
[2] School of Business Administration, Al Akhawayn University, Ifrane, Morocco
m.sarhani@aui.ma

Abstract. Fitness landscape analysis has gained quite some attention in understanding the behaviour of metaheuristics. Swarm intelligence is a type of metaheuristics that has grown considerably on the algorithmic side over the past decade. Nevertheless, only little attention has been paid to understanding the behaviour of algorithms on different fitness landscapes, especially in combinatorial optimization. Our aim in this paper is to re-motivate the importance of this issue. Moreover, by considering particle swarm optimization (PSO), we present a first investigation on its adaptation to three variants of the travelling salesman problem and how its performance is correlated with the ruggedness of the problem instances. The results show that PSO performance deteriorates with the increase in the number of cities and the ruggedness of the instances.

Keywords: Particle swarm optimization · Fitness landscape analysis · Metaheuristics · Travelling salesman problem · Ruggedness

1 Introduction

Many real-world problems require optimization. Optimization algorithms are designed to find the most desirable solution to an optimization problem, taking into account the quality of the solution, the computational cost, or a compromise of both. In particular, metaheuristics have shown promise for solving complex optimization problems where classical exact techniques are computationally expensive and traditional approximate algorithms (e.g. gradient-based methods) perform poorly [7].

In general, work on metaheuristics focuses either on the algorithmic side or on the analysis of the problem itself. Fitness landscape (FL) analysis is an approach that focuses on the latter and provides insights towards the understanding of the performance of different algorithms and the reasons of this performance based on the FL characteristics [50]. This concept was proposed several decades ago in [52] as an abstract notion for understanding biological evolution and was later

© Springer Nature Switzerland AG 2022
D. E. Simos et al. (Eds.): LION 2022, LNCS 13621, pp. 429–444, 2022.
https://doi.org/10.1007/978-3-031-24866-5_31

used to explain the behaviour of optimization algorithms. That is, FL analysis has evolved over the past two decades from a purely theoretical concept to a practical tool supporting optimization [38]. Indeed, characterizing a problem should lead to a better understanding of it and to better choices of optimization algorithms as well as their parameterization, and therefore, have an increased chance of producing better quality solutions.

Before delving into this issue, we note that metaheuristics could often be divided into single solution approaches and population-based algorithms. The former consist of modifying and improving a single candidate solution while the latter work on a collection of solutions in parallel [13, 46]. The first category includes mainly local search approaches and the second could be divided into evolutionary computation approaches and swarm intelligence methods [54]. For a related template we refer to [16].

In general, FL could be defined by a triplet: the fitness function, the solution space and the neighbourhood relation. While the two first elements are algorithm independent, the neighbourhood depends on each algorithm and has been studied primarily for evolutionary computation and local search. In fact, by analyzing the literature, we can notice that most of the work on FL analysis has been devoted to evolutionary algorithms as they share the same origin. Moreover, the neighbourhood of evolutionary operators such as mutation could be easily characterized [36]. In addition, the notion has been included in local search algorithms as they are associated with the concept of neighbourhood. That is, a local search approach usually involves choosing the best solution from a neighbourhood. We refer to [49] for an example of extensive analysis of local search FLs. FL analysis is also useful when hybridizing evolutionary computation and local search (e.g. a memetic algorithm as studied in [29]). The concept of swarm intelligence is different and often consists of the interaction between a number of agents. In fact, most of the work included in the recent surveys on FL analysis fell into the categories of local search and evolutionary computation as can be seen, e.g., in [27] and [33].

On the contrary, we can notice that a large part of the research on the algorithmic side has been devoted to swarm intelligence. In fact, swarm intelligence algorithms had become a major area of research over the last decade and most of the proposed algorithms were part of it [10]. But there is a lack of interest in linking the performance of such algorithms to the characteristics of the problems. This is one of the reasons for the controversy over many of the proposed methods in the past decade, which was exposed, e.g., in [42] and [5].

Recently, de Armas et al. [2], extending [16], proposed a pool template to identify and analyze swarm intelligence algorithms at a component level and to highlight their similarities. In this paper, we aim to shed light on the characterization of swarm intelligence algorithms based on FL features. More specifically, we provide a first step to the FL characterization of a particle swarm optimization (PSO) adaptation to the travelling salesman problem (TSP), which is among the most studied combinatorial optimization problems. In fact, PSO is one of the most adopted swarm intelligence algorithms, and the most studied approach in terms of FL analysis (as highlighted in Sect. 3.2). But, most of the work on this topic has focused on continuous optimization, which is typical for PSO. To the best of our knowledge, the introduction of FL analysis has not yet been explored for discrete and binary PSO adaptations, as well as for other swarm intelligence approaches. Beyond investigating the TSP, we conduct some experiments for two generalizations of the TSP, too.

The rest of the paper is organized as follows: in the next section, we are interested in FL features and in particular the ruggedness. In Sect. 3, we show how PSO could be adopted to the considered problems. Section 4 shows the experiments. Finally, a conclusion is presented.

2   Fitness Ruggedness

2.1   Fitness Landscape Features

The FL of optimization problems has been studied for a while and a number of features have been shown to influence the ability of algorithms to solve problems. An example of an apparent feature of FL is the set of local optima of a given problem. Unimodal functions have only one local optimum, which is also the global optimum while multimodal functions have multiple local optima. For a comprehensive survey of FL features, we refer to [33] and [27]. Here, we note that some features are tailored to evolutionary algorithms such as evolvability and epistatis. The former speculates on the ability of a population to produce offspring (subsequent solutions) fitter than their parents (previous solutions). It could also be defined as the ability of a given search process to move to a place in the landscape of better fitness. The latter refers to the degree of dependency between the genes (bits) of a chromosome (variable). In other words, it reflects the degree of interdependence between variables and how it is possible to adjust one variable to find the optimal value independently of the others. Another analysis of the FL could be carried-out through its basins of attraction. These are the areas which lead to a certain local optimum and are most often sought in local search methods. Additionally, they could be incorporated into swarm intelligence analysis as in [41]. Other measures are based on samples of the search space, such as ruggedness, which is our focus in this paper, and which we describe below.

2.2   Ruggedness Measures

Ruggedness generally refers to the number and distribution of local optima. The most common approach to predict ruggedness is through fitness value samples obtained from the FL using a time series of random walks. The autocorrelation function computed in Eq. (1) and proposed by [51] was previously the most classical approach to measure FL ruggedness.

$$\rho(s) = \frac{E[f_t f_{t+s}] - E[f_t]E[f_{t+s}]}{V[f_t]} \tag{1}$$

where $E[f_t]$ and $V[f_t]$ are the expectation and the variance, respectively, of the time series.

The idea of using this function is that in a rugged landscape, the fitness of neighbouring solutions is less correlated, and thus, it is harder for a search method to infer a search direction from a previous solution. However, this measure has been criticized in some works, which have pointed out its weakness in the characterization of the FL (e.g. [21]). Thus, the measure proposed in [44] to study ruggedness, which is described above, has become the most common over the last decade.

In [44], the authors defined three states to characterize it, which are rugged, smooth and neutral states. The rugged and smooth states correspond to the differences in fitness between neighbouring solutions, while the neutral state is determined by equal neighbouring solutions regarding fitness. The authors adopted the notion of entropy to measure the degree of change between the three states. In this case, the samples are encoded in a certain symbol sequence, and the entropy is measured. The approach was first introduced for combinatorial optimization and then extended to continuous optimization [25]. In this paper, we are interested in combinatorial optimization and we briefly describe the approach as follows.

The time series is represented as a string, $S(\epsilon) = s_1 s_2 s_3 ... s_n$, of symbols $s_i \in \{\bar{1}, 0, 1\}$ and $i \in 1, ..., n$. The encoding function for the time series of a random walk is defined as follows:

$$\Phi(i, \epsilon) = \begin{cases} \bar{1} \text{ if } y_{i+1} - y_i < -\epsilon \\ 0 \text{ if } |y_{i+1} - y_i| \leq \epsilon \\ 1 \text{ if } y_{i+1} - y_i > \epsilon \end{cases} \tag{2}$$

where ϵ is the information sensitivity, which acts as an accuracy parameter of the symbol sequence.

Based on this definition, an entropic measure H is defined as follows:

$$H(\epsilon) = -\sum_{p \neq q} P_{[pq]} log_6 P_{[pq]} \tag{3}$$

where, $P_{[pq]}$ is defined as:

$$P_{[pq]} = \frac{n_{[pq]}}{n} \tag{4}$$

and $n_{[pq]}$ is the number of sub-blocks pq in the string $S(\epsilon)$ (p and q are elements from the set $\{\bar{1}, 0, 1\}$ computed in Eq. (2)). The interested reader is referred to [25] and [34] for a more detailed description of these parameters, which measure the degree of change indicated above. In this paper, we are interested in computing the ruggedness parameter as defined in Eq. (5).

$$R = \max_{\epsilon} H(\epsilon) \tag{5}$$

The analysis of ruggedness has shown efficiency in improving metaheuristic performance. In particular, it is a fruitful way of designing adaptive algorithms that take advantage of FL analysis as studied, e.g., in [18]. In this paper, we aim to use it to analyze the PSO behaviour for the TSP and some generalization.

3   Particle Swarm Optimization for TSP Landscapes

PSO is a swarm intelligence approach that mimics the movement of birds and fish, which was originally proposed in [19]. In the native PSO, the updating of the population of particles is done according to Eq. (6).

$$v_i^j(t+1) = wv_i^j(t) + c_1 r_1 . (p_i^j(t) - x_i^j(t)) + c_2 r_2 . (p_g^j(t) - x_i^j(t)) \qquad (6)$$

where $v_i^j(t)$ and $x_i^j(t)$ correspond to the jth dimension of velocity and position vectors of the particle i. The position $p_i^j(t)$ represents the jth dimension of the best previous position of particle i, while $p_g^j(t)$ represents the jth dimension of the best position among all particles in the population, while the variables r_1 and r_2 are two independently uniformly distributed random variables. The parameters c_1 and c_2 are the acceleration factors and w is the inertia weight. This last parameter was introduced in [39] to balance the exploration and exploitation dilemma.

PSO was mainly adopted for continuous optimization and most of the improvements and extensions proposed in the literature have focused on improving its performance for continuous problems. Nevertheless, some extensions have been proposed to deal with combinatorial optimization and few solvers are implemented (e.g. [45]). Below, we take a look at how PSO could be adapted to the TSP.

3.1   PSO for the TSP

The TSP is a well-known NP-hard combinatorial optimization problem. Metaheuristics have shown to be effective for the TSP, especially for large instances. Most of the adopted approaches are either local searches or evolution computation approaches [43]. Ant colony optimization is the most adopted swarm intelligence approach for the TSP [47]. Regarding PSO, it has been less adopted because it is not easy to adapt its concept to the TSP. This also applies to the classical binary PSO proposed in [20], which consists in transforming continuous solutions using the sigmoid function. Nonetheless, a few approaches have proposed to adapt PSO to this problem and have shown success in this regard. For example, Chen et al. [8] used a set-based representation scheme that allows PSO to characterize the discrete search space. The approach represents TSP solutions as permutations over a number of vertices of a given instance. Another approach was proposed in [48], which introduced the concept of a swap sequence. This concept is a sequence of swap operators that involve exchanging two nodes in a solution. This idea was extended and further illustrated in [15] and [14]. Their idea was inspired by the concept of path relinking [12] and consists of defining velocity operators and choosing from them in a probabilistic way. More precisely, particle's positions are transformed according to a velocity operator. The binary operations take two particles and alter the position of one of them considering the position of the other. Also, in this case, the velocities are updated according to Eq. 7.

$$v_i^j(t+1) = v_i^j(t) \oplus \alpha(p_i^j(t) - x_i^j(t)) \oplus \beta(p_g^j(t) - x_i^j(t)) \qquad (7)$$

We can observe that for this PSO adaptation $w = 1$, α and β correspond to $c_1 r_1$ and $c_2 r_2$, respectively.

3.2 Fitness Landscape Analysis for PSO

As previously noted, FL has been studied less for swarm intelligence than for other types of metaheuristics. Despite this, there are a few papers incorporating it into the PSO analysis. In particular, a funnel is a FL feature that was first introduced for the case of PSO [41]. A funnel in a landscape is a global basin shape that consists of clustered local optima and was also adopted later (e.g. [53]). Other researchers were interested in analyzing the topic more broadly. Malan and Engelbrecht [26] analyzed the correlation between three metrics and the PSO performance for continuous optimization. The three metrics are a ruggedness measure based on entropy (Eq. (3)), a dispersion measure for predicting the presence of funnels and a fitness gradient estimation measure. A similar work with respect to the gradient was proposed in [26] and extended in [24]. The authors were interested in the searchability, which is an adaptation of the evolvability for PSO. Then, a more generic work was conducted in [11]. The authors adopted these different features to compare and analyze the performance of certain PSO variants, in terms of the balance between exploration and exploitation. The measure adopted for this purpose was named the diversity rate-of-change.

As the integration of FL analysis in PSO has so far been dedicated to continuous optimization, a researcher who tries to understand the behaviour of the PSO algorithm on a combinatorial optimization problem, has very little inspiration from previous studies to work on in terms of FL analysis. In this paper, we are interested in the analysis the ruggedness of TSP instances and the analysis of the link between them and the PSO performance. Before diving into the experiments, we briefly present some research concerning the analysis of TSP instances.

3.3 TSP Hardness and Fitness Landscape PSO Assessment

Some papers were interested in the analysis of the FL of TSP instances. An example of a paper which highlighted the concept of ruggedness is [30]. But, most of the papers included other features. For example, Boese et al. [4] asserted that the search space of TSP instances (under 2-opt moves) has a big-valley structure, in which local optima are clustered around one central global optimum. However, this statement has been questioned in multiple papers (e.g. [17]) and its generalization is also an issue of discussion [31]. Indeed, as noted in [32], the TSP structure is not yet fully understood. Moreover, according to [32], many works have revealed the presence of multiple funnels. This means, as stated in Sect. 3.2, that local optima are organized into several clusters, so that a particular local optimum belongs to a particular funnel. Cárdenas-Montes [6] presented a methodology for creating difficult instances of the TSP based on the spatial attributes of previously solved instances. Other papers, such as [40], interested in reducing the difficulty of the TSP. After highlighting these papers, we note that to our knowledge, there is no clear understanding and a quantification of the ruggedness of instances. Also, work on this issue is still ongoing (e.g. [43]) and it depends on each algorithm [28]. From an algorithmic point of view, it is related to the nature of local optima and funnels. In this paper, we investigate the impact of this factor on PSO depending on the number of cities for some randomly generated instances, and then, some benchmark TSP instances are investigated. The next steps are doing the experiments for two special variants of the TSP, namely the clustered TSP (CTSP) and the family TSP (TSP). In the CTSP, the cities are divided into a number of clusters and the salesman must visit all the cities of each cluster contiguously (see [23]). The FTSP includes clusters of cities, too, however, the salesman has to visit a number of cities from each cluster with minimal distance without any limit of finishing with one cluster before starting with another one (see [3]).

On the other hand, Malan and Engelbrecht [28] defined a number of metrics to predict PSO performance. Additionally, the same authors [26] incorporated other metrics such as the success rate (SR) and the success speed (SP).

SR expresses the number of successful runs, which find a solution within the fixed accuracy level of the global optimum, divided by the total number of runs. SR is a value in the range [0,1] where 1 indicates the highest possible rate of success. SP indicates the average number of function evaluations until reaching the global optimum (within the fixed accuracy level) over ns runs. This metric is calculated as follows:

$$SSP_r = \begin{cases} 0 & \text{if run is not successful} \\ \frac{Max(FES)-(FES_r-1)}{Max(FES)} & \text{otherwise} \end{cases} \tag{8}$$

$$SP = \begin{cases} \frac{\sum_{r=1}^{ns} SSP_r}{ns} & if\ ns > 0 \\ 0 & if\ ns = 0 \end{cases} \tag{9}$$

FES_r is the number of function evaluations until success in run r. SSP_r is the success speed of a run r, which is in the range [0,1]. It is 1, if the global optimum is reached in the first function evaluation. (It is noticeable that this metric is related to time-to-target plots [1]).

4   Experiments

4.1   Experimental Setup

In this work, we adopt SR, SP as well as their product (SR·SP). Six sets containing TSP instances of different variants are considered. These sets include random and benchmark TSP, CTSP and FTSP instances.

For PSO, we adopt the values $\alpha = 0.9$ and $\beta = 1$ as they are the values adopted in the papers underlined in Sect. 3.1. The population size and the number of iterations are set to 100 and 200, respectively.

For each instance with the same distance matrices, we perform random walks (2-opt move) 30 times with 200 iterations and calculate the ruggedness factor

(R). In addition, we run the PSO 30 times on the same instances and compute the three factors SR, SP, SR·SP (the best solution is estimated by looking at the different possible solutions). The algorithms are implemented in Python. In particular, we adopt the Mlrose package [35] to implement the random walks. In Sect. 4.2, we depict the results obtained.

4.2   Investigation of PSO Ruggedness on the Sets of Three Different TSP Variants

Regarding the random TSP instances, we use the following number of cities to generate the instances: 10, 20, 50, 100, 200 and 500. The coordinates of the cities are randomly generated in a $[0,1] \times [0,1]$ field. Our motivation in choosing these values is to start with a basic number of cities before studying the PSO performance on larger instances. In Table 1, we show the average results of the random TSP sets obtained by performing the experiments explained above. In the last row, some information about the correlation between ruggedness and the used performance metric $(SR \cdot SP)$ is given. This includes a correlation coefficient (r_s) calculated based on the Spearman's rank method [9] that shows the correlation of the instance ruggedness and the PSO performance. We show also the corresponding pvalue of the null hypothesis: no correlation exists between ruggedness and the performance metric. The relation between the ruggedness and the PSO performance for this set is depicted in Fig. 1

Table 1. PSO performance on the random TSP instances and ruggedness

No. of cities	R	SR	SP	SR·SP
10	0.7183	0.7225	0.7582	0.5478
20	0.7431	0.7012	0.7210	0.5056
50	0.7875	0.6721	0.7092	0.4767
100	0.8626	0.6455	0.6871	0.4435
200	0.8807	0.5512	0.6284	0.3464
500	0.9587	0.1816	0.2268	0.0412
1000	0.9892	0.0067	0.0036	0.0000
1100	0.9936	0.0008	0.0012	0.0000
1200	0.9959	0.0002	0.0008	0.0000
1500	0.9988	0.0000	0.0000	0.0000
Correlation	$r_s = -0.9692$		$pvalue = 0.0000$	

The second set consists of 12 benchmark instances from TSPLIB [37]. The results of this set are shown in Table 2 and Fig. 2.

The next experiments are on 10 CTSP instances, which we build using the data of 10 instances from TSPLIB [37] and clustering the cities by the kmeans method [22]. The obtained results are shown in Table 3 and Fig. 3.

Table 2. PSO performance on the benchmark TSP instances and ruggedness

No. of cities	R	SR	SP	SR·SP
eli51	0.7329	0.7209	0.8325	0.6001
berlin52	0.7451	0.7162	0.8227	0.5892
pr76	0.7608	0.7052	0.8170	0.5761
rat99	0.7745	0.6852	0.7965	0.5458
kroA100	0.7826	0.6733	0.7704	0.5187
lin105	0.7895	0.6682	0.7682	0.5133
pr124	0.7992	0.6542	0.7481	0.4894
pr136	0.8152	0.6206	0.7152	0.4439
pr150	0.8346	0.5720	0.6671	0.3816
rat195	0.8865	0.5347	0.6178	0.3303
kroA200	0.8965	0.5310	0.6008	0.3190
ts225	0.9372	0.5125	0.5820	0.2983
Correlation	$r_s = -1$		$pvalue = 0.0000$	

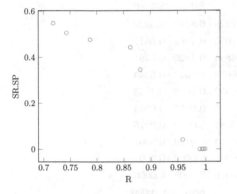

Fig. 1. Ruggedness vs. the PSO performance for the random TSP instances

Fig. 2. Ruggedness vs. the PSO performance for the benchmark TSP instances

Subsequently, we do a similar test on 10 benchmark CTSP instances adopted from [23]. The corresponding results are summarised in Table 4 and Fig. 4.

The last two sets contain random and benchmark FTSP instances. The random ones are used by adapting 10 TSPLIB [37] instances exactly like the FTSP random instances and choosing a random number of required cities to visit from each family within the set $[0.1fm_i, 0.3fm_i]$, where fm_i is the number of members of family i. The benchmark FTSP instances are from [3]. The results related to the two last sets are given by Tables 5 and 6 as well as Figs. 5 and 6.

The results of all the six sets indicate that the performance of the approach deteriorates with the increase of the number of cities. In fact, the approach is

Table 3. PSO performance on the random CTSP instances and ruggedness

No. of cities	R	SR	SP	SR·SP
bays29	0.7012	0.7129	0.8252	0.5883
att48	0.7481	0.6827	0.8092	0.5524
brazil58	0.7663	0.6504	0.7812	0.5081
gr96	0.8015	0.6278	0.7682	0.4823
ch130	0.8572	0.6012	0.7265	0.4368
ch150	0.8962	0.5835	0.6825	0.3982
d198	0.9353	0.5467	0.6132	0.3352
a280	0.9682	0.5178	0.5806	0.3006
d493	0.9892	0.1627	0.2108	0.0343
d657	0.9994	0.0876	0.1502	0.0132
Correlation	$r_s = -0.9970$		$pvalue = 0.0000$	

Table 4. PSO performance on the benchmark CTSP instances and ruggedness

No. of cities	R	SR	SP	SR·SP
49-pcb1173	0.6742	0.7228	0.8190	0.5920
100-pcb1173	0.7506	0.7055	0.7962	0.5617
144-pcb1173	0.7986	0.6894	0.7820	0.5391
10-nrw1379	0.8685	0.6626	0.7612	0.5044
12-nrw1379	0.8892	0.6552	0.7606	0.4983
1500-10-503	0.9678	0.0395	0.0856	0.0034
1500-20-504	0.9716	0.0281	0.0553	0.0016
1500-50-505	0.9775	0.0253	0.0126	0.0003
1500-100-506	0.9865	0.0091	0.0062	0.0001
1500-150-507	0.9982	0.0019	0.0008	0.0000
Correlation	$r_s = -1$		$pvalue = 0.0000$	

Table 5. PSO performance on the random FTSP instances and ruggedness

No. of cities	R	SR	SP	SR·SP
bayg29	0.7061	0.8016	0.7910	0.6341
att48	0.7351	0.7825	0.7792	0.6097
gr48	0.7462	0.7705	0.7208	0.5554
berlin52	0.7628	0.7558	0.7014	0.5301
berlin58	0.7865	0.7413	0.6827	0.5061
gr96	0.8365	0.7206	0.6527	0.4703
bier127	0.8642	0.6862	0.6288	0.4315
ch130	0.8955	0.6570	0.5922	0.3891
ch150	0.9521	0.2636	0.2461	0.0649
brg180	0.9763	0.1826	0.2259	0.0412
Correlation	$r_s = -1$		$pvalue = 0.0000$	

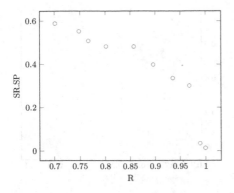

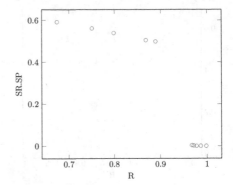

Fig. 3. Ruggedness vs. the PSO performance for the random CTSP instances

Fig. 4. Ruggedness vs. the PSO performance for the benchmark CTSP instances

Table 6. PSO performance on the benchmark FTSP instances and ruggedness

No. of cities	R	SR	SP	SR·SP
a280_1	0.7167	0.8562	0.8447	0.7232
a280_2	0.7289	0.8425	0.8409	0.7085
gr666_1	0.8267	0.7832	0.7613	0.5963
gr666_2	0.8372	0.7725	0.7585	0.5859
gr666_3	0.8416	0.7688	0.7471	0.5744
pr1002_1	0.9724	0.1726	0.0166	0.0029
pr1002_2	0.9766	0.1265	0.0154	0.0019
pr1002_3	0.9814	0.1078	0.0073	0.0008
pr144_4	0.7598	0.7758	0.7245	0.5621
kroA150_3	0.7837	0.7524	0.7031	0.5290
Correlation	$r_s = -0.81818$		$pvalue = 0.00381$	

unable to find the optimal solution in large instances in most executions. The degradation is also correlated with the ruggedness factor.

The ruggedness estimations are negatively correlated with performance for low and high dimensional problems. This can be deduced from the obtained Spearman's correlation coefficients, which are between -0.80 and -1. Most of the problems with high ruggedness values were not solved (SR × SP values of 0).

In the end of this section, it has to be mentioned that the results obtained with the used PSO are not competitive when they are compared to the best-known or optimal results specially for the large-sized instances. The reason is that this PSO is not the best solution methodology for the investigated problems and some enhancements are required such as hybridizing it with other metaheuristics or exact methods.

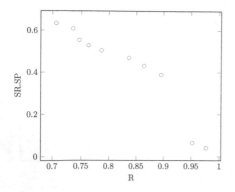

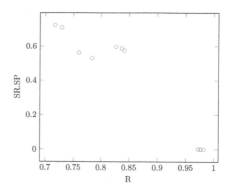

Fig. 5. Ruggedness vs. the PSO performance for the random FTSP instances

Fig. 6. Ruggedness vs. the PSO performance for the benchmark FTSP instances

5   Conclusion and Discussion

In this work, we investigated the fitness landscape analysis of PSO for combinatorial optimization. We started by motivating the topic and presenting the open challenges. Next, we presented a first study of the most common PSO adaptation to the TSP and its two variants. The presented results indicate that the adopted approach could not find satisfactory results for TSP instances out of 100 instances. In addition, we can notice that its performance correlates with the ruggedness. In that sense our work sheds some light on the detailed analysis of PSO as a general solver in combinatorial optimization. This does not mean that PSO is bad; it just indicates additional insights on a popular method and its real impact and behaviour. Note, however, that other FL features also have an impact as PSO does not fully correlate with the ruggedness factor.

Future research should attempt to incorporate other features into the analysis of PSO. Also, it would be important to compare the performance of different swarm intelligence approaches or certain PSO variants, as investigated in [11] for continuous optimization. In addition, PSO could also be studied for other combinatorial optimization problems in order to have a broader overview of its performance.

Acknowledgement. Malek Sarhani was supported by the Alexander von Humboldt Foundation.

References

1. Aiex, R.M., Resende, M.G., Ribeiro, C.C.: TTT plots: a Perl program to create time-to-target plots. Optim. Lett. **1**(4), 355–366 (2007). https://doi.org/10.1007/s00291-020-00604-x

2. de Armas, J., Lalla-Ruiz, E., Tilahun, S.L., Voß, S.: Similarity in metaheuristics: a gentle step towards a comparison methodology. Natural Comput. **21**, 265–287 (2021). https://doi.org/10.1007/s11047-020-09837-9

3. Bernardino, R., Paias, A.: Heuristic approaches for the family traveling salesman problem. Int. Trans. Oper. Res. **28**(1), 262–295 (2021). https://doi.org/10.1111/itor.12771

4. Boese, K.D., Kahng, A.B., Muddu, S.: A new adaptive multi-start technique for combinatorial global optimizations. Oper. Res. Lett. **16**(2), 101–113 (1994). https://doi.org/10.1016/0167-6377(94)90065-5

5. Camacho Villalón, C.L., Stützle, T., Dorigo, M.: Grey wolf, firefly and bat algorithms: three widespread algorithms that do not contain any novelty. In: Dorigo, M., et al. (eds.) ANTS 2020. LNCS, vol. 12421, pp. 121–133. Springer, Cham (2020). https://doi.org/10.1007/978-3-030-60376-2_10

6. Cárdenas-Montes, M.: Creating hard-to-solve instances of travelling salesman problem. Appl. Soft Comput. **71**, 268–276 (2018). https://doi.org/10.1016/j.asoc.2018.07.010

7. Caserta, M., Voß, S.: Metaheuristics: intelligent problem solving. In: Maniezzo, V., Stützle, T., Voß, S. (eds.) Matheuristics. Annals of Information Systems, vol. 10, pp. 1–38. Springer, Boston (2009). https://doi.org/10.1007/978-1-4419-1306-7_1

8. Chen, W.N., Zhang, J., Chung, H., Zhong, W.L., Wu, W.G., Shi, Y.: A novel set-based particle swarm optimization method for discrete optimization problems. IEEE Trans. Evol. Comput. **14**(2), 278–300 (2010). https://doi.org/10.1109/tevc.2009.2030331

9. Daniel, W.W.: Applied Nonparametric Statistics. PWS-KENT Pub, Boston (1990)

10. Dokeroglu, T., Sevinc, E., Kucukyilmaz, T., Cosar, A.: A survey on new generation metaheuristic algorithms. Comput. Industr. Eng. **137**, 106040 (2019). https://doi.org/10.1016/j.cie.2019.106040

11. Engelbrecht, A.P., Bosman, P., Malan, K.M.: The influence of fitness landscape characteristics on particle swarm optimisers. Nat. Comput. (2021). https://doi.org/10.1007/s11047-020-09835-x

12. Glover, F.: A template for scatter search and path relinking. In: Hao, J.-K., Lutton, E., Ronald, E., Schoenauer, M., Snyers, D. (eds.) AE 1997. LNCS, vol. 1363, pp. 1–51. Springer, Heidelberg (1998). https://doi.org/10.1007/BFb0026589

13. Glover, F., Sörensen, K.: Metaheuristics. Scholarpedia **10**(4), 6532 (2015). https://doi.org/10.4249/scholarpedia.6532

14. Goldbarg, E.F.G., Goldbarg, M.C., de Souza, G.R.: Particle swarm optimization algorithm for the traveling salesman problem. In: Greco, F. (ed.) Traveling Salesman Problem, pp. 75–96. InTech (2008). https://doi.org/10.5772/5580

15. Goldbarg, E.F.G., de Souza, G.R., Goldbarg, M.C.: Particle swarm for the traveling salesman problem. In: Gottlieb, J., Raidl, G.R. (eds.) EvoCOP 2006. LNCS, vol. 3906, pp. 99–110. Springer, Heidelberg (2006). https://doi.org/10.1007/11730095_9

16. Greistorfer, P., Voß, S.: Controlled pool maintenance for metaheuristics. In: Rego, C., Alidaee, B. (eds.) Metaheuristic Optimization via Memory and Evolution, pp. 387–424. Kluwer Academic Publishers (2005). https://doi.org/10.1007%2F0-387-23667-8_18

17. Hains, D.R., Whitley, L.D., Howe, A.E.: Revisiting the big valley search space structure in the TSP. J. Oper. Res. Soc. **62**(2), 305–312 (2011). https://doi.org/10.1057/jors.2010.116

18. Huang, Y., Li, W., Tian, F., Meng, X.: A fitness landscape ruggedness multiobjective differential evolution algorithm with a reinforcement learning strategy. Appl. Soft Comput. **96**, 106693 (2020). https://doi.org/10.1016/j.asoc.2020.106693

19. Kennedy, J., Eberhart, R.: Particle swarm optimization. In: Proceedings of ICNN 1995 - International Conference on Neural Networks. IEEE (1995). https://doi.org/10.1109/icnn.1995.488968

20. Kennedy, J., Eberhart, R.: A discrete binary version of the particle swarm algorithm. In: IEEE International Conference on Systems, Man, and Cybernetics. Computational Cybernetics and Simulation. IEEE (1997). https://doi.org/10.1109/icsmc.1997.637339

21. Liefooghe, A., Daolio, F., Verel, S., Derbel, B., Aguirre, H., Tanaka, K.: Landscape-aware performance prediction for evolutionary multi-objective optimization. IEEE Trans. Evol. Comput. **24**(6), 1063–1077 (2019)

22. Lloyd, S.: Least squares quantization in PCM. IEEE Trans. Inf. Theory **28**(2), 129–137 (1982). https://doi.org/10.1109%2Ftit.1982.1056489

23. Lu, Y., Hao, J.K., Wu, Q.: Solving the clustered traveling salesman problem via TSP methods. arXiv preprint arXiv:2007.05254 (2020). https://doi.org/10.48550/arXiv.2007.05254

24. Malan, K.M., Engelbrecht, A.P.: Characterising the searchability of continuous optimisation problems for PSO. Swarm Intell. **8**(4), 275–302 (2014). https://doi.org/10.1007/s11721-014-0099-x

25. Malan, K.M., Engelbrecht, A.P.: Quantifying ruggedness of continuous landscapes using entropy. In: 2009 IEEE Congress on Evolutionary Computation. IEEE (2009). https://doi.org/10.1109/cec.2009.4983112

26. Malan, K.M., Engelbrecht, A.P.: Ruggedness, funnels and gradients in fitness landscapes and the effect on PSO performance. In: 2013 IEEE Congress on Evolutionary Computation. IEEE (2013). https://doi.org/10.1109/cec.2013.6557671

27. Malan, K.M., Engelbrecht, A.P.: A survey of techniques for characterising fitness landscapes and some possible ways forward. Inf. Sci. **241**, 148–163 (2013). https://doi.org/10.1016/j.ins.2013.04.015

28. Malan, K.M., Engelbrecht, A.P.: Fitness landscape analysis for metaheuristic performance prediction. In: Richter, H., Engelbrecht, A. (eds.) Recent Advances in the Theory and Application of Fitness Landscapes. ECC, vol. 6, pp. 103–132. Springer, Heidelberg (2014). https://doi.org/10.1007/978-3-642-41888-4_4

29. Merz, P., Freisleben, B.: Fitness landscape analysis and memetic algorithms for the quadratic assignment problem. IEEE Trans. Evol. Comput. **4**(4), 337–352 (2000). https://doi.org/10.1109/4235.887234

30. Merz, P., Freisleben, B.: Memetic algorithms for the traveling salesman problem. Complex Syst. **13**(4), 297–346 (2001)

31. Ochoa, G., Veerapen, N.: Deconstructing the big valley search space hypothesis. In: Chicano, F., Hu, B., García-Sánchez, P. (eds.) EvoCOP 2016. LNCS, vol. 9595, pp. 58–73. Springer, Cham (2016). https://doi.org/10.1007/978-3-319-30698-8_5

32. Ochoa, G., Veerapen, N.: Mapping the global structure of TSP fitness landscapes. J. Heurist. **24**(3), 265–294 (2017). https://doi.org/10.1007/s10732-017-9334-0

33. Pitzer, E., Affenzeller, M.: A comprehensive survey on fitness landscape analysis. In: Fodor, J., Klempous, R., Suárez Araujo, C.P. (eds.) Recent Advances in Intelligent Engineering Systems. SCI, vol. 378, pp. 161–191. Springer, Heidelberg (2012). https://doi.org/10.1007/978-3-642-23229-9_8

34. Poursoltan, S., Neumann, F.: Ruggedness quantifying for constrained continuous fitness landscapes. In: Datta, R., Deb, K. (eds.) Evolutionary Constrained Optimization. ISFS, pp. 29–50. Springer, New Delhi (2015). https://doi.org/10.1007/978-81-322-2184-5_2

35. PyPI: mlrose, 13 March 2022. https://pypi.org/project/mlrose/

36. Reeves, C.R.: Landscapes, operators and heuristic search. Ann. Oper. Res. **86**, 473–490 (1999). https://doi.org/10.1007/bf01165154
37. Reinelt, G.: TSPLIB—a traveling salesman problem library. ORSA J. Comput. **3**(4), 376–384 (1991). https://doi.org/10.1287/ijoc.3.4.376
38. Richter, H.: Fitness landscapes: from evolutionary biology to evolutionary computation. In: Richter, H., Engelbrecht, A. (eds.) Recent Advances in the Theory and Application of Fitness Landscapes. Emergence, Complexity and Computation, vol. 6, pp. 3–31. Springer, Heidelberg (2014). https://doi.org/10.1007%2F978-3-642-41888-4_1
39. Shi, Y., Eberhart, R.: A modified particle swarm optimizer. In: 1998 IEEE International Conference on Evolutionary Computation Proceedings. IEEE World Congress on Computational Intelligence (Cat. No.98TH8360). IEEE (1998). https://doi.org/10.1109/icec.1998.699146
40. Sun, Y., Ernst, A., Li, X., Weiner, J.: Generalization of machine learning for problem reduction: a case study on travelling salesman problems. OR Spectrum **43**(3), 607–633 (2020). https://doi.org/10.1007/s00291-020-00604-x
41. Sutton, A.M., Whitley, D., Lunacek, M., Howe, A.: PSO and multi-funnel landscapes: how cooperation might limit exploration. In: Keijzer, M. (ed.) Genetic and Evolutionary Computation Conference, pp. 75–82. Association for Computing Machinery, New York (2006). https://doi.org/10.1145/1143997.1144008
42. Sörensen, K.: Metaheuristics-the metaphor exposed. Int. Trans. Oper. Res. **22**(1), 3–18 (2015). https://doi.org/10.1111/itor.12001
43. Varadarajan, S., Whitley, D., Ochoa, G.: Why many travelling salesman problem instances are easier than you think. In: Proceedings of the 2020 Genetic and Evolutionary Computation Conference, pp. 254–262. ACM (2020). https://doi.org/10.1145/3377930.3390145
44. Vassilev, V.K., Fogarty, T.C., Miller, J.F.: Smoothness, ruggedness and neutrality of fitness landscapes: from theory to application. In: Ghosh, A., Tsutsui, S. (eds.) Advances in Evolutionary Computing: Theory and Applications, pp. 3–44. Springer, Heidelberg (2003). https://doi.org/10.1007/978-3-642-18965-4_1
45. Vaz, A.I., Vicente, L.N.: PSwarm: a hybrid solver for linearly constrained global derivative-free optimization. Optim. Methods Softw. **24**(4–5), 669–685 (2009). https://doi.org/10.1080/10556780902909948
46. Voß, S.: Tabu search: applications and prospects. In: Du, D., Pardalos, P.M. (eds.) Network Optimization Problems, vol. 2, pp. 333–353. World Scientific, Singapore (1993). https://doi.org/10.1142/9789812798190_0017
47. Voss, S.: Book review: Marco Dorigo and Thomas Stützle: Ant colony optimization. Math. Methods Oper. Res. **63**(1), 191–192 (2006). https://doi.org/10.1007/s00186-005-0050-4
48. Wang, K.P., Huang, L., Zhou, C.G., Pang, W.: Particle swarm optimization for traveling salesman problem. In: Proceedings of the 2003 International Conference on Machine Learning and Cybernetics (IEEE Cat. No.03EX693), pp. 1583–1585. IEEE (2003). https://doi.org/10.1109/icmlc.2003.1259748
49. Watson, J.P.: Empirical modeling and analysis of local search algorithms for the job-shop scheduling problem. Ph.D. thesis, Colorado State University (2003)
50. Watson, J.P.: An introduction to fitness landscape analysis and cost models for local search. In: Handbook of Metaheuristics, pp. 599–623. Springer, US (2010). https://doi.org/10.1007/978-1-4419-1665-5_20
51. Weinberger, E.: Correlated and uncorrelated fitness landscapes and how to tell the difference. Biol. Cybern. **63**(5), 325–336 (1990). https://doi.org/10.1007/bf00202749

52. Wright, S.: The roles of mutation, inbreeding, crossbreeding and selection in evolution. In: Proceedings of the Sixth International Congress of Genetics, vol. 1, pp. 356–366 (1932)
53. Xin, B., Chen, J., Pan, F.: Problem difficulty analysis for particle swarm optimization. In: Proceedings of the First ACM/SIGEVO Summit on Genetic and Evolutionary Computation - GEC 2009, pp. 623–630. ACM Press (2009). https://doi.org/10.1145/1543834.1543919
54. Yang, X.-S. (ed.): Recent Advances in Swarm Intelligence and Evolutionary Computation. SCI, vol. 585. Springer, Cham (2015). https://doi.org/10.1007/978-3-319-13826-8

A Multi-UAVs' Provider Model
for the Provision of 5G Service Chains:
A Game Theoretic Approach

Giorgia Maria Cappello[1]([✉]), Gabriella Colajanni[2], Patrizia Daniele[2],
Laura Galluccio[1], Christian Grasso[1], Giovanni Schembra[1],
and Laura Rosa Maria Scrimali[2]

[1] Department of Electrical, Electronics and Informatics Engineering (DIEEI)
and CNIT Research Unit, University of Catania,
Viale Andrea Doria, 6 - 95125 Catania, Italy
{giorgia.cappello,christian.grasso,giovanni.schembra}@unict.it
[2] Department of Mathematics and Computer Science, University of Catania,
Viale Andrea Doria,6 - 95125 Catania, Italy
{gabriella.colajanni,patrizia.daniele,laura.scrimali}@unict.it

Abstract. In the last years, the use of Flying Ad-hoc Networks
(FANET) to extend and improve the capability of 5G networks, especially in scenarios characterized by poor or completely inexistent structured networks, has been very successful. The possibility to mount, on
board of an Unmanned Aerial Vehicle (UAV), a Computing Element,
giving it the possibility to host virtual functions (VFs) and provide data
processing services, has allowed 5G networks to be able to extend their
functionalities closer to the user, in the so-called Edge Network. In this
paper, we present a multi-UAVs' providers network based model describing the provisioning of service chains to users and devices on the ground.
The objective of each provider in the proposed model is to establish the
optimal service chain flows to manage and send to, or receive by, other
providers in order to maximize its revenue while minimizing the total execution, execution request and transmission costs, under the constraints
that the total costumer demand, for each service chain, are satisfied, as
well as the capacity constraints. We formulate the nonlinear optimization problem as a non-cooperative game in which each player (provider)
is rational and acts selfishly. In particular, we analyzed the Generalized
Nash Equilibrium Problem (GNEP), and the equivalent formulation of
the GNEP by means of Variational Inequality theory is also provided.
Finally, an illustrative numerical example is presented and analyzed.

Keywords: 5G · Unmanned Aerial Vehicles · Generalized Nash
Equilibrium Problem · Variational formulation

1 Introduction

In the last years, due to the rapid evolution of mobile communication networks,
a plethora of new services has been introduced, for applications ranging from

© Springer Nature Switzerland AG 2022
D. E. Simos et al. (Eds.): LION 2022, LNCS 13621, pp. 445–459, 2022.
https://doi.org/10.1007/978-3-031-24866-5_32

smart agriculture to environmental monitoring, from search and rescue operations during environmental disasters to support for the management of cities and
forest fires [1,2]. However, some typical application scenarios of 5G networks are
characterized by poor or completely inexistent structured networks. In this case,
the use of Unmanned Aerial Vehicles (UAVs) could be of fundamental importance to support and extend traditional wireless networks. 5G [3] is not only the
last step in the evolution of mobile communication networks, but a revolution
due to the introduction of innovative service capabilities. In 5G networks, an
important role is played by the virtualization of physical resources at the edge
of the network, where data are generated by end-users. This is done through the
use of two combined paradigms: network function virtualization (NFV) [4] and
multi-access edge computing (MEC) [5]. In scenarios such as those described
above, at the moment, the data generated by the devices must be sent to remote
clouds, far from the place where these data were produced, for further processing. To address this problem, the possibility of an extension of a 5G network
at the Edge using a Flying Ad-hoc Network (FANET), composed by a certain
number of UAVs, was proposed in [6,7]. More in detail, each UAV is equipped
with a computing element (CE), thus it is able to host one or more Service
Chains (SC), composed by a certain number of Virtual Functions (VFs) linked
together, with the aim of processing data flows coming from end users or ground
devices. However, the main problem when FANETs are used to provide computational resources is represented by the limited flight duration of each UAV.
Indeed, when UAV battery runs out, the FANET loses a component, with a
consequent decrease of the maximum offered computational capabilities. The
problem of charge duration of UAV battery is exacerbated by the presence of
the CE that, depending on the application scenarios, hosts VFs providing a total
power consumption comparable to the power consumption of the engine. All that
said, it is clear that distributing the service chains flows optimally among the
UAVs belonging to different providers in the FANET, in order to minimize the
overall consumption and maximize the FANET lifetime represents an aspect to
pay attention to in these contexts, to achieve a tradeoff that allows to offer as
much service as possible to end users.

In the recent years there has been a growing interest in adopting cooperative
and non-cooperative game theoretic approaches [14–16] to model many communications and networking problems in the context of FANETs. In particular,
game theory has been deployed for the resolution of conflicts among UAV inter-
acting decision makers in wireless network applications [22]. In [20] the authors
tackle the problem of offloading heavy UAV computational tasks with the aim to
achieve the best possible tradeoff between energy consumption, time delay, and
computation cost. They formulate the problem as a non cooperative theoretical
game. In [21] the authors present a framework in which the coordination of a
group of UAVs flying is considered so as to guarantee the maximum network
coverage to mobile-ground units by efficiently utilizing the available on-board
power. A game theoretic and a heuristic approaches are proposed and compared. Furthermore, in [23] the authors explore some new challenges in the multi-
UAV cooperative search, such as collaborative control and search area covering

problems, while in [24] a game-theoretic autonomous decision-making approach for efficient deployment of UAVs in a multi-level and multi-dimensional assisted network is analyzed. The UAVs work in a cooperative manner for achieving the suitable deployment with the optimal coverage values for the candidate region.

In this paper we consider a network-based model for a multi-service chain and multi-provider scenario.

We consider a fleet of UAVs of different providers organized as a FANET. This fleet provides service chains (SCs) on demand to users and devices on the ground, e.g. in a remote geographic area (see [10] and [11]). We formulate a nonlinear optimization problem in order to determine the optimal distribution flow pattern that, for each provider, maximizes its revenue while minimizing the total execution, execution request and transmission costs, under the constraint that the total users' and devices' demand is satisfied, namely the conservation flow constraints, and the capacity constraints.

In our framework we include the possibility for each provider, that manages a service chain, to send chain flows to other providers for the execution, by paying a request execution costs. Therefore, we formulate the optimization problem as a game in which each player (provider) acts in a non cooperative manner. In particular, we analyzed a Generalized Nash Equilibrium Problem (GNEP), because the strategy of a given provider is affected by the strategies of the other providers. We also provide an equivalent formulation of the GNEP by means of Variational Inequality theory. Given the equivalence between the GNEP and the VI problem, conditions guaranteeing the existence of a GNE follow from the existence of a solution of the VI [17]. Furthermore, building on the equivalence between the GNEP and the VI problem, one can borrow solution methods for the GNEP from the vast literature on variational inequalities [18,19].

The rest of the paper is organized as follows. In Sect. 2 we present the network topology and the mathematical formulation of the optimization problem. In Sect. 3 we outline the GNEP, and a variational equilibrium of the GNEP is also provided. Finally, an illustrative numerical example is presented and analyzed in Sect. 4. We summarize our results and present our conclusions in Sect. 5.

2   The Mathematical Formulation

In this section, we present the multi-provider network model. We consider a network topology consisting of the provider layer and service chain layer, as depicted in Fig. 1.

We denote by $k = 1, \ldots, K$ the typical service chain required by customers (users and devices on the ground) and by $q = 1, \ldots, Q$ the typical provider of services. Each provider, for each service chain, can establish the flow of service chain requests it intends to satisfy. In addition, each provider can send (and, obviously, receive) part of the flow to other providers for the execution.

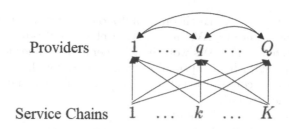

Fig. 1. Network topology

The model variables are described below. Let:

- y_{kq} be the nonnegative flow of service chain k managed by provider q (that is the amount of packet rate requested for the k-th chain and controlled by the provider q); and let us group all these quantities into the vectors $Y_k = (y_{kq})_{q=1,\ldots,Q}$, $Y_q = (y_{kq})_{k=1,\ldots,K}$ and $Y = (y_{kq})_{\substack{k=1,\ldots,K \\ q=1,\ldots,Q}}$;

- $x_{kq\tilde{q}}$ be the nonnegative flow of service chain k managed by provider q and sent to the provider $\tilde{q}$ for the execution with $q \neq \tilde{q}$. It means that a variable $x_{k\overline{q}\overline{q}}$ does not exist because a generic provider $\overline{q}$ cannot send flow to itself. Let us group all these quantities into the vectors $X_{q\tilde{q}} = (x_{kq\tilde{q}})_{k=1,\ldots,K}$, $\forall q, \tilde{q} = 1,\ldots,Q$, $X_q = (x_{kq\tilde{q}})_{\substack{k=1,\ldots,K \\ \tilde{q}=1,\ldots,Q}}$, $\forall q = 1,\ldots,Q$, $\tilde{X}_q = (x_{k\tilde{q}q})_{\substack{k=1,\ldots,K \\ \tilde{q}=1,\ldots,Q}}$, $\forall q = 1,\ldots,Q$ and $X = (x_{kq\tilde{q}})_{\substack{k=1,\ldots,K \\ q,\tilde{q}=1,\ldots,Q}}$.

Costumers, that are users or devices on the ground, require a certain amount of flow (packet rate) for each service chain k, that we denote by D_k.

The flow distribution of each service chain k among the various providers must satisfy the demand for that service chain in the network, namely the following conservation law holds:

$$\sum_{q=1}^{Q} y_{kq} = D_k, \quad \forall k = 1,\ldots,K. \tag{1}$$

For each service chain k a certain amount of execution or computation space, denoted by s_k, is needed. Moreover, we denote by $\overline{C}_q$ the maximum computation capacity of the resources owned by provider q (e.g. CE in the UAVs, etc.).

We now present the demand price and cost functions. Let us introduce:

- the unit demand price of the service chain k provided by q, as the function $\rho_{kq} : \mathbb{R}_+^Q \to \mathbb{R}_+$ that depends on the executed flows, that is:

$$\rho_{kq} = \rho_{kq}(Y_k);$$

the demand price function is assumed to be continuous, continuously differentiable, decreasing with respect to the respective service chain's own demand;

- the execution cost that the provider q has to pay, as the function $c_q^{(E)}$: $\mathbb{R}_+^{K(2Q-1)} \to \mathbb{R}_+$ that depends on the net executed flow by provider q, given by the sum, for each service chain k, of the amount of flows managed by provider q but not sent to other providers and the amount of flows that q receive from others $\sum_{k=1}^{K}\left(y_{kq} + \sum_{\substack{\tilde{q}=1 \\ \tilde{q} \neq q}}^{Q}(x_{k\tilde{q}q} - x_{kq\tilde{q}})\right)$, that is:

$$c_q^{(E)} = c_q^{(E)}\left(\sum_{k=1}^{K}\left(y_{kq} + \sum_{\substack{\tilde{q}=1 \\ \tilde{q} \neq q}}^{Q}(x_{k\tilde{q}q} - x_{kq\tilde{q}})\right)\right) = c_q^{(E)}\left(Y_q, \tilde{X}_q, X_q\right);$$

- the unit request execution cost that provider q has to pay to provider $\tilde{q}$ for the execution of service chains managed by q, as the function $c_{\tilde{q}}^{(S)} : \mathbb{R}_+^{K(Q-1)} \to \mathbb{R}_+$ that depends on the flows sent by all the providers to $\tilde{q}$ for all the service chains, $\tilde{X}_{\tilde{q}}$:

$$c_{\tilde{q}}^{(S)} = c_{\tilde{q}}^{(S)}\left(\tilde{X}_{\tilde{q}}\right);$$

it is clear that $c_{\tilde{q}}^{(S)}$ represents the unit revenue that the provider $\tilde{q}$ receives from executing the service chains managed by other providers;
- the management cost that the provider q has to pay for managing the required flow for the service chain k, as the function $c_{kq}^{(M)} : \mathbb{R}_+ \to \mathbb{R}_+$, that depends on the flow of service chain k managed by provider q, y_{kq}:

$$c_{kq}^{(M)} = c_{kq}^{(M)}(y_{kq});$$

- the transmission cost from the provider q to the provider $\tilde{q}$, as the function $c_{q\tilde{q}}^{(T)} : \mathbb{R}_+^K \to \mathbb{R}_+$ that depends on the flow of all the service chains sent from q to $\tilde{q}$ $\sum_{k=1}^{K} x_{kq\tilde{q}}$:

$$c_{q\tilde{q}}^{(T)} = c_{q\tilde{q}}^{(T)}\left(\sum_{k=1}^{K} x_{kq\tilde{q}}\right) = c_{q\tilde{q}}^{(T)}(X_{q\tilde{q}});$$

In our model we assume that there is not a transmission cost from the provider $\tilde{q}$ to the provider q, incurred by q.

The aim of provider q is to maximize his profit, given by the difference between the revenues and the costs. Particularly, the revenues are obtained by the sum of unit demand price of each service chain (multiplied by the flow of service chain managed by q) and the revenues that the provider q receives from all the providers when it executes service chains managed by them; while the overall cost is defined by the sum of all the execution, request execution (to other providers) and transmission costs.

The nonlinear optimization problem formulation for the provider q is as follows:

$$\text{maximize} \qquad \sum_{k=1}^{K} \rho_{kq}(Y_k)\, y_{kq} - c_q^{(E)}\left(Y_q, \tilde{X}_q, X_q\right) +$$

$$+ \sum_{\substack{\tilde{q}=1 \\ \tilde{q} \neq q}}^{Q} \left(c_q^{(S)}\left(\tilde{X}_q\right) \cdot \left(\sum_{k=1}^{K} x_{k\tilde{q}q}\right) - c_{\tilde{q}}^{(S)}\left(\tilde{X}_{\tilde{q}}\right) \cdot \left(\sum_{k=1}^{K} x_{kq\tilde{q}}\right) \right) +$$

$$- \left[\sum_{k=1}^{K} c_{kq}^{(M)}(y_{kq}) + \sum_{\substack{\tilde{q}=1 \\ \tilde{q} \neq q}}^{Q} c_{q\tilde{q}}^{(T)}(X_{q\tilde{q}}) \right] \qquad (2)$$

subject to:

$$\sum_{\tilde{q}=1}^{Q} y_{k\tilde{q}} = D_k, \quad \forall k = 1, \ldots, K, \qquad (3)$$

$$\sum_{\substack{\tilde{q}=1 \\ \tilde{q} \neq q}}^{Q} x_{kq\tilde{q}} \leq y_{kq}, \quad \forall k = 1, \ldots, K, \qquad (4)$$

$$\sum_{k=1}^{K} s_k \cdot \left(y_{kq} + \sum_{\substack{\tilde{q}=1 \\ \tilde{q} \neq q}}^{Q} (x_{k\tilde{q}q} - x_{kq\tilde{q}}) \right) \leq \overline{C}_q, \qquad (5)$$

$$y_{k\tilde{q}}, \ x_{kq\tilde{q}}, \ x_{k\tilde{q}q} \geq 0, \quad \forall k = 1, \ldots, K, \ \forall \tilde{q} = 1, \ldots, Q. \qquad (6)$$

Constraint (3) represents the conservation law, according to which the demand D_k must be satisfied (without excesses or lacks).

Constraint (4) means that, for each service chain, each provider cannot send to other providers more flow than it manages.

The computational capacity constraint is defined by (5), which establishes that the computational space required by all the service chains for the execution on the provider q's resources must not exceed the maximum computational capacity of q, $\overline{C}_q$.

Finally, constraint (6) defines the domain of the variables.

We assume here that

$$\sum_{k=1}^{K} s_k D_k \leq \sum_{q=1}^{Q} \overline{C}_q. \qquad (7)$$

In other words, we assume that in the network there is sufficient capacity to execute all the requested service chains. We also assume that the demand price terms are continuously differentiable and concave, while all the cost terms are continuously differentiable and convex in the objective function.

3   Generalized Nash Equilibrium Problem Formulation

In this section we provide the Generalized Nash Equilibrium (GNE) problem formulation of the optimization problem (2)–(6), and then we analyze the GNE as variational equilibrium of an appropriate variational inequality.

Let us consider a non cooperative game with Q players that are the providers of the service chains in the network, each of whom are rational and acts selfishly.

Let $\mathbb{K}_q$ denote the feasible strategy vectors set corresponding to provider q, where:

$$\mathbb{K}_q = \{(Y_q, X_q, \tilde{X}_q) \in \mathbb{R}_+^{K(2Q-1)}, \text{ s.t.: } (3) - (5) \ \forall k = 1, \ldots, K; \ \forall \tilde{q} = 1, \ldots, Q\}.$$

Observing that the amount of flow that a provider q sends to another provider $\tilde{q}$ must be equal to the amount that the provider $\tilde{q}$ receives from q, we also define:

$$\mathbb{K} = \{(Y, X) \in \mathbb{R}_+^{KQ^2}, \text{ s.t.: } (3) - (5) \ \forall k = 1, \ldots, K; \ \forall q, \tilde{q} = 1, \ldots, Q \text{ hold}\}$$

in which all the constraints are shared.

As a consequence of the assumption (7), the feasible set $\mathbb{K}$ is nonempty.

Each provider q, $q = 1, \ldots, Q$, hence, seeks to maximize its payoff function $U_q(X, Y) : \mathbb{K} \to \mathbb{R}$ that corresponds to the objective function of the optimization problem (2)-(6), described in the previous section. The utility function U_q of player q depends not only on its own strategy vector, but also on the strategy vector of the other players.

We now state the following definition.

Definition 1. *A Service Chain flow distribution pattern* $(Y^*, X^*) \in \mathbb{K}$, *is said to be a Generalized Nash Equilibrium if for each providers* q, $q = 1, \ldots, Q$,

$$U_q(Y_q^*, X_q^*, \tilde{X}_q^*, \hat{Y}_q^*, \hat{X}_q^*) \geq U_q(Y_q, X_q, \tilde{X}_q, \hat{Y}_q^*, \hat{X}_q^*), \quad \forall (Y_q, X_q, \tilde{X}_q) \in \mathbb{K}_q, \quad (8)$$

where
$$\hat{Y}_q^* \equiv (Y_1^*, \ldots, Y_{q-1}^*, Y_{q+1}^*, \ldots, Y_Q^*), \ \hat{X}_q^* \equiv \left(X_{\tilde{q}\hat{q}}^*\right)_{\substack{\tilde{q}, \hat{q}=1, \ldots, Q \\ \tilde{q}, \hat{q} \neq q \\ \tilde{q} \neq \hat{q}}}.$$

According to the above definition, an equilibrium is achieved if no provider can increase its utility by unilaterally altering the value of its strategy vector, given the strategies of the other providers and the constraints.

We now introduce an equivalent GNE formulation that allows us to analyze and determine the equilibrium solution via a variational inequality through a variational equilibrium (see [8, 9]).

Definition 2. *A vector* $(Y^*, X^*) \in \mathbb{K}$ *is said to be a variational equilibrium of the above GNE if there exist the Lagrange multiplier vectors* $\lambda^{1*} \in \mathbb{R}^K$, $\lambda^{2*} \in \mathbb{R}_+^{KQ}$ *and* $\lambda^{3*} \in \mathbb{R}_+^Q$ *such that the vector* $(Y^*, X^*, \lambda^{1*}, \lambda^{2*}, \lambda^{3*}) \in \mathbb{R}_+^{KQ^2} \times \mathbb{R}^K \times \mathbb{R}_+^{KQ} \times \mathbb{R}_+^Q$ *is a solution to the variational inequality:*

Find $(Y^*, X^*, \lambda^{1*}, \lambda^{2*}, \lambda^{3*}) \in \mathbb{R}_+^{KQ^2} \times \mathbb{R}^K \times \mathbb{R}_+^{KQ} \times \mathbb{R}_+^Q$ *such that:*

$$
\sum_{q=1}^Q \sum_{k=1}^K \left[\frac{\partial c_q^{(E)}\left(Y_q^*, \tilde{X}_q^*, X_q^*\right)}{\partial y_{kq}} + \frac{\partial c_{kq}^{(M)}\left(y_{kq}^*\right)}{\partial y_{kq}} - \rho_{kq}\left(Y_k^*\right) - \frac{\partial \rho_{kq}\left(Y_k^*\right)}{\partial y_{kq}} y_{kq}^* + \right.
$$

$$
\left. + \lambda_k^{1*} - \lambda_{kq}^{2*} + s_k \lambda_q^{3*} \right] \times (y_{kq} - y_{kq}^*) +
$$

$$
+ \sum_{k=1}^K \sum_{q=1}^Q \sum_{\substack{\tilde{q}=1 \\ \tilde{q} \neq q}}^Q \left[\frac{\partial c_q^{(E)}\left(Y_q^*, \tilde{X}_q^*, X_q^*\right)}{\partial x_{kq\tilde{q}}} + \frac{\partial c_{q\tilde{q}}^{(T)}\left(X_{q\tilde{q}}^*\right)}{\partial x_{kq\tilde{q}}} + \lambda_{kq}^{2*} - s_k \lambda_q^{3*} \right] \times (x_{kq\tilde{q}} - x_{kq\tilde{q}}^*) +
$$

$$
+ \sum_{k=1}^K \sum_{q=1}^Q \sum_{\substack{\tilde{q}=1 \\ \tilde{q} \neq q}}^Q \left[\frac{\partial c_q^{(E)}\left(Y_q^*, \tilde{X}_q^*, X_q^*\right)}{\partial x_{k\tilde{q}q}} + s_k \lambda_q^{3*} \right] \times (x_{k\tilde{q}q} - x_{k\tilde{q}q}^*) +
$$

$$
- \sum_{k=1}^K \left[\sum_{q=1}^Q y_{kq}^* - D_k \right] \times (\lambda_k^1 - \lambda_k^{1*}) +
$$

$$
- \sum_{k=1}^K \sum_{q=1}^Q \left[\sum_{\tilde{q}=1}^Q x_{kq\tilde{q}}^* - y_{kq}^* \right] \times (\lambda_{kq}^2 - \lambda_{kq}^{2*}) +
$$

$$
- \sum_{q=1}^Q \left[\sum_{k=1}^K s_k \cdot \left(y_{kq}^* + \sum_{\substack{\tilde{q}=1 \\ \tilde{q} \neq q}}^Q (x_{k\tilde{q}q}^* - x_{kq\tilde{q}}^*) \right) - \overline{C}_q \right] \times (\lambda_q^3 - \lambda_q^{3*}) \geq 0,
$$

$$
\forall (Y, X, \lambda^1, \lambda^2, \lambda^3) \in \mathbb{R}_+^{KQ^2} \times \mathbb{R}^K \times \mathbb{R}_+^{KQ} \times \mathbb{R}_+^Q, \tag{9}
$$

where $\lambda^1 = (\lambda_k^1)_{k=1,\ldots,K}$; $\lambda^2 = (\lambda_{kq}^2)_{\substack{k=1,\ldots,K \\ q=1,\ldots,Q}}$, $\lambda^3 = (\lambda_q^3)_{q=1,\ldots,Q}$ *are the vectors of Lagrange multipliers associated with the constraints* (3), (4) *and* (5), *respectively.*

Note that the terms of the utility functions related to the execution request (both costs and revenues between providers) cancel out because they are opposite.

The existence of a solution to variational inequality (9) is guaranteed from the classical theory of variational analysis, since the feasible set is compact and the function that enters the variational inequality is continuous (see [12]).

4   Illustrative Numerical Example

In this section we present an illustrative numerical example for the GNEP described in the previous section. We consider a simple scenario in which there are $Q = 3$ providers and $K = 4$ service chain requests.

Let us consider the data provided in Table 1, and $\mathbb{K} = \{(Y, X) \in \mathbb{R}^{36}, \text{s.t. } (3) - -(5), \forall k = 1, \ldots, 4; q = 1, 2, 3\}$ the set of the feasible strategies. We solved the VI (9) obtaining the following optimal solutions:

Table 1. Data for the illustrative example.

Demand D_k	$D_1 = 35$, $D_2 = 40$, $D_3 = 20$, $D_4 = 45$
Computation space s_k	$s_1 = 2$, $s_2 = 10$, $s_3 = 3$, $s_4 = 5$
Computational capacity $\hat{C}_q$	$\hat{C}_1 = 370$, $\hat{C}_2 = 290$, $\hat{C}_3 = 200$
Unit demand price functions $\rho_{kq}(Y_k)$	$\rho_{11} = -0.3y_{11} + 2.2y_{12} + 2.3y_{13}$, $\rho_{32} = 1.7y_{31} - 0.5y_{32} + 2y_{33}$ $\rho_{12} = -0.1y_{12} + 1.5y_{11} + 1.5y_{13}$, $\rho_{33} = 1.1y_{31} + 1.2y_{32} - 0.5y_{33}$ $\rho_{13} = -0.2y_{13} + 1.1y_{11} + 1.2y_{13}$, $\rho_{41} = -0.2y_{41} + 1.3y_{42} + 1.5y_{43}$ $\rho_{21} = -0.1y_{21} + 1.1y_{22} + 2.3y_{23}$, $\rho_{42} = 1.2y_{41} - 0.1y_{42} + 1.3y_{43}$ $\rho_{22} = -0.3y_{22} + 1.3y_{21} + 1.1y_{23}$, $\rho_{43} = +1.1y_{41} + 1.3y_{42} - 0.3y_{43}$ $\rho_{23} = -0.1y_{23} + 1.3y_{21} + 1.1y_{22}$, $\rho_{31} = -0.5y_{31} + 1.1y_{32} + 1.2y_{33}$
Execution cost functions $c_q^{(E)}(Y_q, \hat{X}_q, X_q)$	$c_1^{(E)} = 0.3\Big[\sum_{k=1}^{K}(y_{k1} + \sum_{\hat{q}=2}^{Q}(x_{k\hat{q}1} - x_{k1\hat{q}})) \Big]^2 + 2\Big[\sum_{k=1}^{K}(y_{k1} + \sum_{\hat{q}=2}^{Q}(x_{k\hat{q}1} - x_{k1\hat{q}})) \Big]$ $c_2^{(E)} = 0.4\Big[\sum_{k=1}^{K}(y_{k2} + \sum_{\substack{\hat{q}=1 \\ \hat{q}\neq2}}^{Q}(x_{k\hat{q}2} - x_{k2\hat{q}})) \Big]^2 + 3\Big[\sum_{k=1}^{K}(y_{k2} + \sum_{\substack{\hat{q}=1 \\ \hat{q}\neq2}}^{Q}(x_{k\hat{q}2} - x_{k2\hat{q}})) \Big]$ $c_3^{(E)} = 0.25\Big[\sum_{k=1}^{K}(y_{k3} + \sum_{\substack{\hat{q}=1 \\ \hat{q}\neq3}}^{Q}(x_{k\hat{q}3} - x_{k3\hat{q}})) \Big]^2 + 4\Big[\sum_{k=1}^{K}(y_{k3} + \sum_{\substack{\hat{q}=1 \\ \hat{q}\neq3}}^{Q}(x_{k\hat{q}3} - x_{k3\hat{q}})) \Big]$
Management cost functions $c_{kq}^{(M)}(y_{kq})$,	$c_{11}^{(M)} = 0.2y_{11}^2 + 0.1y_{11}$, $c_{12}^{(M)} = 0.1y_{11}^2 + 0.5y_{11}$, $c_{33}^{(M)} = 0.35y_{33}^2 + y_{33}$ $c_{13}^{(M)} = 0.3y_{13}^2 + 0.1y_{13}$, $c_{21}^{(M)} = 0.2y_{21}^2 + 0.2y_{11}$, $c_{41}^{(M)} = 0.25y_{41}^2 + 0.5y_{41}$ $c_{22}^{(M)} = 0.5y_{22}^2 + 0.1y_{22}$, $c_{23}^{(M)} = 0.25y_{22}^2 + 0.2y_{23}$, $c_{42}^{(M)} = 0.25y_{42}^2 + y_{42}$ $c_{31}^{(M)} = 0.2y_{31}^2 + 0.2y_{31}$, $c_{32}^{(M)} = 0.4y_{32}^2 + 0.1y_{32}$, $c_{43}^{(M)} = 0.35y_{43}^2 + 2y_{43}$
Transmission cost functions $c_{q\hat{q}}^{(T)}(\sum_{k=1}^{K} x_{kq\hat{q}})$	$c_{12}^{(T)} = 0.5(\sum_{k=1}^{K} x_{k12})^2 + (\sum_{k=1}^{K} x_{k12})$, $c_{32}^{(T)} = 0.45(\sum_{k=1}^{K} x_{k32})^2 + (\sum_{k=1}^{K} x_{k32})$ $c_{13}^{(T)} = 0.2(\sum_{k=1}^{K} x_{k13})^2 + (\sum_{k=1}^{K} x_{k13})$, $c_{21}^{(T)} = 0.09(\sum_{k=1}^{K} x_{k21})^2 + (\sum_{k=1}^{K} x_{k21})$ $c_{23}^{(T)} = 0.35(\sum_{k=1}^{K} x_{k23})^2 + (\sum_{k=1}^{K} x_{k23})$, $c_{31}^{(T)} = 0.4(\sum_{k=1}^{K} x_{k31})^2 + (\sum_{k=1}^{K} x_{k31})$

$$y_{11}^* = 18.13, \quad y_{12}^* = 9.78, \quad y_{13}^* = 7.08, \quad y_{21}^* = 12.94, \quad y_{22}^* = 15.11, \quad y_{23}^* = 11.95,$$

$$y_{31}^* = 8.80, \quad y_{32}^* = 2.76, \quad y_{33}^* = 8.44, \quad y_{41}^* = 22.83, \quad y_{42}^* = 18.43, \quad y_{43}^* = 3.74$$

$$x_{113}^* = 7.38, \quad x_{123}^* = 3.75, \quad x_{kq\hat{q}}^* = 0, \text{ for all the remaining variables.}$$

These optimal solutions are computed through the Euler Method (see [13] for a detailed description) using the Matlab program on an HP laptop with an AMD compute cores 2C+3G processor, 8 GB RAM. We observe that, in equilibrium, for the providers $q = 1$ and $q = 2$, it is convenient to send part of the flows of the service chain $k = 1$ to the provider $q = 3$ for the execution.

We also analyzed seven different scenarios (instances), Si, $i = 1, \ldots, 7$, with the same network configuration described above, in which we vary the maximum computational capacity of each provider, $\overline{C}_q$, $\forall q = 1, 2, 3$, as reported in Table 2. The optimal solutions are shown in Fig. 2 and Fig. 3. Particularly, in Fig. 2 are reported the optimal variables referred to the flow of each service chain managed by each provider, y_{kq}, $\forall k = 1, 2, 3, 4$, $\forall q = 1, 2, 3$; while in Fig. 3 are reported

Table 2. Scenarios analyzed and maximum computational capacities.

Scenario	$\overline{C}_1$	$\overline{C}_2$	$\overline{C}_3$
S1	370	290	200
S2	350	300	210
S3	340	370	370
S4	320	250	480
S5	300	380	510
S6	250	250	650
S7	220	290	760

the optimal variables referred to the flow of each service chain managed by a provider but executed by another provider, $x_{kq\tilde{q}}$, $\forall k = 1, 2, 3, 4$, $\forall q, \tilde{q} = 1, 2, 3$. Note that, for example, in all seven scenarios, at the equilibrium, the value of the optimal flow variable of the service chain $k = 3$ and the second provider $q = 2$ (y_{32}) is the lowest (see Fig. 2, eighth variable). From Fig. 3 we can observe several null values. They mean that there are no packet rate transfers between some providers. In particular, we underline that there are absolutely no input flows to the second provider and output from the third.

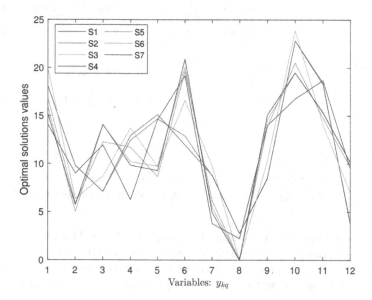

Fig. 2. Optimal solutions: amount of packet rate requested and managed by each provider, for each service chain.

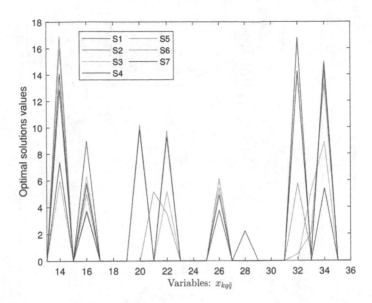

Fig. 3. Optimal solutions: amount of packet rate transferred between providers (one of which manages while the other performs)

We also analyzed, for each scenario, the used computational capacities of each provider, and compared them with the maximum ones. Figures 4, 5 and 6 show, for each provider respectively, the total computational capacity that the provider should have used if he had to run all managed services, $\sum_{k=1}^{K} s_k y_{kq}$, the total computational capacity transferred to or from other providers, $\sum_{k=1}^{K} \sum_{\tilde{q}=1}^{Q} s_k(x_{k\tilde{q}q} - x_{kq\tilde{q}})$, the real total computational capacity to use, given by the sum of the previous ones (that is, $\sum_{k} \left[s_k y_{kq} + \sum_{\tilde{q}=1}^{Q}(x_{k\tilde{q}q} - x_{kq\tilde{q}}) \right]$), and the maximum available computational capacity, $\overline{C}_q$.

In some scenarios, the total computational capacity that a provider should have used if he had to run all managed services, $\sum_{k=1}^{K} s_k y_{kq}$, results to be grater than the maximum available (see the seventh scenario for provider $q = 1$ and the fourth for provider $q = 2$). Note also that the total computational capacity transferred to or from other providers, $\sum_{k=1}^{K} \sum_{\tilde{q}=1}^{Q} s_k(x_{k\tilde{q}q} - x_{kq\tilde{q}})$ could assume a negative value. This means that such a provider (for which this condition occurs) sends flows to other providers rather than receiving them. Therefore, from Figs. 4,

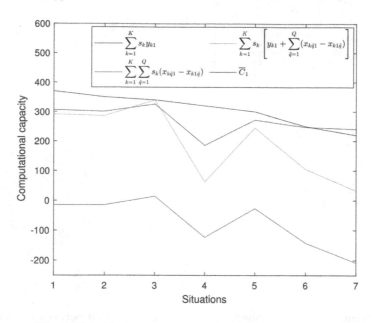

Fig. 4. Provider $q = 1$: computational capacity analysis

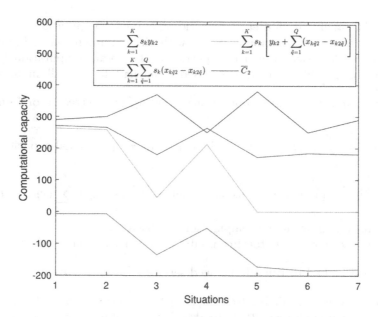

Fig. 5. Provider $q = 2$: computational capacity analysis

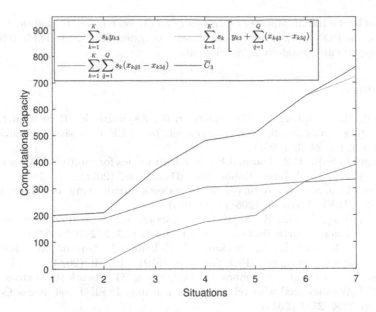

Fig. 6. Provider $q = 3$: computational capacity analysis

5 and 6, we can observe that providers $q = 1$ and $q = 2$ generally send flows to other providers, while provider $q = 3$ generally receives from other providers. Finally, we underline that, except for the last scenario, provider $q = 3$ always uses all his available computational capacity to execute services managed by himself and services received from (and managed by) other providers. These results are justified by the different cost functions.

5 Conclusion and Future Works

In this paper we propose a network based model for covering service chain requests from end users and ground devices on a multi-providers 5G FANET. We consider the scenario in which there is a single FANET, composed by UAVs belonging to different providers whose purpose is to maximize their profits, under the constraint to satisfy all requests for service chains. We studied a GNEP and solved it through a suitable variational inequality.

As future work, it would be interesting to extend the present non-cooperative game considering a scenario in which the different providers also choose how to place Virtual Functions constituting the Service Chains on their own UAVs in the FANET, in order to satisfy the service demand of the users on the ground (see [25] for VF placement optimization problem with one single provider).

Acknowledgement. The research was partially supported by the PIACERI project "OMNIA", the PRIN Project "Liquid Edge", the POR S6 project and the PNRR RESTART SUPER Project. The research of Gabriella Colajanni was also partially supported by the Italian Ministry of University and Research (MUR) and the European Union for the PON project on Research and Innovation 2014-2020, D.M. 1062/2021. These supports are gratefully acknowledged.

References

1. Erdelj, M., Natalizio, E., Chowdhury, K.R., AkyildizI. F.: Help from the Sky: leveraging UAVs for disaster management. In: IEEE Pervasive Computing, vol. 16, no. 1, pp. 24–32 (2017)
2. George, J., Sujit, P.B., Sousa, J.B.: Search strategies for multiple UAV search and destroy missions. J. Intell. Robot. Syst. **61**, 355–367 (2011)
3. Gupta, A., Jha, R.K.: A survey of 5G network: architecture and emerging technologies. IEEE Access **3**, 1206–1232 (2015)
4. Yi, B., Wang, X., Li, K., Sajal, k. D., Huang, M.: A comprehensive survey of network function virtualization. Comput. Netw. **133**, 212–262 (2018)
5. Filali, A., Abouaomar, A., Cherkaoui, S., Kobbane, A., Guizani, M.: Multi-access edge computing: a survey. IEEE Access **8**, 197017–197046 (2020)
6. Faraci, G., Grasso, C., Schembra, G.: Design of a 5G network slice extension with MEC UAVs managed with reinforcement learning. IEEE J. Sel. Areas Commun. **38**, pp. 2356–2371 (2020)
7. Faraci, G., Grasso, C., Schembra, G.: Reinforcement-learning for management of a 5G network slice extension with UAVs. In: IEEE INFOCOM 2019 - IEEE Conference on Computer Communications Workshops (INFOCOM WKSHPS), pp. 732–737 (2019)
8. Facchinei, F., Kanzow, C.: Generalized Nash equilibrium problems. Ann. Oper. Res. **175**, 177–211 (2010)
9. Kulkarni, A.A., Shanbhag, U.V.: On the variational equilibrium as a refinement of the generalized Nash equilibrium. Automatica **48**(1), 45–55 (2012)
10. Colajanni, G., Daniele, P., Sciacca, D.: On the provision of services with UAVs in disaster scenarios: a two-stage stochastic approach. Oper. Res. Forum **3**(1), 1–30 (2022). https://doi.org/10.1007/s43069-022-00127-x
11. Colajanni G., Sciacca D.: An optimization model for service requests management in a 5G network architecture. In: Masone, A., et al. (eds.): Optimization and Data Science: Trends and Applications. AIRO Springer Series, vol. 6, pp. 81–98. Springer, Cham (2021). https://doi.org/10.1007/978-3-030-86286-2_7
12. Kinderlehrer, D., Stampacchia, G.: Variational Inequalities and Their Applications. Academic Press, New York (1980)
13. Dupuis, P., Nagurney, A.: Dynamical systems and variational inequalities. Ann. Oper. Res. **44**, 9–42 (1993)
14. D'Oro, S., Galluccio, L., Palazzo, S., Schembra, G.: A game theoretic approach for distributed resource allocation and orchestration of softwarized networks. IEEE J. Sel. Areas Commun. **35**(3), 721–735 (2017)
15. Nagurney, A., Wolf, T.: A Cournot-Nash-Bertrand game theory model of a service-oriented Internet with price and quality competition among network transport providers. CMS **11**(4), 475–502 (2014)
16. Lasaulce, S., Tembine, H.: Game Theory and Learning for Wireless Networks: Fundamentals and Applications. Academic Press, Cambridge (2011)
17. Scutari, G., Palomar, D.P., Facchinei, F., Pang, J.S.: Convex optimization, game theory, and variational inequality theory. IEEE Sig. Process. Mag. **27**(3), 35–49 (2010)

18. Nagurney, A.: Network Economics: A Variational Inequality Approach, vol. 10. Springer, Dordrecht (1998). https://doi.org/10.1007/978-94-011-2178-1
19. Facchinei, F., Pang, J.S. (eds.): Finite-Dimensional Variational Inequalities and Complementarity Problems. Springer, New York (2003). https://doi.org/10.1007/b97543
20. Messous, M.A., Senouci, S.M., Sedjelmaci, H., Cherkaoui, S.: A game theory based efficient computation offloading in an UAV network. IEEE Trans. Veh. Technol. **68**(5), 4964–4974 (2019)
21. Giagkos, A., Tuci, E., Wilson, M.S., Charlesworth, P.B.: UAV flight coordination for communication networks: genetic algorithms versus game theory. Soft. Comput. **25**(14), 9483–9503 (2021). https://doi.org/10.1007/s00500-021-05863-6
22. Mkiramweni, M.E., Yang, C., Li, J., Zhang, W.: A survey of game theory in unmanned aerial vehicles communications. IEEE Commun. Surv. Tutor. **21**(4), 3386–3416 (2019)
23. Ni, J., Tang, G., Mo, Z., Cao, W., Yang, S.X.: An improved potential game theory based method for multi-UAV cooperative search. IEEE Access **8**, 47787–47796 (2020)
24. Nemer, I.A., Sheltami, T.R., Mahmoud, A.S.: A game theoretic approach of deployment a multiple UAVs for optimal coverage. Transp. Res. Part A Policy Pract. **140**, 215–230 (2020)
25. Cappello, G.M., et al.: Optimizing FANET lifetime for 5G softwarized network provisioning, submitted (2022)

Metabolic Syndrome Risk Forecasting on Elderly with ML Techniques

Elias Dritsas[(✉)], Sotiris Alexiou, and Konstantinos Moustakas

Department of Electrical and Computer Engineering, University of Patras,
26504 Rion, Greece
{dritsase,salexiou}@ceid.upatras.gr, moustakas@ece.upatras.gr
http://www.vvr.ece.upatras.gr/en/

Abstract. Metabolic syndrome is a disorder that affects the overall function of the human body. It is manifested by elevated levels of cholesterol and triglycerides, a significant reduction in energy levels, weight gain with visceral fat deposition in the abdomen, and menstrual disorders while increasing the risk of cardiovascular disease, autoimmune diseases and diabetes. A public dataset is exploited to evaluate the metabolic syndrome (MetS) occurrence risk in the elderly using Machine Learning (ML) techniques concerning Accuracy, Recall and Area Under Curve (AUC). The stacking method achieved the best performance. Finally, our purpose is to identify subjects at risk and promote earlier intervention to avoid the future development of MetS.

Keywords: Metabolic syndrome · Risk prediction · Machine learning

1 Introduction

MetS can be described as a scourge of the modern age, associated with a sedentary lifestyle and poor diet. The rates of people with metabolic syndrome are constantly increasing in the western world. The effects of this increase have already begun to show with the rise of type 2 diabetes from a young age [22]. It is a disease that has no specific symptoms. Hence, its diagnosis is made through laboratory indicators. Insulin resistance (a state of decreased activity and sensitivity, accompanied by increased insulin secretion) is a key feature of the metabolic syndrome that causes a set of symptoms that may not be immediately apparent as being related to the disorder. Under normal conditions, the human body breaks down food into glucose. Insulin is the hormone secreted by the pancreas that helps glucose pass from the blood into the cells [3,13,19].

In people with insulin resistance, the body secretes more and more insulin, which leads to the appearance of [17]:

This work has been supported by the European Union's H2020 research and innovation programme GATEKEEPER under grant agreement No 857223, SC1-FA-DTS-2018-2020 Smart living homes-whole interventions demonstrator for people at health and social risks.

D. E. Simos et al. (Eds.): LION 2022, LNCS 13621, pp. 460–466, 2022.
https://doi.org/10.1007/978-3-031-24866-5_33

- Hypertension which is characterized by high systolic above 130 mmHg or low diastolic blood pressure above 85 mmHg.
- Elevated blood sugar levels, namely, fasting glucose above 100 mg/dl.
- Low levels of good cholesterol HDL less than 40 mg /dl for men and 50 mg/dl for women.
- Elevated triglycerides with values greater than 150 mg/dl.
- Waist fat deposition with development of central obesity and increased visceral fat: waist circumference over 102 cm in men and over 88 cm in women.

In recent years, more and more research works have scientifically seen that sleep remains a major and valuable aspect of human life as, among else, it regulates the proper functioning of the human body. Hence, its absence may have serious consequences. According to [15,21,23], the "incorrect" sleep patterns (short duration <7 h and long duration >9 h) can also affect the risk of developing metabolic syndrome.

Based on the above, a big challenge in the healthcare field is the early forecasting of various chronic conditions, such as diabetes (as classification [8,12] or regression task for continuous glucose prediction [1,5]), hypertension [7], high cholesterol [9,11], COPD [4], CVDs [6], stroke [10] etc. Similar to other conditions, several research studies have been conducted for MetS using ML models. In [2], decision tree (DT) was selected for MetS features selection and data classification. This model was evaluated considering the Youden index, Positive Predicted Value (PPV), Negative Predicted Value (NPV), Sensitivity and Specificity. The key idea in this work was to derive possible rules for the determination of MetS that could enhance its diagnosis. In [14], besides to decision tree, the authors applied the support vector machine (SVM) method to predict MetS. Sensitivity, specificity and accuracy were employed to assess their behavior. In [16], the authors investigated the performance of ML methods using data sampling techniques to generate balanced training sets in order to identify dependencies between diabetes mellitus and metabolic syndrome. For this purpose, they applied DT and Naïve Bayes. In [24], SVM, DT, random forest (RF), artificial neural network (ANN), principal component analysis (PCA) and association analysis (AA) are applied for the modelling and construction of predictive models for metabolic syndrome characterization. In [25], the XGBoost model is the best performing in terms of AUC, Accuracy, Precision, F1-score, Specificity and F2-score.

An ongoing comparative study of various ML techniques is the main contribution of this work. Moreover, the current research models will be integrated into the AI services of the GATEKEEPER[1] system, which aims to improve the independence and ability overtime of the elderly and provide information to professionals to support their decision for implementing personalised prevention and intervention plans (lifestyle changes).

The rest of this paper is organized as follows. Section 2 describes the features of the dataset which are used by experts as diagnostic criteria of MetS. Section 3 presents the methods for data balancing and feature importance ranking. Also, Sect. 4 presents the evaluation of the ML models and Sect. 5 summarizes the paper.

---

[1] https://www.gatekeeper-project.eu/.

2 Dataset Description

Our research was based on a dataset from Kaggle. From this dataset, we focused on participants who are over 50 years old. The number of participants is 396, and all the attributes (11 as input to ML models and 1 for target class) are described as follows

- Age (years): This feature refers to the age of a person who (in this study) is over 50 years old.
- Gender: This feature refers person's gender. The number of men is 193 while the number of women is 203.
- Marital: This feature represents the marital status of the participants and has 5 categories(Widowed, Married, Single, Divorced and Separated).
- WaistCirc -WC (cm): It is the measurement taken around the abdomen at the level of the umbilicus.
- BMI (Kg/m^2): This feature captures the body mass index of a person.
- Albuminuria (mg/g): This feature represents the person's urine albumin level and is categorized as normal to mildly increased (<30 mg/g), moderate increased (30–300 mg/g - microalbumin) and severe increased ($\geq$300 mg/g - macroalbumin) [26].
- UrAlbCr: It captures the urine albumin to creatinine ratio. Its values define the level of albuminuria, as it is explained in the previous feature.
- UricAcid (mg/dL): Uric acid is a chemical created when the body breaks down purines.
- Blood Glucose: This feature captures the person's blood glucose level.
- HDL (mg/dL): High-density lipoprotein absorbs cholesterol and carries it back to the liver.
- Triglycerides (mg/dL): Triglycerides are a type of fat (lipid) found in human blood.
- MetS: This feature represents if a person has Metabolic Syndrome or not.

Table 1. Statistical Characteristics

Features	Min	Max	Mean ± std
Age	50	80	67.23 ± 9.36
BMI	15.7	59.2	29.97 ± 5.93
WaistCirc	66.4	145.6	102.92 ± 14.86
UrAlbCr	1.87	338.54	19.98 ± 42.09
Uric Acid	1.9	9.9	5.69 ± 1.45
Blood Glucose	73	327	113.09 ± 33.
HDL	25	108	55.08 ± 15.89
Triglycerides	41	560	139.29 ± 81.4

In Table 1, we summarize the statistical characteristics of the numerical features in the dataset. The participants are older than 50 years, and their maximum age is 80 years. In the current data, the number of participants who have been diagnosed with MetS is approximately similar in the age groups 55–59 (29), 60–64 (32), 65–69 (35) and 70–64 (32), while about two times greater (57) is the number of participants in the age group of older than 75 years. Moreover, the distribution of patients with MetS and albuminuria severity level is shown in Table 2.

Table 2. Patients with MetS and Albuminuria level per age group

	Level 0	Level 1	Level 2
50–54	16	3	0
55–59	25	3	0
60–64	30	1	0
65–69	20	1	0
70–74	26	4	0
75+	39	21	2

Most of the patients have urine albumin to creatinine ratio in the normal to mild range and even less in the moderate class. In the third class, only patients older than 75 years have occurred. As in [18], here, it is also verified that the prevalence of both metabolic syndrome and albuminuria increases with age.

In this study, 27 women and 24 men suffer from MetS with the simultaneous presence of waist circumference above 88 for females and 102 for males, triglycerides above 150 and HDL lower than 50 and 40, respectively. In Table 3, we see MetS patients distribution for each criterion separately. Also, in the dataset, there are 162 MetS patients with glucose levels above 100. These patients are mainly distributed in the overweight (50) and obese (96) classes. In Table 4, we present the prevalence of MetS in overweight and obesity classes when the waist circumference criterion is also satisfied.

Table 3. WaistCirc, HDL, Triglycerides per class

WC	No	Yes
>102 (male)	34	84
>88 (female)	71	91
HDL	No	Yes
<40 (male)	5	43
<50 (female)	6	42
Triglycerides	No	Yes
>150	12	112

Table 4. WaistCirc vs BMI classes

	No	Yes		No	Yes
Overweight	48	50	Obese I	29	69
WC > 102	20	22	WC > 102	10	39
WC > 88	28	28	WC > 88	19	30
Obese II	8	28	Obese III	4	16
WC > 102	2	15	WC > 102	0	8
WC > 88	6	13	WC > 88	4	8

3   Data Preprocessing and Feature Importance

In this study, random oversampling has been applied to produce a balanced dataset. For the training of the ML-based models, all features were kept, except for race and income. Based on the relevant literature, we focus only on the most important risk factors for metabolic syndrome. To estimate the importance of an attribute x, we employed Gain Ratio (GR) method. The Gain Ratio of an attribute x is calculated as $GR(x) = \frac{IG(x)}{H(x)}$, where $IG(x)$, $H(x)$ capture the Information Gain and entropy of x, respectively [20]. The entropy of x is defined as $H(x) = -\sum_i P(x_i)log_2(P(x_i))$ where $P(x_i)$ captures the probability to have the value x_i by considering all values of an attribute. In the balanced dataset the features' importance and the related weight are as follows: Triglycerides (0.2972), BMI (0.1795), WaistCirc (0.1698) BloodGlucose (0.1526), HDL (0.1065), Age (0.0845), UricAcid (0.0825), UrAlbCr (0.0519), Albuminuria (0.0357), Marital (0.0105) and Sex (0.0000186).

Table 5. Machine learning models performance

	Logistic regression		SVM (linear)		MultiLayer perceptron		Random forest		Stacking	
	Yes	No	Yes	No	Yes	No	Yes	No	Yes	No
Accuracy	0.783		0.785		0.888		0.909		0.919	
Recall	0.773	0.793	0.785	0.785	0.899	0.879	0.919	0.899	0.924	0.914
AUC	0.891	0.891	0.785	0.785	0.927	0.927	0.974	0.974	0.971	0.971

4   Performance Evaluation of ML Models

In this section, the performance of several ML models is evaluated in the WEKA environment using 10-cross validation on the balanced dataset. Logistic Regression (LR), Support Vector Machine (SVM), MultiLayer Perceptron (MLP), Random Forests (RFs) and a Stacking ensemble (using as base classifiers the previous models while as a meta classifier the LR) was applied.

The results in Table 5 indicate that LR and (linear) SVM models present similar satisfactory accuracy and recall 78.5%. The LR model demonstrated higher AUC, and it can discriminate the prevalence of MetS with a higher probability than SVM in populations similar to the dataset. An even higher performance demonstrated RF (as a single classifier) and the superior outcomes were acquired by the stacking method. Stacking performed best concerning the accuracy and recall metrics with a bit (0.3%) lower AUC. Finally, this model is considered powerful for the personalized risk assessment of the MetS in the context of the GATEKEEPER system.

5   Conclusions

In this research work, a publicly available dataset was considered to examine the order of importance of specific risk factors on MetS, aiming at risk prediction in older people living at home. A limitation of the current study is that in the features set was not available the blood pressure in relation to metabolic syndrome. Several ML methods were assessed and the Stacking method was found to yield the best prediction performance against the single classifiers. The results of the stacking method presented consistently high accuracy (0.919), recall (0.919) and AUC (0.971), a fact that seems promising for the discrimination ability of the model regarding possible subjects with MetS.

In future work, we aim to extend the Machine Learning framework through the use of Deep Learning methods by applying Long-Short-term-Memory (LSTM) algorithm and Convolutional Neural Networks (CNN) based on the same dataset and comparing the results concerning the aforementioned metrics.

References

1. Alexiou, S., Dritsas, E., Kocsis, O., Moustakas, K., Fakotakis, N.: An approach for personalized continuous glucose prediction with regression trees. In: 2021 6th South-East Europe Design Automation, Computer Engineering, Computer Networks and Social Media Conference (SEEDA-CECNSM), pp. 1–6. IEEE (2021)
2. Babič, F., Majnarić, L., Lukáčová, A., Paralič, J., Holzinger, A.: On patient's characteristics extraction for metabolic syndrome diagnosis: predictive modelling based on machine learning. In: Bursa, M., Khuri, S., Renda, M.E. (eds.) ITBAM 2014. LNCS, vol. 8649, pp. 118–132. Springer, Cham (2014). https://doi.org/10.1007/978-3-319-10265-8_11
3. Basciano, H., Federico, L., Adeli, K.: Fructose, insulin resistance, and metabolic dyslipidemia. Nutr. metabol. **2**(1), 1–14 (2005)
4. Dritsas, E., Alexiou, S., Moustakas, K.: COPD severity prediction in elderly with ML techniques. In: Proceedings of the 15th International Conference on PErvasive Technologies Related to Assistive Environments, pp. 185–189 (2022)
5. Dritsas, E., Alexiou, S., Konstantoulas, I., Moustakas, K.: Short-term glucose prediction based on oral glucose tolerance test values. In: International Joint Conference on Biomedical Engineering Systems and Technologies - HEALTHINF, vol. 5, pp. 249–255 (2022)
6. Dritsas., E., Alexiou., S., Moustakas., K.: Cardiovascular disease risk prediction with supervised machine learning techniques. In: Proceedings of the 8th International Conference on Information and Communication Technologies for Ageing Well and e-Health - ICT4AWE, pp. 315–321. INSTICC, SciTePress (2022)
7. Dritsas, E., Fazakis, N., Kocsis, O., Fakotakis, N., Moustakas, K.: Long-term hypertension risk prediction with ML techniques in ELSA database. In: Simos, D.E., Pardalos, P.M., Kotsireas, I.S. (eds.) LION 2021. LNCS, vol. 12931, pp. 113–120. Springer, Cham (2021). https://doi.org/10.1007/978-3-030-92121-7_9
8. Dritsas, E., Trigka, M.: Data-driven machine-learning methods for diabetes risk prediction. Sensors **22**(14), 5304 (2022)
9. Dritsas, E., Trigka, M.: Machine learning methods for hypercholesterolemia long-term risk prediction. Sensors **22**(14), 5365 (2022)
10. Dritsas, E., Trigka, M.: Stroke risk prediction with machine learning techniques. Sensors **22**(13), 4670 (2022)

11. Fazakis, N., Dritsas, E., Kocsis, O., Fakotakis, N., Moustakas, K.: Long-term cholesterol risk prediction with machine learning techniques in ELSA database. In: International Joint Conference on Computational Intelligence (IJCCI), pp. 445–450. SCIPTRESS (2021)
12. Fazakis, N., Kocsis, O., Dritsas, E., Alexiou, S., Fakotakis, N., Moustakas, K.: Machine learning tools for long-term type 2 diabetes risk prediction. IEEE Access **9**, 103737–103757 (2021)
13. Freeman, A.M., Pennings, N.: Insulin resistance. StatPearls [Internet] (2021)
14. Karimi-Alavijeh, F., Jalili, S., Sadeghi, M.: Predicting metabolic syndrome using decision tree and support vector machine methods. ARYA Atherosclerosis **12**(3), 146 (2016)
15. Konstantoulas, I., Kocsis, O., Dritsas, E., Fakotakis, N., Moustakas, K.: Sleep quality monitoring with human assisted corrections. In: International Joint Conference on Computational Intelligence (IJCCI), pp. 435–444. SCIPTRESS (2021)
16. Perveen, S., Shahbaz, M., Keshavjee, K., Guergachi, A.: Metabolic syndrome and development of diabetes mellitus: predictive modeling based on machine learning techniques. IEEE Access **7**, 1365–1375 (2018)
17. Raikou, V.D., Gavriil, S.: Metabolic syndrome and chronic renal disease. Diseases **6**(1), 12 (2018)
18. Shih, H.M., Chuang, S.M., Lee, C.C., Liu, S.C., Tsai, M.C.: Addition of metabolic syndrome to albuminuria provides a new risk stratification model for diabetic kidney disease progression in elderly patients. Sci. Rep. **10**(1), 1–9 (2020)
19. Tappy, L., Lê, K.A.: Metabolic effects of fructose and the worldwide increase in obesity. Physiol. Rev. (2010)
20. Trabelsi, M., Meddouri, N., Maddouri, M.: A new feature selection method for nominal classifier based on formal concept analysis. Procedia Comput. Sci. **112**, 186–194 (2017)
21. Troxel, W.M., et al.: Sleep symptoms predict the development of the metabolic syndrome. Sleep **33**(12), 1633–1640 (2010)
22. Vollenweider, P., Eckardstein, A.v., Widmann, C.: HDLS, diabetes, and metabolic syndrome. von Eckardstein, A., Kardassis, D. (eds.) High Density Lipoproteins. Handbook of Experimental Pharmacology, vol. 224, pp. 405–421. Springer, Cham (2015). https://doi.org/10.1007/978-3-319-09665-0_12
23. Wolk, R., Somers, V.K.: Sleep and the metabolic syndrome. Exp. Physiol. **92**(1), 67–78 (2007)
24. Worachartcheewan, A., Schaduangrat, N., Prachayasittikul, V., Nantasenamat, C.: Data mining for the identification of metabolic syndrome status. EXCLI J. **17**, 72 (2018)
25. Yang, H., et al.: Machine learning-aided risk prediction for metabolic syndrome based on 3 years study. Sci. Rep. **12**(1), 1–11 (2022)
26. Zhang, A., et al.: The relationship between urinary albumin to creatinine ratio and all-cause mortality in the elderly population in the Chinese community: a 10-year follow-up study. BMC Nephrol. **23**(1), 1–10 (2022)

Airport Digital Twins for Resilient Disaster Management Response

Evangelia Agapaki$^{(\boxtimes)}$ (iD)

Construction Management, M.E. Rinker, Sr. School of Construction Management,
University of Florida, Gainesville, FL 32603, USA
agapakie@ufl.edu
https://www.evagapaki.com/

Abstract. Airports are constantly facing a variety of hazards and
threats from natural disasters to cybersecurity attacks and airport stakeholders are confronted with making operational decisions under irregular
conditions. We introduce the concept of the *foundational twin*, which can
serve as a resilient data platform, incorporating multiple data sources and
enabling the interaction between an umbrella of twins. We then focus on
providing data sources and metrics for each foundational twin, with an
emphasis on the environmental airport twin for major US airports.

Keywords: Digital twin · Instance segmentation · Graphical neural
network · Geometric deep learning · Point clouds · Computer vision

1 Introduction

The complexity of airport operations regardless of their size extends beyond
the airside side of operations. Natural disasters, climate change threats, high
annual passenger demand, large volumes of cargo and baggage, concessionaires
and vendors as well as other airport tenants may extend an airport's operations
beyond capacity or disrupt operations. Figure shows the airport systems and
stakeholders in an airport. The American Society of Civil Engineers (ASCE)
has rated airport infrastructure with grade "D" and this finding is based on
the anticipated higher passenger demand compared to infrastructure capacity
(Bureau of Transportation Statistics, 2019). Moreover, irregular operations and
disruptions due to internal or external threats can have serious consequences in
cities [24]. For example, in December 2017, one of the busiest airports in the
world, Atlanta's Hartsfield-Jackson airport, suffered an 11hr-long power outage,
which disrupted the airport's operations and also incurred economic losses [34].
These incidents necessitate the need for a resilient airport.

Resilience incorporates the ability to (a) anticipate, prepare for, and adapt to
changing conditions, (b) to absorb, (c) to withstand, respond to, and (d) to recover
rapidly from disruptions. The implementation of resilient solutions in airports can
be performed by preventing or mitigating disruptive events to air traffic operations
[8, 27, 44]. Those events can be either severe weather hazards (e.g., dense fog, flooding, snow, drought, tornado, wildfire, hurricane), threats (e.g., equipment outages,

© Springer Nature Switzerland AG 2022
D. E. Simos et al. (Eds.): LION 2022, LNCS 13621, pp. 467–486, 2022.
https://doi.org/10.1007/978-3-031-24866-5_34

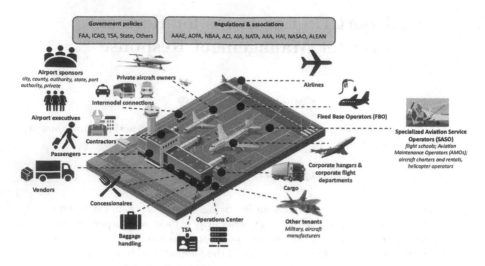

Fig. 1. Airport systems and stakeholders

political changes, economic downturn, pandemics, cyber-attacks, physical attacks) and vulnerabilities (e.g., equipment outages, lack of staff).

Resilience can be quantified by analyzing risks to an airport. We adopt the definition of airport risks by the National Infrastructure Protection Plan (NIPP) [39], where risk is defined by the likelihood and the associated consequences of an unexpected event. Those risks are the hazards most likely to occur, potential threats, and vulnerabilities. Hazards and threats refer to incidents that can damage, destroy, or disrupt a site or asset. The difference between hazards and threats is that the former can happen unexpectedly, typically outside of an airport's control whereas the latter happen purposefully and are usually manmade. Some examples of hazards are natural hazards (e.g., hurricanes, earthquakes, wildfire), technological (e.g., infrastructure failure, poor workmanship, or design), or human-caused threats (e.g., accidents, cyberattacks, political upheaval). The consequences associated with the vulnerabilities of an airport, as a result of a hazard or threat being realized, is one way to measure the impacts associated with risks. Therefore, risk is defined as the multiplication of consequences, probability and vulnerabilities of an airport.

Resilience analysis includes both the time before (planning capability), during (absorbing capability) and after a disruption event occurs (recovery and adaptation capability), including the actions taken to minimize the system damage or degradation, and the steps taken to build the system back stronger than before. Figure 2(a) shows this timeline and the planning [7, 13, 18, 19, 30, 31, 43, 45], absorbing [7, 13, 30, 31, 35, 42, 43, 45], recovering [2, 41, 43, 45, 47], and adapting [7, 13, 41, 43, 45, 47] phases of a resilience event. As shown in Fig. 2, the system initially is in a steady state. After the disruptive event occurs at t_d, the system's performance starts decreasing and then a contingency plan is implementing at time t_c. Then, there are four "recovery" scenarios. In the first scenario (blue line), the performance of the system gradually recovers without any outside intervention until it reaches the original steady state. In the second scenario (purple, dashed line), the system first reaches a new steady state, but eventually returns to its originally state. For example, temporary routes and measures are taken to meet immediate needs of airport operations when the system is damaged due to a hurricane. However, it may take weeks or months for the system to fully recover. The worst scenario is when the system cannot recover (red dashed line). The last scenario is to reach the recovery state earlier by using a holistic Digital Twin (DT) framework (green line), which will be discussed in the last section of this paper. Figure 2(b) showcases the risk assessment adoption framework in airport operations [11].

This paper targets to identify the areas of highest risks for an airport, so that these can serve as indicators to inform policies and investment decisions.

1.1   Background on Airport Resilience

In recent years, there has been a lot of research on resilience in airport operations. Multiple studies have implemented Cost-Benefit Analysis (CBA) tools to estimate costs and benefits after implementing security measures in their security risk assessment policies. However, CBA analysis cannot validate the estimated airport security costs, therefore multiple simulation experiments are needed to investigate the interdependencies between different stakeholders and systems [32]. To overcome these limitations, the ATHENA project investigated a framework to evaluate curbside traffic management control measures and traffic scenarios at the Dallas-Fort Worth International Airport (DFW) [37], optimization of shuttle operations that can lead to 20% energy reductions [29] and traffic demand forecasting [22].

Researchers have also focused on risk assessment models of airport security. Lykou et al. (2019) [23] developed a model for smart airport network security with the objective to mitigate malicious cyberattacks and threats. Zhou and Chen (2020) [46] proposed a method to evaluate an airport's resilient performance under extreme weather events. Their results demonstrated that airport resilience greatly varies based on the level of modal substitution, airport capacity and weather conditions. Previous research has greatly focused on airport security protection [36,44,48]. Agent-based modeling has been used to represent sociotechnical elements of an airport's security system and identify states and behaviors of its agents such as weather, pilots, aircrafts, control tower operators [33]. Recently, Huang et al. (2021) [17] proposed a Bayesian Best Worst Method that identifies the optimal criteria weights with a modified Preference Ranking Organization method for Enrichment evaluations (modified PROMETHEE) to make pairwise comparisons between alternatives for each criterion. They evaluated their method in three airports in Taiwan. This system relies on the judgement of experts for the evaluation of multiple, even overlapping criteria based on pre-determined evaluation scales.

However, a comprehensive framework for resilient management response for airports has not yet been developed. This is a complex and difficult Multiple Cri-

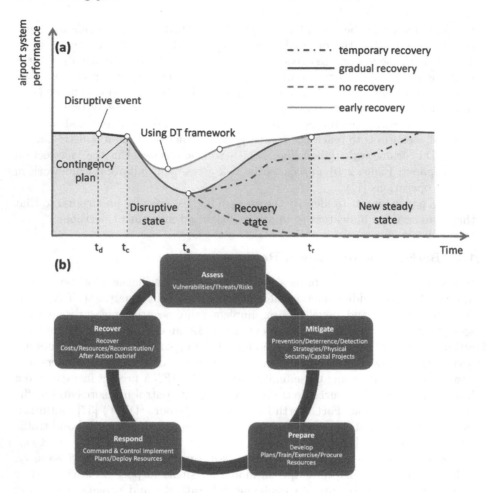

Fig. 2. (a) Resilience framework overview with and without the use of Digital Twins and (b) risk assessment adoption in airports (modified from Crosby et al., 2020).

teria Decision-Making (MCDM) problem. The objective of MCDM is to identify an optimal solution when taking into account multiple overlapping or conflicting criteria. This study intends to develop a framework that leverages recent advances in digital twin technologies and identify the criteria and metrics, which will act as a guidance for a holistic disaster management response.

1.2 Resilience Indexes

Previous literature has focused on identifying multiple metrics (resilience indexes) related to aviation and airport safety and operations [2,4,7,12,13,18, 19,21,24,30,31,35,41,41–43,45,47]. Figure 3 summarizes the most widely used resilience metrics for each resilience phase as described above. These metrics take into account the airport's physical facilities, personnel, equipment and the disaster response phases.

Resilience indexes and web-based platforms have been widely developed for communities. Two widely adopted tools are the National Risk Index (FEMA, 2022) and the ASCE Hazard Tool (ASCE, 2022). The former calculates risk index scores per each US county based on 18 natural hazards by computing the expected annual loss due to natural disasters multiplied by the social vulnerability divided by the community resilience. The ASCE Hazard tool provides reports on natural disasters with widely known parameters. These tools need to be taken into account when considering the exposure of airports to natural disasters.

The uncertainly of disruption occurrences in conjunction with the complexity of airport infrastructure and operations (Fig. 1) necessitates the need for a unified digital platform to integrate information related to airport operations as well as interactions between airport sub-systems and their accurate representation.

2   Airport Digital Twin Framework

Technological innovations have the potential to: (a) capture the detailed geometry of the physical infrastructure and generate the asset's digital twin, (b) enrich the geometric digital twin with real-time sensor data, (c) update, maintain and communicate with the digital twin and (d) leverage the digital twin to monitor the asset's performance and improve decision-making by planning interventions well before the time of need. The goal of this research is to develop a foundational digital twin template that can be implemented across airports of all sizes. This leverages the use of digital twin technologies to explore alternative future scenarios for a more resilient airport. The foundational digital twin will be used as the main framework and specific systems of the digital twin will be investigated. In particular, the key objectives are the following:

- O1: Airport Digital Twin (ADT) definition in the context of operational airport systems.
- O2: Airport Digital Twin (ADT) generation and maintenance. We propose a framework for static and dynamic information curation based on existing sensor data and airport systems.

We introduce the concept of the foundational airport digital twin (Fig. 4), which incorporates an umbrella of twins that could interact with each other; these twins are, but not limited to, the geometric, financial, operations, social and environmental twin. If successful implementation of the foundational twin is achieved, then it can lead to improved efficiency and operations as well as better planning in the presence of irregular events that are of paramount importance for airport executives and stakeholders. The Foundational Digital Twin will serve as a resilient data backbone for airport infrastructure systems and will enable the implementation of more advanced twins such as the adaptive/planning and intelligent twin. As illustrated in Fig. 4, the adaptive twin encapsulates simulated scenarios towards a proactive plan of operating an airport, where planned

Resilience phase	Metrics	References
Plan	Ground crews' safety awareness and risk alertness at work	Zhao et al. (2017); Singh et al. (2019); Ergun and Bulbul (2019)
	Periodic inspection of airport runways	Zhao et al. (2017); Humphries and Lee (2015)
	Accuracy of instruments and system testing at security checkpoints	Zhao et al. (2017); Ergun and Bulbul (2019); Skorupski and Uchronski (2016)
	Safety Management System (SMS) integrity	Zhao et al. (2017); Chen and Li (2016); Ergun and Bulbul (2019)
	Early warning accuracy and timeliness of aviation weather service stations	Zhao et al. (2017); Chen and Li (2016)
	Reliability of airport security and surveillance system	Zhou et al. (2018); Yang et al. (2015); Ergun and Bulbul (2019)
	Reliability of the airport's epidemic prevention system	Huizer et al. (2015)
Absorb	Emergency procedures when flammables and explosives are identified	Singh et al. (2019)
	Ground crew training on airport safety	Singh et al. (2019); Zhao et al. (2017); Ergun and Bulbul (2019)
	Number of security personnel and aviation police officers	Yang et al. (2015); Skorupski and Uchronski (2016)
	Comprehensive physical drainage system	Birgani and Yazdandoost (2018)
	Planning & management of tarmacs, number of collisions between vehicles, machinery and aircraft	Chen and Li (2016)
	Earthquake prevention measures for terminals	Singh et al. (2019)
Recover	Adequacy of fire protection resources inside and outside terminals	Zhao et al. (2017); Bao and Zhang (2018)
	Stability of communication systems in various departments at an airport	Zhao et al. (2017); Yang et al. (2018)
	Comprehensive emergency evacuation measures and clear escape instructions	Bao and Zhang (2018)
	Cost of maintenance work for terminals	Yang et al. (2015); Wallace and Webber (2017)
	Morale of airport staff for post-disaster reconstruction	Yang et al. (2015); Zhou et al. (2018)
Adapt	Timeliness of airport runway repair operations	Humphries and Lee (2015)
	Spare power generation equipment to ensure uninterrupted Information Systems	Yang et al. (2015); Wallace and Webber (2017)
	Spare Management Information Systems (MIS)	Yang et al. (2015); Wallace and Webber (2017)

Fig. 3. Literature review on metrics for each airport resilience phase

interventions will be more sophisticated than ever before. The data collection, modeling and intervention will become increasingly automated. That level of automation will lead to the Intelligent Twin, where we envision an informed decision-making system with minimal human intervention.

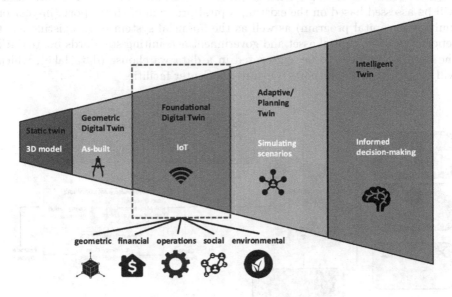

Fig. 4. Digital Twin framework

We expand on the definition of the Foundational Digital Twin in Fig. 5. The geometric twin entails spatial data collection as well as their intelligent processing, Building Information Modeling (BIM) and validation with laser scanned data and GIS data integration. The process of laser scanning to BIM has already been applied to other complex infrastructure assets and is named as geometric digital twinning [1]. The Financial twin should have the capacity to: (a) simulate the allocation of funding from a variety of sources and visualize it in different physical assets at the airport, (b) visualize potential conflicts in funding utilization in real-time and (c) facilitate the fiscal management of an airport expansion and renewal projects. The Social twin should visualize the human demand on infrastructure and predict social behaviors based on historical data. In particular, it should: (a) integrate geospatial and airport-specific data (e.g., area and airport infrastructure reachability in correlation to the number of runways, taxiways), (b) integrate and process demographic data such as urban indexes and population around the airport to predict human demand and (c) integrate geographic and urban data related to the existing built environment surroundings that can affect passenger demand. Lastly, the environmental twin should account for natural hazards, energy consumption, occupancy rates, pollution and air volume for airports to be on track to achieve net-zero carbon infrastructure by 2050 (UN Environment Program, 2020).

As presented in Fig. 5, the foundational digital twin will be integrated with existing infrastructure Asset Management (AM) software and the configuration parameters (asset lifecycle, risk management, consequence of failure, probability of failure) will be predicted based on the capabilities of the twin. The asset will be registered in a Common Data Environment (CDE) and the infrastructure needs will be assessed based on the existing capital program of the airport (in-year or multi-year capital program) as well as the financial system (e.g., existing asset reports, tangible capital asset and government accounting standards board). All the data is expected to be aggregated in a data warehouse (data lake), which will be hosted in the airports' Operations Center facilities.

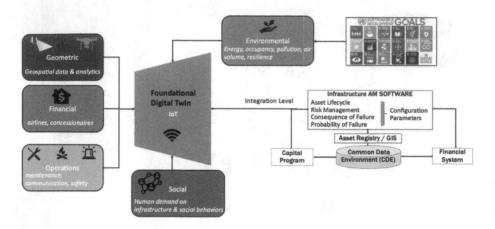

Fig. 5. Definition of foundational digital twin

2.1   Exploration of Threats and Hazards for Airport Digital Twins

The threats and hazards are grouped into each Foundational Twin and are summarized in Fig. 6. Each group is then analyzed below.

Geometric. The majority of aging industrial facilities lacks accurate drawings and documentation [1]. Without capturing the existing geometry of an asset accurately, the incurred information loss throughout an asset's lifecycle would be immense.

Financial. Budget reductions and overruns, economic downturns, inefficient funding allocations are some of the financial threats airport authorities may be encountered with [15,38].

Operational. Equipment maintenance is critical for airport operations (SMS Pilot Studies; FAA, 2019). The condition of passenger bridge boarding and ancillary equipment is another critical component that needs to be reliably assessed and maintained.

Social. Airports are at risk of malicious events that can put their operations in jeopardy. Such incidents have been reported in the Los Angeles International airport in 2002 and 2013, when assailants killed airport personnel during attacks [10,25]. In 2014 an intoxicated passenger attacked another passenger in a hate crime at the Dallas airport [28], indicating the need for vigilance at all levels. Incidents of cyberattacks continue to increase and airport infrastructure is a target for those. Zoonotic diseases[1] will likely increase in their extent and frequency in the future [20]. We need to look into the diseases that have the highest likelihood of infecting regions where the airports operate. For example, in Texas, the most common vectors include ticks and mosquitoes carrying spotted fever rickettsiosis and West Nile virus, respectively [6]. Other unknown zoonotic diseases could arise in the future in the same way that the novel coronavirus is thought to have emerged as a zoonotic disease and then rapidly spread around the world, primarily via airports and air travel [40]. In such scenarios, the airport's operations could facilitate the spread of new zoonotic diseases from one person to another, including passengers, flight crews, airport workers, transportation personnel, and others. Other societal threats are demographic changes (e.g., sudden population growth) and unprecedented industrial or staff accidents.

Environmental. These threats include natural hazards and impacts of climate change that can significantly alter meteorological conditions, which may affect airport operations. Extreme natural events can lead to power system and flight disruptions. Studies have estimated that these events will be exacerbated to a modest degree by climate change [14,26]. Although current research cannot definitively conclude whether climate change will increase or decrease the frequency of tornadoes in every situation, overall warmer temperatures will likely reduce the potential for wind shear conditions that lead to tornadoes [5,16]. Severe wind could become more common because previous research has shown that climate change is responsible for a gradual increase in world average wind speeds [9]. Fewer and/or less powerful tornadoes could help improve operations, because airport structures and visitors are less likely to be damaged or harmed by airborne debris caused by tornadoes. On the other hand, stronger overall winds could lead to more difficult weather conditions for pilots of arriving or departing aircraft.

Climate change also has the potential to increase the annual average temperatures and to make the temperature swings more extreme. This can result in an increase in electricity demand as HVAC systems respond to temperature changes. This increased electricity demand can stress system components in both daily operations and in heat wave and cold snap events.

There are three features of the twin that have the potential to assist decision making better than any other technology tool. Those are: (a) interoperability between software tools, which facilitates better communication between multiple stakeholders, (b) relationship mapping between both static (e.g., static infrastructure) and dynamic entities (e.g., humans, moving infrastructure) and

---

[1] A zoonotic disease is one in which an animal acts as intermediary for disease transmission between the vector and the infected human (e.g., Lyme disease occurs in white-footed mice, but it is transmitted to humans via ticks).

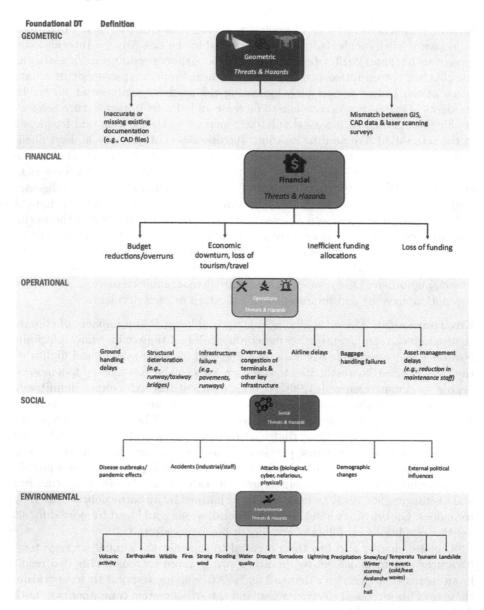

Fig. 6. Definition of foundational DTs in relation to hazards and threats

(c) semantics that allow Artificial Intelligence (AI) tools and data analytics to forecast future scenarios. An example of airport stakeholders involved in the operational twin is presented in Fig. 7. Figure 8 shows semantics in the geometric digital twin applied to heavy industrial facilities. We will also investigate potential data sources for the generation and maintenance of DTs in the next section.

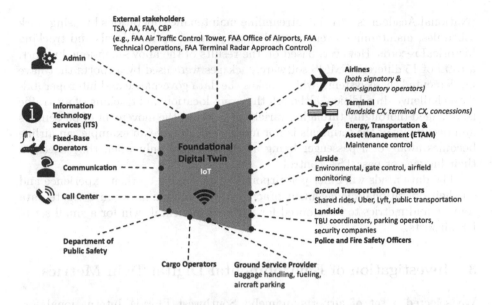

Fig. 7. Airport stakeholders of the operational twin

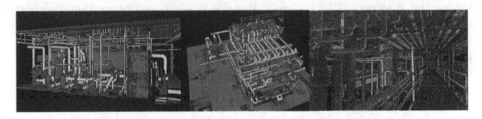

Fig. 8. Automated geometric digital twinning on industrial assets (Agapaki, 2020)

2.2 Data Sources for the Generation and Maintenance of Airport DTs

A sample of data sources and existing datasets for each Foundational Twin is summarized in Figs. 9 and 10. Existing literature provides the insight that there is an abundance of systems related to air traffic control (i.e., ADS-B, SWIM, ERAM). FAA's goal of the NEXT Generation Air Transportation System (NextGen) is a national upgrade of the air traffic control systems. The System Wide Integration Management (SWIM) is the first implementation of this vision, and it integrates a variety of data systems, including weather, communication radar information, traffic flow management systems, en route flight plan changes, arrival and departure procedures, microburst information, NOTAMs, storm cells, wind shear and terminal area winds aloft. However, the systems related to operating, maintaining and predicting failures and irregular operations in the landside and terminal operations are limited. Airport Computer Maintenance Management Systems (CMMS) have been proposed for use in airports

(National Academies, 2015) to streamline maintenance operations by using work schedules, maintaining inventory and spare parts at optimal levels and tracking historical records. However, based on the results of the above-mentioned report, a total of 15 different CMMS software packages were used by airports that have implemented these systems, which makes the data governance and interoperability a challenge. In addition, although the exact location and tracking of air traffic is managed thoroughly through a variety of systems, the movement of passengers and occupants of the terminals is not monitored. To give an example, an airline becomes aware of a passenger being at the terminal only when they check in their luggage or pass TSA control.

The data needs will differ per airport size depending on their experience and on their operational management practices. At a minimum, we investigate data sources and metrics to be adopted for the environmental twin for a small set of US airports.

3   Investigation of Environmental Digital Twin Metrics

We selected a set of airports, namely: Southwest Florida International airport (RSW), Hartsfield-Jackson International airport (ATL), Charlotte Douglas International airport (CLT), Ronald Reagan Washington International airport (DCA), William Hobby airport (HOU), Dalls/Fort Worth International airport (DFW), Dallas Love Field airport (DAL), Austin-Bergstrom International airport (AUS) and Orlando international airport (MCO). The reason for our selection was based on their number of enplanements, being in the top-10 busiest US airports for large and medium hub airports [3].

We investigated the natural hazard risk index scores for selected counties, where the identified airports operate. Figure 11 presents the relative distribution of hazard type risk index scores for each county, which can be used as one of the data sources for the environmental airport DT.

Another indicator of irregular airport operations is the number of cancelled and/or delayed flights arriving or departing at the airport. We collected data from TranStats (Bureau of Transportation Statistics, 2022) for the on-time arrival and departure flights. Figure 12 shows the arrivals and departures in five of the above-mentioned airports and their distribution per year for the last ten years. Then, we looked at extreme weather events that have occurred and could affect the operation of these airports. In particular, we investigated Hurricane Irma. Hurricane Irma occurred in September 2017 affecting MCO, RSW, ATL, CLT and DCA airports and primarily ATL airport with 3.76 and 3.78% cancelled arrival and departure flights respectively. A power outage also affected the ATL airport with nearly 5% cancelled flights in December 2017. Similarly, RSW, CLT, DCA and MCO were affected with the majority of their canceled flights being in September 2017 (Fig. 13). Another important aspect that should be investigated is the recovery time of airport operations after these events occurred, which is part of future work.

Foundational Twin	Data Source	Existing Datasets
GEOMETRIC	Spatial data: Light Detection and Ranging (LIDAR) data & photogrammetric surveys of infrastructure (improving ACRP Report 39)	Limited availability
	GIS data, surveillance-broadcast	Automatic Dependent Surveillance-Broadcast (ADS-B) by FAA
	3D modeling data: BIM models (improving ACRP Report 39)	Limited availability
OPERATIONS	Flight data	SWIM Flight Data Publication Service (SWIM)
		En Route Automation Modernization (ERAM)
		Runway surface friction datasets
		Terminal Flight Data Manager (TFDM)
		Terminal Automation Modernization and Replacement (TAMR)
		Operations Network (OPSNET)
	Ground handling	Aircraft turn times (gate occupancy)
		Common Use Passenger Processing Platforms (CUPPS)
		Common Use Self Service (CUSS) Platform
	Baggage handling	Time to load & offload baggage
	Passenger terminal operations	Airport Operational Database (AODB)
		Common Use Terminal Equipment (CUTE)
	Airport security	Security Screening Checkpoint (SSC)
	Pavement surface conditions	Airfield Pavement Management System (APMS)
	Safety	Aviation Safety Reporting System (ASRS)
		Aviation Safety Information Analysis and Sharing System (ASIAS)
		NTSB Accident Reports
		FAA Aviation Incident Data System (AIDS)
	Customer satisfaction	Airport Service Quality (ASQ)
FINANCIAL	Annual financial reports	FAA Certification Activity Tracking System (CATS) database
	Terminal concessions	Sales per passenger
SOCIAL	Geospatial & airport-specific datasets	OurAirports database
		Open Street Maps geometric data
		GlobalAirports database
	Demographic data	United Nations demographic data
	Urban data	INFORM Index
		Urban risk index Economy
		Logistic Performance Index

Fig. 9. Data sources and existing datasets for the geometric, operations, financial and social foundational twin

Foundational Twin	Data Source	Existing Datasets
Environmental	Noise monitoring system	National Transportation Noise map
		Airport Noise Contours
		Noise complaint data
	Energy management system	Utility usage
		HVAC system data (e.g., Morpheus)
	Airport traffic	DFW real-time airport traffic demand of North and South entrances (e.g., ATHENA project)
		DFW shuttle routes (e.g., ATHENA project)
	Weather (wind, rain, snow)	Weather reports (NCEI)
	Natural Hazards	FEMA, ASCE Hazard Tools
	Climate change	USGCRP indicator

Fig. 10. Data sources and existing datasets for the environmental foundational twin

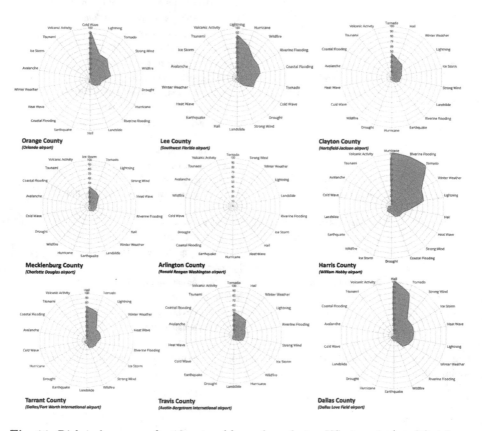

Fig. 11. Risk index scores for 18 natural hazards and nine US airports (modified from FEMA).

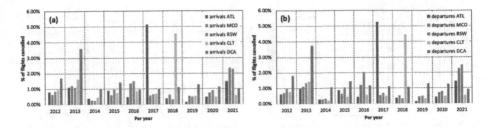

Fig. 12. Distribution of percentage of cancelled (a) arrival and (b) departure flights for ATL, MCO, RSW, CLT and DCA airports.

Climate change metrics are another significant factor that needs to be taken into account when generating and maintaining a digital twin. The National Climate Assessment Representative Concentration Pathways (RCP) provide key indicators for quantifying climate change. Those are: cooling degree days, heating degree days, average daily mean temperature, annual number of days with minimum and maximum temperatures beyond threshold values, annual precipitation, dry days, and days with more than 2" precipitation. Cooling and heating degree days refer to the number of hours per year and the degrees above or below 65 °F as detailed by the NOAA methodology (US Department of Commerce 2022). This methodology means that the cooling degree days and heating degree days can exceed 365 as they are multiplied with the number of hours per year and the magnitude of temperatures above or below 65 °F.

Temperature has the potential to impact human health. It can also affect power, transport, and water system resilience. For example, at the DFW airport, the number of cooling degree days is projected to increase from 2600 to 3400 and the number of heating degree days is projected to decrease from 2400 to 1900 (both according to the low emissions scenario). Cooling degree days will increase electricity consumption from air conditioning, increase on site water consumption, and can exacerbate the airport's peak electricity demand which was analyzed in a subsequent section. Both maximum and minimum ambient temperature can affect electricity demand. The average high and low temperatures at 2 m above the ground, the average precipitation and average wind speed are compared for ATL, CLT, RSW, DFW and HOU. The average precipitation is computed by accumulating rainfall over the course of a sliding 31-day period and the average wind speed is computed as the mean hourly wind speed at 10 m above the ground.

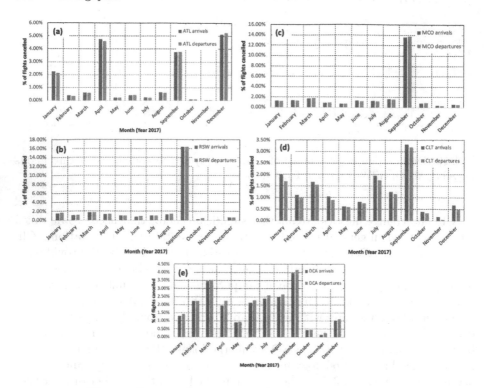

Fig. 13. Per month percentage of cancelled flights in 2017.

Aside from the average temperature, wind and precipitation values, heat waves and cold spells can have an acute effect on airport operations. Using the SAFRAN methodology (EPA, 2021), those can be identified in future work.

The overall amount of precipitation, dry days, and days with significant rainfall (>2" per day) is expected to change very little from present time through 2050. We used DFW as an example, where there have been several historical events where extreme precipitation has led to power system and flight disruptions. These events have included snowfall (US Department of Commerce 2021a; US Department of Commerce 2021b; Narvekar 2011; NBCDFW 2021; Lindsey 2021; L'Heureux 2021), droughts (Runkle and Kunkle 2017; Hegewisch and Abatzoglou 2021; Centers for Disease Control and Prevention National Environmental Public Health Tracking 2020), and floods (USGCRP 2018; First Street Foundation 2021) (Fig. 14).

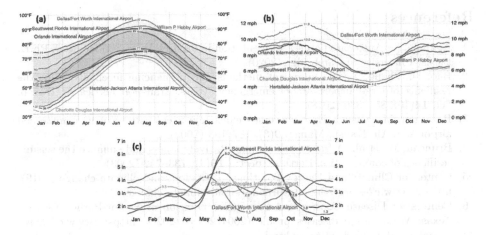

Fig. 14. (a) Average high and low temperature (b) average wind speed and (c) average monthly rainfall.

4 Discussion

When dealing with risk, an airport organization has the choice to either accept the risk, avoiding the risk by planning and preparing for interrupted activities or working to eliminate the risk through mitigation. This research reviewed the existing risks in the context of gathering available data sources, grouped them into categories (geometric, financial, social and environmental) and developed a unified framework for the assessment of risks using multiple criteria. We particularly emphasized environmental threats and provided metrics as well as open-source existing databases for the evaluation of those risks.

5 Conclusion

There have been many studies investigating airport resilience frameworks, however a unified framework that identifies and then combines multiple data sources has not yet been investigated. The aim of this study was to use a foundational digital twins in order to identify key threats and hazards. We then suggested metrics and data sources for the environmental digital twin that can be used as guidance for the development of a unified and integrated data framework for the DT development. Future research directions include the foundational digital twin implementation on airport case studies.

Acknowledgments. We thank the Florida Space Institute (NASA) for sponsoring this research. We gratefully acknowledge the collaboration of all academic and industrial project partners. Any opinions, findings and conclusions or recommendations expressed in this material are those of the authors and do not necessarily reflect the views of the institutes mentioned above.

References

1. Agapaki, E.: Automated object segmentation in existing industrial facilities. Ph.D. thesis, University of Cambridge (2020)
2. Bao, D., Zhang, X.: Measurement methods and influencing mechanisms for the resilience of large airports under emergency events. Transportmetrica A: Transp. Sci. **14**(10), 855–880 (2018)
3. Bazargan, M., Vasigh, B.: Size versus efficiency: a case study of us commercial airports. J. Air Transp. Manag. **9**(3), 187–193 (2003)
4. Bruneau, M., et al.: A framework to quantitatively assess and enhance the seismic resilience of communities. Earthq. Spectra **19**(4), 733–752 (2003)
5. Center for Climate and Energy Solutions: Tornadoes and climate change (2019). https://www.c2es.org/content/tornadoes-and-climate-change/
6. Centers for Disease Control and Prevention Division of Vector-Borne Diseases: Texas: Vector-borne diseases profile (2004–2018) (2020). https://www.cdc.gov/ncezid/dvbd/vital-signs/texas.html
7. Chen, W., Li, J.: Safety performance monitoring and measurement of civil aviation unit. J. Air Transp. Manag. **57**, 228–233 (2016)
8. Clark, K.L., Bhatia, U., Kodra, E.A., Ganguly, A.R.: Resilience of the us national airspace system airport network. IEEE Trans. Intell. Transp. Syst. **19**(12), 3785–3794 (2018)
9. Climate Central: Extreme weather and climate change (2021). https://www.climatecentral.org/library/climopedia/extreme-weather-and-climate-change-the-southeast
10. CNN: Cnn.com - Los Angeles airport shooting kills 3 - July 5, 2002 (2004). https://web.archive.org/web/20041204221915/, http://archives.cnn.com/2002/US/07/04/la.airport.shooting/
11. Crosby, Missouri, L.: Airport security vulnerability assessments. Program Appl. Res. Airport Secur. (2020)
12. Damgacioglu, H., Celik, N., Guller, A.: A route-based network simulation framework for airport ground system disruptions. Comput. Industr. Eng. **124**, 449–461 (2018)
13. Ergün, N., Bülbül, K.G.: An assessment of factors affecting airport security services: an AHP approach and case in turkey. Secur. J. **32**(1), 20–44 (2019)
14. Future Climate Dashboard' Web Tool. Climate Toolbox: Future climate dashboard: Location: 32.9025o n, 97.0433o w. (2021). https://climatetoolbox.org/tool/Future-Climate-Dashboard
15. Graham, A., Morrell, P.: Airport Finance and Investment in the Global Economy. Routledge, London (2016)
16. Hausfather, Z.: Tornadoes and climate change: what does the science say? (2019). https://www.carbonbrief.org/tornadoes-and-climate-change-what-does-the-science-say-2
17. Huang, C.N., Liou, J.J., Lo, H.W., Chang, F.J.: Building an assessment model for measuring airport resilience. J. Air Transp. Manag. **95**, 102101 (2021)
18. Huizer, Y., Swaan, C., Leitmeyer, K., Timen, A.: Usefulness and applicability of infectious disease control measures in air travel: a review. Travel Med. Infect. Dis. **13**(1), 19–30 (2015)

19. Humphries, E., Lee, S.J.: Evaluation of pavement preservation and maintenance activities at general aviation airports in Texas: practices, perceived effectiveness, costs, and planning. Transp. Res. Rec. **2471**(1), 48–57 (2015)
20. James, M., Gage, K., Khan, A.: Potential influence of climate change on vector-borne and zoonotic diseases: a review and proposed research plan. Environ. Health Perspect. **118**, 1507–14 (2010). https://doi.org/10.1289/ehp.0901389
21. Janić, M.: Modeling the resilience of an airline cargo transport network affected by a large scale disruptive event. Transp. Res. Part D: Transp. Environ. **77**, 425–448 (2019)
22. Lunacek, M., et al.: A data-driven operational model for traffic at the Dallas fort worth international airport. J. Air Transp. Manag. **94**, 102061 (2021). https://doi.org/10.1016/j.jairtraman.2021.102061, https://www.sciencedirect.com/science/article/pii/S0969699721000442
23. Lykou, G., Anagnostopoulou, A., Gritzalis, D.: Smart airport cybersecurity: threat mitigation and cyber resilience controls. Sensors **19**(1), 19 (2018)
24. Metzner, N.: A comparison of agent-based and discrete event simulation for assessing airport terminal resilience. Transport. Res. Procedia **43**, 209–218 (2019)
25. News, N.: AX shooting: TSA officer Hernandez bled for 33 minutes at scene - report - U.S. news (2013). https://web.archive.org/web/20131118070335/, http://usnews.nbcnews.com/_news/2013/11/15/21471203-lax-shooting-tsa-officer-hernandez-bled-for-33-minutes-at-scene-report
26. NOAA National Centers for Environmental Information: Texas - key findings (2017). https://statesummaries.ncics.org/downloads/TX-screen-hi.pdf
27. Pishdar, M., Ghasemzadeh, F., Maskeliūnaitė, L., Bražiūnas, J.: The influence of resilience and sustainability perception on airport brand promotion and desire to reuse of airport services: the case of Iran airports. Transport **34**(5), 617–627 (2019)
28. Post, W.: Man tackled at Dallas airport after homophobic attack (2014). https://www.washingtonpost.com/news/morning-mix/wp/2014/10/27/man-tackled-at-dallas-airport-after-spewing-homophobic-slurs/
29. Sigler, D., et al.: Route optimization for energy efficient airport shuttle operations - a case study from Dallas fort worth international airport. J. Air Transp. Manag. **94**, 102077 (2021). https://doi.org/10.1016/j.jairtraman.2021.102077, https://www.sciencedirect.com/science/article/pii/S0969699721000600
30. Singh, V., Sharma, S.K., Chadha, I., Singh, T.: Investigating the moderating effects of multi group on safety performance: the case of civil aviation. Case Stud. Transp. Policy **7**(2), 477–488 (2019)
31. Skorupski, J., Uchroński, P.: A fuzzy system to support the configuration of baggage screening devices at an airport. Expert Syst. Appl. **44**, 114–125 (2016)
32. Stewart, M.G., Mueller, J.: Cost-benefit analysis of airport security: are airports too safe? J. Air Transp. Manag. **35**, 19–28 (2014)
33. Stroeve, S.H., Everdij, M.H.: Agent-based modelling and mental simulation for resilience engineering in air transport. Saf. Sci. **93**, 29–49 (2017)
34. Sun, X., Wandelt, S., Zhang, A.: Resilience of cities towards airport disruptions at global scale. Res. Transport. Bus. Manag. **34**, 100452 (2020)
35. Tahmasebi Birgani, Y., Yazdandoost, F.: An integrated framework to evaluate resilient-sustainable urban drainage management plans using a combined-adaptive MCDM technique. Water Resour. Manag. **32**(8), 2817–2835 (2018)
36. Thompson, K.H., Tran, H.T.: Operational perspectives into the resilience of the us air transportation network against intelligent attacks. IEEE Trans. Intell. Transp. Syst. **21**(4), 1503–1513 (2019)

37. Ugirumurera, J., et al.: A modeling framework for designing and evaluating curbside traffic management policies at Dallas-Fort worth international airport. Transport. Res. Part A: Policy Pract. **153**, 130–150 (2021). https://doi.org/10.1016/j.tra.2021.07.013, https://www.sciencedirect.com/science/article/pii/S0965856421001956

38. Urazova, N., Kotelnikov, N., Martynyuk, A.: Infrastructure project planning. In: IOP Conference Series: Materials Science and Engineering, vol. 880, p. 012105. IOP Publishing (2020)

39. US Department of Homeland Security: National infrastructure protection plan, pp. 29–33 (2013)

40. Van Beusekom, M.: Studies trace Covid-19 spread to international flights (2020). https://www.cidrap.umn.edu/news-perspective/2020/09/studies-trace-covid-19-spread-international-flights

41. Wallace, M., Webber, L.: The disaster recovery handbook: a step-by-step plan to ensure business continuity and protect vital operations, facilities, and assets. Amacom (2017)

42. Willemsen, B., Cadee, M.: Extending the airport boundary: connecting physical security and cybersecurity. J. Airport Manag. **12**(3), 236–247 (2018)

43. Yang, C.L., Yuan, B.J., Huang, C.Y.: Key determinant derivations for information technology disaster recovery site selection by the multi-criterion decision making method. Sustainability **7**(5), 6149–6188 (2015)

44. Yanjun, W., Jianming, Z., Xinhua, X., Lishuai, L., Ping, C., Hansen, M.: Measuring the resilience of an airport network. Chin. J. Aeronaut. **32**(12), 2694–2705 (2019)

45. Zhao, J.N., Shi, L.N., Zhang, L.: Application of improved unascertained mathematical model in security evaluation of civil airport. Int. J. Syst. Assur. Eng. Manag. **8**(3), 1989–2000 (2017)

46. Zhou, L., Chen, Z.: Measuring the performance of airport resilience to severe weather events. Transp. Res. Part D: Transp. Environ. **83**, 102362 (2020)

47. Zhou, L., Wu, X., Xu, Z., Fujita, H.: Emergency decision making for natural disasters: an overview. Int. J. Disaster Risk Reduct. **27**, 567–576 (2018)

48. Zhou, Y., Wang, J., Yang, H.: Resilience of transportation systems: concepts and comprehensive review. IEEE Trans. Intell. Transp. Syst. **20**(12), 4262–4276 (2019)

Strategies for Surviving Aggressive Multiparty Repeated Standoffs (Extended Abstract)

Evangelos Kranakis[1,2]($\boxtimes$) (iD)

[1] School of Computer Science, Carleton University, Ottawa, ON, Canada
[2] Research Supported in Part by NSERC Discovery Grant, Ottawa, Canada
kranakis@scs.carleton.ca

Abstract. A multiparty standoff system, arises from a confrontation involving quarrelling parties for which no strategy may exist for any party to achieve victory and the parties cannot withdraw from the conflict without suffering a loss (which may even include their demise). In a repeated standoff the parties are pursuing in rounds an aggressive attack strategy, which may involve selecting opponents and "repeatedly firing shots" whose success depends on a random adversary. Each participating party has been pre-assigned a probability indicating the chance a given shot will succeed against an opponent.

Eventually, every standoff terminates with the resulting system consisting in the limit (of the number of rounds) of isolated nodes. We consider confrontation strategies and analyze the resulting survivability of the participating parties. We investigate what strategy should co-operating parties follow so as to maximize the number of surviving nodes. We are also interested in what strategy should a given party (or group of parties) follow so as to maximize its chance of survival.

Keywords: Duel · Graph · Node · Parties · Round · Shooting · Standoff · Survival

1 Introduction

A confrontation involving quarrelling parties for which there may not exist a strategy for any party to achieve victory and in which the parties cannot withdraw from the conflict without suffering a significant loss is called a standoff. In this research investigation, we consider multiparty, repeated standoffs involving many peers and study the survival characteristics of the participating peers involved under the effect of aggressive behaviour in an interconnected system which is modelled as a graph with (un)directional links.

Repeated standoffs are modelled as synchronous, interconnected systems represented by graph. The parties are nodes which are acting aggressively attacking each other according to a certain strategy and causing their neighbors to perish (or retreat or be eliminated) with a certain probability which may depend on

© Springer Nature Switzerland AG 2022
D. E. Simos et al. (Eds.): LION 2022, LNCS 13621, pp. 487–504, 2022.
https://doi.org/10.1007/978-3-031-24866-5_35

the capabilities of the parties involved. The strategy is executed in synchronous rounds and may change from round-to-round. In general, we are interested in developing strategies that optimize the overall probability of survival of the parties as well as the expected number of survivals in the graph.

Consider a synchronous system of nodes forming a connected graph with (un)directional edges. Assume the nodes of the graph are occupied by shooters, one shooter per vertex. Shooters may shoot synchronously and in rounds. All shooters follow their own attack strategy which involves shooting at a chosen neighboring opponent. They have unlimited ammunition and can shoot at will anyone of their outgoing neighbors. In a given round a shooter u may attack a single neignboring node and either eliminate it with probability p_u (independent of the neighbor chosen by u) or fail to eliminate it with probability $1 - p_u$. The shooting continues as long as there are non-isolated nodes in the graph, i.e., that have not been eliminated and have at least one non-eliminated neighbor. A node survives if it is not eliminated by any of its neighbors.

Given such a directed graph, we are interested in proposing and analyzing attack strategies and investigate the following questions.

1. What is an optimal survival strategy of a given node as measured by its maximum survival probability?
2. How do the nodes of a coalition set S (of two or more nodes) in a graph coordinate their attack strategies so as to maximize the survival probability of the nodes in S?
3. What is the expected number of surviving nodes for a given strategy?
4. Can we give a node strategy which maximizes the expected number of survivals in a given graph?

In the sequel, we give strategies for a variety of topologies, including circular unidirectional and bidirectional ring and complete bipartite graphs, and study their node and group survival characteristics.

1.1 Model, Notation, and Terminology

In the sequel we describe the basic elements of the model and give necessary definitions of concepts to be used throughout the paper.

Consider a directed (not necessarily) strongly connected graph $G = (V, E)$ with directed edges. For each vertex $u \in V$, let $N_+(u)$ and $N_-(u)$ be the set of outgoing and ingoing neighbors of u, respectively, and $d_+(u)$ and $d_-(u)$ the corresponding out- and in-degree of u.

Each vertex of the graph has the capabilities of a shooter in that it is equipped with an inexhaustible amount of ammunition and can fire a single bullet at a time (in a round) to any of its outgoing neighbors which it may select according to a strategy. A node u (referred to as the *aggressor*) may attack node v (referred to as the *victim*) if (u, v) is a directed arrow (edge) with tail the node u and head the node v. A shooter at node $u \in V$ is associated with a probability $0 \le p_u \le 1$ such that when it is attacking a targeted neighbor v, u eliminates v

with probability p_u and fails to eliminate v with probability $1-p_u$. The parameter p_u is given to each node in advance and cannot be altered during the execution of an algorithm.

Shooting is done synchronously by all shooters and independently of each other at random in the given graph. After completing round $t-1$ all nodes move to round t. At round t all shooters that are alive in this round may shoot at their (newly chosen) targets. Following an algorithm, in a round, a node that is alive chooses to attack a single neighbor. More than one shooter may choose to attack the same victim. A shooter fatally shot in a given round t will be considered eliminated (or dead) and will therefore be out of action and withdraw from all subsequent rounds t', where $t' > t$. Shooters have an inexhaustible supply of ammunition. At the end of a round t nodes emerge either dead or alive. A shooter that survives in a round t will shoot again at a target of its own choice (as this is determined by its own strategy) in the next round $t+1$ as long as there exist nodes within its own target set which have survived.

A *repeated standoff* (or standoff, for short) is a multiparty confrontation in which none of the opponents appears to have a measurable advantage. In our setting a standoff is defined by a directed (not necessarily) strongly connected graph $G = (V, E)$. As the standoff evolves the original graph is also evolving and changing. At round $t = 0$, let $G_0 := G$ be the original graph representing an interconnected system of shooters. If G_t is the graph that has survived in round t then G_{t+1} is the graph resulting from G_t after removing the nodes that were eliminated after round t as well as their adjacent edges.

During the standoff the in- and out-degree of a node in the graph evolves. A node u in the current graph is *isolated* if it can nether attack (either because it has no outgoing neighbors or because $p_u = 0$) or be attacked by another node in the graph (either because it has no ingoing neighbors or because every ingoing node $v \in N_-(u)$ has attack probability $p_v = 0$).

An attack strategy for shooter u is an algorithm $\mathcal{A}_u = \{\mathcal{A}_u(t) : t \geq 0\}$ which is executed by node u (as long as it has not been eliminated) and describes at which neighbors u should shoot at each round t. Shooters are interested in designing an algorithm $\{\mathcal{A}_u(t) : u \in V, t \geq 0\}$ that will maximize their survival probability or alternatively minimize the probability that they will be eliminated in a round.

A *duel* is a confrontation between two parties, while *standoffs* are multiparty confrontations. We use the notation $u \to_t v$ to indicate that node u attacks node v in round t and $u \to v$ to indicate that $u \to_t v$, for some $t \geq 0$. When $u \to_t v$ then the notation $u \Rightarrow_t v$ (resp. $u \not\Rightarrow_t v$) means that u eliminates (resp. does not eliminate) v in round t. Similarly, $u \Rightarrow v$ means that $u \Rightarrow_t v$, for some $t \geq 0$.

1.2   Related Work

There are many references of duels in the literature. A historical development can be found in [2]. In the present paper, a duel can be thought of as a standoff involving two "aggressive" peers. There are numerous studies on aggression models in social networks which focus on the causes of aggression and its effects in various segments of the population (e.g, education, work-place, etc), e.g., [5,16,21]. Additional studies are focusing on analyzing how existing (social) network topologies and network relations between peers are causing verbal aggressive behaviour, e.g., [3,4]. Ways to reduce aggression and competition in computer games is studied in [22].

Duelling is a known form of resolution in the animal kingdom and plays an important role in the evolution of species. The repeated standoffs described resemble confrontation of battling rams as can be seen in the following video [17]. In addition, there are numerous studies in Behavioural Biology and Ecology where one is interested in the evolutionary origins and history of animal contests, and conflict resolution, as well as patterns arising. For example, [1] investigates models of fight escalation, a game theoretic study on animal contests [19] can be found in the edited volume [8,13] discusses dyadic (two-party) and multiparty contests, and [14] discusses patterns in conflicts.

Winkler in [23][page 33] mentions the related problem of Group Russian Roulette. There are n angry and armed people in a room. At each chime of the clock each person synchronously spins around and shoots a random other person. Persons shot fall dead and the survivors spin and shoot again at the next round; eventually, either everyone is dead or there is a single survivor. The question posed is what is the limiting probability that there will be a survivor? This probability does not tend to a limit but rather varies according to the fractional part of $\ln n$ (see also [11]). A related problem has been proposed by Prodinger [18] in which n people select a loser by flipping coins (whose outputs are 0 or 1) and recursively selecting one of their peers. It is shown that this process stops on the average in about $\log_2 n$ steps. Interestingly, this problem has an interesting application in randomized leader election whereby at every stage those peers that survived so far flip a biased coin (whose outcome is Head or Tail), and those who received, say a Tail. survive for the next round. The process continues until only one peer remains [15]. An example of a Mexican standoff (three nodes in a triangle with bidirectional edges) is mentioned in [10]. For the definition of the geometric distribution and other related probability concepts see any standard textbook, e.g., [20].

It will be seen that finding an optimal (in terms of peer survival) strategy in aggressive repeated standoffs is anything but a simple problem. In the sequel we will restrict our attention merely to standoffs on some special graphs, including rings and complete bipartite. To the best of our knowledge the model proposed is new and there is no related mathematical literature on the standoff problem on a graph discussed in our paper.

1.3   Outline of the Paper

In Sect. 2 we begin with observations on general graphs and in Subsect. 2.2 we analyze the standard dyadic (two-person) duel. In Sect. 3 we consider rings; in Subsect. 3.1 we analyze standoffs for a unidirectional ring, while in Subsect. 3.2 we analyze the triadic (Mexican) standoff. In Sect. 4 we remind the reader of a well-known result about structural properties of balanced networks considered in social network theory which motivates the next section. In Sect. 5 we consider standoffs for complete bipartite graphs. We conclude the paper with Sect. 6 where we discuss additional open problems and possibilities for extension and further research. Due to space limitations this is an extended abstract and all missing proofs will appear in a forthcoming full paper.

2 Observations on Standoffs

We begin with some observations on standoffs for general graphs and also discuss the simple two-person duel.

2.1 General Graphs

Consider a connected graph with n nodes, denoted $1, 2, \ldots, n$, all of which are shooters. A standoff in this graph is non-trivial if not all the p_i are 0. Two simple observations are valid for all non-trivial standoff situations in which all nodes which are alive eventually shoot at all their neighbors at repeated rounds.

Observation 1: If $p_i > 0$, for some $i = 1, 2, \ldots, n$, then the probability that all nodes survive is 0.

Indeed, the probability that all nodes of the graph survive the t-th round will be at most $(1 - p_1)^{t_1}(1 - p_2)^{t_2} \cdots (1 - p_n)^{t_n}$, where $t_1, t_2, \ldots, t_{n-} \le t$ and t_i is the number of rounds that the i-th node failed to kill a neighbor, which converges to 0, if at least one of $t_1, t_2, \ldots, t_n$ converges to ∞.

Observation 2: If $p_i > 0$, for all $i = 1, 2, \ldots, n$, then with probability 1 in the limit as $t \to \infty$ the resulting graph consists only of isolated nodes.

Indeed, consider the final standoff graph and assume that some pair of nodes, say i, j remain connected for ever. Clearly this is impossible because they will keep shooting at each other until at least one of the nodes is eliminated.

2.2 Dyadic (One-to-One) Duel

We begin our study by discussing a basic standoff known as "one-to-one duel" involving only two nodes. Unlike traditional duels in which the two peers send only a single shot against each other synchronously, we consider a model of duel in which the peers shoot in synchronous rounds.

Consider a duel with two shooters as depicted in Fig. 1. In this standoff there is a simple shooting strategy since shooters are not given a choice which neighbor to shoot at and instead keep shooting synchronously at each other. A shooter will keep shooting as long as it and its opponent are alive. Eventually, either

Fig. 1. Two parties u and v in a duel.

both shooters may get eliminated at the same round or a shooter survives the duel if he is the first to eliminate its opponent and is not eliminated at the same round. In this setting we can prove the following theorem.

Theorem 1. *Consider a one-to-one standoff with two nodes u, v. Then*

1. $\Pr[u\ declared\ winner] = \frac{p_v}{p_u+p_v-p_up_v}$
2. $\Pr[v\ declared\ winner] = \frac{p_u}{p_u+p_v-p_up_v}$
3. $\Pr[both\ u, v\ are\ eliminated] = \frac{p_up_v}{p_u+p_v-p_up_v}$

Remark 1. The quantity $p_u + p_v - p_up_v$ in the denominator above is the probability of the event that in a shootout at least one of the two shooters eliminates its opponent. Therefore $\frac{p_u}{p_u+p_v-p_up_v}$, $\frac{p_v}{p_u+p_v-p_up_v}$, and $\frac{p_up_v}{p_u+p_v-p_up_v}$ are the probabilities of the events that in a duel, shooter u survives, v survives, and no shooter survives, respectively, under the condition that at least one of the two shooters eliminates its opponent.

3   Standoffs on Rings

In this section we analyze standoffs on unidirectional rings and also analyze the case of the three-node bidirectional ring which is known as the Mexican standoff.

3.1   Standoffs on Unidirectional Rings

First we consider unidirectional rings. Consider a unidirectional ring of n nodes labeled $0, 1, \ldots, n-1$ and placed in this order along the ring. Consider labeling the nodes in the counter clockwise orientation as depicted in Fig. 2 (the clockwise orientation is similar). Throughout this section we abbreviate $i \pm 1 \bmod n$ with $i \pm 1$, respectively. Suppose that node i can shoot node $i + 1$ and the probability of success (resp. failure) is p_i (resp. $1 - p_i$),

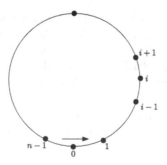

Fig. 2. Standoff in a unidirectional ring of n nodes $0, 1, \ldots, n-1$ orinted counter clockwise.

For simplicity, from now on assume that $0 < p_i < 1$, for all i. For each i, let K_i be the random variable that counts the number of rounds until node i is eliminated by node $i-1$, Note that $\Pr[K_i = k_i] = (1-p_i)^{k_i-1}p_i$. Now we look at the probability that a node *survives the shootout* (i.e., never gets eliminated). It is clear that node i survives iff $i-1$ is eliminated by $i-2$ before i is eliminated by $i-1$. It follows that $\Pr[i \text{ survives}] = \Pr[K_{i-1} < K_i]$.

Theorem 2. *Consider a ring of n nodes labeled $0, 1, \ldots, n-1$, where $n \geq 2$.*

1. *In clockwise orientation, $\Pr[i \text{ survives}] = 1 - \frac{p_{i-1}}{1-(1-p_{i-1})(1-p_{i-2})}$, and*
2. *In counter clockwise orientation, $\Pr[i \text{ survives}] = 1 - \frac{p_{i+1}}{1-(1-p_{i+1})(1-p_{i+2})}$,*

for all i.

Proof. Consider only the clockwise direction. The counter clockwise orientation is similar. Indeed, elementary calculations show that

$$\Pr[i \text{ survives}] = \Pr[K_{i-1} < K_i]$$

$$= \sum_{k_{i-1} < k_i} \Pr[K_{i-1} = k_{i-1}] \cdot \Pr[K_i = k_i]$$

$$= \sum_{k_i=1}^{\infty} \sum_{k_{i-1}=1}^{k_i-1} (1-p_{i-2})^{k_{i-1}-1}p_{i-2}(1-p_{i-1})^{k_i-1}p_{i-1}$$

Simplifying the righthand side of the last equation above we derive the following

$$\Pr[i \text{ survives}] = p_{i-2}p_{i-1} \sum_{k_i=1}^{\infty} (1-p_{i-1})^{k_i-1} \sum_{k_{i-1}=1}^{k_i-1} (1-p_{i-2})^{k_{i-1}-1}$$

$$= p_{i-2}p_{i-1} \sum_{k_i=1}^{\infty} (1-p_{i-1})^{k_i-1} \cdot \frac{1-(1-p_{i-2})^{k_i-1}}{1-(1-p_{i-2})}$$

$$= p_{i-1} \sum_{k_i=1}^{\infty} (1-p_{i-1})^{k_i-1} \cdot (1-(1-p_{i-2})^{k_i-1})$$

$$= p_{i-1} \sum_{k_i=1}^{\infty} (1-p_{i-1})^{k_i-1} - p_{i-1} \sum_{k_i=1}^{\infty} ((1-p_{i-1})(1-p_{i-2}))^{k_i-1}$$

$$= 1 - \frac{p_{i-1}}{1-(1-p_{i-1})(1-p_{i-2})}. \tag{1}$$

This proves Theorem 2.   □

The probability that i gets eliminated in the clockwise (and counter clockwise) orientation is a simple corollary.

Corollary 1. *Consider a ring of n nodes $0, 1, \ldots, n-1$. If $p_i > 0$ then*

$$\Pr[\text{both } i \text{ and } i+1 \text{ survive}] = 0.$$

Proof. Without loss of generality assume a clockwise orientation. Consider the event S_i that node i survives. Now observe that

$$\begin{aligned}
&\Pr[\text{both } i \text{ and } i+1 \text{ survive}] \\
&= \Pr[\text{all shots by } i \text{ do not eliminate } i+1 \ \& \ S_i] \\
&= \Pr[\text{all shots by } i \text{ do not eliminate } i+1 | S_i] \cdot \Pr[S_i] \\
&= \left(\lim_{k \to \infty} (1 - p_i)^k \right) \cdot \Pr[S_i] = 0.
\end{aligned}$$

This proves Corollart 1. □

Corollary 2. *Consider a ring of n nodes $0, 1, \ldots, n-1$. For each i we can compute the probability that for a given i, nodes i and $i+l$ survive while all nodes $i+1, \ldots, i+l-1$ between i and $i+l$ die.*

Proof. Without loss of generality assume a clockwise orientation. The event describing this probability is the following

$$(i \text{ and } i+l \text{ survive}) \ \& \ (i+1 \text{ and } i+2 \text{ and } \cdots \text{ and } i+l-1 \text{ get eliminated})$$

The probability of each of the events above can be computed from Formula (1). This proves the proof of Corollary 2. □

To conclude, consider the simpler case that all nodes have the same shooting capability, namely $p_i = p$, for all i. We observe the following facts for a unidirectional ring.

- By Lemma 2, a given node in the ring survives with probability $\frac{1-p}{2-p}$ and gets eliminated with probability $\frac{1}{2-p}$, and
- By Lemma 2, the expected number of surviving nodes is $\frac{1-p}{2-p}n$ and the number of of eliminated nodes $\frac{1}{2-p}n$,
- By Corollary 1, with probability 1, no two consecutive nodes survive.

Remark 2. One can study in a similar manner an oriented line of n nodes. Details are similar to the oriented ring and are left to the reader.

3.2 Triadic (Mexican) Standoff

In the Mexican standoff the three shooters $0, 1, 2$ are arranged in a triangle whose three edges are bidirectional as depicted in Fig. 3 so that shooter 0 (respectively, 1 and 2) can decide which one of 1 and 2 (respectively, 0 and 2, and 0 and 1) may shoot (one at a time) and can eliminate the chosen one with probability p_0 (respectively, p_1 and p_2) or fail to eliminate with probability $1-p_0$ (respectively, $1-p_1$ and $1-p_2$).

Consider the following *Change-Orientation Algorithm* which is in two phases. In the first CCW (Counter Clockwise)-phase each node i which is alive attacks its CCW neighbor $i+1$ as long as the latter is alive. If node i eliminates node

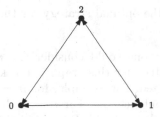

Fig. 3. Depiction of a Mexican standoff represented as a triangle graph with bidirectional links.

$i+1$ then i enters its second CW (Clockwise)-phase and attacks its CW neighbor $i-1$, as long as the letter is alive.

Observe that in this algorithm, although the shooting is synchronous the nodes may not necessarily be in the same phase at the same time. Consider the first node i which changes phase (from CCW to CW) in the above algorithm. For some t, such a node i will enter the second CW phase of the algorithm at round $t+1$ if it eliminated its CCW opponent $i+1$ at round t. The probability that node i enters the second phase at time $t+1$ ie given by the formula

$$\sum_{s=0}^{t-1}(1-p_i)^s(1-p_{i-1})^s p_i = p_i \frac{1-(1-p_i)^t(1-p_{i-1})^t}{1-(1-p_i)(1-p_{i-1})}$$

$$= p_i \frac{1-(1-p_i)^t(1-p_{i-1})^t}{p_i+p_{i-1}-p_i p_{i-1}}.$$

Now the two nodes $i-1$ and i will be running a dyadic duel which is starting at round $t+1$ with nodes i and $i-1$ attacking each other. In this dyadic duel $i-1$ (resp. i) is a winner with probability $\frac{p_{i-1}}{p_i+p_{i-1}-p_i p_{i-1}}$ (resp. $\frac{p_i}{p_i+p_{i-1}-p_i p_{i-1}}$), while both are eliminated with probability $\frac{p_i p_{i-1}}{p_i+p_{i-1}-p_i p_{i-1}}$. Multiplying out it follows that $i-1$ survives w.p.

$$p_i p_{i-1} \frac{1-(1-p_i)^t(1-p_{i-1})^t}{(p_i+p_{i-1}-p_i p_{i-1})^2} \tag{2}$$

and i survives w.p.

$$p_i^2 \frac{1-(1-p_i)^t(1-p_{i-1})^t}{(p_i+p_{i-1}-p_i p_{i-1})^2} \tag{3}$$

Moreover, asymptotically as $t \to \infty$ we have that $i-1$ survives w.p.

$$\frac{p_i p_{i-1}}{(p_i+p_{i-1}-p_i p_{i-1})^2} \tag{4}$$

and i survives w.p.

$$\frac{p_i^2}{(p_i+p_{i-1}-p_i p_{i-1})^2} \tag{5}$$

Remark 3. We note that the optimal strategy for the triadic Mexican standoff is not known.

Remark 4. More generally, one could consider the n-node bidirectional ring. However, the optimal strategy for this graph is not known. Nevertheless, some special cases can be analyzed. For example, for $n = 2$ this is the same as the dyadic duel discussed in Subsect. 2.2. For $n = 3$ this leads to the triadic (or Mexican) standoff previously analyzed. For $n = 4$ this is the complete bidirectional, bipartite $K_{2,2}$ which is analyzed in Sect. 5 as part of the more bipartite graph $K_{m.n}$.

4   Relationships

There is a strong connection between standoffs and Positive/Negative relationships whereby positive represents "friendship" and negative "antagonism;; whose structure is studied extensively in social network theory, e.g., see [9][Chapter 5]. Consider Peaceful/Aggressive relationships whereby in a Peaceful (resp., Aggressive) relationship opponents have no incentive (resp. attack) each other.

As depicted in Fig. 4, one could distinguish four possible relationships, namely Peaceful (P) and Aggressive (A), between three entities (forming a triangle). The two triangles to the left have an odd number of Peaceful relationships while the two triangles to the right have an even number of Peaceful relationships.

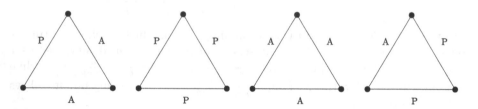

Fig. 4. Triangles representing relationships between three entities: the left two relationships are considered stable while the right two are unstable.

As in [9][Chapter 5], it can be argued that the two triangles to the left represent stable social relationships in that the entities which are connected with a Peaceful relationship have no (or little) incentive to flip to an Aggressive one. However, the two triangles to the right represent unstable social relationships in that the entities which are connected with an Aggressive relationship have an incentive to flip to a Peaceful one and thus change the overall structure of the relationship between the three entities. Based on this reasoning, one refers to triangles with one or three P's as balanced, since they are free of these sources of instability, while triangles with zero or two P's are referred to as unbalanced.

The (un)balanced property defined above refers only to three nodes forming a triangle. More generally, a complete graph of n nodes all of whose edges are labeled either with P or A is called balanced if every one of its triangles is balanced. An interesting structural characterization of balanced complete graphs is due to [6,12] and is presented as a theorem below (for a proof see the book [9][Chapter 5]).

Theorem 3 ([6,12]). *If every triangle in a complete graph is Stable then there is a bipartition of the nodes into two parts X, Y such that 1) every pair of nodes in X is Peaceful, 2) every pair of nodes in Y is Peaceful, 3) every pair of nodes one from X and the other from Y is Aggressive, and vice versa.*

As depicted in Fig. 5, Theorem 3 gives a structural characterization of a balanced complete graph with a Bipartition.

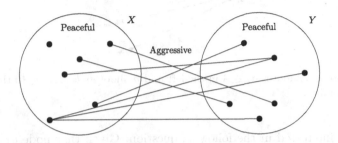

Fig. 5. Bipartition of a balanced complete graph.

We note that Theorem 3 can be extended in many ways. First to non-complete graphs in which only the edges that are present have Peaceful or Aggressive relationships. Also, a similar theorem holds for approximately balanced networks whereby an asymptotically large number of the edges form a balanced network. Further, one can also study the concept of weakly stable in which no triangle can have exactly two Peaceful relationships: such graphs are called weakly stable and can be proved that they correspond to multipartite graphs [7]. For additional details and proofs of the results listed above the interested reader may consult the book [9][Chapter 5] by Easley and Kleiberg.

5   Standoffs on Complete Bipartite Graphs

Motivated from the structural balance property presented in Sect. 4, in this section we consider standoffs on complete bipartite graphs, a graph whose vertex set is partitioned into two parts A, B so that each pair of vertices $\{u, v\}$ with $u \in A, v \in B$ are connected with a bidirectional edge.

5.1   One-to-Many Standoff

Consider the bipartite configuration depicted in Fig. 6 in which $A = \{a_1\}$ is a singleton and $B = \{b_1, b_2, \ldots, b_n\}$. Shooter a_1 has probability p_1 (resp. $1-p_1$) for a successful (resp. unsuccessful) attack, while shooter $b_i \in B$ has probability q_i (resp. $1-q_i$) for a successful (resp. unsuccessful) attack. In the sequel we analyze the survival probabilities of the nodes under a variety of attack algorithms.

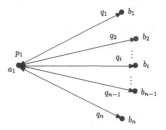

Fig. 6. There is a shooter $a_1 \in A$ in the left and n shooters $b_1, \ldots, b_n$ to the right in a bipartite standoff.

We are interested in the following question: Given that node a_1 knows the attack probabilities of all the nodes in B, what strategy will maximize its survival probability? Consider the following algorithm.

Algorithm 1. AttackInOrder-a_1 Shooting Algorithm

1: **for** $t = 0, 1, \ldots$; **do**
2:     as long as there are nodes alive in round t not all of whose outgoing neighbors are dead;
3:         **if** a_1 is alive; **then**
4:             attack the surviving node in B with the smallest index i;
5:         **else**
6:             stop;

Intuitively, in Algorithm 1 node a_1 in round t, where $t = 1, 2, \ldots$, attacks the node i which has the smallest index among the nodes in B surviving in this round. We can prove the following lemma.

Lemma 1. *If node a_1 follows Algorithm 1 then the probability that it eliminates all opponent nodes in B is given by the formula*

$$\frac{P}{1-P} \cdot \left(\frac{p_1}{1-p_1}\right)^n \prod_{i=1}^{n-1} \frac{\frac{P}{(1-q_i)\cdots(1-q_1)}}{1 - \frac{P}{(1-q_i)\cdots(1-q_1)}}, \tag{6}$$

where $P = (1-p_1)(1-q_1)\cdots(1-q_n)$.

Proof. Consider the event E which describes that node a_1 eliminates all its opponents in the standoff. For each k define the random variable $K_{1,\ldots,k}$ so that that $K_{1,\ldots,k} = t$ iff node a_1 eliminates all its k opponents $b_1, \ldots, b_k$ in this order and is not itself eliminated in round t or before, i.e., b_1 is eliminated first, followed by b_2 second, etc., until b_k is eliminated last, and no one eliminates a_1 in round t or before. Let $E_{1,\ldots,k}$ be the event $\bigcup_{t \geq 1}(K_{1,\ldots,k} = t)$. Since by definition, Algorithm 1 attacks the nodes of B in the order $b_1, b_2, \ldots, b_n$, the event E is equivalent to the event $E_{1,\ldots,n}$.

Now we proceed with the calculation of the related probabilities. Let us define $P := (1 - p_1)(1 - q_1) \cdots (1 - q_n)$. Due to the independence of the actions of the parties involved, elementary calculations show that

$$\Pr[K_1 = 1] = P\frac{p_1}{1 - p_1},$$

and more generally for all $t \geq 1$,

$$\Pr[K_1 = t] = P^t\frac{p_1}{1 - p_1}$$

and therefore we conclude from the discussion above that

$$\Pr[E_1] = \frac{p_1}{1 - p_1} \sum_{t=1}^{\infty} P^t = \frac{p_1}{1 - p_1}\frac{P}{1 - P}.$$

Using conditional probabilities, the last formula above, and elementary calculations we see that

$$\Pr[E_{1,2}] = \Pr[E_2|E_1] \cdot \Pr[E_1]$$

$$= \frac{\frac{P}{1-q_1}}{1 - \frac{P}{1-q_1}} \left(\frac{p_1}{1 - p_1}\right)^2 \frac{P}{1 - P}, \tag{7}$$

where the last identity is valid because after eliminating node $b_1 \in B$, node $a_1 \in A$ is only shooting at node b_2 among the surviving nodes $b_2, \ldots, b_k \in B$.

Using Formula (7), a similar formula can be derived for the subsequence $1, 2, 3$, namely

$$\Pr[E_{1,2,3}] = \Pr[E_3|E_{1,2}] \cdot \Pr[E_{1,2}]$$

$$= \frac{\frac{P}{(1-q_2)(1-q_1)}}{1 - \frac{P}{(1-q_2)(1-q_1)}}\frac{p_1}{1 - p_1} \cdot \Pr[E_{1,2}]$$

$$= \frac{\frac{P}{(1-q_2)(1-q_1)}}{1 - \frac{P}{(1-q_2)(1-q_1)}}\frac{\frac{P}{1-q_1}}{1 - \frac{P}{1-q_1}} \left(\frac{p_1}{1 - p_1}\right)^3 \frac{P}{1 - P} \tag{8}$$

Finally, using induction on k and arguing as above we derive the more general formula

$$\Pr[E_{1,2,\ldots,k}] = \frac{\frac{P}{(1-q_{k-1})\cdots(1-q_1)}}{1 - \frac{P}{(1-q_{k-1})\cdots(1-q_1)}} \cdots \frac{\frac{P}{1-q_1}}{1 - \frac{P}{1-q_1}} \left(\frac{p_1}{1 - p_1}\right)^k \frac{P}{1 - P}. \tag{9}$$

This proves Formula (6) in Lemma 1.   □

Now instead of Algorithm 1, consider the following algorithm in which node $a_1 \in A$ in round $t = 1, 2, \ldots$ attacks the node which has the highest attack probability among the nodes surviving in this round.

Algorithm 2. AttackMax Shooting Algorithm

1: **for** $t = 0, 1, \ldots$; **do**
2:     as long as there are nodes $i = 1, 2, \ldots, n$ alive in round t not all of whose outgoing neighbors are dead;
3:     **if** a_1 is alive; **then**
4:         attack a surviving node i whose attack probability q_i is highest among all the nodes that survived this round;
5:     **else**
6:         stop;

Theorem 4. *Algorithm 2 under which node a_1 in round $t = 1, 2, \ldots$ attacks the node which has the highest attack probability among the nodes surviving in this round gives the highest overall survival probability for node a_1. Moreover, if $i_1, i_2, \ldots, i_n$ is a permutation of the nodes $1, 2, \ldots, n$ so that the eliminating probabilities in descending order are $q_{i_1} \geq q_{i_2} \geq \cdots \geq q_{i_n}$ then the survival probability of node a_1 is equal to*

$$\left(\frac{p_1}{1 - p_1}\right)^n \cdot \frac{P}{1 - P} \cdot \frac{\frac{P}{(1 - q_{i_{n-1}}) \cdots (1 - q_{i_1})}}{1 - \frac{P}{(1 - q_{i_{n-1}}) \cdots (1 - q_{i_1})}} \cdots \frac{\frac{P}{(1 - q_{i_2})(1 - q_{i_1})}}{1 - \frac{P}{(1 - q_{i_2})(1 - q_{i_1})}} \cdot \frac{\frac{P}{1 - q_{i_1}}}{1 - \frac{P}{1 - q_{i_1}}},$$

where $P = (1 - p_1)(1 - q_1) \cdots (1 - q_n)$.

Proof. As shown in the proof of Lemma 1, Formula (6) is valid if node a_1 attacks the remaining nodes $b_1, b_2, \ldots, b_n$ of the graph in the order $1, 2, \ldots, n$. Lets use the notation $A(1, 2, \ldots, n)$ for this product formula. Clearly, for any permutation $i_1, i_2, \ldots, i_n$ of the nodes $b_1, b_2, \ldots, b_n$, if node a_1 attacks the nodes $b_1, b_2, \ldots, b_n$ in the order $i_1, i_2, \ldots, i_n$ then the resulting formula for the survival probability of node a_1 will be $A(i_1, i_2, \ldots, i_n)$, by which we mean that one replaces in the formula each index k by i_k while it is assumed that node a_1 attacks the nodes in the order $i_1, i_2, \ldots, i_n$.

We would like to prove that the product $A(i_1, i_2, \ldots, i_n)$ is maximized when the permutation $i_1, i_2, \ldots, i_n$ is chosen so that $q_{i_1} \geq q_{i_2} \geq \cdots \geq q_{i_n}$, Indeed, it is easy to observe the following monotonicity: since $q_{i_1} \geq q_1$ we must have that

$$\frac{\frac{P}{1 - q_{i_1}}}{1 - \frac{P}{1 - q_{i_1}}} \geq \frac{\frac{P}{1 - q_1}}{1 - \frac{P}{1 - q_1}}$$

Similarly, since $q_{i_1} \geq q_1$ and $q_{i_2} \geq q_2$ we must have that

$$\frac{\frac{P}{(1 - q_{i_2})(1 - q_{i_1})}}{1 - \frac{P}{(1 - q_{i_2})(1 - q_{i_1})}} \geq \frac{\frac{P}{(1 - q_2)(1 - q_1)}}{1 - \frac{P}{(1 - q_2)(1 - q_1)}}$$

The reader can easily verify that the same monotonicity argument will work for any $n \geq 3$. This proves Theorem 4. $\square$

Next we look at the survival probability of the remaining nodes, namely b_i, for $i \geq 1$. All these nodes do not have any attack flexibility because they are not connected to each other and are only connected to the node $a_1 \in A$. Let the random variable I be defined as the first $i = 1, 2, \ldots, n$ such that node b_{i+1} is not eliminated by node a_1 when following Algorithm 1. Thus, $I = i$ denotes the event that all nodes b_j, for $1 \leq j \leq i$, are eliminated by node a_1 but b_{i+1} is not, while $I = n$ denotes the event that all nodes b_i, for $i = 1, 2, \ldots, n$, are eliminated by node a_1.

In the sequel we indicate how to compute $\Pr[I = i]$ as well as derives a formula for the expectation $E[I] = \sum_{i=1}^{n} i \Pr[I = i]$. To this end, observe that

$$I = i \text{ iff } (a_1 \Rightarrow b_1) \, \& \, (a_1 \Rightarrow b_2) \, \& \, \cdots \, \& \, (a_1 \Rightarrow b_i) \, \& \, (a_1 \not\Rightarrow b_{i+1}), \qquad (10)$$

where $a_1 \Rightarrow b_k$ denotes the event that node a_1 eliminates node b_k. Let $P = (1 - p_1)(1 - q_1) \cdots (1 - q_n)$. Recall that according to the algorithm eliminated nodes are withdrawn. Therefore using conditional probabilities we can compute

$$\Pr[a_1 \Rightarrow b_1] = \sum_{t=1}^{\infty} P^{t-1} p_1 = \frac{1}{1 - P} p_1$$

$$\Pr[a_1 \Rightarrow b_1 \, \& \, a_1 \Rightarrow b_2] = \Pr[a_1 \Rightarrow b_2 | a_1 \Rightarrow b_1] \Pr[a_1 \Rightarrow b_1] = \frac{1}{1 - \frac{P}{1-q_1}} \frac{1}{1 - P} p_1^2$$

$$\vdots$$

$$\Pr[a_1 \Rightarrow b_1 \, \& \, \cdots \, \& \, a_1 \Rightarrow b_i] = \frac{1}{1 - \frac{P}{(1-q_{i-1})\cdots(1-q_1)}} \cdots \frac{1}{1 - \frac{P}{1-q_1}} \frac{1}{1 - P} p_1^i$$

Combining the last formula with Equation (10) we compute that

$$\Pr[I = i] = \frac{1}{1 - \frac{P}{(1-q_i)\cdots(1-q_1)}} \frac{1}{1 - \frac{P}{(1-q_{i-1})\cdots(1-q_1)}} \cdots \frac{1}{1 - \frac{P}{1-q_1}} \frac{1}{1 - P} p_1^i q_{i+1}$$

Clearly, a formula for $E[I] = \sum_{i=1}^{n} i \Pr[I = i]$ easily follows from the last identity for $\Pr[I = i]$ derived above.

5.2 Many-to-Many Complete Bipartite Standoff

This standoff forms a complete bipartite graph consisting of two parts, the first with m nodes and the second with n nodes (see Fig. 7). The two parts consist of the sets $A = \{a_1, a_2, \ldots, a_m\}$ and $B = \{b_1, b_2, \ldots, b_n\}$ with corresponding attack probabilities p_i, q_j, for $i = 1, 2, \ldots, m$ and $j = 1, 2, \ldots, n$.

Consider the following shooting Algorithm 3 (which is a generalization of Algorithm 1) in which nodes attack nodes in their opposite part in order.

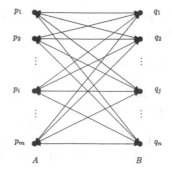

Fig. 7. A bipartite standoff with m shooters $a_1, a_2, \ldots, a_m$ in the left part A and n shooters $b_1, b_2, \ldots, b_n$ in the right part B.

Algorithm 3. AttackInOrderAll Shooting Algorithm

1: **for** $t = 0, 1, \ldots;$ **do**
2:     as long as there are nodes alive in round t not all of whose outgoing neighbors are dead;
3:         **if** two alive nodes a_i, b_j, where $i \leq m$ and $j \leq n$ exist; **then**
4:             the first such node $a_i \in A$ attacks the surviving node in $b_j \in B$ with the smallest index j and the first such node $b_j \in B$ attacks the surviving node in $a_i \in A$ with the smallest index i;
5:         **else**
6:             stop;

Similarly, one is led to the following Algorithm 4 which is a generalization of Algorithm 2.

Algorithm 4. AttackMaxAll Shooting Algorithm

1: **for** $t = 0, 1, \ldots;$ **do**
2:     as long as there are nodes alive in both parts A and B;
3:         **if** $a_i \in A$ (resp. $b_j \in B$) are alive; **then**
4:             attack a surviving node $b_j \in B$ (resp. $a_i \in A$) whose attack probability q_j (resp. p_i) is highest among all the nodes in B (resp. A) that survived this round;
5:         **else**
6:             stop;

We can prove the following theorem.

Theorem 5. *Algorithm 4 under which nodes a_i, b_j in round $t = 1, 2, \ldots$ attack the node which has the highest attack probability among the nodes surviving in this round gives the highest overall survival probability for each of the nodes of the graph.* □

Moreover, we can compute the survival expectations of the nodes.

Theorem 6. *The expected number of survivors in the Algorithm 4 in the complete bipartite graph with two parts A, B of size m, n, respectively, is given by the formulas $E[X_A] = \sum_{i=1}^{m} p_A(i)$ and $E[X_B] = \sum_{j=1}^{n} q_B(j)$, where $p_A(i), q_B(j)$ are given by closed Formulas. The AttackMax algorithm has the maximum number of survivors among all attack-in-order algorithms.* $\square$

6 Conclusion

We considered attack algorithms for aggressive standoffs on a graph in which the nodes shoot at a single neighbor per round for a variety of topologies including rings, and complete bipartite. We also studied the survivability of a node resulting from various attack strategies. We looked at various graphs and analyzed the unidirectional ring on n nodes with arbitrary attack probabilities. In addition, for the case of complete bipartite graphs we determined the optimal strategy for surviving the standoff.

There are many interesting additional questions to be investigated concerning various graphs. One may want to look at standoffs on specific classes of graphs, like bidirectional rings of n nodes (the special case of which for $n = 3$ is known as the Mexican standoff) as well as trees and complete graphs. Interesting problems concern the survivability on general social graphs with strong and weak links, as well as on small world graphs.

Additional questions concern the attack model used. All the algorithms considered were offline in that the peers had complete knowledge of the abilities of their opponents (e.g., they are aware of their attack probabilities). However, very little is known about oblivious strategies in which peers have incomplete knowledge of their opponents. Moreover, in the strategies discussed in this paper, the attack probability of a peer u is independent of the opponent v as well as the current round. An interesting generalization would be for the attack probability of u, v to depend either on both neighbors u and v or the current round. Additional questions include survivability for peers with limited amount of ammunition or when peers are allowed to withdraw from the standoff. Finally, one could also consider standoffs in which peers may form collisions.

References

1. Archer, J., Huntingford, F.: Game theory models and escalation of animal fights. The dynamics of aggression: biological and social processes in dyads and groups, pp. 3–31 (1994)
2. Baldick, R.: The Duel: a history of duelling. Hamlyn Publ Group Ltd., (1965)
3. Bekiari, A., Deliligka, S., Koustelios, A.: Examining relations of aggressive communication in social networks. Soc. Netw. **6**(1), 38–52 (2016)
4. Bekiari, A., Pachi, V., Hasanagas, N.: Investigating bullying determinants and typologies with social network analysis. J. Comput. Commun. **5**(7), 11–27 (2017)

5. Cairns, R.B., Cairns, B.D., Neckerman, H.J., Gest, S.D., Gariepy, J.-L.: Social networks and aggressive behavior: peer support or peer rejection? Dev. Psychol. **24**(6), 815 (1988)

6. Cartwright, D., Harary, F.: Structural balance: a generalization of Heider's theory. Psychol. Rev. **63**(5), 277 (1956)

7. Davis, J.A.: Clustering and structural balance in graphs. Hum. Relat. **20**(2), 181–187 (1967)

8. Dugatkin, L.A., Reeve, H.K.: Game Theory and Animal Behavior. Oxford University Press on Demand, Oxford (2000)

9. Easley, D., Kleinberg, J.: Networks, Crowds, and Markets: Reasoning About a Highly Connected World. Cambridge University Press, Cambridge (2010)

10. Eastaway, R., Standoff, M.: Electronic Edition of The New Scientist, Puzzle Section, 21 Apr 2020

11. Fill, J.A., Mahmoud, H.M., Szpankowski, W.: On the distribution for the duration of a randomized leader election algorithm. Ann. Appl. Probab. **6**(4), 1260–1283 (1996)

12. Harary, F.: On the notion of balance of a signed graph. Michigan Math. J. **2**(2), 143–146 (1953)

13. Hardy, I.C.W., Briffa, M.: Animal Contests. Cambridge University Press, Cambridge (2013)

14. Huntingford, F.A.: Animal conflict. Springer Science & Business Media (2013). https://doi.org/10.1007/978-94-009-3145-9

15. Janson, S., Szpankowski, W.: Analysis of an asymmetric leader election algorithm. Electron. J. Comb. **4**, R17–R17 (1997)

16. Kolanowski, A.: Aggressive behavior in institutionalized elders: a theoretical framework. Am. J. Alzheimer's Disease **10**(2), 23–29 (1995)

17. YouTube Video of Duelling Rams. These rams go head to head - literally. https://www.youtube.com/watch?v=5NTvMDEyDpg. Accessed 29 May 2021

18. Prodinger, H.: How to select a loser. Discrete Math. **120**(1–3), 149–159 (1993)

19. Riechert, S.E.: Game theory and animal contests. In: Dugatkin, L.A., Reeve, H.K. (eds.) Game Theory and Animal Behavior, pp. 64–93. Oxford University Press, Oxford, New York (1998)

20. Ross, S.M.: Probability Models for Computer Science. Harcourt Academic Press, San Diego (2002)

21. Huesmann, L.R.: An information processing model for the development of aggression. Aggressive Behav. **14**(1), 13–24 (1988)

22. Williams, R.B., Clippinger, C.A.: Aggression, competition and computer games: computer and human opponents. Comput. Hum. Behav. **18**(5), 495–506 (2002)

23. Winkler, P.: Mathematical Puzzles: a Connoisseur's Collection. CRC Press, Boca Raton (2003)

A Hybridization of GRASP and UTASTAR for Solving the Vehicle Routing Problem with Pickups and Deliveries and 3D Loading Constraints

Themistoklis Stamadianos, Magdalene Marinaki, Nikolaos Matsatsinis, and Yannis Marinakis(✉)

School of Production Engineering and Management, Technical University of Crete, Chania, Greece

tstamadianos@tuc.gr, magda@dssl.tuc.gr, nikos@dpem.tuc.gr, marinakis@ergasya.tuc.gr

Abstract. As urban centers grow, demand for goods transportation grows as well. The emergence of e-commerce has been a great catalyst, with online sales placing a big load on transportation companies. Current social conditions further amplify the effect. An unnoticed segment has been the delivery of large-size items in urban centers, where restrictions of different kinds impose the use of small vehicles. This research presents a novel combination of UTASTAR with Vehicle Routing Problem with Pickups and Deliveries and three-dimensional loading constraints to provide solutions. Scenarios of demand exceeding capacity are considered. A Decision Support System (DSS) is created to assist Decision Makers (DMs) of logistics companies get routing suggestions based on their priorities. The considerable size and weight of the items require careful handling of the smaller vehicles. The utilization of heuristic methods for routing expedites the solution process, enabling the formation of multiple solutions, which get ranked by the UTASTAR method on four criteria. The criteria values and thresholds are set indirectly by the DM. The model is tested on modified instances from the literature and a case study.

Keywords: Vehicle routing problem · GRASP · UTASTAR · Multi-criteria VRP · Three dimensional loading

1 Introduction

Planning deliveries in urban environments is a demanding task. Unpredictable traffic, small roads, and weight restrictions are some of the most commonly faced issues. Most of the research has been on the public transport scope, with freight remaining largely unregulated and neglected. The research that exists on urban deliveries, excluding research solely on vehicle routing, is on the subject of sustainability [2,3].

© Springer Nature Switzerland AG 2022
D. E. Simos et al. (Eds.): LION 2022, LNCS 13621, pp. 505–520, 2022.
https://doi.org/10.1007/978-3-031-24866-5_36

This research paper aims to solve a Vehicle Routing Problem (VRP) in an urban context, with the potential for real-life application, and provide tools to assist Decision Makers in creating delivery plans that correspond to their current needs. For example, different parts of the city, days of the week, or even types of customers, may require different handling.

The discussed problem arises when the transportation of large items needs to take place in a restrictive environment, like an urban one. The problem is viewed from the perspective of a carrier company specializing in large item transportation. They get hired by businesses and tasked with picking up items from their premises and transporting them within city limits. The scenario of transportation demand exceeding capacity is explored. While, many customers are interested in delivery services, the fleet of the company cannot provide service for everyone.

A formulation of the problem as a Vehicle Routing Problem (VRP) with Pickups and Deliveries and three-dimensional loading constraints (3L-VRPPD) is proposed and a novel combination of the UTASTAR method for preference disaggregation analysis and of the Greedy Randomized Adaptive Search Procedure (GRASP) for solving the routing problem, helps solve the described problem. GRASP algorithm is used to generate multiple solutions for the VRP. Then, UTASTAR takes over and ranks them according to the preferences of the Decision Maker (DM). By ranking a few dummy solutions, UTASTAR can quickly and accurately determine the DM's preferences as long as the DM's choices remain consistent. The described process allows the DM to shift policies and get tailored suggestions for the needs of the business.

The rest of the paper is organized as follows. In Sect. 2 the Vehicle Routing Problem is discussed. UTASTAR is described in Sect. 3. The instances and the computational results are given in Sect. 4. Finally, conclusions and future research are presented in Sect. 5.

2   Vehicle Routing Problem with Pickups and Deliveries and 3D Loading Constraints

2.1   Problem Description

Designing a viable loading plan would be an essential element for a business model such as the one described. The items to be transported are large in size; therefore, their dimensions and total weight need to be taken into account. An item may also get classified as fragile. The VRP solved combines the Pickup and Delivery problem, and the three-dimensional loading problem. The problem presented is well suited to be solved as a Vehicle Routing Problem with some necessary extensions. A number of vehicles are employed to serve the customers' transportation needs. The items get picked up at a certain location and delivered to another one. Each location is either a pickup point or a delivery point. A routing example is provided in Fig. 1.

The limited size of the vehicles used for urban transportation is an important factor, not just for item volume, but for item weight as well. Since the concept of split delivery is not considered, and the size of the vehicles is small, only the total customer weight is needed. Further weight limitations may also be imposed by local authorities in order to protect older roads (i.e. roads of historic centers) and other roads susceptible to weight damages ([6,7,10]).

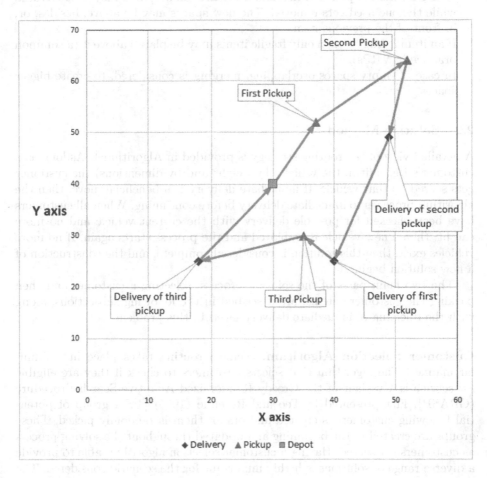

Fig. 1. Pickup and delivery example

Some of the most important three-dimensional loading rules are described in short in the following list.

- Initially, the only available space has the dimensions of the loading bay of the vehicle.
- The items being placed must be positioned as far back and as left as possible.
- When placing new items, the first spaces to be checked should be those above other items.

- Available spaces closest to the cabin will be preferred.
- A slight box overhang is allowed, however, modifications will have to be made to other spaces being affected. At least 75% of both the width and length must be supported in order to consider the space as a viable placement option.
- When an item is placed a maximum of three new spaces may be generated while the one used gets removed. The new spaces may be above, besides, or, in front of the previous item.
- If an item is fragile, then only fragile items may be placed above it (a common practice in VRPs).
- In case of empty spaces overlapping, merging is considered, to create bigger spaces.

2.2   Solution Method

A detailed view of the routing strategy is provided in Algorithm 1. As long as a customer's items fit in the vehicle (by weight and by dimensions) the customer gets served by that vehicle. If immediate delivery is a beneficial move, then the algorithm proceeds to immediate delivery before continuing. When all customers have been checked for possible delivery with the current vehicle and no more can fit, then a new vehicle is employed and the process starts again. If no more vehicles exist, then this solution is considered complete, and the construction of a new solution begins.

The two main parts of the solution process, selecting a customer and then packing the customer's items, are described in the following subsections, along with the dilemma of immediate delivery faced in this problem.

Customer Selection Algorithm. Vehicle routing takes place in a familiar manner. The algorithm that selects customers to check if they are eligible for service is a variant of the Greedy Randomized Adaptive Search Procedure (GRASP), first presented by Feo and Resende (1995) [4]. A group of potential following customers is created and one of them is randomly picked. These groups are created at the beginning and updated throughout the solving process as customers get served. Having a customer selection algorithm, able to provide a diverse range of solutions is highly important for the scenario considered. The criterion upon which the groups are created is a metric of distance. Algorithm 2 describes the process.

Packing Algorithm. To load the vehicles, a packing algorithm is developed. Each time a new customer is considered, the packing algorithm has to check if a feasible packing plan that includes this customer exists. A common concept in three-dimensional packing problems is the use of maximal spaces. At each step of the loading process, a number of spaces are available for loading. The plans must follow the Last In First Out (LIFO) concept, meaning that the last items to be loaded belong to the customer that will be served first. To serve a customer, no additional movement of items should take place. This principle is

---

Algorithm 1: VRP solver

Data: Customer Information (pickup coordinates, delivery coordinates,
number of items, dimensions, fragility, weight), Vehicle Information
(number of vehicles, loading bay dimensions, maximum payload),
Depot coordinates.

Result: Feasible VRP Solution

First customer = pickAcustomer();

while *There are unserved customers* **do**

    Pick the next vehicle.;

    if *The current customer is served* **then**

        customer = pickAcustomer();

    while *Termination condition is not met* **do**

        if *Adding customer's weight does not exceed vehicle capacity* **then**

            if *Customer's items fit in the loading bay* **then**

                The customer gets served by the current truck;

        if *Delivering the items that were just picked is beneficial* **then**

            Immediately deliver the items that were picked;

            Termination condition is met.;

        customer = pickAcustomer();

        if *Too many checks have been made* **then**

            Termination condition is met.;

    Add the delivery points to the route.;

Algorithm 2: pickAcustomer

Data: Restricted Candidate List (RCL), Customers

Result: A potential customer to be served with the current vehicle.

Pick a random and eligible customer.

if *The customer belongs to the RCL* **then**

    Pick the customer.

else

    Pick the closest customer.

Mark the chosen customer as an already explored customer for the vehicle.

---

exceptionally important when dealing with large items. The algorithm employed
in this research makes an exhaustive search. When a feasible packing plan is
found, the algorithm immediately stops.

The Delivery Dilemma. A critical decision is made just after the items of
customer 1 are loaded and before the vehicle leaves to visit the pickup point
of customer 2. There is the option to deliver the items of customer 1 before
visiting customer 2. If it is considered to be a beneficial move, then a diversion
is made, and the delivery takes place immediately. If not, the vehicle proceeds
to the pickup point of customer 2. This takes place after every customer pickup,

in an effort to free space and serve more customers along the way, without the need for extra vehicles.

Whether immediate delivery is beneficial or not, is determined partly by the ratio between the distance with delivery and without delivery and partly by the number of items of customer 1. Intuitively, close deliveries and a higher number of items commend a higher chance of immediate delivery.

3   UTASTAR

After the algorithm has generated the predefined number of solutions, the UTAS-TAR method is used in order to evaluate them. UTA (UTility Additives), originally developed by Jacquet-Lagreze and Siskos (1982) [8], is a method of Multi-Attribute Utility Theory (MAUT). Various methods can be employed in order to attain information about the DM's value system. These may include previous decisions of the DM or the creation of a set of representative alternatives for the DM to rank. In this case, the provided alternatives are the routing plans. This information is used to create a function or functions, able to provide rankings according to the DMs values. These functions are referred to as additive utility functions and when combined they form a model which can be used to evaluate future options. In this research paper, an extension of the UTA method was used, UTASTAR (or UTA*), introduced by Siskos and Yannacopoulos (1985) [11].

UTASTAR is the tool employed to rank the different solutions provided by the VRP algorithm. A number of solutions, differing significantly in many cases, are the result of the VRP algorithm. This differentiation is allowing for a greater chance to generate and select a good solution as opposed to a static approach which would provide moderate results in a predictable manner. The solutions are judged upon four different points of quality. The level of customer service is a very important factor. Being able to provide service to as many customers as possible yields a higher income. To ensure that the vehicles are used efficiently, the total weight transferred and the total volume of items are also observed. The last criterion, the total traveled distance, is a good indicator of operational costs. All the above help provide a proper customer experience.

The most important part of UTASTAR is the data related to the DM's preferences, regardless of the method employed to acquire them. The DM has to be consistent to avoid errors. In this case, the DM is presented with a total of 8 different plans to rank, as shown in Table 1. The first column enumerates the alternatives. The Distance column refers to the total travel distance, the Serviced column to the total number of served customers, and the next two refer to the total Load and Volume of the items. The last column contains a hypothetical Ranking of the alternatives by the DM, from best (1) to worst (8). This ranking is the required input for the UTASTAR algorithm.

Table 1. UTASTAR alternatives

Alternative	Distance	Serviced	Load	Volume	Ranking (DM)
1	100	12	225	525	8
2	100	19	340	660	7
3	250	19	260	830	6
4	250	19	390	720	5
5	250	26	660	750	4
6	400	37	700	1080	2
7	400	19	870	920	1
8	400	26	660	940	3

The Linear Program (LP) concerning UTASTAR is presented below. The objective function is provided in Eq. 1, under constraints 2 to 10. The objective function minimizes the sum of all overestimation and underestimation errors of the utility functions.

The constraints 2 to 8 describe the maximum possible utility difference between the alternatives in the parenthesis. Multiple tests were carried to define the level of difference needed to set apart two of the alternatives. The smallest possible difference was found to be 0.00005, and this value was used throughout all the experiments. Such a small delta will pick apart even the smallest differences, but it bears the danger of failure if the DM gets inconsistent. Constraint 9 denotes that the sum of the weights must be one. These weights are the transformed utilities. Lastly, constraints 10 ensure that all weights and errors are equal or greater than zero.

All delta functions are described in Table 2. The variables are presented in the first column, while the rest of the columns correspond to their values in the delta function each column describes.

$$min : F = \Sigma[\sigma(a)^+ + \sigma(a)^-] \tag{1}$$

Under Constraints:

$$\Delta(7,6) \geq 0.00005 \tag{2}$$

$$\Delta(6,8) \geq 0.00005 \tag{3}$$

$$\Delta(8,5) \geq 0.00005 \tag{4}$$

$$\Delta(5,4) \geq 0.00005 \tag{5}$$

$$\Delta(4,3) \geq 0.00005 \tag{6}$$

$$\Delta(3,1) \geq 0.00005 \tag{7}$$

$$\Delta(2,1) \geq 0.00005 \tag{8}$$

$$\Sigma(w_{ij}) = 1 \tag{9}$$

$$w_{ij}, \sigma(a)^+, \sigma(a)^- \geq 0, \forall i, j, a \qquad (10)$$

Table 2. Delta values

Var.	$\Delta(7,6)$	$\Delta(6,8)$	$\Delta(8,5)$	$\Delta(5,4)$	$\Delta(4,3)$	$\Delta(3,2)$	$\Delta(2,1)$
w_{11}	0	0	−1	0	0	0	0
w_{12}	0	0	0	0	0	−1	0
w_{21}	0	0	0	0	0	0	1
w_{22}	−1	0	0	1	0	0	0
w_{23}	−1	1	0	0	0	0	0
w_{31}	0	0	0	0.2326	0.6047	−0.3721	0.5349
w_{32}	0	0	0	1	0	0	0
w_{33}	0.7907	0.1860	0	0.0233	0	0	0
w_{41}	0	0	0	0	0	0.2703	0.7297
w_{42}	0	0	0.7838	0.1622	−0.5946	0.6486	0
w_{43}	−0.8649	0.7568	0.2432	0	0	0	0
σ_1^-	0	0	0	0	0	0	1
σ_1^+	0	0	0	0	0	0	−1
σ_2^-	0	0	0	0	0	1	−1
σ_2^+	0	0	0	0	0	−1	1
σ_3^-	0	0	0	0	1	−1	0
σ_3^+	0	0	0	0	−1	1	0
σ_4^-	0	0	0	1	−1	0	0
σ_4^+	0	0	0	−1	1	0	0
σ_5^-	0	0	1	−1	0	0	0
σ_5^+	0	0	−1	1	0	0	0
σ_6^-	1	−1	0	0	0	0	0
σ_6^+	−1	1	0	0	0	0	0
σ_7^-	−1	0	0	0	0	0	0
σ_7^+	1	0	0	0	0	0	0
σ_8^-	0	1	−1	0	0	0	0
σ_8^+	0	−1	1	0	0	0	0

After solving the LP, the objective function was equal to zero, meaning no errors were present, hinting at the existence of multiple optimal solutions.

A new LP was created with four new objective functions, one for each criterion, as presented in Eqs. 11 to 14. Constraints 2 to 8 have to be modified. The sigma variables are no longer necessary; therefore, they are not included in the delta constraints of the new LP. Constraints 9 and constraints 10 remain the same.

The average solution of these objective functions is presented in Table 3. It is called the centroid and it forms the utility function, representing the DM. In essence, to compute the utility of a solution, the distance, number of served customers, total load and total volume are multiplied to the respective coefficients from Table 3, and then summed.

$$min : F_1 = w_{11} + w_{12} \tag{11}$$

$$min : F_2 = w_{21} + w_{22} + w_{23} \tag{12}$$

$$min : F_3 = w_{31} + w_{32} + w_{33} \tag{13}$$

$$min : F_4 = w_{41} + w_{42} + w_{43} \tag{14}$$

Table 3. UTASTAR Utility function

Utility	Distance	Served	Load	Volume
$u(g) = 0.0559 * u_1(g_1)$		$+0.217 * u_2(g_2)$	$+0.3576 * u_3(g_3)$	$+0.3696 * u_4(g_4)$

4   Computational Experiments

4.1   Instances

The VRP [12] variant addressed in this paper is a combination of Capacitated VRP with three-dimensional loading constraints, and the one-to-one Pickup and Delivery VRP. Each of these have publicly available benchmarks; however, no combination of the above exists. Subsequently, the creation of appropriate instances is a necessity in this case. The item dimensions and fragility represent a great part of the problem; thus, instances from the 3L-CVRP literature are the best candidates.

In the literature, 3L-CVRP instances were created by Gendreau, Iori, Laporte, and Martello (2006) [5] by adding item dimensions, fragility, and vehicle dimensions in the classic CVRP benchmark instances. It is worth noting that the item dimensions were produced randomly, in relevance to the size of the vehicle. Furthermore, about a quarter of the items to be delivered in each instance are fragile, with one instance having about 40% fragile items. These instances were also modified and used in [1,9].

In this research, instances of [5] are used as well as a basis for the new instances. The data provided by these instances contains the set of customer coordinates, the dimensions (height, width, and length), fragility, and weight of the items to be transported, the total weight per customer, the number of available vehicles (homogeneous fleet) and the dimensions (height, width, and length), as well as the maximum capacity of the vehicles.

To create appropriate instances, the following modifications and additions were necessary, the included set of customer coordinates were used as the pickup locations. For each customer, new coordinates were generated and used as delivery locations. The number of available vehicles was lowered by one to create a scenario in which serving all customers is not possible. It is worth noting, that the new delivery coordinates were created randomly, within the same area of the original ones. Designing instances with the LIFO aspect in mind is deliberately avoided, since it would not be a realistic approach to the problem.

4.2 Case Study

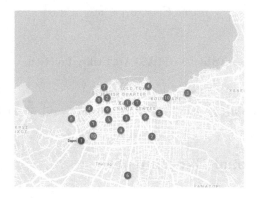

Fig. 2. Chania Case Study, Source: Google Maps

To further study the feasibility of such a plan and analyze how effective the proposed strategies are, a Case study was conducted. The city of Chania, Greece, was chosen as a test area. Chania is densely populated with small roads and generally bad accessibility, especially during summer months, when tourists visit the city. Most of the businesses are located close to or within the center and the historic part of the city, which is comprised of small roads and heavy traffic. While pickup points may lay outside the city center and be easier to reach, the majority of the delivery points will be inside the city center. Figure 2 includes all the Pickup and Delivery points and the location of the depot. These points represent realistic transport scenarios within the city and are presented in Table 4. The coordinates of the depot are provided in the last row of the table.

The vehicles used in this case study are assumed to carry a maximum of 700 kg and have a cargo space of 2000 mm in length, 1500 mm in width, and 1350 mm in height, for 4.05 cubic meters of space. The dimensions and the capacity are similar to most small commercial cargo vans for sale today. According to the total weight of the items, a minimum of four vehicles would be necessary to fully satisfy the demand. To make the scenario infeasible, only three vehicle will be used for the case study. The items to be transferred are described in Table 5.

Table 4. Pickup and delivery coordinates

Customer	Pickup		Delivery	
	X	Y	X	Y
1	35,50913	24,00842	35,51385	24,01825
2	35,50623	24,0252	35,51502	24,01245
3	35,51039	24,01795	35,51608	24,03528
4	35,51259	24,0073	35,51764	24,02412
5	35,5115	24,0273	35,51007	24,01292
6	35,49736	24,01837	35,51028	24,00224
7	35,51388	24,02104	35,51744	24,01159
8	35,51454	24,01005	35,50763	24,0164
9	35,51227	24,01254	35,51063	24,02335
10	35,51503	24,02949	35,5063	24,00846
Depot	35,50529	24,00503		

4.3 Computational Results

Both the solution algorithm and UTASTAR were implemented in MATLAB and tested on a Laptop computer equipped with an Intel i3 8130u and 6.0 GB of DDR4 RAM. A total of five runs for each of the instances took place. In each run, four solutions were created and then evaluated by the UTASTAR method. Only the best suited solution from each run is kept, according to the ranking of UTASTAR. The results presented are the average of these five runs per instance.

The results of the modified instances from the literature are presented in Table 6. The number of vehicles employed and the customers are provided in the parenthesis in the first column. Relative standard deviation is a better way to compare results, due to the different orders of magnitude used between instances. This way, the deviation is expressed as a percentage rather than a simple difference of values. The results can be found in Table 7. The second column of Table 7 refers to the average computational value, which can be described as reasonable. However, it may significantly deviate due to the randomness of the algorithm. The rest of the values described are more predictable. The total carried load, the total distance traveled, and the number of served customers have a relative standard deviation below 10%, except for a single case. In some cases, the relative standard deviation between the number of served customers is 0%, meaning the same number of customers were served in every run of the algorithm, providing a reassuring sense of stability.

Table 5. Items characteristics

Customer	Item	Height (Meters)	Width (Meters)	Length (Meters)	Fragile	Total Weight (Kilograms)
1	1	0,65	0,8	0,5	No	124
2	1	0,5	0,5	1,2	No	533
	2	0,4	0,3	0,3	No	
3	1	0,25	0,6	1	No	104
4	1	0,3	1	0,5	No	80
5	1	1	0,8	0,7	No	332
	2	0,3	0,8	0,8	No	
6	1	0,8	0,9	0,9	Yes	184
7	1	0,5	1	0,8	No	362
8	1	0,65	0,8	0,5	No	205
	2	0,7	0,5	1	No	
	3	0,25	0,3	0,3	Yes	
9	1	1,2	0,75	1	Yes	260
10	1	1	1	0,3	No	319
Sum:	14					2503

Table 6. Average results

Instance	Time (sec)	Load	Distance	Served
1 (3,15)	11,3241	166,6	447,701	12,2
2 (4,15)	0,1831	218,8	524,69136	13
3 (4,20)	26,68202	299,6	811,12284	16
4 (5,20)	12,3129	332,8	966,93202	18,2
5 (5,21)	114,7398	181,6	914,64212	19
6 (5,21)	14,50276	178	940,39068	19,2
7 (5,22)	49,64232	358,7	1603,48	19,6
8 (5,22)	20,95414	309,64	1645,62	19,6

Table 7. Relative standard deviation

Instance	Time (sec)	Load	Distance	Served
1 (3,15)	44,39%	8,28%	2,97%	3,28%
2 (4,15)	38,66%	2,79%	7,82%	0,00%
3 (4,20)	68,06%	4,85%	5,23%	0,00%
4 (5,20)	75,18%	2,49%	3,21%	2,20%
5 (5,21)	34,35%	2,88%	11,18%	0,00%
6 (5,21)	61,20%	4,42%	3,18%	2,08%
7 (5,22)	31,68%	5,60%	2,97%	2,50%
8 (5,22)	39,64%	7,06%	4,17%	2,50%

Although the VRP could not provide solutions that fully satisfy the demand, the results can be described as acceptable. The deviation in number of served customers is exceptionally low, while the deviation in vehicle load demonstrates that different customers are being checked, and the algorithm is not trapped in any local optimum. The execution of the solution algorithm along with the UTASTAR ranking takes place quickly, enabling the DM to add new customers, and possibly vehicles too, and generate new solutions in a matter of seconds.

There are many cases, where pickup and delivery take place separately, one after the other. An algorithm that follows this notion was created and tested against the more complex system employed in the present research article, for comparison purposes only. The simpler algorithm variation, will pick up all the items to be transported and then plan routes to deliver them, as opposed to the more advanced system, which considers the delivery of loaded items between pickups. Table 8 shows the differences between the two methods. Computational time seems to be the most volatile aspect of the problem, albeit, still in a reasonable time frame. The number of served customers marginally improved, despite the lower carried load. This can be attributed to the selection of different customers to serve, since not all customers can be served, due to the removal of a vehicle. The most impacting change can be observed in the total traveled distance, which on average was lowered by 16% without any drawback in the number of served customers.

Table 8. Performance difference to sequential Pickup and Delivery

Instance	Time (sec)	Load	Distance	Served
1 (3,15)	55,50%	−26,02%	−26,43%	−5,43%
2 (4,15)	−57,03%	−3,01%	−16,61%	0,00%
3 (4,20)	−1,21%	6,58%	−15,58%	−5,88%
4 (5,20)	1,94%	11,23%	−10,37%	−1,62%
5 (5,21)	−23,46%	−7,16%	−13,56%	5,56%
6 (5,21)	4,33%	−13,30%	−313,29%	2,13%
7 (5,22)	−22,89%	−9,09%	−20,66%	4,26%
8 (5,22)	−38,88%	−4,68%	−13,76%	2,62%
Average	−10,21%	−5,68%	−16,28%	0,20%

Figure 3 is a visual representation of a vehicle from a solution of the second problem instance. On the left side (a) is the solution provided by the algorithm developed in this paper. The other side (b) is solved with the algorithm that completes all the pickups before the deliveries. An arrow is present in both sides to indicate the direction of travel. As demonstrated, the items of customer 5 are the first to be picked up, and they are delivered immediately. There is a twofold benefit with this method. First, the vehicle will have more free space (the maximum possible in this example), allowing for the possibility to serve more customers with this vehicle. Secondly, when compared to the routing of the same vehicle on side (b), it is obvious that the immediate delivery had a positive effect on the total travelled distance without changing the order of visits or the number of customers. Figure 3 is, also, a visual representation of the restrictive LIFO order of service, imposed by the significant size of the transported items.

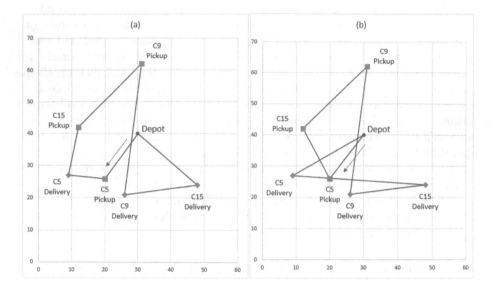

Fig. 3. Visual comparison of the two pickup and delivery variants.

Case Study. A separate analysis was carried out for the case study at Chania, Greece. The problem described in Sect. 4.2 was solved using the same algorithm that was used for the larger instances derived from literature. The least number of vehicles needed according to weight were four, whereas it was just two when judging by volume. However, to create an infeasible scenario, the number of vehicles was lowered by one, hence the least number of vehicles is three in this case. The fact that only two vehicles would be needed when judging only by volume does not mean that there will be lots of available space in the cargo space of the vehicles. The shape of the items to be delivered could be of a peculiar shape, rendering part of the cargo space unusable. Also, the presence of fragile items should be kept in mind.

Thorough experimentation resulted in the same single best solution every time, which is presented in Table 9. The small size of the problem commended more runs, therefore, 100 runs were made in just about a second. By making deliveries between pickups, the algorithm eliminated the need for one of the four trucks, originally presumed to be needed. This way, operational costs are proven to get lower while all customers get served.

Table 9. Average results of case study

Instance	Time (sec)	Load (kg)	Distance	Served
Case Study	1,0298	2503	2,2	10

In contrast to the previous tests, in the case study, the algorithm was able to provide a solution in which all customers get served, despite having a vehicle less than the ideal number.

5   Conclusion

Vehicle routing in urban spaces is a delicate matter. It is of high importance for businesses and residents alike and a determining factor to uneventful daily activities. A unique scenario where transportation demand exceeds the transportation capacity of the carrier is explored.

In this research paper, the Vehicle Routing Problem with Pickup and Delivery, and three-dimensional Loading was solved in an urban context. The inclusion of three-dimensional loading makes for a strong approach towards managing urban transportation logistics. New instances of appropriate characteristics were created through a combination of instances from the literature and new elements where needed. A case study for the city of Chania, Greece, was conducted as well. A total of 10 customer requests were fulfilled using vehicles modeled after the most common commercial vans, and pairs of origin and destination points were created.

To solve these routing problems heuristic methods were employed. A GRASP-inspired algorithm was developed to select customers to serve, while a savings algorithm was used to locate the closest customers. The solution process allows for delivery of the loaded items between pickups in order to free up cargo space, lower the travel distance, and subsequently, cost. Fewer vehicles may be needed as a side effect, as observed in the case study. Multiple solutions are created on each run for the sake of variety. Each of them will have different strengths and weaknesses, stemming from the randomness of the customer selection process. The goal is to provide the DM with a single solution, fitted to his priorities. To solve this issue, UTASTAR was involved. UTASTAR has the ability to analyze previous choices of an individual and take decisions based upon the same principles, effectively eliminating the need for manual evaluation of the provided solutions, unless the DM's priorities change.

Numerous tests were carried to assess the quality of the proposed algorithm. Average results were presented along with relative standard deviation. Computational time was the most volatile, albeit short and reasonable. When compared to a sequential pickup and delivery plan, the plan of this research was superior. Total driving distance was substantially lowered, about the same amount of customers were served; however, it was customers with lower demands.

As for the case study, the small size of the problem allowed for more runs, thus, more experimentation. Different thresholds were used when exploring the option to deliver immediately or later. Having a higher tolerance on a slightly longer trip gives the ability to serve more customers on a single route with a single vehicle. Both the case study and the rest of the instances were carried using the same UTASTAR inputs.

Many opportunities for future research exist. Further refinement to the routing algorithm could provide solutions better suited to each specific problem, instead of taking a generic approach, able to generate results for each and every case but with a differing rate of success. Another important factor to consider altering would be the method through which decisions are made when considering to deliver the items loaded last or not. In this case, a rule of thumb was used through experimentation. At last, it is worth mentioning the potential for the adoption of Electric Vehicles, which are well-suited for urban use and the future of mobility, in general.

References

1. Bortfeldt, A.: A hybrid algorithm for the capacitated vehicle routing problem with three-dimensional loading constraints. Comput. Oper. Res. **39**(9), 2248–2257 (2012)
2. Chaovalitwongse, W., Furman, K. C., Pardalos, P.M., (Eds.) Optimization and logistics challenges in the enterprise, Springer-Verlag (2009).https://doi.org/10. 1007/978-0-387-88617-6
3. Cinar, D., Gakis, K., Pardalos, P.M. (eds.): Sustainable Logistics and Transportation. SOIA, vol. 129. Springer, Cham (2017). https://doi.org/10.1007/978-3-319-69215-9
4. Feo, T.A., Resende, M.G.: Greedy randomized adaptive search procedures. J. Glob. Optim. **6**(2), 109–133 (1995)
5. Gendreau, M., Iori, M., Laporte, G., Martello, S.: A tabu search algorithm for a routing and container loading problem. Transp. Sci. **40**(3), 342–350 (2006)
6. Heng, K., Li, R., Li, Z., Wu, H.: Dynamic responses of highway bridge subjected to heavy truck impact. Eng. Struct. **232**, 111828 (2021)
7. Henning, T., Alabaster, D., Greenslade, F., Fussell, A., Craw, R.: The relationship between vehicle axle loadings and pavement wear on local roads June 2017 (2017). (Technical Report)
8. Jacquet-Lagreze, E., Siskos, J.: Assessing a set of additive utility functions for multicriteria decision-making, the UTA method. Eur. J. Oper. Res. **10**(2), 151–164 (1982)
9. Lacomme, P., Toussaint, H., Duhamel, C.: A GRASP× ELS for the vehicle routing problem with basic three-dimensional loading constraints. Eng. Appl. Artif. Intell. **26**(8), 1795–1810 (2013)
10. Pais, J.C., Amorim, S.I., Minhoto, M.J.: Impact of traffic overload on road pavement performance. J. Transp. Eng. **139**(9), 873–879 (2013)
11. Siskos, Y., Yannacopoulos, D.: Utastar: an ordinal regression method for building additive value functions. Investigaçao Operacional **5**(1), 39–53 (1985)
12. Toth, P., Vigo, D.: The vehicle routing problem. SIAM (2002)

Packing Hypertrees and the k-cut Problem in Hypergraphs

Mourad Baïou[1] and Francisco Barahona[2]($\boxtimes$)

[1] Université Clermont-Auvergne, CNRS, Mines de Saint-Étienne,
Clermont-Auvergne-INP, LIMOS, 63000 Clermont-Ferrand, France
[2] IBM Research AI, New York, USA
barahon@us.ibm.com

Abstract. We give a combinatorial algorithm to find a maximum packing of hypertrees in a capacitated hypergraph. Based on this we extend to hypergraphs several algorithms for the k-cut problem, that are based on packing spanning trees in a graph. In particular we give a γ-approximation algorithm for hypergraphs of rank γ, extending the work of Ravi and Sinha [24] for graphs. We also extend the work of Chekuri, Quanrud and Xu [7] in graphs, to give an algorithm for the k-cut problem in hypergraphs that is polynomial if k and the rank of the hypergraph are fixed. We also give a combinatorial algorithm to solve a linear programming relaxation of this problem in hypergraphs.

Keywords: k-cut · Packing hypertrees · Hypergraphic matroids

1 Introduction

Hypergraphic matroids were introduced by Lorea [18] and later studied by Frank, Kiraly and Kriesell [11]. They showed that the notion of circuit-matroid of graphs can be generalized to hypergraphs. The notion of spanning trees generalizes to hypertrees. Then Frank et al. [11] extended a theorem of Tutte [27] and Nash-Williams [23] about spanning trees, to a similar theorem giving the maximum number of disjoint hypertrees contained in a hypergraph. Based on this we give an algorithm to find a maximum packing of hypertrees in a capacitated hypergraph.

Spanning tree packing has been used to study the k-cut problem in graphs. It was used by Naor and Rabani [21] to derive a linear programming relaxation, and by Thorup [26] to develop an algorithm that is polynomial for fixed k. The linear programming relaxation was further studied by Chekuri, Quanrud and Xu [7], shedding light on the connections among several of these results. Here we show that hypertree packing and other algorithms for hypergraphic matroids, can be used to extend to the k-cut problem in hypergraphs, several of the results mentioned above.

Below we describe previous work, then we give more details about our contribution, and we outline the organization of this paper.

© Springer Nature Switzerland AG 2022
D. E. Simos et al. (Eds.): LION 2022, LNCS 13621, pp. 521–534, 2022.
https://doi.org/10.1007/978-3-031-24866-5_37

1.1 Previous Work

Based on the Theorem of Tutte [27] and Nash-Williams [23], polynomial algorithms for packing spanning trees in a graph have been given by Barahona [2] and Gabow and Manu [14]. The k-cut problem in graphs is NP-hard if k is part of the input, see [16]. If k is fixed, Goldschmidt and Hochbaum [16] gave the first polynomial algorithm. Later other algorithms improving the running time have been found. For a graph $G = (V, E)$, let $n = |V|$ and $m = |E|$. Thorup [26] gave an $O(mn^{2k-2})$ algorithm. Chekuri, Quanrud and Xu [7] improved the running time to $O(mn^{2k-3})$. Also they presented a framework that unifies the tree packing approach of Thorup [26] and the linear programming approach of Naor and Rabani [21]. When k is part of the input, several 2-approximation algorithms have been developed. Saran and Vazirani [25], gave a $(2 - 2/k)$ approximation. Nagamochi and Kamidoi [20], and Kapoor [17], found a similar approximation. Narayanan, Roy and Patkar [22] used submodular functions to give a $2(1 - 1/n)$ approximation algorithm. Naor and Rabani [21] used a linear programming relaxation to obtain a $2(1 - 1/n)$ approximation. Ravi and Sinha [24] used Lagrangian relaxation to also give a $2(1 - 1/n)$ approximation. Under the Small Set Expansion hypothesis [19], a factor of 2 is the best possible approximation. For more references on the k-cut problem see [7].

For a hypergraph $H = (V, E)$, let $n = |V|$, $m = |E|$, and $\gamma = \max\{|e| : e \in E\}$. This last number is called the *rank* of H. For the k-cut problem in hypergraphs, Fukunaga [13] extended Thorup's algorithm and gave an $O(m^2 n^{\gamma k-1})$ algorithm. Chandrasekaran, Xu and Yu [5] obtained a randomized algorithm that runs in $\tilde{O}(pn^{2k-1})$ time, where $p = \sum_{e \in E} |e|$. Fox, Panigrahi and Zhang [10] improved the randomized run-time to $\tilde{O}(mn^{2k-2})$. Recently Chandrasekaran and Chekuri [4] gave two deterministic algorithms with complexities $O(n^{3k(k-1)/2})$ and $O(n^{8k})$. When k is part of the input, the k-cut problem in hypergraphs is hard to approximate to within large factors under the Exponential Time Hypothesis, see [6], [19]; recall that for graphs there are several 2-approximation algorithms.

1.2 Our Contribution

We give a combinatorial algorithm for packing hypertrees in a hypergraph. Then we use this algorithm and the tools developed in [1], to study the k-cut problem in hypergraphs. First we extend the algorithm of Ravi and Sinha [24] for graphs, to obtain a γ-approximation algorithm for hypergraphs of rank γ. Then we study a linear programming relaxation for hypergraphs similar to the one used by Naor and Rabani [21] for graphs. We give a combinatorial algorithm to solve this linear relaxation. Then we extend to hypergraphs the analysis done by Cheruki et al. [7] for graphs, and show that the integrality gap is γ. We also build on their analysis to show that a maximum hypertree packing gives an $O(mn^{\gamma k-3})$ algorithm for k-cut in hypergraphs of rank γ. This improves by a factor of $O(mn^2)$ the time of the algorithm of [13], that is based on an approximate hypertree packing. In summary, our work shows that the use of hypergraphic matroids leads to natural extensions of several algorithms for the k-cut problem, that were initially developed for graphs.

1.3 Organization

Section 2 contains definitions, notation and some preliminary results. In Sect. 3 we give the algorithm for packing hypertrees. Section 4 contains lower and upper bounds for the value of a minimum k-cut. In Sect. 5 we study a linear programming relaxation. Section 6 contains a polynomial algorithm for fixed k and fixed rank.

2 Preliminaries

Let $H = (V, E)$ be hypergraph. For a non-empty set $X \subset V$ and $F \subseteq E$, $F[X]$ denotes the set of hyperedges in F contained in X. For $S \subset V$ we use $H(S)$ to denote the hypergraph $(S, E[X])$. Let $\mathcal{P} = \{V_1, \ldots, V_k\}$ be a family of non-empty subsets of V with $V_i \cap V_j = \emptyset$ for $i \neq j$, we denote by $\delta_F(\mathcal{P})$ the set of hyperedges in F included in $\cup_i V_i$ and that intersect at least two sets in $\mathcal{P}$. Notice that $\mathcal{P}$ is not necessarily a partition of V. For a hypergraph $H' = (V, E')$ sometimes we use $\delta_{H'}(\mathcal{P})$ instead of $\delta_{E'}(\mathcal{P})$. Also when there is no confusion we use $\delta(\mathcal{P})$ instead of $\delta_F(\mathcal{P})$. We say that H is *connected* if $\delta(S, V \setminus S) \neq \emptyset$, for all $S \subset V$, $\emptyset \neq S \neq V$. For $S \subset V$, *shrinking* S means creating a new hypergraph $H' = (V', E')$. Here $V' = (V \setminus S) \cup \{s\}$, where s is a new node that represents S. And $E' = E_1 \cup E_2$, where $E_1 = \{e \in E : e \cap S = \emptyset\}$, and $E_2 = \{(e \setminus S) \cup \{s\} : e \in E, e \cap S \neq \emptyset, e \cap (V \setminus S) \neq \emptyset\}$. For a vector $x \in \mathbb{R}^E$, and $S \subseteq E$ we use $x(S)$ to denote $\sum_{e \in S} x(e)$.

Let $D = (V, A)$ be a directed graph, for $S \subseteq V$, we denote by $\delta^+(S)$ the set $\delta^+(S) = \{(u, v) \in A \mid u \in S, v \notin S\}$. Given two distinguished vertices s and t, for a set $S \subset V$, with $s \in S$, $t \notin S$, the set of arcs $\delta^+(S)$ is called an *st-cut*. Given a capacity vector $c \in \mathbb{R}^A_+$, a *minimum st-cut* is an *st-cut* $\delta^+(S)$ such that $c(\delta^+(S))$ is minimum. A minimum *st-cut* can be found in $O(|V|^3)$ time with the push-preflow algorithm of [15].

For a hypergraph $H = (V, E)$, a *hyperforest* is a set $F \subseteq E$ such that $|F[X]| \leq |X| - 1$ for every non-empty $X \subseteq V$. A hyperforest F is called a *hypertree* of H if $|F| = |V| - 1$. If H is a graph, $F \subseteq E$ is a hypertree if and only if F is a spanning tree. It was proven by Lorea [18] that the hyperforests of a hypergraph form the family of independent sets of a matroid. These are called *hypergraphic matroids*. Frank et al. [11] further studied hypergraphic matroids. In particular they gave the following formula for the rank $r(F)$ of $F \subseteq E$, $r(F) = \min\{|V| - |\mathcal{P}| + |\delta_F(\mathcal{P})| : \mathcal{P} \text{ is a partition of } V\}$. Notice that matroid rank and rank of a hypergraph are completely different concepts.

Remark 1. *It follows from the formula above that if T is a hypertree then $|\delta_T(\mathcal{P})| \geq |\mathcal{P}| - 1$, for every partition $\mathcal{P}$ of V.*

Frank et al. [11] extended a theorem of Tutte [27] and Nash-Williams [23] giving the maximum number of spanning trees in a graph. They gave a similar formula for the maximum number of disjoint hypertrees contained in a hypergraph. This is in the theorem below.

Theorem 2 *[11]. A hypergraph contains k disjoint hypertrees if and only if*

$$|\delta(\mathcal{P})| \geq k(|\mathcal{P}| - 1) \tag{1}$$

holds for every partition $\mathcal{P}$ of V.

Based on this theorem we give an algorithm to find a maximum packing of hypertrees in a hypergraph. For that we use three algorithms mentioned below, that were developed in [1].

2.1 Separation of Partition Inequalities

Since there is an exponential number of inequalities (1), we need a polynomial algorithm to test if there is any of them that is violated. For that we assume that $\bar{x} \in \mathbb{R}_+^E$ is an input vector and we solve

$$\text{minimize} \quad \bar{x}(\delta(\mathcal{P})) - \beta(|\mathcal{P}| - 1). \tag{2}$$

where the minimum is taken among all partitions $\mathcal{P}$ of V, and $\beta > 0$ is a fixed number. Since $\mathcal{P} = \{V\}$ is a partition, the minimum is always less than or equal to zero. This gives us a most violated inequality, if there is any. In [1] this was reduced to a sequence of $|V|$ minimum cut problems in a graph with $O(|V|+|E|)$ nodes.

2.2 Strength of a Network

Given a hypergraph $H = (V, E)$, with a capacity vector $c \in \mathbb{R}_+^E$, we also need to find the maximum value of k so that $c(\delta(\mathcal{P})) \geq k(|\mathcal{P}| - 1)$, for all partitions $\mathcal{P}$ of V. We compute $s = \min \frac{c(\delta(\mathcal{P}))}{|\mathcal{P}|-1}$, where the minimum is taken over all partitions $\mathcal{P}$ of V, with $|\mathcal{P}| \geq 2$. For graphs this was called *Network Strength* in [8]. Thus we call the value s, the *strength* of H. Then $k = \lfloor s \rfloor$. The strength can be found with the same asymptotic complexity as $|V|$ applications of the push-preflow algorithm [15], in a graph with $O(|V| + |E|)$ nodes, see [1].

2.3 Network Reinforcement

The following problem was studied in [8]. Given a graph, a number k and a set of candidate edges, each of them with an associated cost, find a minimum cost set of candidate edges to be added to the network so it has strength equal to k. We need to solve a similar problem for a hypergraph. An algorithm for it was given in [1]. It requires to solve $|E||V|$ minimum cut problems in a graph with $O(|V| + |E|)$ nodes.

3   Packing Hypertrees

Here we give an algorithmic proof of Theorem 2. If H contains k disjoint hypertrees then it follows from Remark 1 that (1) holds for every partition. So we have to prove the other direction.

A partition is called *tight* if (1) holds as equation. We proceed by induction on $|V| + |E|$. So we assume that the statement is true for any hypergraph $H' = (V', E')$ with $|V'| + |E'| < |V| + |E|$. If an edge does not belong to any tight partition, we remove it and we apply the induction hypothesis. So we assume that every edge appears in a tight partition.

Case 1. Suppose that the partition $\{S_1, \ldots, S_p\}$ is tight, and at least one set, S_1 say, has $|S_1| > 1$. We shrink S_1 to form the hypergraph H'. From the induction hypothesis we know that there are k disjoint hypertrees in H'. Now consider $H'' = H(S_1)$. If there is a partition $T_1, \ldots, T_l$ of S_1 with $\delta_{H''}(T_1, \ldots, T_l) < k(l-1)$, then $\delta_H(T_1, \ldots, T_l, S_2, \ldots, S_p) < k(l-1) + k(p-1) = k(l+p-2)$, a contradiction. Thus by our induction hypothesis there are k disjoint hypertrees in $H(S_1)$. Lemma 3 below shows that any hypertree of H' can be combined with a hypertree of H'' to obtain a hypertree of H.

Lemma 3. *Let T' be a hypertree of H', and T'' a hypertree of H''. Then $T = T' \cup T''$ is a hypertree of $H = (V, E)$.*

Proof. Suppose that for a partition $\{U_1, \ldots, U_p\}$ of V, we have $|\delta_T(U_1, \ldots, U_p)| < (p-1)$. After renumbering the sets $\{U_i\}$, we can assume that $S_1 \subseteq \cup_{i=1}^{i=r} U_i$, and $S_1 \cap U_i \neq \emptyset$ for $i = 1, \ldots, r$. Also $|\delta_T(U_1, \ldots, U_p)| < (r-1) + (p-r) = (p-1)$. But this is not possible because $|\delta_T(U_1, \ldots, U_r)| \geq (r-1)$, and $|\delta_T(\cup_{i=1}^{i=r} U_i, U_{r+1}, \ldots, U_p)| \geq (p-r)$. Thus $|\delta_T(\mathcal{P})| \geq |\mathcal{P}| - 1$, for every partition $\mathcal{P}$ of V. Since $r(T) = \min\{|V| - |\mathcal{P}| + |\delta_T(\mathcal{P})| : \mathcal{P}$ a partition of $V\}$, we have $r(T) = |V| - 1$. In view of $|T| = |V| - 1$, we conclude that T is a hypertree of H.

Case 2. Now we assume that the partition $\{S_1, \ldots, S_p\}$ is tight, and all sets $\{S_i\}$ are singletons. We have

$$|E| = k(|V| - 1). \tag{3}$$

If every hyperedge has exactly two elements we have a graph. Then the result follows from the Theorem of Tutte [27] and Nash-Williams [23]. In this case the algorithms of [2] or [14] give the packing of spanning trees.

Suppose that a hyperedge e has at least three elements. We remove one element v from e, and test if all inequalities (1) are satisfied. If so we keep working with the new hypergraph. If not, there is a tight partition $\mathcal{P}$ whose inequality becomes violated after removing v from e. This is not the partition of all singletons, because (3) is not violated after removing v from e. Then we can treat $\mathcal{P}$ as in Case 1. This completes the proof.

3.1 Integral Packing

Based on the above proof, now we derive an algorithm. We assume that for a
hypergraph $H = (V, E)$, we have a capacity vector $c \in \mathbb{Z}_+^E$. We denote by $\mathcal{T}(H)$
the set of hypertrees of H. A *maximum integral packing of hypertrees* is a solution
of the following.

$$\max \sum_{T \in \mathcal{T}(H)} y_T; \quad \sum_{T : e \in T} y_T \le c(e), \text{ for each edge } e; \; y \ge 0, \text{ integer valued.} \quad (4)$$

The algorithm has several stages as follows.

- First we compute

$$k = \min \left\lfloor \frac{c(\delta(\mathcal{P}))}{|\mathcal{P}| - 1} \right\rfloor, \quad (5)$$

over all partitions $\mathcal{P}$ of V. This is the maximum value of k such that $c(\delta(\mathcal{P})) \ge k(|\mathcal{P}| - 1)$ for every partition $\mathcal{P}$ of V. This is the strength of the hypergraph,
as defined in Sub-section 2.2. For that we use the algorithm in [1]. This
requires $|V|$ applications of the push-preflow algorithm [15], in a graph with
$O(|V| + |E|)$ nodes. This gives the value of the maximum in (4), but not the
values for the variables y.
- Once the value k is known, we should adjust the capacities so that every
hyperedge appears at least in a tight partition. For that we solve the linear
program below.

$$\min x(E) \quad (6)$$
$$x(\delta(\mathcal{P})) \ge k(|\mathcal{P}| - 1), \text{ for all partitions } \mathcal{P} \text{ of } V, \quad (7)$$
$$0 \le x(e) \le c(e). \quad (8)$$

This is called Network Reinforcement, and as mentioned in Sub-section 2.3,
it reduces to $|E||V|$ minimum cut problems in a graph with $O(|V| + |E|)$
nodes, see [1]. If the capacities are integers, this algorithm produces an integer
solution. Denote by $\bar{x}$ the solution obtained. We lower the capacities c to $\bar{x}$,
i.e., we set $c \leftarrow \bar{x}$.
- Now we have to treat Cases 1 and 2 above. We have to find a tight partition.
For that we pick an edge e, we decrease by one its capacity $c(e)$, and find
a most violated partition inequality, with the algorithm of [1]. This involves
$|V|$ minimum cut problems in a graph with $O(|V| + |E|)$ nodes. A violated
inequality is a tight inequality if we do not decrease $c(e)$. Let $\{S_1, \ldots, S_p\}$ be
the associated partition of V.
In Case 1 we assume that one set, S_1 say, has $|S_1| > 1$. We shrink S_1 to a
single node and we denote by H' the resulting hypergraph. Then we look for
a packing of hypertrees of value k in H'. We also denote by $H'' = H(S_1)$, and
look for a packing of hypertrees of value k in H''. Then we combine hypertrees
in H' with hypertrees in H'' to obtain a packing of hypertrees of value k in
H. This is done as below.

Let $S' = \{T'_1, \ldots, T'_r\}$ be a set of hypertrees of H' with positive weights $\{\alpha_1, \ldots, \alpha_r\}$, and let $S'' = \{T''_1, \ldots, T''_s\}$ be a set of hypertrees of H'' with positive weights $\{\beta_1, \ldots, \beta_s\}$. Pick any hypertree in the first set, T'_1 say, and any hypertree in the second set, T''_1 say, and form a hypertree in H, $T = T'_1 \cup T''_1$ with weight $\gamma = \min\{\alpha_1, \beta_1\}$. Subtract γ from α_1 and β_1, and remove from S' and S'' any hypertree with zero weight. Continue until S' and S'' are empty. This procedure produces at most $r + s$ hypertrees of H. If the weights $\{\alpha_i\}$ and $\{\beta_j\}$ are integers, then the new weights are also integer.

In Case 2 we assume that $\{S_1, \ldots, S_p\}$ is the partition of all singletons. If all hyperedges have exactly two elements, we have a graph. Then we can apply the algorithms of [2] or [14]. If the capacities are integer these two algorithms produce an integral packing. The algorithm of [14] has complexity $O(|V|^3|E|\log(|V|^2/|E|))$ and produces at most $2|E| + 2|V| - 2$ spanning trees. Assume now that there is a hyperedge e with at least three nodes and with capacity $c(e)$. We remove a node v and look for a most violated partition inequality. If there is none, we keep working with the new hypergraph. Otherwise, let α be the violation. We create a new hyperedge $e' = e \setminus \{v\}$ with capacity $c(e) - \alpha$, and give a capacity α to e. Then we have a tight partition that is treated as in Case 1. Notice that Case 1 arises at most $|V|$ times, therefore during the entire execution of this algorithm, at most $|V|$ new hyperedges are created.

Since all arithmetic operations are additions and subtractions, and the capacities are integer, this algorithm produces an integral packing. Now we analyze the complexity of this algorithm. Notice that it requires finding at most $|V|$ violated partition inequalities. This amounts to at most $|V|^2$ minimum cut problems in a graph with $O(|V| + |E|)$ nodes. So the complexity of this part is $O(|V|^2(|V| + |E|)^3)$. This dominates the complexity of the algorithm for finding a packing of spanning trees that is $O(|V|^3|E|\log(|V|^2/|E|))$.

After some transformations our algorithm requires finding packings of spanning trees in graphs. For each of these trees, each edge is contained in a hyperedge of the original hypergraph. The algorithm of [14] produces $2m+2n$ different spanning trees for a graph with n nodes and m edges. Thus we conclude that our algorithm produces at most $O(|E| + |V|)$ hypertrees.

If the capacities are all equal to one, packing hypertrees is a special case of Matroid Partition, and can be solved as Edmonds showed in [9]. However it is not clear to us how to use Edmonds approach when the capacities are nonnegative integers, or for fractional packing. This last case is the key for the algorithm in Sect. 5.

3.2 Fractional Packing

Now we consider problem (4), but without requiring integrality of the variables y. The only difference here is that formula (5) is replaced by $k' = \min \frac{c(\delta(\mathcal{P}))}{|\mathcal{P}|-1}$. Then k' might be a non-integer number. All other steps of the algorithm remain the same. Notice that in Case 2 when a new hyperedge is created, it receives a fraction of the capacity of the original hyperedge. We illustrate this below with a simple example.

Consider $H = (V = \{a, b, c, d\}, E = \{V\})$. Let $c(V) = 1$. Then using formula above we get $k' = 1/3$, given by the partition of all singletons. Thus in our algorithm, this partition is tight. As in Case 2, removing node a from the hyperedge V gives a violation of $1/3$ for the partition inequality associated with $\{\{a\}, \{b, c, d\}\}$. Thus we give capacity $1/3$ to V and capacity $2/3$ to the new hyperedge $\{b, c, d\}$; then the partition $\{\{a\}, \{b, c, d\}\}$ is tight. When we shrink $\{b, c, d\}$ to single node we obtain a graph with one edge. Then we have to keep working with $H(\{b, c, d\})$. Here the partition of all singletons is tight, so we remove b from the hyperedge $\{b, c, d\}$. Then the partition inequality associated with $\{\{b\}, \{c, d\}\}$ is violated by $1/3$. Thus we give capacity $1/3$ to $\{b, c, d\}$, and capacity $1/3$ to $\{c, d\}$. When we shrink $\{c, d\}$, we obtain a graph with one edge, and the hypergraph $H(\{c, d\})$ is also a graph with one edge. In both cases the packing is trivial to find. In summary, the algorithm made three copies of the hyperedge V (or subsets of it), each of them with capacity $1/3$, and gave the weight $1/3$ to the resulting hypertree.

4   A Relaxation of the k-cut Problem

Consider a hypergraph $H = (V, E)$, with a weight vector $w \in \mathbb{R}_+^E$, and a fixed number k. The k-cut problem consists of finding a partition $\{S_1, \ldots, S_k\}$ of V that minimizes $w(\delta(S_1, \ldots, S_k))$. Let $\lambda_k(H)$ denote the value of the minimum. For a non-negative number b, a lower bound of $\lambda_k(H)$ is

$$l(b) = \min_{p \geq 1} \left\{ w(\delta(S_1, \ldots, S_p)) - b(p - k) \right\}. \tag{9}$$

Here the minimum is taken over all partitions of V, and b is a fixed non-negative number. The function $l(\cdot)$ is concave and piece-wise linear, and it has at most n break-points. The maximum of l is found at a break-point $\bar{b}$. In what follows we study how to find all break-points $b_i \leq \bar{b}$. For graphs, a similar lower bound was proposed in [3] and independently in [24].

4.1   Break-Points of l

We start with $b = 0$, then the trivial partition $\mathcal{P} = \{V\}$ gives the minimum in (9). The following lemma gives us a way to generate the subsequent break-points.

Lemma 4. Let $\mathcal{P}' = \{S_1, \ldots, S_p\}$ be a solution of (9) for $b = b'$. Assume that for $b = b''$, $b'' \geq b'$, $\mathcal{P}'$ and $\mathcal{P}'' = \{T_1, \ldots, T_q\}$ are both solutions of (9), with $q > p$. Then b'' is the strength of one of the sets $\{S_i\}$. Recall that the strength was defined in Sub-section 2.2.

Proof. Since $q > p$, we can assume that after renumbering, $S_1 \subseteq \cup_{i=1}^r T_i$ and $Q_i = T_i \cap S_1 \neq \emptyset$, for $i = 1, \ldots, r, r > 1$.

Since $\mathcal{P}'$ is a solution for $b = b''$, we have $w(\delta(Q_1, \ldots, Q_r)) \geq b''(r-1)$. If this is not the case, we would have $w(\delta(Q_1, \ldots, Q_r)) < b''(r-1)$, and we could improve the solution $\mathcal{P}'$ by removing S_1 and adding the sets $\{Q_i\}$. With the same argument we conclude that a similar inequality holds for any partition of S_1.

Since $\mathcal{P}''$ is also a solution for $b = b''$ we cannot have $w(\delta(Q_1, \ldots, Q_r)) > b''(r-1)$. If that was the case, we could improve the solution $\mathcal{P}''$ replacing the sets T_i, $1 \leq i \leq r$, with their union.

Therefore $w(\delta(Q_1, \ldots, Q_r)) = b''(r-1)$, and $w(\delta(R_1, \ldots, R_t)) \geq b''(t-1)$, for any partition $\{R_1, \ldots, R_t\}$ of S_1. Thus b'' is the strength of S_1.

This suggests the following procedure.

Algorithm 1

Step 0. Start with $\mathcal{P}_0 = \{V\}$, $b = \bar{b} = 0$, $j = 0$.

Step 1. Compute the strength of each set in $\mathcal{P}_j$. Among them, let S_q be a set with minimum strength s_q.

Step 2. Update $\bar{b} \leftarrow s_q$, and to obtain $\mathcal{P}_{j+1}$, replace S_q in $\mathcal{P}_j$ with a partition of S_q giving its strength. Set $j \leftarrow j+1$, if $|\mathcal{P}_j| < k$ go to Step 1, otherwise stop.

The sequence of partitions produced here can be studied in the context of submodular functions as in [12].

4.2   The Maximum of l

Consider the last value j, we have $|\mathcal{P}_j| \geq k$. Let $\bar{b}$ be the associated value of the parameter. The corresponding lower bound is

$$\mu = w(\mathcal{P}_{j-1}) - \bar{b}(|\mathcal{P}_{j-1}| - k) = w(\mathcal{P}_j) - \bar{b}(|\mathcal{P}_j| - k). \tag{10}$$

If $|\mathcal{P}_j| = k$, this is a solution of the k-cut problem. Now we treat the case when $|\mathcal{P}_j| > k$.

Since $\bar{b} = \frac{w(\mathcal{P}_j) - w(\mathcal{P}_{j-1})}{|\mathcal{P}_j| - |\mathcal{P}_{j-1}|}$, we obtain the expression below that is needed in the next sub-section.

$$\mu = \frac{|\mathcal{P}_j| - k}{|\mathcal{P}_j| - |\mathcal{P}_{j-1}|} w(\delta(\mathcal{P}_{j-1})) + \frac{k - |\mathcal{P}_{j-1}|}{|\mathcal{P}_j| - |\mathcal{P}_{j-1}|} w(\delta(\mathcal{P}_j)). \tag{11}$$

4.3   An Upper Bound

Now we produce an approximate solution for the k-cut problem. Let $l = k - |\mathcal{P}_{j-1}|$ and $\gamma = \max\{|e| : e \in E\}$. Let $\{T_1, \ldots, T_r\}$ be the partition of the last set S_q obtained in Step 2 of Algorithm 1. We number the sets $\{T_i\}$ so that $w(\delta(T_i)) \leq w(\delta(T_{i+1}))$, for $i = 1, \ldots, r-1$. We choose $\{T_1, \ldots, T_l\}$, and combine $\{T_{l+1}, \ldots, T_r\}$ into one set. A similar procedure was proposed for graphs in [24].

Theorem 5. *The value of this solution is at most* $\gamma(1 - \frac{1}{n})\lambda_k(H)$.

Proof. We have

$$\sum_{i=1}^{l} w(\delta(T_i)) \leq \frac{l}{r} \sum_{i=1}^{r} w(\delta(T_i)) \leq \gamma\frac{l}{r}w(\delta(T_1, \ldots, T_r)) =$$

$$\gamma\frac{r-1}{r} \frac{l}{r-1} w(\delta(T_1, \ldots, T_r)).$$

Thus the value of this solution is at most

$$w(\mathcal{P}_{j-1}) + \gamma(1 - \frac{1}{r})\frac{l}{r-1}w(\delta(T_1, \ldots, T_r)) =$$

$$w(\mathcal{P}_{j-1})\left(1 - \gamma(1 - \frac{1}{r})\frac{l}{r-1}\right) + \gamma(1 - \frac{1}{r})\frac{l}{r-1}w(\mathcal{P}_j) \leq$$

$$\gamma(1 - \frac{1}{r})\, w(\mathcal{P}_{j-1})(\frac{r-1-l}{r-1}) + \gamma(1 - \frac{1}{r})\frac{l}{r-1}w(\mathcal{P}_j) =$$

$$\gamma(1 - \frac{1}{r})\mu \leq \gamma(1 - \frac{1}{r})\lambda_k(H) \leq \gamma(1 - \frac{1}{n})\lambda_k(H).$$

Thus we have a γ-approximation algorithm for hypergraphs of rank γ. Recall that the k-cut problem in hypergraphs is hard to approximate to within large factors under the Exponential Time Hypothesis, see [6,19]. Also recall that for $\gamma = 2$, under the same hypothesis, a factor of 2 is the best possible approximation, cf. [19].

5   A Linear Programming Relaxation for k-cut

Let $H = (V, E)$ be a connected hypergraph, and $w \in \mathbb{Z}_+^E$. Let $\mathcal{T}(H)$ denote the set of hypertrees of H. We study the linear program

$$\min \sum w(e)x(e) \tag{12}$$

$$\sum_{e \in T} x(e) \geq k - 1 \text{ for } T \in \mathcal{T}(H) \tag{13}$$

$$0 \leq x(e) \leq 1 \text{ for } e \in E \tag{14}$$

An integer solution of this gives a solution of the k-cut problem. For graphs a similar linear program was proposed in [21].

Now we extend to hypergraphs the analysis used in [7] for graphs. Let $\mathcal{P}_j$ be the last partition produced by Algorithm 1, and let $\{T_1, \ldots, T_r\}$ be the partition of the last set S_q obtained in Step 2 of Algorithm 1. The vector $\bar{x}$ defined below is a feasible solution of (12)–(14).

- $\bar{x}(e) = 1$ for $e \in \delta(\mathcal{P}_{j-1})$; $\bar{x}(e) = 0$ for $e \in E \setminus \delta(\mathcal{P}_j)$.
- $\bar{x}(e) = \alpha$ for $e \in \delta(\mathcal{P}_j) \setminus \delta(\mathcal{P}_{j-1})$, where $\alpha = \frac{k - |\mathcal{P}_{j-1}|}{|\mathcal{P}_j| - |\mathcal{P}_{j-1}|}$.

Its value is

$$w(\delta(\mathcal{P}_{j-1})) + \frac{k - |\mathcal{P}_{j-1}|}{|\mathcal{P}_j| - |\mathcal{P}_{j-1}|} w(\delta(T_1, \ldots, T_r)) = \tag{15}$$

$$\frac{|\mathcal{P}_j| - k}{|\mathcal{P}_j| - |\mathcal{P}_{j-1}|} w(\delta(\mathcal{P}_{j-1})) + \frac{k - |\mathcal{P}_{j-1}|}{|\mathcal{P}_j| - |\mathcal{P}_{j-1}|} w(\delta(\mathcal{P}_j)).$$

This is the value μ as in (11).

Now consider the dual problem.

$$\max(k-1) \sum_{T \in \mathcal{T}(H)} y_T - \sum_{e \in E} z(e) \tag{16}$$

$$\sum_{T : e \in T} y_T \le w(e) + z(e) \forall e \in E \tag{17}$$

$$y \ge 0, \; z \ge 0 \tag{18}$$

To obtain a dual solution define

$$w(e) + \bar{z}(e) = \begin{cases} (b_j/b_i)w(e) & \text{for } e \in \delta(\mathcal{P}_i) - \delta(\mathcal{P}_{i-1}), i < j, \\ w(e) & \text{otherwise.} \end{cases}$$

Using $w + \bar{z}$ as capacities leads to a hypergraph whose strength is b_j. Thus this set of capacities yields a maximum (fractional) hypertree packing $\bar{y}$ of value b_j. Now we compute the objective value for the dual vector $(\bar{y}, \bar{z})$. This is

$$(k-1)b_j - \sum_{i=1}^{j-1} (\frac{b_j}{b_i} - 1)(w(\mathcal{P}_i) - w(\mathcal{P}_{i-1})) =$$

$$(k-1)b_j - \sum_{i=1}^{j-1} (\frac{b_j - b_i}{b_i})(w(\mathcal{P}_i) - w(\mathcal{P}_{i-1})) =$$

$$(k-1)b_j - \sum_{i=1}^{j-1} (b_j - b_i)(|\mathcal{P}_i| - |\mathcal{P}_{i-1}|) =$$

$$(k-1)b_j - (|\mathcal{P}_{j-1}| - 1)b_j + w(\mathcal{P}_{j-1}) =$$

$$\frac{k - |\mathcal{P}_{j-1}|}{|\mathcal{P}_j| - |\mathcal{P}_{j-1}|} w(\delta(T_1, \ldots, T_r)) + w(\mathcal{P}_{j-1}).$$

Here we obtained expression (15). Thus $\bar{x}$ and $(\bar{y}, \bar{z})$ have the same value, therefore they are optimal solutions. Hence we have polynomial combinatorial algorithms to produce optimal primal and dual solutions of (12)–(14). For graphs, other authors have suggested the use of the ellipsoid method, or the use of approximate tree packing, see [7,13,21,26]. Having fast algorithms to produce lower and upper bounds gives the possibility of embedding this in a branch and bound procedure.

The optimal value of this linear program is exactly the lower bound μ defined in (10). Hence from Theorem 5 we obtain the following.

Theorem 6. *The integrality gap of this linear program is at most* $\gamma(1 - \frac{1}{n})$.

Notice that as long as the hypergraph is connected, the algorithm from Subsect. 3.2 makes fractional copies of the hyperedges to produce a fractional packing of hypertrees. Consider the example in Subsect. 3.2, with $k = 2$. The value of the linear program is $1/3$ and the value of a minimum 2-cut is 1. Here we have exactly the gap given by Theorem 6. We can extend this example to a hypergraph with n nodes, and one hyperedge with weight 1, containing all nodes. Then for $k = 2$ the lower bound is $1/(n - 1)$. Again we have exactly the gap given by Theorem 6.

Consider now a non-connected hypergraph. Assuming that the weights are integer, we propose the following. We multiply by n all the weights, and add a minimal set of artificial edges to make the hypergraph connected. We give the weight 1 to each artificial edge. Then minimum k-cuts in the new hypergraph correspond to minimum k-cuts in the original one.

6   A Polynomial Algorithm for Fixed γ and k

Now we show that the algorithm for graphs given in [7] can be extended to hypergraphs. The lemma below was proved in [7] for $\gamma = 2$, the proof for larger values of γ is similar.

Lemma 7. *Let* $(\bar{y}, \bar{z})$ *be an optimal solution of (16)–(18). Let* E' *be any set of hyperedges such that* $w(E') \leq \alpha \lambda_k(H)$ *for some* $\alpha \geq 1$. *For each hypertree* T *let* $l_T = |E' \cap E(T)|$. *Let* $\tau = \sum_T \bar{y}_T$ *and* $p_T = \bar{y}_T/\tau$. *For an integer* $h \geq (k - 1)$, *let* $q_h = \sum_{T:l_T \leq h} p_T$. *Then*

$$q_h \geq 1 - \frac{\gamma\alpha(k - 1)(1 - \frac{1}{n})}{h + 1}$$

Corollary 8. *Let* $(\bar{y}, \bar{z})$ *be an optimal solution of (16)–(18). Define the support of* $\bar{y}$ *as* $\{T : \bar{y}_T > 0\}$. *For every optimum* k-cut $A \subseteq E$ *there is a hypertree* T *in the support of* $\bar{y}$ *such that* $|E(T) \cap A| \leq \gamma k - 3$.

Proof. We apply Lemma 7 with $h = \gamma k - 3$ and $\alpha = 1$. We obtain

$$q_h \geq 1 - \frac{(\gamma k - \gamma)(1 - \frac{1}{n})}{\gamma k - 2},$$

and for $\gamma \geq 2$ we have $q_h > 1$.

This suggest the following algorithm: For each hypertree in the support of $\bar{y}$, choose $\gamma k - 3$ hyperedges, contract the remaining hyperedges, and find a minimum k-cut in the resulting hypergraph. This has to be repeated for every choice of $\gamma k - 3$ hyperedges. Recall that the packing algorithm produces $O(m+n)$ hypertrees, so this leads to an $O((m + n) n^{\gamma k-3})$ algorithm that enumerates all minimum k-cuts. Fukunaga has given an $O(m^2 n^{\gamma k-1})$ algorithm based on a greedy packing of hypertrees. Using an optimal packing leads to decrease the complexity by a factor of $O(mn^2)$, and to a simpler derivation. Chandrasekaran and Chekuri [4] gave two algorithms with complexities $O(n^{3k(k-1)/2})$ and $O(n^{8k})$, (that are independent of γ). Thus the hypertree packing approach seems to be of interest for hypergraphs of small rank.

The theorem below was proved in [7] for $\gamma = 2$, a similar proof works for larger values of γ, so we omit it.

Theorem 9. *For $\alpha \geq 1$ the number of α-approximate minimum k-cuts is $O(n^{\lfloor \gamma\alpha(k-1)\rfloor})$.*

7 Concluding Remarks

We have given an algorithm for (fractional) packing of hypertrees. This and other algorithms for hypergraphic matroids allow us to use in hypergraphs, many algorithms originally developed for the k-cut problem in graphs.

References

1. Baiou, M., Barahona, F.: On some algorithmic aspects of hypergraphic matroids. arXiv 2111.05699 (2021)
2. Barahona, F.: Packing spanning trees. Math. Oper. Res. **20**(1), 104–115 (1995)
3. Barahona, F.: On the k-cut problem. Oper. Res. Lett. **26**(3), 99–105 (2000)
4. Chandrasekaran, K., Chekuri, C.: Hypergraph k-cut for fixed k in deterministic polynomial time. In: 2020 IEEE 61st Annual Symposium on Foundations of Computer Science (FOCS), pp. 810–821. IEEE (2020)
5. Chandrasekaran, K., Xu, C., Yu, X.: Hypergraph k-cut in randomized polynomial time. Math. Program. **186**(1), 85–113 (2021)
6. Chekuri, C., Li, S.: A note on the hardness of approximating the k-way hypergraph cut problem. Manuscript, http://chekuri.cs.illinois.edu/papers/hypergraph-kcut.pdf (2015)
7. Chekuri, C., Quanrud, K., Xu, C.: LP relaxation and tree packing for minimum k-cut. SIAM J. Discrete Math. **34**(2), 1334–1353 (2020)
8. Cunningham, W.H.: Optimal attack and reinforcement of a network. J. of ACM **32**, 549–561 (1985)
9. Edmonds, J.: Lehman's switching game and a theorem of Tutte and Nash-Williams. J. Res. Nat. Bur. Standards Sect. B **69**, 73–77 (1965)
10. Fox, K., Panigrahi, D., Zhang, F.: Minimum cut and minimum k-cut in hypergraphs via branching contractions. In: Proceedings of the Thirtieth Annual ACM-SIAM Symposium on Discrete Algorithms, pp. 881–896. SIAM (2019)
11. Frank, A., Király, T., Kriesell, M.: On decomposing a hypergraph into k connected sub-hypergraphs. Discrete Appl. Math. **131**(2), 373–383 (2003)
12. Fujishige, S.: Theory of principal partitions revisited. In: Research Trends in Combinatorial Optimization, pp. 127–162. Springer, Berlin, Heidelberg (2009). https://doi.org/10.1007/978-3-540-76796-1_7
13. Fukunaga, T.: Computing minimum multiway cuts in hypergraphs. Discrete Optim. **10**(4), 371–382 (2013)

14. Gabow, H.N., Manu, K.: Packing algorithms for arborescences (and spanning trees) in capacitated graphs. Math. Program. **82**(1), 83–109 (1998)
15. Goldberg, A.V., Tarjan, R.E.: A new approach to the maximum-flow problem. J. Assoc. Comput. Mach. **35**(4), 921–940 (1988)
16. Goldschmidt, O., Hochbaum, D.S.: A polynomial algorithm for the k-cut problem for fixed k. Math. Oper. Res. **19**, 24–37 (1994)
17. Kapoor, S.: On minimum 3-cuts and approximating k-cuts using cut trees. In: Cunningham, W.H., McCormick, S.T., Queyranne, M. (eds.) IPCO 1996. LNCS, vol. 1084, pp. 132–146. Springer, Heidelberg (1996). https://doi.org/10.1007/3-540-61310-2_11
18. Lorea, M.: Hypergraphes et matroides. Cahiers Centre Etudes Rech. Oper. **17**, 289–291 (1975)
19. Manurangsi, P.: Inapproximability of maximum edge biclique, maximum balanced biclique and minimum k-cut from the small set expansion hypothesis. In: 44th International Colloquium on Automata, Languages, and Programming (ICALP 2017). Schloss Dagstuhl-Leibniz-Zentrum fuer Informatik (2017)
20. Nagamochi, H., Kamidoi, Y.: Minimum cost subpartitions in graphs. Inf. Process. Lett. **102**(2–3), 79–84 (2007)
21. Naor, J., Rabani, Y.: Tree packing and approximating k-cuts. In: SODA. vol. 1, pp. 26–27 (2001)
22. Narayanan, H., Roy, S., Patkar, S.: Approximation algorithms for min-k-overlap problems using the principal lattice of partitions approach. J. Algorithms **21**(2), 306–330 (1996)
23. Nash-Williams, C.S.J.A.: Edge-disjoint spanning trees of finite graphs. J. London Math. Soc. **36**, 445–450 (1961)
24. Ravi, R., Sinha, A.: Approximating k-cuts using network strength as a Lagrangean relaxation. Eur. J. Oper. Res. **186**(1), 77–90 (2008)
25. Saran, H., Vazirani, V.V.: Finding k cuts within twice the optimal. SIAM J. Comput. **24**(1), 101–108 (1995)
26. Thorup, M.: Minimum k-way cuts via deterministic greedy tree packing. In: Proceedings of the Fortieth Annual ACM Symposium on Theory of Computing, pp. 159–166 (2008)
27. Tutte, W.T.: On the problem of decomposing a graph into n connected factors. J. London Math. Soc. **36**, 221–230 (1961)

Maximizing the Eigenvalue-Gap and Promoting Sparsity of Doubly Stochastic Matrices with PSO

Panos K. Syriopoulos[✉], Nektarios G. Kalampalikis,
and Michael N. Vrahatis[iD]

Computational Intelligence Laboratory, Department of Mathematics, University of
Patras, 26110 Patras, Greece
{p.syriopoulos,nkalamp,vrahatis}@math.upatras.gr

Abstract. The eigenvalue-gap of doubly stochastic matrices with sparsity constraints is maximized using the unified particle swarm optimizer. This is possible through the use of an iterative normalization procedure that maps the search space of the swarm to the set of doubly stochastic matrices with given sparsity pattern. We extend the method to the problem of finding doubly-stochastic matrices of given dimensions that are as sparse as possible, and attain a given eigenvalue-gap target.

Keywords: Eigenvalue-gap · Particle swarm optimization · Sparse matrix identification · Distributed averaging

1 Introduction

The study of complex networks is a fascinating subject with far reaching results both in terms of their practical applications and in their theoretical foundations. Understanding complex networks is crucial for the cooperation and control of autonomous agents. It is fair to say that complex networks occur in various aspects of life. Interestingly, one of the pioneers of the study of complex networks was a psychiatrist, and founder of group-therapy, Jacob Moreno (1889–1974), with his paper "Statistics of social configurations" [18]. The study of social networks still remains an active area of research [3]. In engineering the applications of complex networks are numerous. For example, after the introduction of the multiprocessor, consensus networks were quickly applied for the task of dynamic load balancing [4]. Modern data centers require more sophisticated solutions to be scalable and reliable [14,25]. Among other applications, energy micro grids is a recent area for distributed control and presents new challenges [11,13]. Additionally, sensor networks have several modern day applications [12], while significant attention has been given to autonomous vehicle control [22]. Complex networks also model many natural processes such as bird flocking [10]. In general, one has to consider the dynamics of the network (how the network structure

© Springer Nature Switzerland AG 2022
D. E. Simos et al. (Eds.): LION 2022, LNCS 13621, pp. 535–548, 2022.
https://doi.org/10.1007/978-3-031-24866-5_38

changes) and the dynamics in the network (how nodes interact). The major concepts can be found, among others, in [1]. A lot of the theoretical interest is on emergent phenomena such as the unidimensionality of beliefs [6].

A basic but integral part of distributed control is *consensus*. The purpose is for all agents in the network to agree on an underlying state of their environment. A special case is distributed averaging. In this instance, the nodes have to agree on the average of their initial measurements by communicating through the network [8]. We consider linear update rules. In each communication round, each node updates its estimate by taking a weighted sum of the estimates of its neighbors. To mention a few applications, the distributed Kalman filter can be seen as a distributed averaging problem [19], while, distributed averaging is heavily leveraged in modern distributed optimisation applications [7].

We are concerned with two key problems related to (a) the *speed of convergence to the average* and (b) the problem of *optimizing the network itself*. In other words, for problem (a) we want to minimize the number of communication rounds required for all the nodes to approximate the average within an error bound. It turns out that the asymptotic speed of convergence is bounded above by the second largest eigenvalue in magnitude. The problem is to optimize the network weights for speed, while respecting the network connectivity constraints. For problem (b), given a target for the asymptotic speed of convergence, we aim to find a network structure that is as *sparse* as possible. We tackle these problems using the *unified particle swarm optimization UPSO* (see [21], the corresponding MATLAB code and references therein), which is an effective and efficient evolutionary algorithm, and can also be considered as a complex network. The algorithm consists of particles that perform random search locally, and exchange information in the framework of their network connectivity in order to minimize a given objective function. The ability of the algorithm to optimize arbitrary functions is an emergent phenomenon of the dynamics between the particles. There are also several advantages. Firstly, the algorithm is easily applied and widely accessible. Secondly, it is incredibly capable in solving hard problems: the objective function can be non-differentiable and even discontinuous, and the algorithm can be applied to spaces with non connected regions. These characteristics appear frequently in difficult real life applications and UPSO can effectively tackle these issues. As a result, we are interested to study how the algorithm performs in aforementioned problems (a) and (b). In any case, we believe that this work is a useful demonstration of the ability of evolutionary algorithms in optimization. Our contribution is in the use of an iterative normalization scheme, which enables the application of evolutionary algorithms in this context.

The paper is structured as follows. Section 2 contains the background material. In Sect. 3 the objective functions are formulated. Finally, Sect. 4 contains experiments, while, in Sect. 5, a synopsis and concluding remarks are given.

2   Preliminaries

This section provides a basic understanding of the background material. To motivate the problem of eigenvalue gap maximization, we examine the effect of eigenvalues in the context of consensus networks [8]. We exhibit that the consensus speed depends on the magnitude of the second largest eigenvalue in modulus (for the diagonalizable case). In general, the consensus problem is grounded in the theory of Markov chains. Details can be found in [15]. Alternatively, we refer the interested reader to [17] for further mathematical treatments of bounds for consensus speed. Furthermore, we describe the unified PSO algorithm (UPSO) [21] with which we tackle the eigenvalue gap maximization problem.

Throughout the remainder of the paper, the structure of a network is represented by a graph $G(V, E)$ where $V = \{1, 2, \ldots, n\}$ is the vertex set and E is the ordered set containing the edges $E = \{(i, j) : \text{node } i \text{ listens to node } j\}$. The set of weights that nodes give to their neighbors can be represented by a matrix W with elements $W_{i,j}$ such that:

$$W_{i,j} = \begin{cases} W_{i,j} > 0, & \text{if } \{i, j\} \in E, \\ 0, & \text{if } \{i, j\} \notin E. \end{cases} \tag{1}$$

2.1   Consensus and Eigenvalues

Suppose $W \in \mathbb{R}^{n \times n}$ is a stochastic matrix with non-negative entries so that each row sums to one, i.e. $\sum_j^n W_{i,j} = 1$, for all $i \in \{1, 2, \ldots, n\}$. Given any $x(0) \in \mathbb{R}^n$, we are concerned with the convergence speed of the sequence $x(t)$, $t = 0, 1, \ldots$ defined by:

$$x(t + 1) = Wx(t) = W^{t+1}x(0). \tag{2}$$

This is known as the *consensus model of DeGroot* and was initially formulated as a method for pooling probability distributions [5]. Intuitively $x(0)$ can be seen as the vector of beliefs of each node at time 0 (their initial approximations), and W is the matrix of the network's edge weights. The application of W updates the belief of the nodes by taking a linear combination of the beliefs of their neighbors. The *consensus vector* denoted by x_c is given by the following limit, provided it exists:

$$x_c = \lim_{t \to \infty} x(t) = \lim_{t \to \infty} W^t x(0). \tag{3}$$

It can be seen that the existence of the limit in Eq. (3) depends on the existence and uniqueness of a left eigenvector with eigenvalue 1. Suppose $\pi \in \mathbb{R}^n$ such that $\pi X = \pi$ is the unique eigenvector (up to normalization) with eigenvalue 1. Then we have that:

$$\pi x(t + 1) = \pi(Wx(t)) = (\pi W)x(t) = \ldots = \pi x(0). \tag{4}$$

Taking the limit as $t \to \infty$ above we obtain:

$$\pi x_c = \pi x(0).$$

At this point it can be seen that if W is doubly-stochastic (rows and columns sum to one), then $\pi = [1/n, 1/n, \ldots, 1/n]^\top$ is a left eigenvector, and the system converges to the average of the values of $x(0)$.

We can derive sufficient conditions for the existence of the limit in Eq. (3) using linear algebra. It is known that the dominant eigenvalue of any stochastic matrix has modulus one. Additionally, we know from the Perron-Frobenius theorem that strictly positive matrices have simple dominant eigenvalues. Since the eigenpairs of the matrices W^t are the same for all $t = 1, 2, \ldots$, we simply require that W^{t_0} is strictly positive for some $t_0 \in \mathbb{N}$. In terms of the graph induced by W, the interpretation of this condition is that for all $t \geqslant t_0$, any vertex can be reached from any other vertex in exactly t steps. This is equivalent to requiring that the graph of the network is irreducible and aperiodic. These conditions are also necessary for consensus. To this end, consider, for example, the following periodic matrix:

$$T = \begin{pmatrix} 0 & 1 \\ 1 & 0 \end{pmatrix}.$$

The beliefs of the nodes will swap in every iteration, thus, never reaching consensus. If, on the other hand, the network is reducible, the independent strongly connected components of the network can reach consensus on their respective averages, but not on the total average of all nodes in the network (since information does not flow from some independent component to others).

We showcase that the convergence speed depends directly on the second largest eigenvalue for diagonalizable matrices. We do this by considering the spectral decomposition. Supposing W is diagonalizable, we can write:

(a) $W = PDP^{-1}$,
(b) $W^t = PD^tP^{-1}$,

where D is a diagonal matrix $D = \text{diag}\{\lambda_1, \lambda_2, \ldots, \lambda_n\}$, $P = [f_1, f_2, \ldots, f_n]^\top$ where f_i's are the right eigenvectors of W, and $P^{-1} = [\pi_1, \pi_2, \ldots, \pi_n]$ where π_i's are the left eigenvectors of W. Assuming, W is generic (i.e. non-singular) we have that:

$$f_i \pi_j = \begin{cases} 0, & \text{if } i \neq j, \\ 1, & \text{if } i = j. \end{cases}$$

We can define matrices M_i for $i = 1, 2, \ldots, n$ such that:

$$M_k = f_k \pi_k^\top = \begin{bmatrix} f_k(1)\,\pi_k(1) & \cdots & f_k(1)\,\pi_k(n) \\ \vdots & \ddots & \vdots \\ f_k(n)\,\pi_k(1) & \cdots & f_k(n)\,\pi_k(n) \end{bmatrix}.$$

Then we can check that:

$$M_i M_j = \begin{cases} 0, & \text{if } i \neq j, \\ M_i, & \text{if } i = j. \end{cases} \tag{5}$$

with these definitions, one can see that the above relation $W = PDP^{-1}$ is equivalent to:

$$W = PDP^{-1} = \lambda_1 M_1 + \lambda_2 M_2 + \cdots + \lambda_n M_n.$$

By taking powers of the above equation, and due to property (5), we can see that:

$$W^t = \lambda_1^t M_1 + \lambda_2^t M_2 + \cdots + \lambda_n^t M_n. \tag{6}$$

Eq. (6) shows that the convergence in Eq. (2) directly depends on the eigenvalues of W. For simplicity, index the eigenvalues so that:

$$|\lambda_1| \geqslant |\lambda_2| \geqslant \cdots \geqslant |\lambda_n|.$$

Since we are working with doubly stochastic matrices, $\lambda_1 = 1$ is the dominant eigenvalue. By the Perron-Frobenius theorem, λ_1 is simple, i.e. $|\lambda_2| < |\lambda_1| = 1$, and so $\lim_{t \to \infty} |\lambda_2|^t = 0$. Clearly, the smaller the magnitude $|\lambda_2|$, the faster the decay, and the faster the convergence of Eq. (2). The works cited in the beginning of this section provide several proofs for the general case of irreducible and aperiodic Markov Chains. In general, the second largest eigenvalue in magnitude bounds the convergence speed of Eq. (2) from above.

2.2 Relevant Approaches

The previous subsection demonstrated that in order to maximize the asymptotic speed of convergence of the DeGroot model, one has to minimize the second largest eigenvalue in modulus. In the literature, the second largest eigenvalue magnitude is often abbreviated as SLEM and its minimization solves the *Fastest Mixing Markov Chain* (FMMC) problem. It has been solved for medium and large, symmetric problems by Boyd *et al.* using a sub-gradient method [2,26]. In fact, they show that SLEM minimization can be cast as a *Semi-Definite Program* (SDP). To see the formulation of the problem one can simply observe that the complete graph with equal weights yields the fastest possible network. It corresponds to full connectivity (all nodes communicate with all other nodes), and it converges to the average in one iteration. Obviously this particular case is of no practical interest. We would like to impose the network's communication restrictions. The maximisation of the asymptotic convergence speed and the per-step convergence speed are defined by the following two problems:

$$
\begin{array}{llcll}
\min & \varrho(W - \mathbf{1}\mathbf{1}^\top n^{-1}), & & \min & \|W - \mathbf{1}\mathbf{1}^\top n^{-1}\|, \\[2mm]
\text{s.t.} & W \in S, & & \text{s.t.} & W \in S, \\[2mm]
& \mathbf{1}^\top W = \mathbf{1}^\top, & \text{and} & & \mathbf{1}^\top W = \mathbf{1}^\top, \\[2mm]
& W\mathbf{1} = \mathbf{1}. & & & W\mathbf{1} = \mathbf{1}.
\end{array}
\tag{7}
$$

respectively. The matrix $\mathbf{1}\mathbf{1}^\top n^{-1}$ is known as the *averaging matrix* and corresponds to a complete graph with all weights equal to $1/n$. The first problem minimises the spectral radius (i.e., the largest absolute value of the eigenvalues of the matrix) and is generally hard to solve because it is non-convex and not Lipschitz continuous [20]. The second problem minimises the spectral norm, that is, $\|W\|$ is the largest singular value of W. If W is constrained to be symmetric, then the two problems coincide. The set S is the network's communication restrictions and it corresponds to Eq. (1).

The problem of optimizing the network itself is concerned with finding the sparsest possible communication graph, on a given set of nodes, for a given asymptotic convergence speed target (given in terms of the modulus of the second largest eigenvalue). Our solution to this problem is inspired by the work of authors in [16]. Their work is concerned with continuous time distributed consensus, but it is easily relatable to the discreet equivalent. Their aim is to solve the following problem:

$$\min J(W) + \gamma \, \mathbf{card}(W), \tag{8}$$

where $J(W)$ is a measure of performance, $\mathbf{card}(W)$ is the number of non-zero elements of the matrix W, and γ is a scalar factor. They do this by considering a relaxation of the $\mathbf{card}(\cdot)$ function and forming an SDP. They solve problem (8), and follow with a "polishing step" (of the sparse network) to optimize for convergence speed. Our work differs in the fact that the "sparsification" of the network is done simultaneously with the convergence speed optimization. Moreover, we treat the convergence speed as an input: a target for asymptotic convergence speed given in terms of SLEM, and the final output is a network optimized for sparsity with given speed.

2.3 Unified Particle Swarm Optimization

Particle Swarm Optimization (PSO) is a stochastic optimization method inspired by the aggregating behaviors of populations. Given an objective function to be minimized, a population of candidate solutions (particles) is moved around in the search space in accordance with a set of mathematical formulas which dictate the particles' positions and velocities. The movement of each particle of the swarm is influenced by its local best-known position, and also by the best-known positions of the swarm. Iterating the process is expected to shift the swarm towards a good candidate solution. We can distinguish between two approaches: the local approach and the global approach. In the global approach, particles take into account the overall best position ever found by all particles in the swarm. Known as *global PSO variant* (gbest), this approach has a good convergence ability towards the global best positions found during the optimization process, and is distinguished for its exploitation abilities. On the other hand, in the *local PSO variant* (lbest), particles take into account the best position ever found by neighboring particles only, thus, having better exploration ability. The local variant is better at detecting the most favorable regions of the search space. The *Unified Particle Swarm Optimization* (UPSO) [21] variant harnesses both properties. Through the use of a new parameter $u \in (0,1)$, called *unification factor*, it controls the impact of exploitative and exploratory characteristics. Suppose N is the swarm size, n is the dimension of the problem at hand, V_i is the velocity of the particle x_i, X_i is the position of the particle x_i, P_i is the best position ever visited by particle x_i. Also, let $G_i(t+1)$ denote the velocity update of the i-th particle, x_i, for the global PSO variant with constriction coefficient χ, which is defined as:

$$G_i(t+1) = \chi \left(V_i(t) + c_1 r_1 (P_i(t) - X_i(t)) + c_2 r_2 (P_g(t) - X_i(t)) \right), \qquad (9)$$

and $L_i(t+1)$ denote the corresponding velocity update for the local PSO variant:

$$L_i(t+1) = \chi \left(V_i(t) + c_1' r_1' (P_i(t) - X_i(t)) + c_2' r_2' (P_{l_i}(t) - X_i(t)) \right), \qquad (10)$$

where r_1, r_2, r_1', r_2' are stochastic parameters uniformly distributed within the range $[0,1]$, χ is a parameter called constriction coefficient or constriction factor (with typical value $\chi = 0.729$), c_1, c_2, c_1', c_2' are weighting constants called cognitive and social parameter respectively (usually set to 2.05), g denotes the index of the overall best position, and l_i denotes the best position in the neighborhood of x_i. Then, the velocity V_i and the position of the particle X_i are updated as follows:

$$V_i(t+1) = u\, G_i(t+1) + (1-u)\, L_i(t+1), \qquad (11)$$

$$X_i(t+1) = X_i(t) + V_i(t+1). \qquad (12)$$

Eqs. (11) and (12) indicate that the new shifted position of a particle in UPSO is made up of a weighted combination of the Global and Local PSO position shifts. Consequently, both the local's inherent exploration capabilities and the global's inherent exploitation capabilities contribute. In the special cases where the unification factor $u = 0$ and $u = 1$, UPSO coincides with the original local and global PSO variant, correspondingly. Values around the middle point, $u = 0.5$, are expected to produce more balanced behaviors with respect to the exploration/exploitation abilities. Several additional PSO methods, together with standard parameter settings have been proposed (see e.g., [21]).

3 Problem Formulation

In this section we formulate objective functions for the key problems presented in the introduction: (a) eigenvalue gap maximization, and (b) the maximally sparse network given asymptotic convergence speed target. Our approach is relatively straight forward. We utilize the UPSO algorithm because it can be effectively applied to difficult problems, including among others, problems with a discontinuous function. It is also worth noting that an effective solution of aforementioned problem (b) is enabled by an effective solution of the problem (a).

Start by fixing a graph $G(V, E)$, and a corresponding doubly-stochastic weight matrix $W \in \mathbb{R}^{n \times n}$. We index the eigenvalues of W so that:

$$1 = |\lambda_1| \geqslant |\lambda_2| \geqslant \cdots \geqslant |\lambda_n|.$$

Note that $|\lambda_1| = |\lambda_2|$ occurs in three cases: (i) W induces a periodic graph, (ii) W induces a graph with more than one strongly connected components, (iii) both (i) and (ii). As indicated in Subsect. 2.1 these cases are of no interest to us. Denote the eigenvalue gap of W as:

$$e_{\text{gap}}(W) = ||\lambda_1| - |\lambda_2||, \tag{13}$$

we define the function:

$$L(W) = \begin{cases} -e_{\text{gap}}(W), & \text{if } e_{\text{gap}}(W) \neq 0, \\ \infty, & \text{if } e_{\text{gap}}(W) = 0. \end{cases} \tag{14}$$

There are several things to note. First, eigenvalues need to be calculated, a task which is computationally expensive. We can ease the computational burden by using Lanczos algorithm for the approximation of dominant eigenvalues (see for example [23]). Secondly, the loss function disregards reducible and/or aperiodic matrices by giving them maximum penalty. This should not be a problem for evolutionary algorithms. Last but not least, the loss function (14) has to be minimized for weight matrices satisfying the following restrictions:

(a) $W\mathbf{1} = \mathbf{1}^{\top}W = \mathbf{1}$ (doubly-stochastic).
(b) $W_{i,j} = 0$ if $\{i, j\} \notin E$.

The second restriction can be satisfied naturally by a vectorization of the sparse matrix W. If there are $k = |E|$ edges (here $|E|$ denotes the cardinality of the edge set E), then $\mathbf{vec}(W) \in \mathbb{R}^k$. The first restriction is crucial and needs to be discussed. There are several ways by which one might try to restrict the attention of the particles to doubly-stochastic matrices. One of them is to penalize deviations of the sum of the rows and columns from unity. According to our experience this procedure is not efficient. Allowing the particles to search the k-dimensional Euclidean space (where $k = |E|$) makes it very difficult to randomly come across a doubly stochastic matrix. Even if that happens, the evolution of the velocity vectors makes it very difficult to reach a different doubly-stochastic matrix. To get around this issue, we find a map from the Euclidean space to the set of doubly-stochastic matrices with a given sparsity pattern. This way, the particles will be able to search the Euclidean space, then map their position into the set of doubly-stochastic matrices, and retain the relevance of the velocity vectors. The loss function, then, essentially becomes a composition of Eq. (14) with that mapping.

We present an iterative normalization scheme for constructing a doubly stochastic matrix. The idea is to normalize the rows and the columns of W iteratively until the matrix becomes doubly stochastic. The proposed iterative normalization scheme is exhibited in Algorithm 1.

Iterative normalization is related to the Sinkhorn-Knopp algorithm [24]. It is shown that it converges in linear time and the only requirement is that the initial matrix W has support. It should also be seen that the sparsity pattern of W remains unchanged.

Denoting the iterative normalization function as $N(\cdot)$ there are two ways to incorporate iterative normalization in the UPSO algorithm and both of them

---

Algorithm 1. Iterative Normalization I

1: **function** NORMALIZE(W, $k = 10$)
2:     dim ← the dimension of W, i.e. n
3:     **for** $i = 0, 1, \ldots, k$ **do**
4:         **for** $j = 0, 1, \ldots, $ dim **do**
5:             $W_j \leftarrow W_j / \left(\sum_{k=1}^{\text{dim}} W_{j,k} \right)$
6:         $W \leftarrow W^{\top}$
7:     **if** W is doubly-stochastic **then**
8:         **return** W
9:     **else**
10:        Normalize(W)

---

are viable. One way is to incorporate it into the loss function. Then, the loss for problem (a) becomes:

$$L_a(W) = L \circ N(W) = L(N(W)). \tag{15}$$

The other is to incorporate $N(\cdot)$ into the UPSO step. In that case the UPSO equations for particle i become:

$$
\begin{aligned}
G_i(t+1) &= \chi \left(V_i(t) + c_1 r_1 (P_i(t) - X_i(t)) + c_2 r_2 (P_g(t) - X_i(t)) \right), \\
L_i(t+1) &= \chi \left(V_i(t) + c_1' r_1' (P_i(t) - X_i(t)) + c_2' r_2' (P_{li}(t) - X_i(t)) \right), \\
V_i(t+1) &= u\, G_i(t+1) + (1-u)\, L_i(t+1), \\
X_i^{\text{temp}}(t+1) &= X_i(t) + V_i(t+1), \\
X_i(t+1) &= N\left(X_i^{\text{temp}}(t+1) \right).
\end{aligned}
\tag{16}
$$

In our experience, without iterative normalization, the velocity vector is tasked with determining the absolute change necessary of the state vector $X_i = $ **vec**(W_i). In contrast, the output of $N(W_i)$ is determined by the relative magnitudes of the elements of W_i. By composing the loss function (14) with $N(\cdot)$, the velocity vectors are now tasked with determining favorable changes for the relative magnitudes of the elements of $X_i(t+1)$ instead. To elaborate a little further, the elements of the velocity vectors may contain positive elements only (although not necessarily). The relative magnitudes of these elements translate to changes in the relative magnitudes of the elements of the state vectors X_i^{temp}. These, in turn, determine a doubly-stochastic matrix through iterative normalization. Both schemes (15) or (16), results in effective solutions for eigenvalue-gap maximization with UPSO and other evolutionary algorithms.

For the problem of optimizing the network itself, we would like to provide the number of nodes n, and a target eigenvalue gap denoted by $e_{\text{gap}}^{\text{tgt}} \in (0,1)$. The task is to find a positive weight matrix $W^* \in \mathbb{R}^{n \times n}$ which satisfies the following:

(a) It is doubly-stochastic: $W^* \mathbf{1} = \mathbf{1}^{\top} W^* = \mathbf{1}$.

(b)   It attains the target eigenvalue gap: $L(W^*) = e_{\text{gap}}^{\text{tgt}}$.

(c)   It is as sparse as possible, i.e. $\mathbf{card}(W^*)$ is as small as possible.

We achieve the above by allowing for a trade-off between the deviation from the target eigenvalue gap, and the number of non-zero elements of the corresponding weight matrix W. When the trade-off is favorable, particles in UPSO will remove an edge of W, and continue subsequent iterations by improving the loss until the next trade-off is favorable. Note that this is happening with a single initialization, a single UPSO swarm. Doing this is enabled by UPSO's ability to perform well in problem (a). Furthermore, to facilitate the process, we find a suitable relaxation for the $\mathbf{card}(\cdot)$ function that penalizes small entries of W. First we add a safety loop in the iterative normalization scheme. If an entry of W is sufficiently small, we turn it to zero. To this end we give the following Algorithm 2:

Algorithm 2. Iterative Normalization II

1: **function** NORMALIZE(W, $k = 10$, threshold $= 0.01$)

2:      dim $\leftarrow$ the dimension of W, i.e. n

3:      **for** every entry of W **do**

4:          **if** $W_{i,j} <$ thresholds **then** $W_{i,j} \leftarrow 0$

5:      **for** $i = 0, 1, \ldots, k$ **do**

6:          **for** $j = 0, 1, \ldots, \text{dim}$ **do**

7:              $W_j \leftarrow W_j / \left(\sum_{k=1}^{\text{dim}} W_{j,k} \right)$

8:          $W \leftarrow W^\top$

9:      **if** W is doubly-stochastic **then**

10:          **return** W

11:     **else**

12:          Normalize(W)

This will enable us to reliably identify the graph associated with the weight matrix W. For the remainder of the paper at hand, for the sake of readability, we simply write W for the matrix that results after iterative normalization (c.f., Algorithm 2). We define the adjacency matrix A^W associated with W whose entries are given by:

$$A_{i,j}^W = \begin{cases} 0, & \text{if } W_{i,j} = 0, \\ 1. & \text{if } W_{i,j} > 0. \end{cases}$$

Then, the relaxation of the $\mathbf{card}(\cdot)$ function for a particular W matrix is given by:

$$C(W) = \sum_{i,j} (A_{i,j}^W - W_{i,j}). \tag{17}$$

Clearly, $C(\cdot)$ is positive and increasing with the number of non-zero elements of W. Additionally, the smaller the entry $W_{i,j}$, the larger its contribution to $C(W)$. Thus, C essentially penalizes small entries. When a small entry $W_{i,j}$ falls below the threshold value, i.e. $W_{i,j}$ < threshold, Algorithm 2 turns it to zero, and $C(W') \approx C(W) - (1 - \text{threshold})$. Of course, $C(\cdot)$ is a discontinuous function.

To control the eigenvalue gap, we define the following function:

$$D(W) = \min\left(\frac{|L(W)|}{e_{\text{gap}}^{\text{tgt}}}, \frac{e_{\text{gap}}^{\text{tgt}}}{|L(W)|}\right). \tag{18}$$

That is, $0 < D(W) < 1$ if the eigenvalue gap of W deviates from the target, while $D(W) = 1$ if W attains the target eigenvalue gap. Then, the loss function can be written as follows:

$$L_b(W) = \frac{C(W)}{D(W)^2}. \tag{19}$$

In the loss function (19), the denominator is maximized when W attains the target eigenvalue-gap, and the numerator is minimized when W is as sparse as possible. The square in the denominator guarantees that reaching the target gap is prioritized over removing edges from the graph associated with W. In our experience, when a certain entry of the W matrix approaches the user defined threshold value (in Algorithm 2), the particle swarm faces a trade-off between the numerator and the denominator. If the edge associated with that entry of W is removed, then the numerator is reduced by approximately $(1 - \text{threshhold})$, however, a deviation from the eigenvalue-gap target is induced, and is reflected in the denominator. The next section shows that UPSO performs nicely in both problems: eigenvalue gap maximization, and, maximization of network sparsity given an asymptotic convergence speed target.

A final note concerns the restriction of search to symmetric matrices. The simplest way to do this is to augment the Eqs. (16) with an extra equation to "symmetrize" the doubly-stochastic matrix. That is, supposing $W_i(t + 1)$ is the matrix corresponding to vector $X_i(t + 1)$, the following step is taken directly after iterative normalization:

$$W_i'(t + 1) = \left(W_i(t + 1) + W_i^\top(t + 1)\right)/2. \tag{20}$$

The new vectorized position of the particle is $X_i'(t + 1) = \mathbf{vec}(W_i'(t + 1))$.

4   Experiments

The experiments carried out are restricted to symmetric edge weights. We consider two 5×5 graphs with known optimal eigenvalue-gaps. The graphs are shown in Fig. 1. We use loss function in Eq. (14) with the augmented UPSO Eqs. (16) together with Eq. (20) for "symmetrization". In our experiments we find optimal or near-optimal eigenvalue-gaps. We employ the widely used convention of $4|E|$ particles. Then, we test objective (19) on the sparse matrix identification problem. We do this on a graph of 30 nodes, with eigenvalue-gap target of 0.2. The

result is comparable to that of [16], however, in contrast to [16], we have chosen
the asymptotic speed of convergence. In addition we have observed that UPSO
on the "sparsification" problem with the objective function of Eq. (19) usually
finds effective solutions with way less than $4\,|E|$ particles. In all cases, the swarm
particles communicate with 4 neighbors in a cyclic manner.

With optimal SLEM values of 0.4286 and 0.25, the optimal eigenvalue-gaps
of the graphs in Fig. 1 are 0.5714 and 0.75 respectively [2]. Our experiments
indicate that UPSO can find the optimal, or near-optimal values. The results
can be seen in Table 1.

For the considered sparsification problem, on a graph of 30 nodes, UPSO
with objective function of Eq. (19) and $e_{\text{gap}}^{\text{tgt}} = 0.2$ produced a graph with 57
bidirectional edges. The number of particles used was 200, and the number of
iterations was 400. The final eigenvalue-gap is exactly 0.2 (Fig. 2).

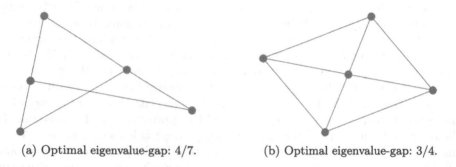

(a) Optimal eigenvalue-gap: 4/7.   (b) Optimal eigenvalue-gap: 3/4.

Fig. 1. Two small networks with known optimal eigenvalue-gaps e^* and optimal edge
weights given in [2].

Table 1. Maximal eigenvalue-gaps produced by UPSO for several unification parameter
values u. The optimal eigenvalue gap of Graph (a) is 0.5714 and the optimal eigenvalue
gap of Graph (b) is 0.5. Number of iterations: 400. Bold-face entries indicate optimal
values.

u	Graph (a)	Graph (b)
0	0.5634	**0.7500**
0.25	**0.5714**	**0.7500**
0.5	0.5660	0.7489
0.75	**0.5714**	0.7492
1	**0.5714**	0.7498

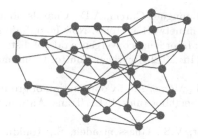

Fig. 2. Graph on 30 nodes with symmetric weight matrix. It contains 57 bidirectional edges and 5 non-zero diagonal elements.

In a future correspondence, we will also examine the case of asymmetric weight matrices. A relevant work can be found in [9].

5   Synopsis and Concluding Remarks

We tackled two difficult problems using unified particle swarm optimization (UPSO). Namely, (a) eigenvalue-gap maximization on doubly-stochastic matrices with sparsity constraints, and, (b) sparse doubly-stochastic matrix identification with given eigenvalue-gap target. The intricacy we faced is that the domains of these problems are hard to navigate. We get around this issue by using an iterative normalization procedure that maps the Euclidean space to the domain of the respective problem. UPSO seems to provide optimal or near optimal solutions to problem (a) with symmetric edge weights, and yields effective solutions to problem (b). In a future correspondence, we would like to study the properties of the iterative normalization scheme, and assess the ability of UPSO to find optimal eigenvalue-gaps for matrices with asymmetric edge weights, and compare with other evolutionary algorithms.

References

1. Boccaletti, S., Latora, V., Moreno, Y., Chavez, M., Hwang, D.-U.: Complex networks: structure and dynamics. Phys. Rep. **424**(4–5), 175–308 (2006)
2. Boyd, S., Diaconis, P., Xiao, L.: Fastest mixing Markov chain on a graph. SIAM Rev. **46**(4), 667–689 (2004)
3. Chandrasekhar, A.G., Larreguy, H., Xandri, J.P.: Testing models of social learning on networks: evidence from two experiments. Econometrica **88**(1), 1–32 (2020)
4. Cybenko, G.: Dynamic load balancing for distributed memory multiprocessors. J. Parallel Distrib. Comput. **7**(2), 279–301 (1989)
5. DeGroot, M.H.: Reaching a consensus. J. Am. Stat. Assoc. **69**(345), 118–121 (1974)
6. DeMarzo, P.M., Vayanos, D., Zwiebel, J.: Persuasion bias, social influence, and unidimensional opinions. Q. J. Econ. **118**(3), 909–968 (2003)
7. Duchi, J.C., Agarwal, A., Wainwright, M.J.: Dual averaging for distributed optimization: convergence analysis and network scaling. IEEE Trans. Autom. Control **57**(3), 592–606 (2011)

8. Hadjicostis, C.N., Domínguez-García, A.D., Charalambous, T.: Distributed averaging and balancing in network systems. Now Foundations (2018)

9. Hao, H., Barooah, P.: Improving convergence rate of distributed consensus through asymmetric weights. In: 2012 American Control Conference (ACC), pp. 787–792. IEEE, (2012)

10. Jadbabaie, A., Lin, J., Morse, A.S.: Coordination of groups of mobile autonomous agents using nearest neighbor rules. IEEE Trans. Autom. Control **48**(6), 988–1001 (2003)

11. Jirdehi, M.A., Tabar, V.S., Ghassemzadeh, S., Tohidi, S.: Different aspects of microgrid management: a comprehensive review. J. Energ. Storage **30**, 101457 (2020)

12. Kandris, D., Nakas, C., Vomvas, D., Koulouras, G.: Applications of wireless sensor networks: an up-to-date survey. Appl. Syst. Innov. **3**(1), 14 (2020)

13. Khan, M.R.B., Jidin, R., Pasupuleti, J.: Multi-agent based distributed control architecture for microgrid energy management and optimization. Energ. Convers. Manag. **112**, 288–307 (2016)

14. Koponen, T., et al.: Onix: a distributed control platform for large-scale production networks. In: 9th USENIX Symposium on Operating Systems Design and Implementation (OSDI 10), (2010)

15. Levin, D.A., Peres, Y.: Markov chains and mixing times. Am. Math. Soc. 107, (2017)

16. Lin, F., Fardad, M., Jovanović, M.R.: Identification of sparse communication graphs in consensus networks. In: 2012 50th Annual Allerton Conference on Communication, Control, and Computing (Allerton), pp. 85–89. IEEE, (2012)

17. Montenegro, R., Tetali, P.: Mathematical aspects of mixing times in Markov chains. Foundations and Trends® in Theoretical Computer Science 1(3), 237–354 (2006). https://doi.org/10.1561/0400000003

18. Moreno, J.L., Jennings, H.H.: Statistics of social configurations. Sociometry, pp. 342–374 (1938)

19. Olfati-Saber, R.: Distributed Kalman filter with embedded consensus filters. In: Proceedings of the 44th IEEE Conference on Decision and Control, pp. 8179–8184. IEEE (2005)

20. Overton, M.L., Womersley, R.S.: On minimizing the special radius of a nonsymmetric matrix function: optimality conditions and duality theory. SIAM J. Matrix Anal. Appl. **9**(4), 473–498 (1988)

21. Parsopoulos, K.E., Vrahatis, M.N.: Particle swarm optimization and intelligence: advances and applications. Information Science Publishing (IGI Global) (2010)

22. Peng, Z., Wang, J., Wang, D., Han, Q.-L.: An overview of recent advances in coordinated control of multiple autonomous surface vehicles. IEEE Trans. Ind. Inform. **17**(2), 732–745 (2020)

23. Saad, Y.: Numerical methods for large eigenvalue problems: revised edition. SIAM (2011)

24. Sinkhorn, R., Knopp, P.: Concerning nonnegative matrices and doubly stochastic matrices. Pacific J. Math. **21**(2), 343–348 (1967)

25. Thramboulidis, K., Perdikis, D., Kantas, S.: Model driven development of distributed control applications. Int. J. Adv. Manuf. Technol. **33**(3), 233–242 (2007)

26. Xiao, L., Boyd, S.: Fast linear iterations for distributed averaging. Syst. Control Lett. **53**(1), 65–78 (2004)

Value of Information in the Mean-Square Case and Its Application to the Analysis of Financial Time-Series Forecast

Roman V. Belavkin[1]([📧])(iD), Panos Pardalos[2](iD), and Jose Principe[3](iD)

[1] Department of Computer Science, Middlesex University, London NW4 4BT, UK
r.belavkin@mdx.ac.uk
[2] Department of Industrial and Systems Engineering, University of Florida, P.O. Box 116595, Gainesville, FL 32611–6595, USA
pardalos@ufl.edu
[3] Department of Electrical and Computer Engineering, University of Florida, P.O. Box 116130, Gainesville, FL 32611–6130, USA
principe@cnel.ufl.edu

Abstract. The advances and development of various machine learning techniques has lead to practical solutions in various areas of science, engineering, medicine and finance. The great choice of algorithms, their implementations and libraries has resulted in another challenge of selecting the right algorithm and tuning their parameters in order to achieve optimal or satisfactory performance in specific applications. Here we show how the value of information $(V(I))$ can be used in this task to guide the algorithm choice and parameter tuning process. After estimating the amount of Shannon's mutual information between the predictor and response variables, $V(I)$ can define theoretical upper bound of performance of any algorithm. The inverse function $I(V)$ defines the lower frontier of the minimum amount of information required to achieve the desired performance. In this paper, we illustrate the value of information for the mean-square error minimization and apply it to forecasts of cryptocurrency log-returns.

Keywords: Value of information · Shannon's information · Mean-square error · Time-series forecast

1 Introduction

The value of information $V(I)$ is the maximum gain in performance one can achieve due to receiving the amount I of information (mathematical meaning of 'performance' and 'information' will be clarified later). This concept was discussed in various settings in the literature, but the main advances of the theory behind it were made by Ruslan Stratonovich and his colleagues in the 1960 s s [10,15–17,19,20]. Inspired by Shannon's rate-distortion theory [12], Stratonovich first extended the ideas to more general class of Bayesian systems and various

© Springer Nature Switzerland AG 2022
D. E. Simos et al. (Eds.): LION 2022, LNCS 13621, pp. 549–563, 2022.
https://doi.org/10.1007/978-3-031-24866-5_39

types of information. He then used original techniques and some methods of statistical physics to derive very deep results on asymptotic equivalence of the value functions for different types of information. Stratonovich and his colleagues also studied the value of information in different settings, from the simplest Boolean and Gaussian systems to stochastic processes in continuous time. Many of these examples are covered in the classical monograph [18], which has recently been published in English [14].

Recent advances of intelligent and learning systems combined with exponential growth of the size and dimensionality of datasets facilitated by the growth in computer performance has prompted a new interest in the value of information theory and its applications. Some results of the theory have facilitated better understanding of the role of randomization in machine learning algorithms [1,2,5]. For example, the value of information was used to derive optimal control functions of mutation rates in genetic algorithms [3,4,8]. It was shown also that the value of information theory is closely related to optimal transport [7] and can have unexpected applications in explaining some decision-making paradoxes in behavioural economics [6].

The purpose of this paper is to demonstrate how the value of information can be used to evaluate the performance and tune parameters of different data-driven models with a specific focus on the mean-square error criterion. In the next section, we briefly overview the VoI theory for the case of translation invariant objective functions, such as the mean-square deviation. We derive a simple expression for the smallest root-mean-square error (RMSE) as a function of Shannon's mutual information between the predictor and response variables. This function is then used in Sect. 3 as performance frontier for several models attempting to forecast daily log-returns of some cryptocurrencies. We conclude by the discussion of these results, the importance of correct estimation of the amount of information in data as well as the choice of objective functions to evaluate the models.

2 Value of Information for Translation Invariant Objective Functions

Let us review some of the main ideas of the value of information theory in the context of optimal estimation, although the context of optimal control is also relevant. Let $(\Omega, P, \mathcal{A})$ be a probability space, and let $x \in X$ be a random variable (i.e. a measurable function $x = x(\omega)$ on a probability space, and $P(X) = P\{\omega : x(\omega) \in X\}$ is the corresponding push-forward measure). Consider the problem of finding an element $y \in Y$ maximizing the expected value of *utility* function $u : X \times Y \to \mathbb{R}$. Let us denote the corresponding optimal value as follows:

$$U(0) := \sup_{y \in Y} \mathbb{E}_{P(x)}\{u(x, y)\}$$

where zero in $U(0)$ designates the fact that no information about specific value of $x \in X$ is given, only the prior distribution $P(x)$. At the other extreme, full information entails that there is an invertible function $z = f(x)$ such that $x \in X$ is determined uniquely $x = f^{-1}(z)$ by the 'message' $z \in Z$. The corresponding optimal value is

$$U(\infty) := \mathbb{E}_{P(x)}\{\sup_{y(z)} u(x, y(z))\}$$

where optimization is over all mappings $y(z)$ (i.e. $y : Z \to Y$). In the context of estimation, variable x is the *response* (i.e. the variable of interest), and z is the *predictor*. The mapping $y(z)$ represents a model with output $y \in Y$.

Let us denote by $U(I)$ the intermediate values in the interval $[U(0), U(\infty)]$ for all information amounts $I \in [0, \infty]$. The value of information is then defined as the following difference [14]:

$$V(I) := U(I) - U(0)$$

There are, however, different ways in which information amount I and the quantity $U(I)$ can be defined leading to different types of function $V(I)$. For example, suppose that $z \in Z$ partitions X into a finite number of subsets. This corresponds to a mapping $z : X \to Z$ with a constraint on the cardinality of its image $|Z| \le e^I < |X|$. Then, given such a partition $z : X \to Z$, one can find optimal $y(z)$ maximizing the conditional expected utility $\mathbb{E}_{P(x|z)}\{u(x, y) \mid z\}$ for each subset $f^{-1}(z) \ni x$. The optimal value $U(I)$ is then defined by repeating the above and optimizing over all partitions $z(x)$ satisfying the cardinality constraint $\ln |Z| \le I$:

$$U(I) := \sup_{z(x)} \left[\mathbb{E}_{P(z)} \left\{ \sup_{y(z)} \mathbb{E}_{P(x|z)}\{u(x, y) \mid z\} \right\} : \ln |Z| \le I \right] \tag{1}$$

Here $P(z) = P\{x \in f^{-1}(z)\}$. The quantity $I = \ln |Z|$ is called *Hartley's information*, and the difference $V(I) = U(I) - U(0)$ in this case is the value of Hartley's information.

Example 1. Let $X \equiv \mathbb{R}^n$ and $u(x, y) = -\frac{1}{2}\|x - y\|^2$. Then the optimal estimator is the expected value $y = \mathbb{E}\{x\}$, which is found from the stationary condition:

$$\frac{\partial}{\partial y} \mathbb{E}_{P(x)} \left\{ -\frac{1}{2}\|x - y\|^2 \right\} = y - \mathbb{E}\{x\} = 0$$

The optimal value is $U(0) = -\frac{1}{2}\sigma_x^2$, where σ_x^2 is the variance of x. Given a partition $z : X \to Z$ of X into $k = |Z|$ subsets, one can compute k estimators given by conditional expectations $y(z) = \mathbb{E}\{x \mid z\}$. The value $U(\ln k)$ can be estimated by computing and minimizing the average of conditional variances $\sigma_x^2(z)$ over several partitions.

One can see from Eq. (1) that the computation of the value of Hartley's information is quite demanding, and Example 1 suggests that it might involve a procedure such as the k-means clustering algorithm or training a multilayer neural network. Indeed, computing the error at the output layer of a perceptron and adjusting the output weights corresponds to finding optimal output function $y(z)$ in equation (1); back-propagation of the error into hidden layers and adjusting their weights corresponds to finding optimal partition $z(x)$ in (1). Although there exist efficient algorithms for such optimization, it is clear that using the value of Hartley's information is not practical due to high cost of the computations involved. The main result of the theory [14] is that the value of Hartley's information (1) is asymptotically equivalent to the value of Shannon's information, which is much easier to compute.

Recall the definition of Shannon's mutual information [12]:

$$I(X,Y) := \mathbb{E}_{W(x,y)}\left\{\ln\frac{P(x \mid y)}{P(x)}\right\} = H(X) - H(X \mid Y)$$
$$= H(Y) - H(Y \mid X)$$

where $W(x,y) = P(x \mid y)Q(y)$ is the joint probability distribution on $X \times Y$, and $H(\cdot) = -\mathbb{E}_P\{\ln P(\cdot)\}$ is the entropy function. The following inequality is valid:

$$0 \leq I(X,Y) \leq \min\{H(X), H(Y)\} \leq \min\{\ln|X|, \ln|Y|\}$$

The value of Shannon's information is defined using the quantity:

$$U(I) := \sup_{P(y|x)} \left[\mathbb{E}_W\{u(x,y)\} : I(X,Y) \leq I\right] \tag{2}$$

where optimization is over all conditional probabilities $P(y \mid x)$ (or joint measures $W(x,y) = P(y \mid x)P(x)$) satisfying the information constraint $I(X,Y) \leq I$. Contrast this with $U(I)$ for Hartley's information (1), where optimization is over the mappings $y(x) = y \circ z(x)$. As was pointed out in [7], the relation between functions (1) and (2) is similar to that between optimal transport problems in the Monge and Kantorovich formulations.

Function $U(I)$ defined in (2) is strictly increasing and concave, and it has the following inverse:

$$I(U) := \inf[I(X,Y) : \mathbb{E}_W\{u(x,y)\} \geq U] \tag{3}$$

It is a proper convex and strictly increasing function, where it is finite. The strictly increasing and concave (resp. convex) properties of $U(I)$ (resp. $I(U)$) can be shown in more general settings, when information is defined by any closed functional (see Proposition 3 in [5]). This means that solutions to these conditional extremum problems can be found by the standard method of Lagrange multipliers (see [5,14] for details). Thus, the optimal joint distributions belong to the following exponential family:

$$W(x,y;\beta) = P(x)Q(y)e^{\beta u(x,y) - \gamma(x;\beta)} \tag{4}$$

where P and Q are the marginal distributions of W, and function $\gamma(x;\beta)$ is defined by the normalization condition $\int_{X \times Y} dW(x,y;\beta) = 1$. Parameter β is called the *inverse temperature*, and it is the Lagrange multiplier associated to the constraint $\mathbb{E}\{u\} \geq U$ in (3). The temperature β^{-1} is associated respectively to the constraint $I(X, Y) \leq I$ in (2). Their values are defined by the following conditions:

$$\beta^{-1} = U'(I), \qquad \beta = I'(U)$$

In fact, this can also be seen from the following considerations. Function $U(I)$ is a proper concave function, and therefore it is the Legendre-Fenchel dual (see [11, 21]) of some proper concave function $F(\beta^{-1})$:

$$U(I) = \inf\{\beta^{-1}I - F(\beta^{-1})\} \quad \Longleftrightarrow \quad I = F'(\beta^{-1}) \quad \Longleftrightarrow \quad \beta^{-1} = U'(I)$$

Function $I(U)$ is a proper convex function, and therefore it is the Legendre-Fenchel dual of some proper convex function $\Gamma(\beta)$:

$$I(U) = \sup\{\beta U - \Gamma(\beta)\} \quad \Longleftrightarrow \quad U = \Gamma'(\beta) \quad \Longleftrightarrow \quad \beta = I'(U)$$

Convex function $\Gamma(\beta)$ is the cumulant generating function of distribution (4). In particular, $U(\beta) = \Gamma'(\beta)$ is the expected value $\mathbb{E}_{W(\beta)}\{u(x, y)\}$. Concave function $F(\beta^{-1})$ is sometimes referred to as *free energy*, and $I(\beta^{-1}) = F'(\beta^{-1})$ is equal to Shannon's mutual information $\mathbb{E}_{W(\beta)}\{\ln W - \ln(P \otimes Q)\}$ of distribution (4). Functions F and Γ have the following relation:

$$F(\beta^{-1}) = -\beta^{-1}\Gamma(\beta)$$

The following procedure can be used to obtain the dependencies $U(I)$ or $I(U)$ and the value of Shannon's information $V(I) = U(I) - U(0)$. Optimal solution (4) is used to define the expression for function $\Gamma(\beta)$, which is then used to derive two functions:

$$U(\beta) = \Gamma'(\beta), \qquad I(\beta) = \beta\,\Gamma'(\beta) - \Gamma(\beta)$$

The dependency $U(I)$ (or $I(U)$) is then obtained either parametrically from $U(\beta)$ and $I(\beta)$ or explicitly by excluding β from one of the equations. Alternatively, one can use free energy $F(\beta^{-1})$ and define $U(I)$ from $I(\beta^{-1}) = F'(\beta^{-1})$ and $U(\beta^{-1}) = \beta^{-1}I(\beta^{-1}) - F(\beta^{-1})$.

Let us now consider function $\Gamma(\beta)$ for distribution (4). Taking partial traces of solution (4) and using the law of total probability leads to the following system of integral equations:

$$\int_X dW(x, y) = dQ(y) \qquad \Longrightarrow \qquad \int_X e^{\beta u(x,y) - \gamma(x;\beta)}\, dP(x) = 1 \quad (5)$$

$$\int_Y dW(x, y) = dP(x) \qquad \Longrightarrow \qquad \int_Y e^{\beta u(x,y)}\, dP(y) = e^{\gamma(x;\beta)} \quad (6)$$

If the linear transformation $T(\cdot) = \int_X e^{\beta u(x,y)}(\cdot)$ has inverse, then from (5) we have $e^{-\gamma(x;\beta)}dP(x) = T^{-1}(1)$ or

$$\gamma(x; \beta) = -\ln \int_Y b(x, y)\, dy + \ln[dP(x)/dx] = \gamma_0(x; \beta) - h(x)$$

where $b(x, y)$ is the kernel of the inverse linear transformation T^{-1}, $\gamma_0(x; \beta) := -\ln \int_Y b(x, y) \, dy$, and $h(x) = -\ln[dP(x)/dx]$ is random entropy or *surprise*. Integrating the above with respect to measure $P(x)$ we obtain

$$\Gamma(\beta) := \int_X \gamma(x; \beta) \, dP(x) = \Gamma_0(\beta) - H(X)$$

where $\Gamma_0(\beta) := \int_X \gamma_0(x; \beta) \, dP(x)$. Notice that $\Gamma'(\beta) = \Gamma_0'(\beta) = U(\beta)$, and therefore

$$I(\beta) = \beta \, \Gamma'(\beta) - \Gamma(\beta) = H(X) - [\Gamma_0(\beta) - \beta \, \Gamma_0'(\beta)]$$

Function $\Gamma_0(\beta) - \beta \, \Gamma_0'(\beta)$ is clearly the conditional entropy $H(X \mid Y)$, because $I(X, Y) = H(X) - H(X \mid Y)$.

Further analysis is complicated by the dependency of solution (4) on marginal distribution $P(x)$. Generally, $P(x)$ influences not only the output distribution $Q(y)$ (i.e. as $dP(x) \mapsto \int_X dP(y \mid x) \, dP(x) = dQ(y)$), but also the conditional probability $P(x \mid y) = P(x)e^{\beta u(x,y) - \gamma(x;\beta)}$. However, as was shown in [14], this dependency on $P(x)$ disappears, if the product $e^{-\gamma(x;\beta)}P(x)$ is independent of x. Indeed, let $e^{-\Gamma_0(\beta)} = e^{-\gamma(x;\beta)} \, dP(x)/dx = \text{const}$. Then from equation (5) we obtain

$$e^{-\Gamma_0(\beta)} \int_X e^{\beta u(x,y)} \, dx = 1 \quad \Longrightarrow \quad \Gamma_0(\beta) = \ln \int_X e^{\beta u(x,y)} \, dx$$

It turns out that $e^{-\gamma(x;\beta)} \, dP(x)/dx = \text{const}$, if the objective function is translation invariant: $u(x, y) = u(x + z, y + z)$. Indeed, using translation invariance and equation (5) gives

$$\int_X e^{\beta u(x+z, y+z) - \gamma(x+z;\beta)} \, dP(x + z) = \int_X e^{\beta u(x,y) - \gamma(x+z;\beta)} \, dP(x + z) = 1$$

Combining this with equation (5) implies that

$$e^{-\gamma(x+z;\beta)} dP(x + z)/dx = e^{-\gamma(x;\beta)} dP(x)/dx = \text{const}$$

Many objective functions $u(x, y)$ are defined using the difference $x - y$, which means they are translation invariant.

Example 2 (Squared error and Gaussian case). Let $u(x, y) = -\frac{1}{2}(x - y)^2$. Then $u(x, y) = u(x + z, y + z)$, and

$$\Gamma_0(\beta) = \ln \int_{-\infty}^{\infty} e^{-\frac{1}{2}\beta(x-y)^2} \, dx = \ln \sqrt{\frac{2\pi}{\beta}}$$

$$U(\beta) = \Gamma_0'(\beta) = -\frac{1}{2\beta}$$

$$I(\beta) = -\frac{1}{2} - \Gamma(\beta) = -\frac{1}{2} + H(X) - \Gamma_0(\beta) = H(X) - \frac{1}{2}[\ln(2\pi) + 1 - \ln \beta]$$

The latter expression allows us to express $\beta = 2\pi e^{2[I-H(X)]+1}$ and write explicit dependency

$$U(I) = -\frac{1}{4\pi}e^{2[H(X)-I]-1} \tag{7}$$

The value of information in this case is

$$V(I) = U(I) - U(0) = \frac{1}{4\pi}e^{2H(X)-1}\left(1 - e^{-2I}\right)$$

For Gaussian density $dP(x)/dx = \frac{1}{\sqrt{2\pi\sigma_x^2}}e^{-\frac{x^2}{2\sigma_x^2}}$ we have

$$H(X) = \frac{1}{2}\left[\ln(2\pi\sigma_x^2) + 1\right], \qquad e^{2H(X)-1} = 2\pi\sigma_x^2$$

and in this case

$$U(I) = -\frac{1}{2}\sigma_x^2 e^{-2I}, \qquad V(I) = \frac{1}{2}\sigma_x^2(1 - e^{-2I})$$

Example 3 (Root-mean-square error). The root-mean-square error (RMSE or standard error) is one of the most important criteria to evaluate data-driven models. The result from Example 2 can be used to compute the smallest RMSE as a function of information. Indeed, $\mathrm{RMSE}(I) = \sqrt{-2U(I)}$, where $U(I)$ is given by equation (7):

$$\mathrm{RMSE}(I) = \frac{1}{\sqrt{2\pi e}}\,e^{H(X)-I}$$

If x is assumed to have normal distribution with variance σ_x^2, then $e^{H(X)} = \sigma_x\sqrt{2\pi e}$ and

$$\mathrm{RMSE}(I) = \sigma_x\, e^{-I} \tag{8}$$

If the amount of information I can be estimated from data (e.g. as mutual information $I(X, Z)$ between the predictors and response variables), then the functions above define the smallest possible standard error.

3   Application: Analysis of Forecasts of Cryptocurrency Log-Returns

In this section, we illustrate how the value of information can facilitate the analysis of performance of data-driven models. Here we use time-series forecasts applied to daily log-returns of cryptocurrency exchange rates.

The dataset used contains daily prices $s(t)$ of several cryptocurrency pairs during the period between Jan 1, 2019 and Jan 11, 2021. Figure 1 shows an example of prices of Bitcoin in US$s (BTC/USD) and the corresponding log-returns, which are defined as

$$r(t + 1) := \ln\left[\frac{s(t + 1)}{s(t)}\right]$$

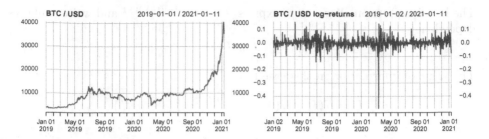

Fig. 1. Close day prices of BTC/USD (left) and the corresponding log-returns (right).

Figure 2 shows the distribution of log-returns $r(t)$ for BTC/USD. They are approximately zero-mean with $r(t) > 0$ corresponding to a price increase and vice versa. Although it is quite common to model log-returns by a Gaussian distribution, it is easy to see that the distribution has heavy tails (see the QQ-plot on Fig. 2 comparing the distribution with a Gaussian), and some extreme price changes are not unusual (e.g. notice the significant price decrease on March 12, 2020, which was caused by the announcements related to the COVID-19 pandemic).

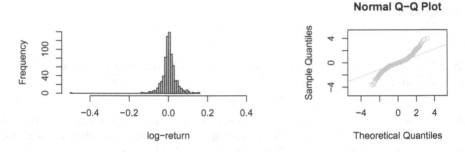

Fig. 2. Distribution of BTC/USD log-returns (left) and its comparison with normal distribution (right).

Predicting price changes is very challenging. In fact, the existence of such forecasts would create an arbitrage, which should quickly disappear in an open market. The left chart on Fig. 3 plots log-returns for two consecutive days: $r(t)$ (abscissa) and $r(t + 1)$ (ordinates). One can see that there is no obvious relation between $r(t)$ and $r(t + 1)$, and they are often assumed to be independent (and hence prices $s(t)$ are often modelled by a Markov process).

On the other hand, in continuous time independence of log-returns would mean that $\{r(t)\}$ is a so-called δ-correlated stochastic process (i.e. its autocorrelation function is proportional to the Dirac δ-function). It is well-known that

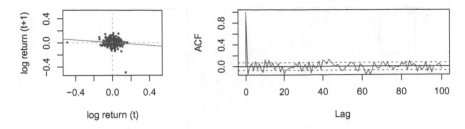

Fig. 3. Relation between log-returns on two consecutive days (left) and the autocorrelation function (right).

such processes are unphysical, because any δ-correlated stochastic process must have infinite variance σ^2 (indeed, one can show that σ^2 is the integral of spectral density, which is the Fourier transform of the autocorrelation function; the Fourier image of the δ-function is a constant function [13]). Therefore, there must be some small information about future log-return $r(t+1)$ contained in the past values $r(t), r(t-1), \ldots, r(t-n)$. This can be seen from the plot of the autocorrelation function for BTC/USD shown on the right chart of Fig. 3.

The idea of autoregressive models is to use the small amount of information between the past and future values for forecasts. Here, we shall employ several techniques to learn models $y = f(z)$, where the predictor $z = (r(t), r(t-1), \ldots, r(t-n))$ is a vector of previous values of log-returns, and the model output $y(z)$ is the forecast of the unknown future log-return $x = r(t+1)$ (the response). The hypothesis is that increasing the number n of lags should increase the amount of information used for the forecasts.

In addition to autocorrelations (correlations between the values of $\{r(t)\}$ at different times), information can be increased by using cross-correlations (correlations between log-returns of different symbols in the dataset). Thus, the vector of predictors is an $m \times n$-tuple, where m is the number of symbols used, and n is the number of time lags. In this paper we report result of predicting log-returns of BTC/USD using the range $m \in \{1, 2, \ldots, 5\}$ of symbols (BTC/USD, ETH/USD, DAI/BTC, XRP/BTC, IOT/BTC) and $n \in \{2, 3, \ldots, 20\}$ of lags. This means that the models used predictors $(z_1, \ldots, z_{m \times n})$, where $m \times n$ ranged from 2 to 100.

In order to analyse the performance of models using the value of information, one has to estimate the amount of information between the predictors $z_1, \ldots, z_{m \times n}$ and the response variable x. Here we employ the following Gaussian formula for Shannon's mutual information [14]:

$$I(X, Z) \approx \frac{1}{2} \left[\ln \det K_z + \ln \det K_x - \ln \det K_{z \oplus x} \right]$$

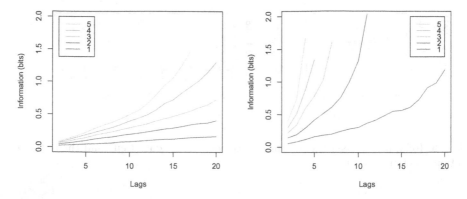

Fig. 4. The average amount of mutual information between predictors and response in the training sets (left) and test sets (right). Abscissa shows the number n of lags, and different curves correspond to different numbers m of symbols used.

where K_z is the covariance matrix of predictors $z \in \mathbb{R}^{m \times n}$, K_x is the covariance of response x (for one dimension $\det K_x = \sigma_x^2$), and $K_{Z \oplus X}$ is the covariance of $Z \oplus X$. We use the approximate sign $\approx$, because the distributions of log-returns are generally not Gaussian (in fact, the above formula gives a lower bound for non-Gaussian random variables). Natural logarithm corresponds to measuring information in 'nats'; for 'bits' one has to use $\log_2$.

For each collection of predictors $(z_1, \ldots, z_{m \times n})$ and response x, the data was split into multiple training and testing subsets using the following rolling window procedure. Here we used 100 and 25 d data windows for training and testing respectively. After training and testing the models, the windows were moved forward by 25 d. Thus, the data of approximately 700 d (Jan 2019 to Jan 2021) was split into $(700 - 100)/25 = 24$ pairs of training and testing sets. The results reported here are the average results from these 24 subsets.

Figure 4 shows the average amounts of information $I(X, Z)$ in the training sets (left) and testing sets (right). Information (ordinates) is plotted against the number n of lags (abscissa) and for $m \in \{1, 2, \ldots, 5\}$ symbols (different curves). The data was used to train and test the following types of models:

1. Multiple mean-square linear regression (LM).
2. Partial least squares regression (PLS).
3. Feed-forward neural network (NN).

The first model has no hyperparameters; the PLS regression used here employed SIMPLS algorithm [9] with 3 components; NN used here had just one hidden layer with 3 logistic units and trained for 30 epochs. This is admitably not an optimal choice of models, but finding the best model or a set of hyperparameters was not the purpose of this study. The models were used to illustrate their performance from the point of the value of information theory.

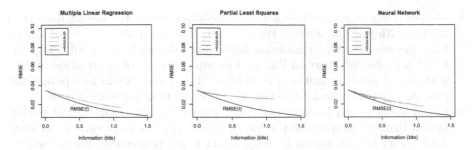

Fig. 5. RMSE results of fitted values of three types of models on training data as functions of information in the training data. Theoretical RMSE(I) curve (8) is plotted for standard deviation of response $\sigma_x \approx .0386$ estimated from the training sets.

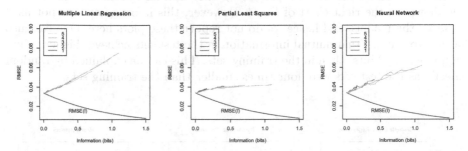

Fig. 6. RMSE results of predicted values from three types of models on testing data as functions of information in the training data. Theoretical RMSE(I) curve (8) is plotted for standard deviation of response $\sigma_x \approx .0361$ estimated from the testing sets.

Figures 5 and 6 show standard errors (RMSE) of the models as function of the information amount I contained in the training data. Different curves are plotted for different numbers of symbols $m \in \{1, \ldots, 5\}$. Theoretical lower bounds are shown by the RMSE(I) curves computed using formula (8) with standard devition of response x estimated from the training and testing sets. Figure 5 shows RMSE of the models fitting the training data after training, while Fig. 6 shows the errors of prediction on testing data. The following observations can be made from the results shown on Figs. 5 and 6:

1. Errors of fitting the training data closely follow theoretical curve RMSE(I). One can see that LM and NN achieve errors on the training data close to theoretical. PLS has higher errors, which can be explained by the fact that the aim of the PLS algorithm is not to minimize squared errors, but to maximize covariance between predictors and reponse [9].
2. All models show higher errors on the testing data. PLS achieved smaller and more stable errors in forecasts than LM or NN in this experiment.
3. Increasing information leads to decreasing errors on the training data, but not necessarily on new data (testing or prediction).

4. Models using $m > 1$ symbols achieve smaller errors on the testing data than models with just one symbol. We note also that when using $m = 4$ or 5 symbols, the amount of information of say $I = .1$ bits can be achieved using only $n \leq 5$ lags (see left chart on Fig. 4). The same amount of information in data with $m = 1$ symbol requires $n > 20$ lags. Thus, cross-correlations potentially provide more valuable information for forecasts than autocorrelations.

5. Linear models, and in particular PLS, appear to have more robust performance than the simple neural network used here. The large variance of standard errors for NN shown on Figs. 5 and 6 are potentially due to random initialization and higher uncertainty in the setting of hyper-parameters (e.g. hidden nodes, the number of epochs to train, activation functions).

Remark 1. RMSE can also be plotted against mutual information in the test set shown on the right chart of Fig. 4. However, this information was not used to learn the models, and hence we do not report these plots here. One can also notice from Fig. 4 that mutual information in the test sets achieves higher values (approaching 2 bits) than in the training sets. This can be explained by random effects, as the test sets were four times smaller than the training sets.

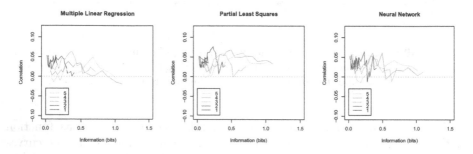

Fig. 7. Correlation between predicted values from models and desired response in the test data as functions of information in the training sets.

Let us point out that RMSE is a general, but certainly not the only and potentially not the most useful measure to assess model's performance. Figure 7 reports correlations between the predicted and the desired log-returns (i.e. correlation between the model output $y(z)$ and the desired response x). One may notice that the best linear models (LM and PLS) are those using $m \in \{2, 3\}$ symbols, and the maximum correlations are generally achieved at higher amounts of information than those achieving the minimums of RMSE.

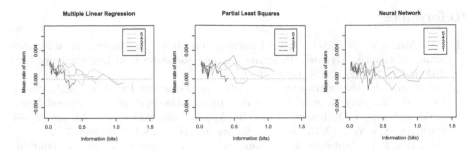

Fig. 8. Mean rates of return as functions of information for different models.

Finally, we estimated the mean rates of return (MRR) from the model forecasts, if they were used for trading. Here, we used the following formula:

$$MRR := e^{\mathbb{E}\{\text{sign}(y(z))\,\text{sign}(x)|x|\}} - 1$$

where $y(z)$ is the predicted log-return, x is the 'true' log-return from the test data, and sign is the signum function. Thus, when the signs of $y(z)$ and x coincide, then the log-return from trading is positive $|x|$; otherwise, the log-return is $-|x|$. The expected value $\mathbb{E}\{\text{sign}(y(z))\,\text{sign}(x)|x|\}$ is the mean log-return from trading $\langle r \rangle$, which is converted into the effective rate of return by the formula $e^{\langle r \rangle} - 1$. Thus, the value of MRR $= .01$ means 1% return per day without taking into account trading fees. Figure 8 reports the estimated mean rates of return for the three types of models. Some models achieve mean rates of return .3% and .4% per day, which is slightly higher than the average rate of return of .26% from BTC/USD in the testing sets. Note also that the mean rate of return from the models can also be as low as $-.5\%$ per day.

4 Discussion

We have reviewed the main mathematical ideas of the value of information theory in the context of translation invariant objective functions. These functions are important for data-driven models, such as the mean-square cost or standard error. We have derived simple expressions for the lower bound of RMSE as a function of mutual information and applied it to the analysis of performance of time-series forecasts using cryptocurrency data. We showed how these information-theoretic ideas can enrich our understanding of data and the models and potentially lead to a more intelligent learning and optimization of model parameters.

Acknowledgements. Stefan Behringer is deeply acknowledged for additional discussion of the example, Roman Tarabrin is deeply acknowledged for providing a MacBookPro laptop used for the computational experiments. This research was funded in part by the ONR grant number N00014-21-1-2295.

References

1. Belavkin, R.V.: Bounds of optimal learning. In: 2009 IEEE International Symposium on Adaptive Dynamic Programming and Reinforcement Learning, pp. 199–204. IEEE, Nashville, TN, USA (2009)
2. Belavkin, R.V.: Information trajectory of optimal learning. In: Hirsch, M.J., Pardalos, P.M., Murphey, R. (eds.) Dynamics of Information Systems: Theory and Applications, Springer Optimization and Its Applications Series, vol. 40, pp. 29–44. Springer, New York (2010). https://doi.org/10.1007/978-1-4419-5689-7_2
3. Belavkin, R.V.: Mutation and optimal search of sequences in nested Hamming spaces. In: IEEE Information Theory Workshop. IEEE (2011)
4. Belavkin, R.V.: Dynamics of information and optimal control of mutation in evolutionary systems. In: Sorokin, A., Murphey, R., Thai, M.T., Pardalos, P.M. (eds.) Dynamics of Information Systems: Mathematical Foundations, Springer Proceedings in Mathematics and Statistics, vol. 20, pp. 3–21. Springer, New York (2012). https://doi.org/10.1007/978-1-4614-3906-6_1
5. Belavkin, R.V.: Optimal measures and Markov transition kernels. J. Global Optim. **55**, 387–416 (2013)
6. Belavkin, R.V.: Asymmetry of risk and value of information. In: Vogiatzis, C., Walteros, J.L., Pardalos, P.M. (eds.) Dynamics of Information Systems. SPMS, vol. 105, pp. 1–20. Springer, Cham (2014). https://doi.org/10.1007/978-3-319-10046-3_1
7. Belavkin, R.V.: Relation between the Kantorovich–Wasserstein metric and the Kullback–Leibler divergence. In: Ay, N., Gibilisco, P., Matúš, F. (eds.) IGAIA IV 2016. SPMS, vol. 252, pp. 363–373. Springer, Cham (2018). https://doi.org/10.1007/978-3-319-97798-0_15
8. Belavkin, R.V., Channon, A., Aston, E., Aston, J., Krašovec, R., Knight, C.G.: Monotonicity of fitness landscapes and mutation rate control. J. Math. Biol. **73**(6), 1491–1524 (2016)
9. de Jong, S.: Simpls: an alternative approach to partial least squares regression. Chemom. Intell. Lab. Syst. **18**(3), 251–263 (1993)
10. Grishanin, B.A., Stratonovich, R.L.: Value of information and sufficient statistics during an observation of a stochastic process. Izvestiya of USSR Academy of Sciences, Technical Cybernetics **6**, 4–14 (1966). in Russian
11. Rockafellar, R.T.: Conjugate Duality and Optimization, CBMS-NSF Regional Conference Series in Applied Mathematics, vol. 16. Society for Industrial and Applied Mathematics, PA (1974)
12. Shannon, C.E.: A mathematical theory of communication. Bell System Technical Journal 27, 379–423 and 623–656 (July and October 1948)
13. Stratonovich, R.L.: Topics in the Theory of Random Noise, vol. 1. Martino Fine Books (2014)
14. Stratonovich, R.L. Theory of Information and its Value. Springer, Cham (2020). https://doi.org/10.1007/978-3-030-22833-0
15. Stratonovich, R.L.: On value of information. Izvestiya of USSR Academy of Sciences, Technical Cybernetics **5**, 3–12 (1965). in Russian
16. Stratonovich, R.L.: Value of information during an observation of a stochastic process in systems with finite state automata. Izvestiya of USSR Academy of Sciences, Technical Cybernetics **5**, 3–13 (1966). in Russian
17. Stratonovich, R.L.: Extreme problems of information theory and dynamic programming. Izvestiya of USSR Academy of Sciences, Technical Cybernetics **5**, 63–77 (1967). in Russian

18. Stratonovich, R.L.: Theory of Information. Sovetskoe Radio, Moscow, USSR (1975). in Russian
19. Stratonovich, R.L., Grishanin, B.A.: Value of information when an estimated random variable is hidden. Izvestiya of USSR Academy of Sciences, Technical Cybernetics **3**, 3–15 (1966). in Russian
20. Stratonovich, R.L., Grishanin, B.A.: Game-theoretic problems with information constraints. Izvestiya of USSR Academy of Sciences, Technical Cybernetics **1**, 3–12 (1968). in Russian
21. Tikhomirov, V.M.: Analysis II, Encyclopedia of Mathematical Sciences, vol. 14, chap. Convex Analysis, pp. 1–92. Springer-Verlag (1990)

Author Index